PROBLEMS OF SOLAR AND STELLAR OSCILLATIONS

Problems of Solar and Stellar Oscillations

Proceedings of the 66th IAU Colloquium held at the Crimean Astrophysical Observatory, U.S.S.R., 1-5 September, 1981

Edited by

D. O. GOUGH

Institute of Astronomy, Cambridge

Reprinted from

Solar Physics, Vol. 82, Nos. 1/2

D. Reidel Publishing Company

Dordrecht : Holland / Boston : U.S.A.

ISBN-13: 978-94-009-7090-8 e-ISBN-13: 978-94-009-7088-5
DOI: 10.1007/978-94-009-7088-5

Softcover reprint of the hardcover 1st edition 1983

TABLE OF CONTENTS

Problems of Solar and Stellar Oscillations

(Proceedings of the 66th IAU Colloquium)

FOREWORD

D. O. GOUGH

Institute of Astronomy, Madingley Road, Cambridge, U.K.

IAU Colloquium 66 on 'Problems of Solar and Stellar Oscillations' was held at the Crimean Astrophysical Observatory, U.S.S.R., on 1–5 September, 1981. The principal purpose of the colloquium was to study the low-amplitude oscillations of the Sun and, to a lesser extent, to consider similar oscillations of other stars.

Much of the emphasis of the discussions was on the diagnostic value of the oscillations. In the last few years we have become aware that the frequencies of the five-minute modes of high degree, which constitute the major component of the oscillations discovered twenty years ago by Evans and Michaud, can be used to put quite tight bounds on the stratification of the solar convection zone. These permit a calibration of solar models computed from so-called standard evolution theory. Modes of low degree penetrate beneath the convection zone to the core of the Sun, and can in principle test the evolution theory. Therefore there was considerable interest in the reports of the latest observations of such modes. Broadly speaking, those observations confirm the calibration by the high-degree modes, but there remain some systematic discrepancies that demand some revision of the theory.

Besides the gross aspects of evolution theory, there are also more intricate details to be understood. For example, how are oscillations influenced by sunspots, or the large-scale convective flow? And, of course, what is the 160^{m} oscillation?

The mechanism by which dynamical oscillations are excited is of considerable interest. It is likely that nonlinear interactions between the modes and the convection, and between the modes themselves, play a crucial role in determining the oscillation amplitudes. Some progress in this difficult area of investigation is reported in these proceedings.

The aim of the observer is to resolve some aspects of the dynamics of the stellar atmosphere. When many modes are present, this can be a formidable task, demanding careful diagnostic techniques. Advances in the study of the interaction between radiative transfer and the dynamics of the oscillations, which has a direct bearing on the interpretation of the data, and in the statistical analyses of the raw data, are reported here.

In principle, full-disk measurements, whether they be of the radiation intensity or of spectrum-line shifts, can be made for other stars. There is already observational evidence that other late-type main sequence stars undergo oscillations similar to the five-minute oscillations of the Sun, but the individual modes have never been resolved. Issues concerning stellar evolution theory upon which the eventual resolution of such modes may bear are discussed in these proceedings. Furthermore, theoretical estimates of the amplitudes of the oscillations are reported, to guide the observers in their initial selection of suitable stars. Finally, there is a look to the future, in a summary of plans for observations, and what one might hope to be achieved.

Solar Physics **82** (1983) 7–8. 0038–0938/83/0821–0007$00.30.

I am very grateful to my colleagues on the Scientific Organising Committee: R. M. Bonnet, F.-L. Deubner, W. A. Dziembowski, I. Iben, Jr., P. Ledoux, A. B. Severny, and J. M. Wilcox, for their support. Every visiting participant of the Colloquium is indebted to the Local Organising Committee: A. B. Severny, E. Ergma, V. S. Imshennik, V. A. Kotov, A. Massevich, and N. V. Steshenko, for their efforts in arranging our interesting visit to the Crimea. We are also grateful to A. Selina, who was always available to cope with administrative problems. Financial support was given by the International Astronomical Union, the Crimean Astrophysical Observatory, and the Soviet Academy of Sciences.

MANIFESTATION OF THE 160-min SOLAR OSCILLATIONS IN VELOCITY AND BRIGHTNESS (OPTICAL AND RADIO OBSERVATIONS)*

(Invited Review)

V. A. KOTOV, A. B. SEVERNY, T. T. TSAP,
I. G. MOISEEV, V. A. EFANOV, and N. S. NESTEROV

Crimean Astrophysical Observatory, P.O. Nauchny, Crimea 334413, U.S.S.R.

Abstract. All evidence of the solar origin of 160-min period oscillations is collected, and the present state of observations of this oscillation in optical and radio-ranges is considered. The main results are summarized: (a) the 160-min oscillation was observed in 1981 as well as before, (b) an attempt to find a nonradial component with $l = 2$ has failed, (c) the intensity and circular polarization of radioemission show with statistical significance the presence of this 160-min periodicity.

1. On the Solar Origin of 160-min Oscillations

The 160-min periodicity has the same phase at different sites of the Earth's globe: Crimea, Stanford, and South Pole (see Figure 1); the last observations lasting for about 5 days without interruptions caused by night-time intervals as is inevitable in usual ground-based observations show also that the 1-day sampling effect does not play any role. Moreover, the 160-min peak in the power spectrum (computed for 1974–1980) is the highest (see Figure 2). A possible influence of the differential transparency of the terrestrial atmosphere (difference in transparency between east and west limbs) can, in principle, bring some modulation, but this effect, on average, was reported to be at least 10 times smaller than the observed one (see Severny *et al.*, 1980) and does not produce the systematic shift of the phase from year to year. Besides that, the changes of transparency do not show the 160-min variations, as follows from the work by Clarke (1980). The apparent absence of the 160-min oscillations in velocity when one uses a telluric line, as well as the indication for the 27.2-day periodicity in the variations of an amplitude of this 160-min oscillation (Kotov *et al.*, 1982), are also worth noticing. One should also note that the synchronous, in parallel with the line-of-sight velocity, variations in the intensity and circular polarization of radioemission from the Sun are observed (see below), which can not be ascribed to fluctuations in the ionosphere of the Earth. The 160-min variations in IR limb-darkening were observed by Koutchmy's and Kotov (1980). A full analysis of possible non-solar sources and errors is in Kotov *et al.* (1982).

* Proceedings of the 66th IAU Colloquium: *Problems in Solar and Stellar Oscillations*, held at the Crimean Astrophysical Observatory, U.S.S.R., 1–5 September, 1981.

Solar Physics **82** (1983) 9–19. 0038–0938/83/0821–0009$01.65.

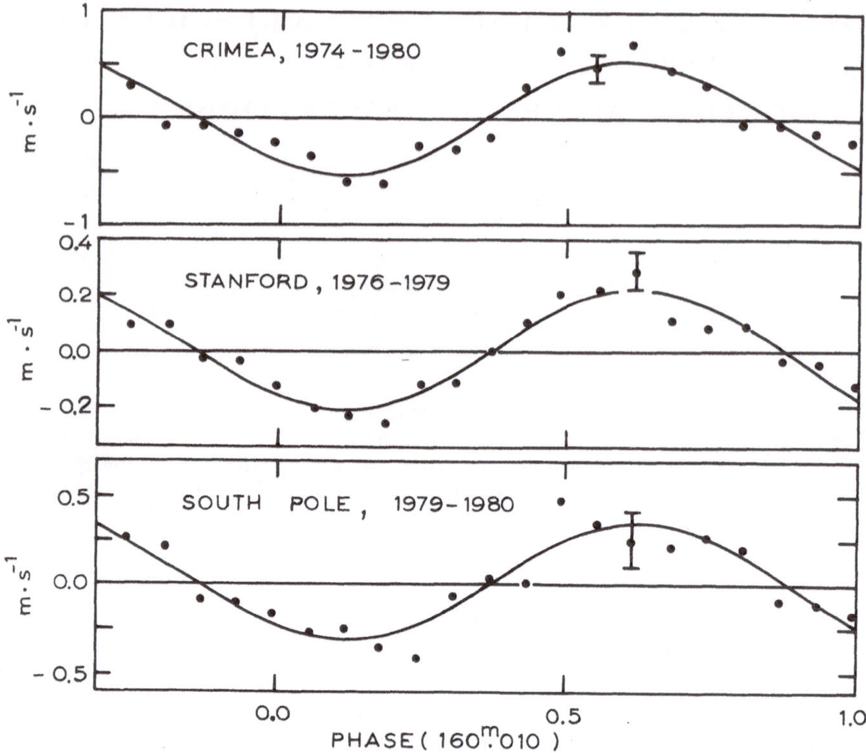

Fig. 1. Superposed epoch plot of the Crimean, Stanford, and South Pole observations with a folding period of 160.010 min. Zero phase corresponds to UT 00^h00^m on 1 January, 1974. The sinusoids represent the best fitted harmonic waves computed for three observatories separately. Vertical lines indicate rms error for each point.

2. Present State of Optical Observations (1974–1980)

A summary of the Crimean measurements of solar velocity is presented in Table I.

The results of superposed epoch analysis for the Crimea, Stanford, and South Pole are illustrated by Figure 1 showing good agreement in phases of mean line-of-sight velocity curves; some difference in amplitudes can be easily ascribed to the differences in methodes of measurements (different areas for averaging of velocity over the solar disk) and, in some part, – to difference in calibration. There is also a good agreement between the dependences of the moments of observed maximum (of outward velocity) upon the time (year) of observation (Figure 3); both Crimea and Stanford clearly show a year-to-year progressive drift of this moment provided that the period of the oscillations, in superposed epoch plots, is precisely 160.000 min; this phase drift is found to be in quite a good agreement with South Pole result too (see label N on Figure 3). This drift of the velocity maximum, determined for each individual year with exactly 160.000 min period, points to a true period of 160.010 min, corresponding to a yearly shift of the phase by about 32 min, on average.

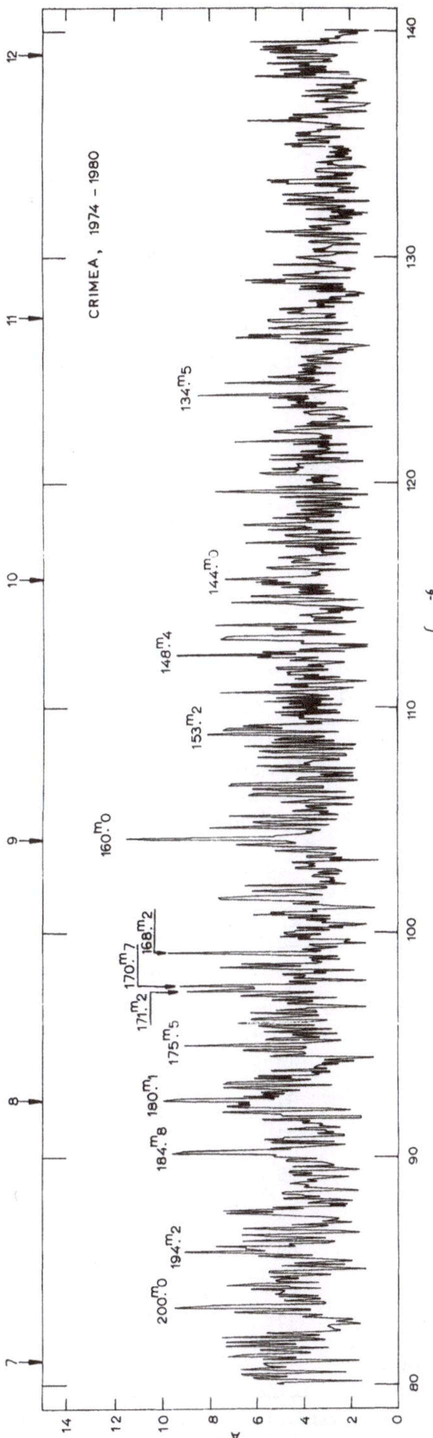

Fig. 2. Power spectrum of the Crimean measurements of solar differential velocity in 1974–1980. The power spectrum for each individual year was computed with a resolution of about 6×10^{-8} Hz (about 0.1 min in periods), normalized, then 7 spectra were averaged. The arrows and digits on the top indicate even harmonics of a day.

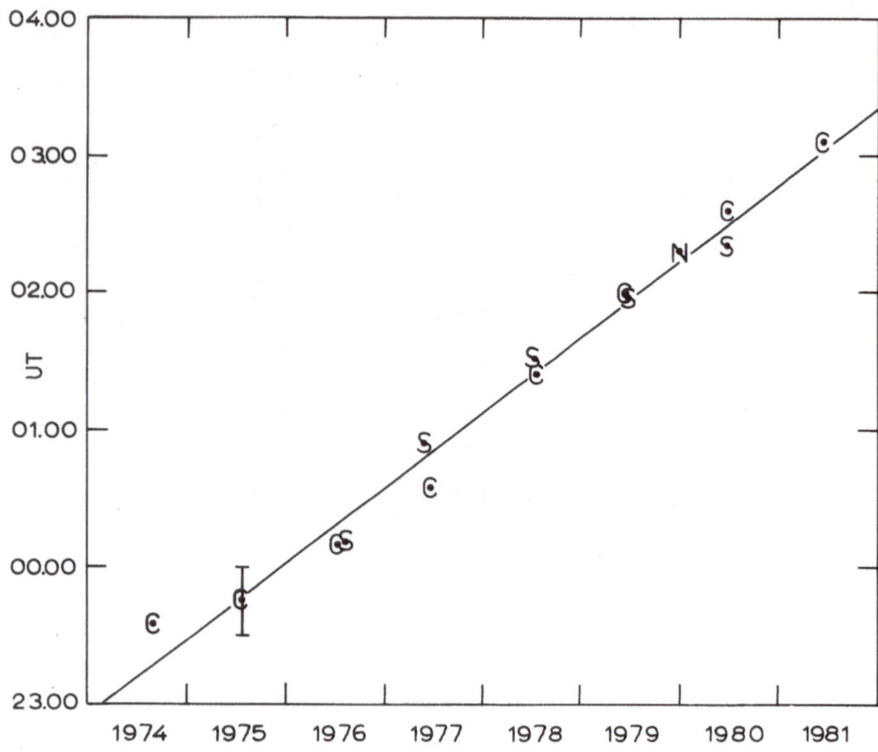

Fig. 3. Phase-shift diagram showing the progressive drift of the velocity maximum, by about 32 min per year, according to observations from different sites: C – Crimea, S – Stanford, N – geographic South Pole (see Scherrer *et al.*, 1980; Grec *et al.*, 1980). The phase shift was computed with respect to a 160 min period.

TABLE I

Summary of Crimean observations of solar velocity

Year	Dates	Days	Hours	Harmonic amplitude (m s^{-1})	UT[a] of maximum velocity
1974	4 August – 9 October	14	76	1.42	23^h35^m
1975	16 March – 14 October	32	171	0.81	23^h45^m
1976	26 April – 15 October	32	163	0.36	00^h09^m
1977	11 March – 24 October	59	280	0.78	00^h35^m
1978	22 April – 21 October	78	471	0.48	01^h24^m
1979	20 March – 30 December	60	381	0.56	01^h59^m
1980	14 January – 26 July	50	304	0.45	02^h36^m
1981	17 June – 31 July	37	248	0.55	03^h07^m
Total	4 August 1974 – 31 July 1981	362	2094	–	–

[a] UT means the time of maximal velocity with respect to zero moment taken at UT 00^h00^m; superposed epoch analysis for each year is made with the period 160.000 min.

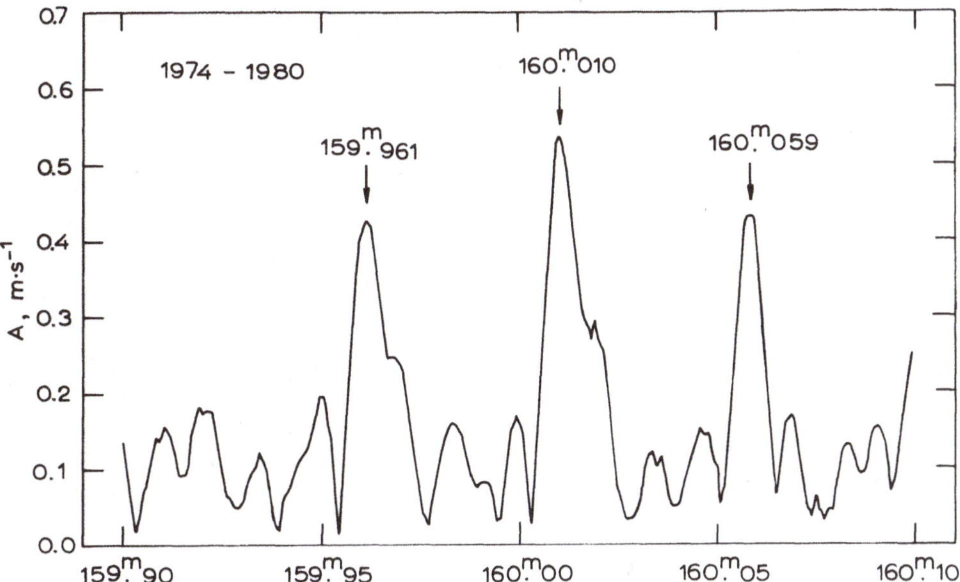

Fig. 4. High resolution power spectrum of the Crimean 1974–1980 measurements calculated near 160-min period (amplitude, in m s^{-1}, versus period, in min).

There is no power in the velocity data at exactly 1/9th of a day (= 160 min), as demonstrated by fine power spectrum on Figure 4; here the dominating peak is definitely located at the 160.010 min period, with two side bands produced by annual regularity of observational seasons. Comparison of the power spectra of long-period oscillations obtained in the Crimea (1974–1979) and at Stanford (1977–1980) shows good agreement of two observatories, especially with regard to the appearance of this 160-min oscillation (see the paper of Scherrer and Wilcox, 1983). However, in addition to the common and dominant maximum at the 160-min period, one can see other significant maxima. Particularly, in the power spectrum of the Crimean data 1974–1980 shown in Figure 2, the periods near 134, 148, 168, and 171 min (besides, of course, the artifact periods near 144 min and 180 min which are side bands of the 160.010 min period) are seen. When considering all these maxima one should keep in mind that in addition to some Earth-originated peaks, we may have also the transient, irregular effects of supergranulation and the newly discovered phenomenon of 'mesagranulation' (with a characteristic life-time of several hours) superposed on global oscillations of the Sun (November *et al.*, 1981).

3. Velocity Observations in 1981

The persistent 160-min oscillation was observed again in 1981, as may be judged from Figure 3 where the phase (more precisely: initial phase) of maximal velocity according to reduction of the 1981 year observations agrees nicely with regression line computed

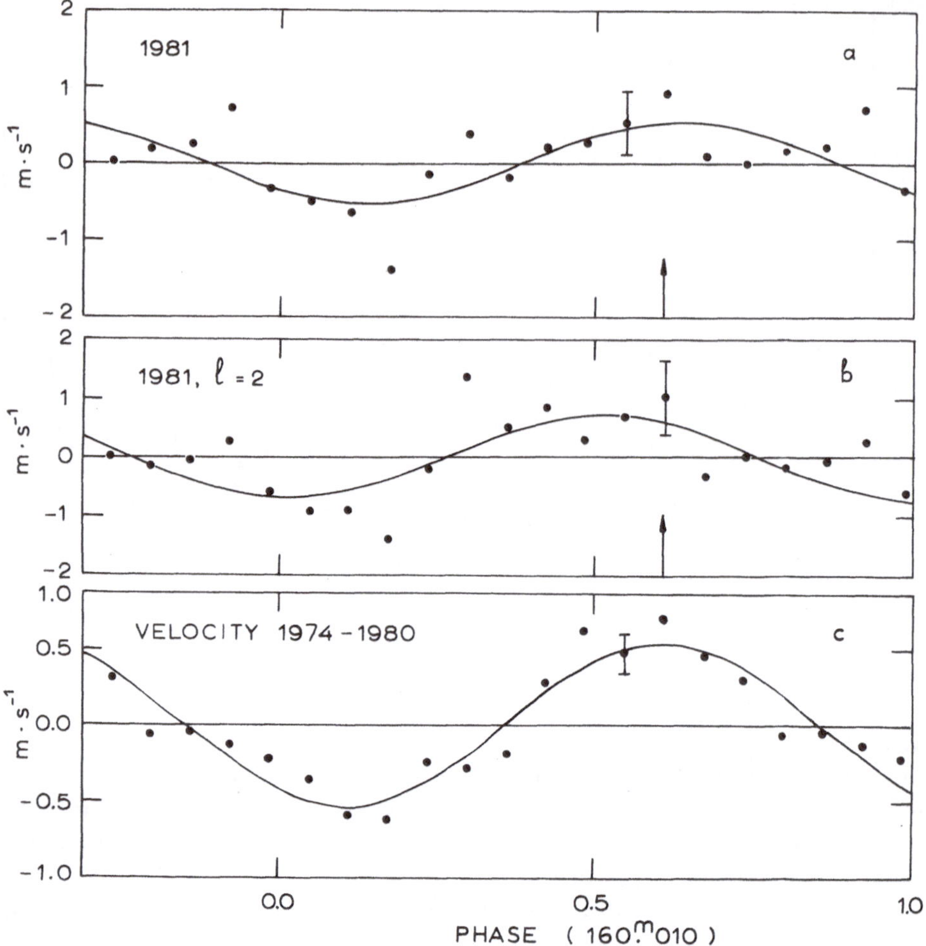

Fig. 5. (a) Superposed epoch plot of the total set of the velocity observations in 1981 including July measurements intended to look for $l = 2$ oscillations (see Figure 6b). (b) Result of July measurements for detection of quadrupols ($l = 2$) motion. (c) Mean velocity curve according to previous, 1974–1980, Crimean observations. Vertical arrows show the position of velocity maxima predicted for 1981 on the basis of 1974–1980 observations.

from previous observations in 1974 to 1980, and implying a 160.010 min period. The superposed epoch plot of these measurements made during 37 days in the summer of 1981 (from 17 June through 31 July, 248 hr in total) is shown in Figure 5 (top). The phase of the best-fitted harmonic curve corresponds to UT 01^h41^m, being in close agreement with the previous Crimean observations 1974 through 1980 (UT 01^h37^m), and Stanford measurements in 1976–1979 (UT 01^h37^m), as well as with the observation at the South geographic pole (UT 01^h41^m; Grec *et al.*, 1980); zero phase everywhere is taken at UT 00^h00^m, 1 January, 1974.

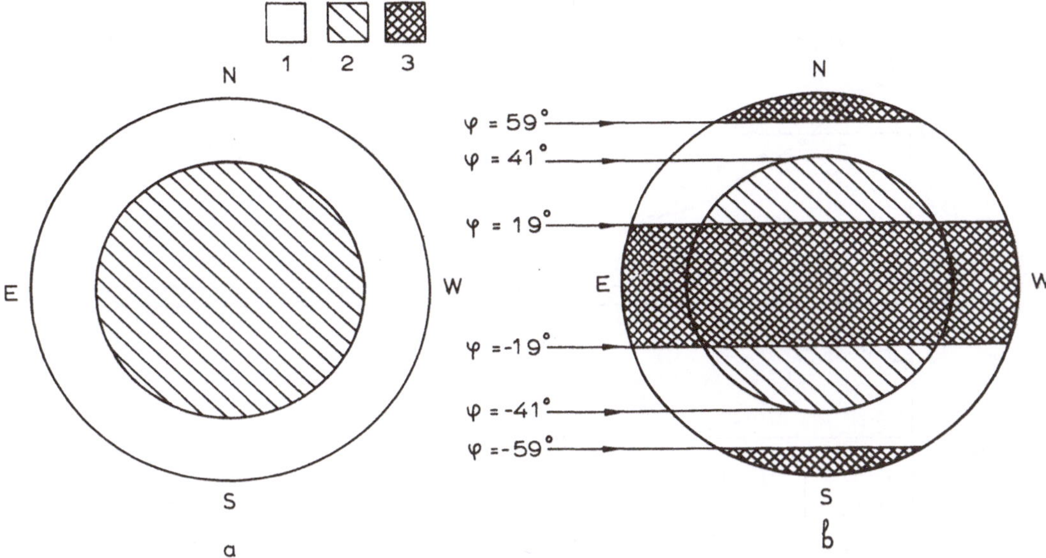

Fig. 6. A sketch showing the different areas of the solar disk used in the Crimea to measure differential solar velocity. (a) For the usual observations in 1974–1981 when the instrument detects the difference between the outer annulus and a weighted full disk. The weighting is such that the outer annulus (unshaded area) has full transmission and the inner circular area (shaded area of 0.66 $D_\odot$ in diameter) has approximately half transmission. (b) For July 1981 observations to detect a quadrupole motion on the solar surface, when the equatorial and polar regions were screened by an opaque shield. The numbers 1, 2, and 3 on top shows three different kinds of areas: 1 – non-screened area; 2 – the area covered by circular polarizer; 3 – completely screened area. By φ we denote different solar latitudes.

Our understanding of the 160-min oscillation, together with other modes of long periods ($\gtrsim 1^{\rm h}$) can suffer from a lack of knowledge about the spatial distribution of amplitudes. This year we made an attempt to determine the type of 160-min oscillation with the use of the same observational technique but with quite different, than earlier, averaging of the velocities over the solar disk. Namely, to look for the quadrupole mode ($l = 2$) given by harmonic P_2° $(\cos\theta)$ having two nodes on the solar surface (at heliolatitudes $\varphi \approx 55°$ and $\approx -55°$ where the radial component of the velocity $V_r = 0$), we screened three portions of the solar disk: two polar regions for $|\varphi| > 59°$ and an extensive equatorial region limited by $|\varphi| < 19°$, see Figure 6. In other respects the optical geometry for measurements of the differential $(V_c - V_l)$ velocity was retained the same as in all previous Crimean observations (see Kotov *et al.*, 1982).

In these new series of observations (performed from 11 to 31 July, 1981; in all 19 days, 129 hr) we measured the difference between the line-of-sight velocities of two large areas located near heliolatitudes $|\varphi| \approx 25°$ and the rest of non-screened areas $59° > |\varphi| > 19°$ (see Figure 6). The result of these observations is plotted in the middle of Figure 5; we see that in these measurements purposed to make discrimination between the radial and quadrupole – types of oscillation, almost the same 160-min wave

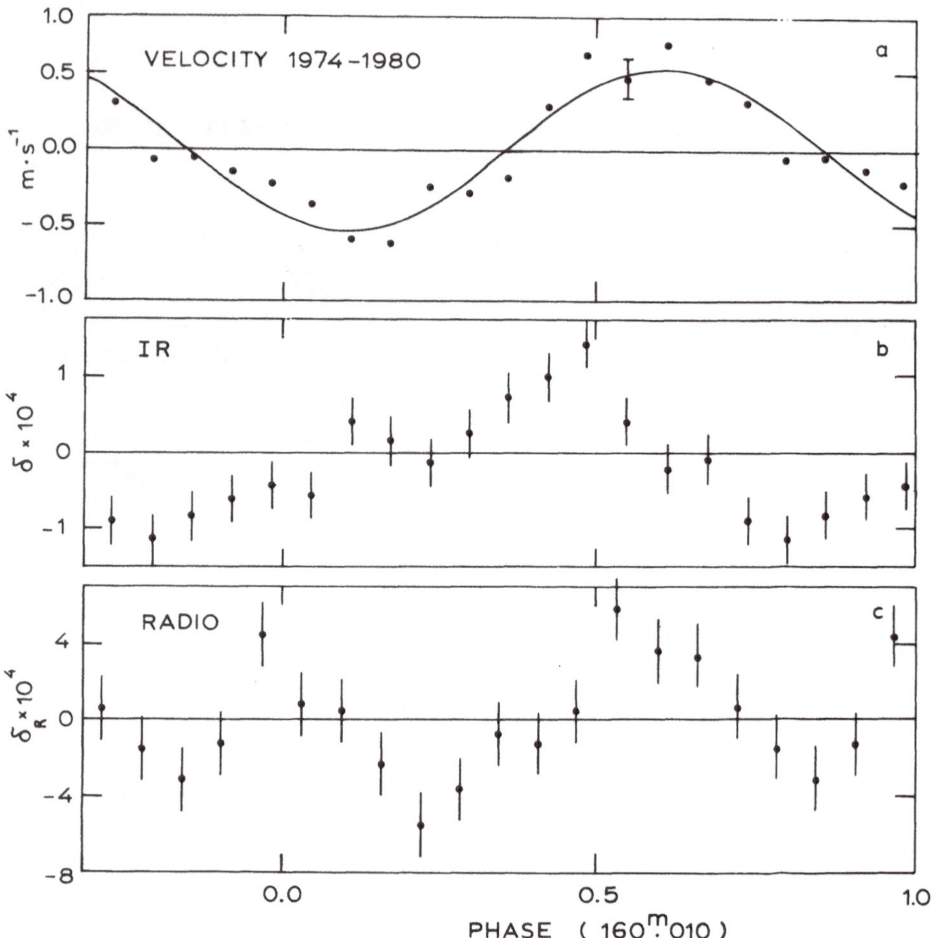

Fig. 7. Variations of velocity (a), IR brightness (b), and radioemission (c) with 160.010 min period. Phase zero is taken at UT 00^{h}00^m, 1 January, 1974 (the same is for Figure 8).

has appeared as in the previous 1974–1980 observations (Figure 5, bottom). From this preliminary result one may conclude that 160-min oscillation shows mostly radial motion.

4. Radio Observations of 160-min Oscillations

To detect the changes in the center-to-limb difference of radiotemperature for the quiet Sun, and following the work by Eryushev *et al.* (1979), we measure alternately, with 5-min integration time, the brightness temperature in two quiet regions of area of about 4 arc min square on the solar disk; one is selected near the disk center and the other is located about 10 arc min to the north from the center. The first measurements of the ratio $(T_l - T_c)/(T_l + T_c)$ were made during 48 days in 1977–1980 with the use of the Crimean 22-m dish antenna (see Ivanov *et al.*, 1967) at four wavelengths: 1.9, 2.25, 2.5, and 3.5 cm (mostly at 2.25 cm; in total about 400 hr of measurements were collected). The first results were reported by Eryushev *et al.* (1979).

Diurnal drifts were eliminated by the usual procedure of 2nd order polinomials approximation (similarly to the analysis of velocity records). The residuals then were subjected to superposed epoch analysis with the 160.010 min period according to the optical measurements.

Figure 7c shows the result of superposed epoch analysis for the difference $\delta_R = 2(I_l - I_c)/(I_l + I_c)$ for the whole data set of 1977–1980. In the middle of Figure 7 we have plotted the mean curve of near infrared brightness changes according to measurements undertaken in 1977–1978 by Koutchmy *et al.* (1980), and at the top – the mean velocity curve for 1974–1980; the phase was set to be 0 at UT 00^h00^m on 1 January, 1974. Comparing all these curves one may conclude that there is an indication of the 160-min waves in solar intensity, IR and radio, with high statistical significance ($2A/\sigma \gtrsim 5$). The average power spectrum of the radio emission (not shown here) calculated for the total set of observations in 1977–1980 also exhibits a significant peak at 160^m.

There also exists a strong overtone with 80–90 min period in the differential (center-to-limb) radiobrightness. The delay in the phase of radio oscillations (see Figure 7) as compared with velocity variations might be caused by the time needed for the oscillations to propagate from the photosphere to the low corona; anyway, this question about time delay and correlation between the velocity and radio data might present a great interest for the theory called up to explain the 160-min oscillations.

Measurements of polarization of the radioemission were made at 1.35 cm wavelength in June–July 1980 and June 1981, during 38 clear days in total. The degree of circular polarization was monitored successively in five different quiet regions on the solar disk: northern (N), central (C), southern (S), eastern (E), and western (W) locations, choosen far away from active centers. The Sun was observed for about 9 hr per day on average (on each of 38 days in 1980 and 1981). Slow drifts again were subtracted from the data by best-fitting a parabola, and detrended residuals have formed a time series which have been processed using the superposed epoch technique and power spectrum analysis. Some data on the observations of these five locations and results – amplitudes and UT of observed maxima of polarization at the 160.010 min period – are given in Table II.

All five quiet regions on the Sun have shown the presence of statistically significant ($2A/\sigma \approx 2$–3) variations with the period 160.010 min and nearly with the same phase (see Table II, last column). The harmonic amplitude is about 10^{-5} (see Table II and Figure 8). The power spectrum averaged for these five sets of measurements of radio-polarization exhibited also a strong 160-min peak.

The polarization of the solar radioemission, as is well known, should be connected with the solar magnetic field. Hence, the 160-min variations of the degree of radio-polarization, shown in Figure 8 (bottom), might mean synchronous changes in the solar magnetic field. This offers a possibility to estimate the amplitude of the magnetic field variations. For the cm-wavelength radioemission induced by thermal bremstrahlung there is a simple relation between degree of polarization and magnetic field strength (see

TABLE II

Radio-polarization, λ1.35 cm, P = 160.010 min (measurements in June–July 1980 and June 1981[a])

Location	Number of			$A \times 10^5$	$2A'/\sigma$	t
	days	hours	5^m-points			
North	26	235	1105	2.4	3.7	02^h10^m
Center	38	335	2122	1.2	4.2	01^h59^m
South	28	246	957	1.1	2.8	01^h54^m
East	13	104	688	1.0	2.2	01^h50^m
West	3	28	182	5.4	3.8	02^h00^m
All	38		5054	1.5	6.9	02^h01^m

[a] A denotes a harmonic amplitude, $2A'$ means 'peak-to-peak' amplitude (see Kotov *et al.*, 1982); t – time of maximal polarization in superposed epoch plots for 160.010 min period and zero moment taken at UT 00^h00^m, 1 January, 1974.

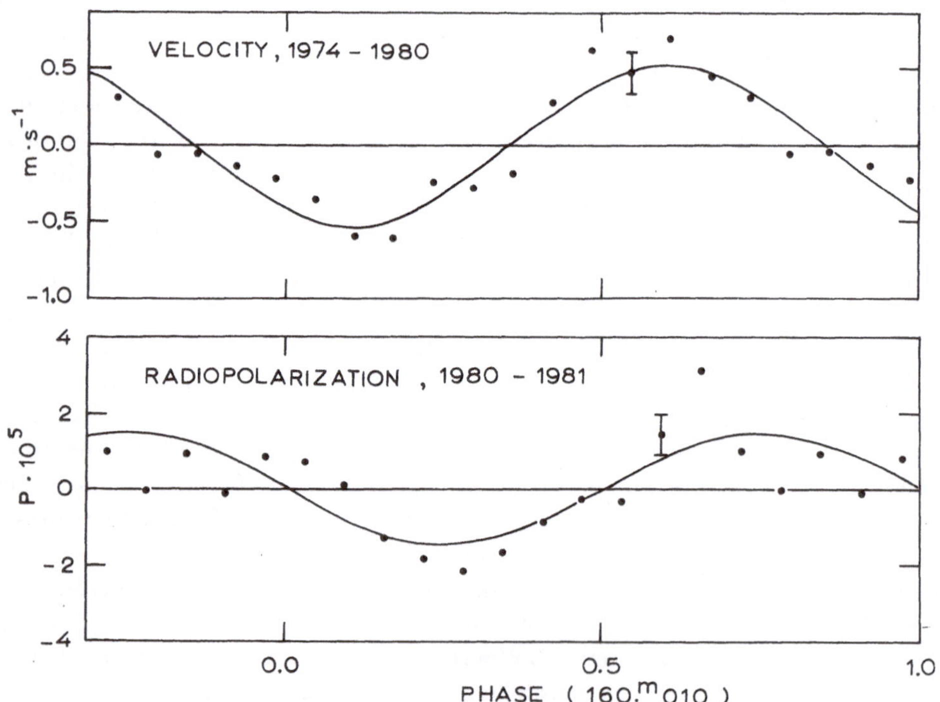

Fig. 8. Mean variations of the velocity and the degree of radio-polarization (at λ1.35 cm) with the period of 160.010 min.

Bogod and Gelfreikh, 1980):

$$p \approx n \frac{f_H}{f} ,$$

where $n = \partial(\lg T_b)/\partial(\lg f)$ is the spectral index of radio-emission, $f_H = eH/2\pi$ cm is the gyrofrequency, f is the frequency of the observations and H is the magnetic field strength. Substituting $\lambda = 1.35$ cm and $n \approx 1$, we get for the field strength:

$$H \approx 0.79 \times 10^4 p ,$$

and for our mean value $p \approx 1.5 \times 10^{-5}$ the amplitude of magnetic field variations attains the value of about 0.12 G. This amplitude exceeds the directly estimated variations in the solar mean magnetic field inferred from optical observations (see Severny, 1969; Kotov *et al.*, 1982) by a factor of ≈ 5. However, this apparent discrepancy might be readily explained by the strongly inhomogeneous structure of the chromosphere and low corona, as well as by the fact that when one observes the Sun-as-a-star field one measures optically some average longitudinal component of the magnetic field for the solar disk as a whole, in contrast to the radio polarization measurements, where we deal with several small areas on the disk, and where the field strength may considerably exceed the mean field value.

It should be emphasized that the values of circular polarization and the amplitude of fluctuations of radioemission can not be attributed to ionospheric effects, where the optical depth for our wavelength (1.35 cm) is about 5×10^{-7} (for the measurements of radioflux) and $\sim 10^{-11}$ for the polarization.

References

Bogod, V. M. and Gelfreikh, G. B.: 1980, *Solar Phys.* **67**, 29.

Clarke, D.: 1980, *Monthly Notices Roy. Astron. Soc.* **190**, 641.

Eryushev, N. N., Kotov, V. A., Severny, A. B., and Tsvetkov, L. I.: 1979, *Pis'ma v Astron. Zh. (USSR)* **5**, 546.

Grec, G., Fossat, E., and Pomerantz, M.: 1980, *Nature* **288**, 541.

Ivanov, V. N., Moiseev, I. G., and Monin, Y. G.: 1967, *Izv. Krymsk. Astrofiz. Obs.* **38**, 141.

Kotov, V. A., Severny, A. B., and Tsap, T. T.: 1982, *Izv. Krymsk. Astrofiz. Obs.* **66**, 3.

Koutchmy, S., Koutchmy, O., and Kotov, V. A.: 1980, *Astron. Astrophys.* **90**, 372.

November, L. J., Toomre, J., Gebbie, K. B., and Simon, G. W.: 1981, *Astrophys. J.* **245**, L123.

Scherrer, P. H. and Wilcox, J. M.: 1983, *Solar Phys.* **82**, 37 (this volume).

Scherrer, P. H., Wilcox, J. M., Severny, A. B., Kotov, V. A., and Tsap, T. T.: 1980, *Astrophys. J.* **237**, L97.

Severny, A.: 1969, *Nature* **224**, 53.

Severny, A. B., Kotov, V. A., and Tsap, T. T.: 1980, *Astron. Astrophys.* **88**, 317.

OBSERVATION OF GLOBAL 160-MIN INFRARED (DIFFERENTIAL) INTENSITY VARIATION OF THE SUN*

V. A. KOTOV

Crimean Astrophysical Observatory, Nauchny, Crimea 334413, U.S.S.R.

S. KOUTCHMY

S.A.S. Institut d'Astrophysique, CNRS, 98 bis BD Arago, F 75014 Paris, France

and

O. KOUTCHMY

Laboratoire d'Analyse Numerique, Université P&M Curie, F 75006 Paris, France

Abstract. The method developed and the instrument designed for detecting variations of the solar limb darkening at the atmospheric transparency window of the solar opacity minimum region of λ 1.65 μ are described. This differential technique proved to be successful in rejecting undesirable low frequency noises due to the atmosphere and to the instrument. Analysis of observations made in 1977, 1978, and 1981 indicates the persistance of global fluctuations of the IR differential, center-to-limb intensity at the well-known 160 min period with an average amplitude of about $\pm 2 \times 10^{-4}$ in units of the 'average Sun' intensity near 1.65 μm.

1. Introduction

One of the principal activities of solar physicists in the last decade has been to tackle the measurements of solar oscillations. In addition to the well-known 5-min oscillation of the Sun, others, of longer periods (7 to 70 min), with small amplitudes were found by Hill *et al.* (1978) in measurements of the apparent solar diameter, which have been lately interpreted in terms of fluctuations in the limb darkening (Hill and Caudell, 1979). Using sensitive Doppler velocity measurements Severny *et al.* (1976, 1979) have found clear evidence for the existence of global oscillation of the Sun with a surprisingly stable period of 160.010 min which was also found to be highly phase coherent in time. This finding has been supported by observations of other groups (see Brookes *et al.*, 1976; Scherrer *et al.*, 1980; Grec *et al.*, 1980) and the period is now determined with an accuracy of about ± 0.002 min. It is worthwhile to note that irrespective of the history of this 160 min oscillation prior to 1974 both Crimean and Stanford data (Scherrer *et al.*, 1980) demonstrate the long term (over at least 8 yr) stability of an underlying 'clock' mechanism, whatever its' nature.

* Proceedings of the 66th IAU Colloquium: *Problems in Solar and Stellar Oscillations*, held at the Crimean Astrophysical Observatory, U.S.S.R., 1–5 September, 1981.

Solar Physics **82** (1983) 21–35. 0038–0938/83/0821–0021$02.25.

For the purpose of the present paper, however, we stress the claim that this 160 min persistent oscillation is accompanied by synchronous variations in the solar radio emission and radio polarization at cm-wavelengths and, presumably, in the differential (central portion of the solar disk with respect to outer annulus) optical intensity of the Sun (see Kotov *et al.*, 1982, 1983).

Nevertheless, the 160 min oscillation has not been uniformly accepted by solar physicists, mainly because it is close to being an integral division ($\frac{1}{9}$th) of a day (see, for example, Dittmer, 1977; Grec and Fossat, 1979). Therefore, it becomes clear that at least a partial answer to the question on the nature of 160-min oscillation lies in the acquisition of a long reliable series of data together with a proper analysis of the observations.

In ground based measurements of the solar output and its variations at any wavelength, the major uncertainty arises in correcting for atmospheric transparency changes. In addition observations of long-period oscillations require very high instrumental stability. These difficulties may be reduced by the use of relative measurements, for instance, those of center-to-limb ratio of the solar intensity, especially in the near infrared.

The near IR spectral range seems to be much more favorable than the optical one for measurements of small variations in the solar intensity. This is because of the weakness of terrestrial atmospheric influences at the IR transparency windows, and also because the wavelengths observed (1.6–1.7 μm) correspond to the minimum of solar atmospheric opacity. In order to minimize further the remaining atmospheric and instrumental disturbances, we measure relative, center-to-limb variations in the IR brightness. In other words, we attempt to detect small changes in the limb darkening function supposed to be measurable at definite periods of solar global oscillations (see Hill *et al.*, 1978; Kotov *et al.*, 1982).

3. Instrumental Set-up

Our instrument uses an image of the Sun, 9 mm in diameter which is periodically scanned across the PbS detector [N.E.P. $\lesssim 10^{-12}$ (W Hz^{-1})] with a frequency of $2\omega/2\pi = 40$ Hz. The instrument is attached to the Crimean Solar Tower Telescope and has been used successfully since August 1977. The first results obtained by this differential technique were previously published by Koutchmy *et al.* (1980).

Figure 1 shows the Solar Tower and a general view of our device and Figure 2 shows the action of this device schematically. The entrance aperture, 1×1 mm^2 in size, is formed by crossing two narrow, rectangular slits denoted by S1 and S2 in Figure 2a, each of 1×10 mm in size; these slits are illuminated by a 'parallel' beam of solar light. This 'parallel' beam is directed into our instrument by two flat coelostat mirrors (see Figures 1b, c) which have a photoguiding system ensuring a pointing accuracy of about 1 arc sec (note the same beam is simultaneously used by Crimean observers for Doppler velocity measurements). At 1-m distance below the aperture (see Figures 1c and 2a) we get a pin-hole image of the solar disk feeding the PbS detector (with an open area of

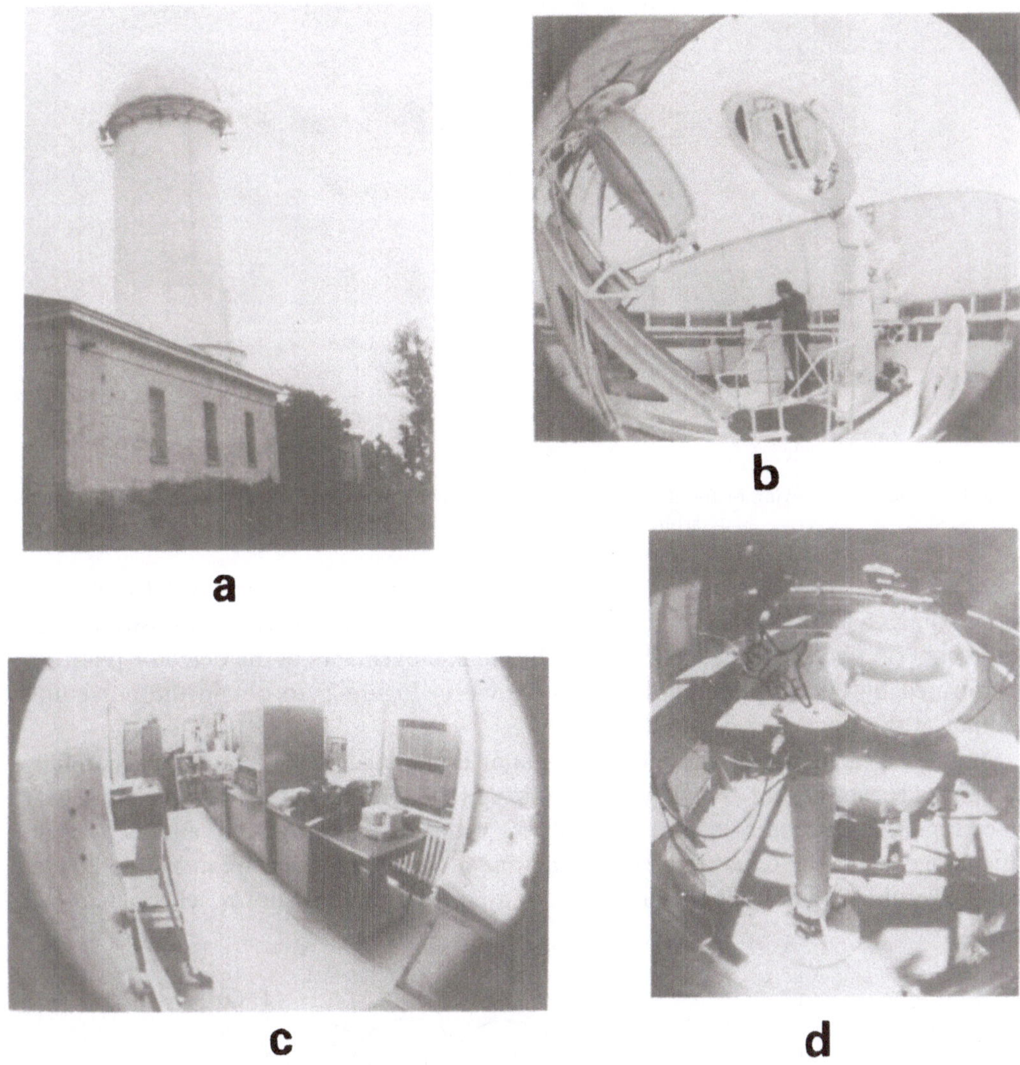

Fig. 1. General view of the Crimean Solar Tower Telescope and the instrumentation. (a) Solar Tower; (b) coelostat mirrors (about 24 m above the ground level); (c) the solar magnetograph (used also for Doppler velocity measurements and, partially, for the recording of IR observations) and the computer; (d) the diagonal flat mirror used for the Doppler velocity measurements and the IR device (its entrance aperture is indicated) illuminated by a parallel beam.

1×0.4 mm) which is coupled to the analog electronics; the working voltage of the detector is given by a stable 90 V battery. Just above the detector we place a Ge interference filter with 60% transmission at the 1.65 μm wavelength with FWHM = 0.06 μm (for some of the 1977 measurements the filter was set for the wavelength 1.75 μm).

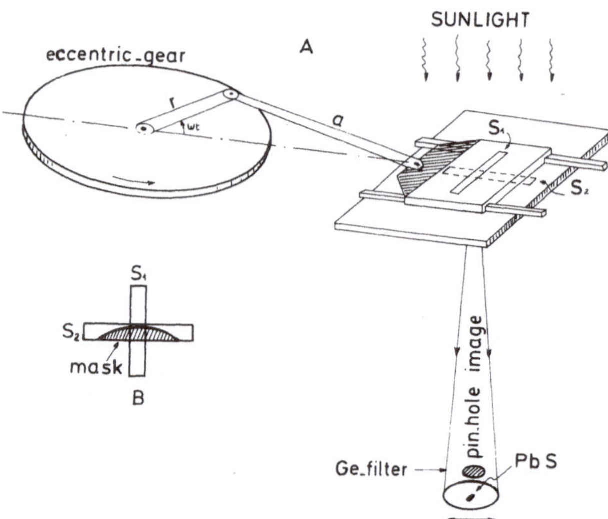

Fig. 2. Schematic drawing of the device used for the IR intensity measurements (dimensions are quite arbitrary); for the explanation see text.

Before each observation the PbS detector was positioned by hand to place it approximately in the center of a light beam; then the remaining errors in position were corrected (in two coordinates, X and Y) by small movements of the detector-platform performed by two stepping-motors (not shown in Figure 2) to obtain the maximum output at the frequency of modulation.

In these measurements, the linear modulation is obtained by the rotation of a carefully balanced eccentric gear allowing long scans, up to 8 mm to be used. The eccentric gear (see Figure 2a) periodically moves the slit S1 within the range ± 4 mm and as a result, the image scans the detector twice during one rotation. The gear frequency is 20 Hz, so the resulting scan frequency is 40 Hz in a periodic pattern shown schematically in

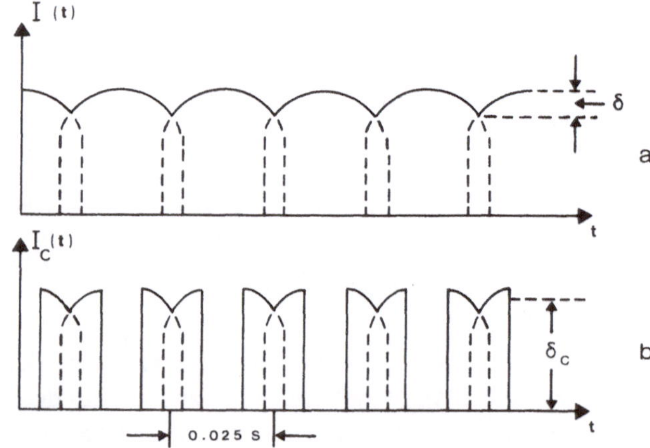

Fig. 3. Schematic plots of the PbS output for the recording of the differential intensity signal (a) and the calibration (b). Frequency of modulation is 40 Hz.

Figure 3a. It is clear that the amplitude of the 40 Hz signal in the detector output $I(t)$ well reflects the limb darkening function (which is somewhat distorted, in reality, by diffraction):

$$I(\rho) = 1 - U_1 + U_1(1 - \rho^2)^{1/2} \tag{1}$$

within the scanning range $|\rho| < 0.9$; $U_1 = 0.23$ for $\lambda 1.65$ μm. Here ρ is a heliocentric distance which is determined, in our case, by a function of the position of the modulator; omitting the terms proportional to $(r/a)^n$, $n \geq 2$ (parameters of the eccentric gear r and a are indicated in Figure 2a; in our case $r/a \approx 0.06$), we get $\rho \sim \cos \omega t$.

However, what we actually measure is a harmonic amplitude of $I(t)$-modulation of the detector output, which is linearly detected within a specially designed Shebyshev-type narrow passband analog filter at the 40 Hz frequency:

$$\delta'(t) = \alpha(t) A\, FT^{-1} \left\{ \int_{2\pi}^{\infty} FT[I(t)]\, G(\omega)\, d\omega \right\}, \tag{2}$$

where FT means Fourier transform, A is some instrumental constant, $\alpha(t)$ is slowly variable atmospheric transparencies (much smaller than the frequency of modulation); $I(t)$ is PbS output which includes not only variations caused by the limb darkening function (see Figure 3a and Expression (1)) but also fast variations produced by Earth's atmospheric effects (seeing, turbulence, etc. ...) at higher frequencies. $G(\omega)$ describes the frequency reponse of the detection system including the electronic filter. The effective passband was equal to ≈ 0.05 Hz, so the function $G(\omega)$ is well peaked at 40 Hz.

Further, in accordance with the linear modulation produced by the eccentric gear, we have $\rho(t) \approx 0.9 \cos \omega t$; therefore, substituting (1) into (2) we get for the measured differential (center-to-limb) signal we are looking for:

$$\delta'(t) \approx \alpha(t) B u_1(t), \tag{3}$$

where $B = \text{const.}$, $u_1(t)$ the limb darkening parameter which is thought to be slowly variable (at frequencies much smaller than the modulation frequency 40 Hz). It is clear that any periodic variation in the center-to-limb ratio function will result in a similar variation of the measured electric output at the 40 Hz frequency.

To correct (3) for $\alpha(t)$-slow changes in atmospheric transparency, we detect also a 'calibration' signal recorded during certain time intervals when the central portion of the solar disk (as it is viewed from the PbS-cell) is screened by an opaque shield, ≈ 3 mm in width, inserted into the light beam (by chopping electromagnet) just under the stationary slit S2 and giving rise to a modulation pattern of the PbS output shown in Figure 3b. In mathematical form this type of modulation can be written as follows:

$$I_c(t) = I(\rho) H(|\rho|), \tag{4}$$

where again $\rho = 0.9 \cos \omega t$, $I(\rho)$ the limb darkening function (1) and

$$H(|\rho|) = \begin{cases} 1, & |\rho| \geq 0.67, \\ 0, & |\rho| < 0.67. \end{cases} \tag{5}$$

It is easy to understand that in this last case the signal I_c' (calibration) being the result of an application of the Expression (2) to the $I_c(t)$-pattern shown in Figure 3b and given by (4) and (5), is equal to

$$I_c' \approx \alpha(t) A \frac{I_\odot}{K} , \tag{6}$$

where $K \approx 1.33$ for our type of modulation, $I_\odot$ the intensity of the 'average Sun' supposed to be constant (in the frame of our investigation) because the instrument is not aimed, in principle, to detect fluctuations of the entire Sun's luminosity. Accordingly, taking the ratio δ'/I_c' we presumably completely reject the influence of atmospheric transparency changes on our measurements.

The observations in 1977 and 1978 were carried out with the use of a specially profiled mask (see Figure 2b) set onto the unmovable slit S2. The corresponding modulation pattern, really recorded as the PbS-output, is shown in Figure 4. Its shape is much more smoothed than the schematic patterns in Figure 3, due to the low spatial resolution employed: we use a 0.4 mm width of an open area of the PbS corresponding to a resolution of about (1/22.5) $D_\odot$; the solar image also is diffraction limited with a resolution of $\approx (1/10) D_\odot$.

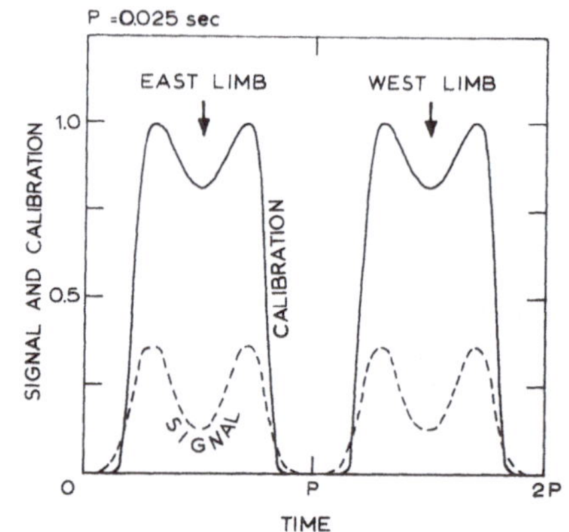

Fig. 4. Modulation pattern of the PbS output with the use of a profiled mask (see text).

The aim of the profiled mask (which reduces by a factor of two intensity of the central part of the solar disk, as it is viewed by PbS) was to minimize the off-set signal, i.e. to make the difference 'center minus limb' near zero. It was believed that any errors, of atmospheric or instrumental origin, are proportional to the off-set signal, i.e. to the 'centerlimb' difference; hence, the use of this mask seemed to be capable to diminish many undesirable noises. In practice, however, the use of the mask significantly alters the short-time noise and diurnal drifts neither for δ' nor for the ratio δ'/I_c', because (1) the mask did not influence the calibration signal I_c' and its errors, (2) the main sources of errors for the differential signal δ' (and for I_c' too) are guiding error and inhomogeneity of atmospheric transparency; all these sources are not affected significantly when we use the profiled mask. Therefore, in 1981 we made observations without using the mask.

The voltage of the PbS output is detected using the same 40 Hz electric filter and synchronous amplifier for both δ' and I_c' signals, then digitally recorded on a computer M6000 with integration time 30 (or 50) s, with 10 s dead time, in the following succession: $I_c' - \delta' - I_c' - \delta' - I_c' - \ldots$, to get 90 (or 60) values of δ' and I_c' per 1 hr of observing time. Then, for each δ'-value we compute the ratio

$$\Delta' = \frac{\delta'(t)}{KI_c''(t)} \approx \frac{0.56\, U_1(t)}{I_\odot} \approx \frac{I_c - I_L}{I_\odot}\,, \tag{7}$$

where $I_c''(t)$ is the average of two adjacent values of $I_c'(t)$, I_c and I_L are the intensities near the center and near the limb of solar disk.

It is obvious that by computing the ratio $\delta'(t)/I_c''(t)$ we (a) suppress considerably the influence of all spurious disturbances of instumental or atmospheric origin and (b) get the simplest way to convert our differential intensity signal Δ' into the intensity of the 'average Sun'. Some higher frequency noise and long-time diurnal drifts appearing in the original data series are due to (a) electronic noise, (b) nonhomogeneity of the atmospheric transparency, and (c) guiding errors.

The most serious error which might be thought as the source of persistent periodicities (in particular, at the 160-min period) in our IR measurements is, of course, the guidance error. Measurements showed that a 1 arc sec shift of the solar beam gives a decrease of about 0.2% and 0.04% of the δ' and I_c' signals, respectively, i.e. it produces approximately 1.7×10^{-4} error in the measured relative intensity. This is close to the mean amplitude of 160-min variations (see below); however, it is unlikely that such a guidance error might be a source of the 160-min periodicity because (1) the sign of a misalignment (guidance) error in our intensity measurements is distributed at random from day to day and (2) careful account of the guidance error of the telescope did not show the presence of any significant (and persistent for many days) periodicity at 160-min in excess of 0.1 arc sec (Kotov *et al.*, 1982).

3. Observations and Analysis

Each observation lasted at least one 2^h40^m cycle while many observations covered 2–3 successive cycles (i.e. $\approx$ 5–8 hr). Both signals, δ' and I_c' could be displayed on graphs to demonstrate the quality of the observations (see, for instance, Figures 5 and 6). Poor observations and offending points (well displaced from the record) could be identified and deleted from the analysis if desired (without information about exact UT of the observation). Then the series of Δ' data is treated using a computer data reduction program developed by Severny *et al.* (1979) and Koutchmy *et al.* (1980) to derive the superposed epoch plots or a power spectrum using a fast Fourier transform.

The slow drifts produced mainly by diurnal variation of the guidance system can reach the value of about 10–15% per 4–6 hr of observing time for the ratio $\Delta' = \delta'/I_c''$. These drifts, as a rule, can be easily approximated by quadratic polynomials $\Delta_0(t)$ and the residuals $\Delta = \Delta' - \Delta_0$ averaged over each 5-min interval of time form our basic data series subjected to further analysis.

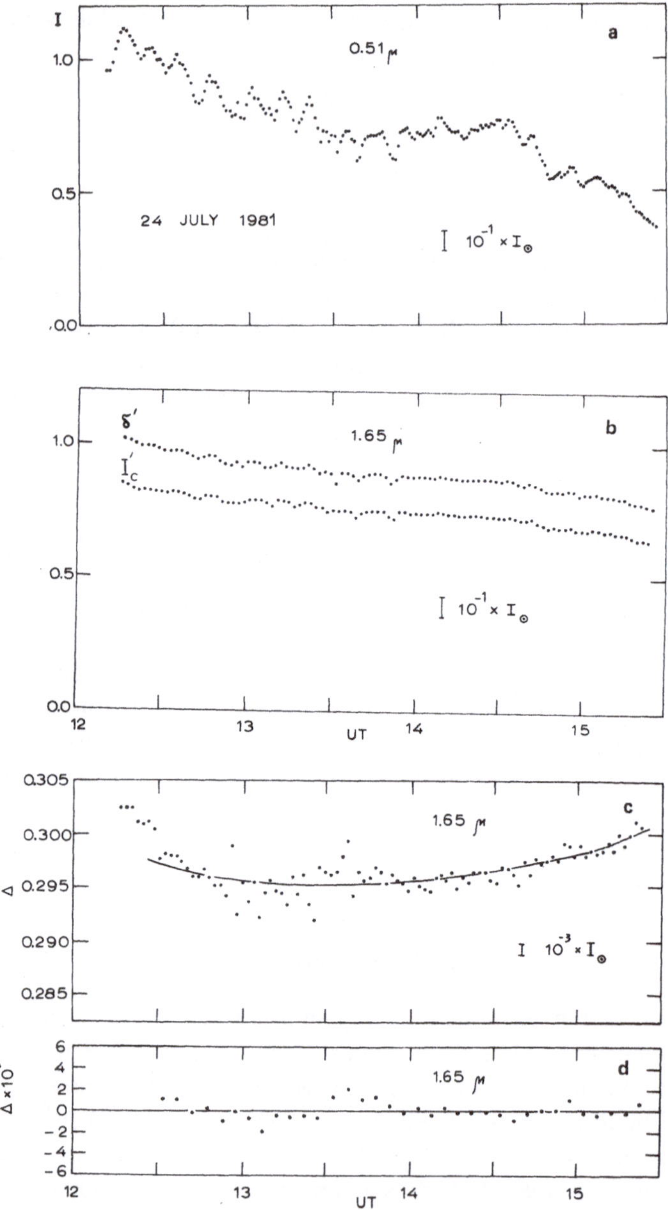

Fig. 5. An example of intensity records obtained on 24 July 1981 under poor atmospheric conditions (hazy sky): (a) the run of solar intensity measured at λ0.51 μm; (b) the same for λ1.65 μm (for the two IR signals, δ' and I_c'); (c) the ratio of two IR signals, $\Delta' = \delta'/I_c''$; (d) the plot of residuals $\Delta = \Delta' - \Delta_0$ (5-min averages).

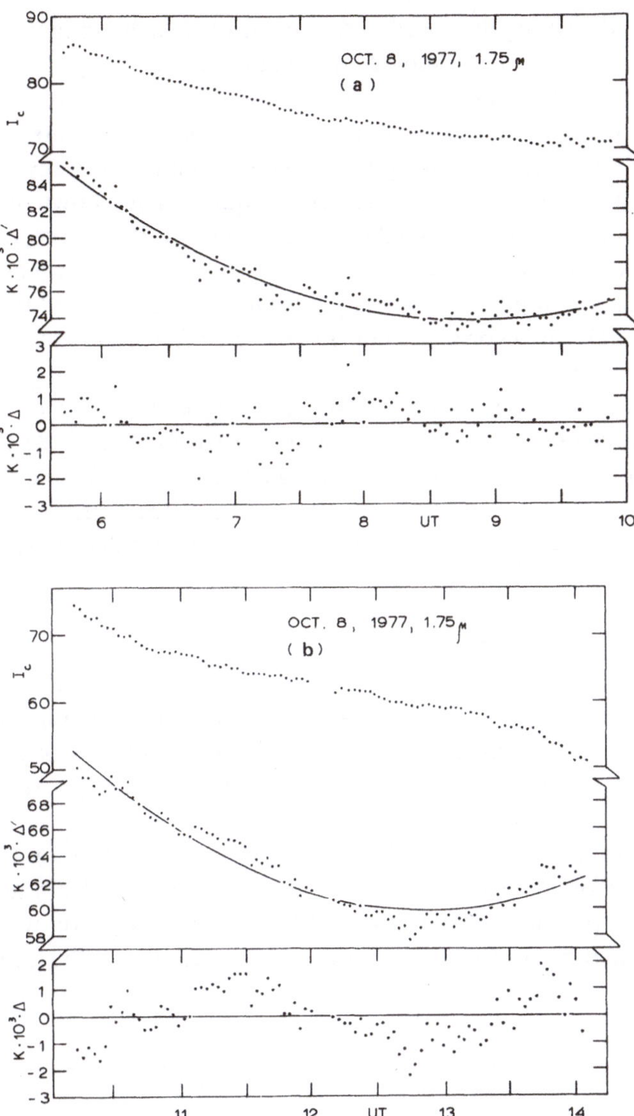

Fig. 6a–b. The plot of originally recorded calibration signal I_c' (top), of the ratio $K\Delta' = \delta'/I_c''$ together with the least squares fit by parabola (middle) and of the residuals $k\Delta$ (bottom), vs UT.

Figure 5 demonstrates an advantage of the near IR spectral range (here λ 1.65 μm) for a detection of solar brightness variations as compared with optical (λ 0.51 μm) observations; fluctuations of both IR signals, δ' and I_c' are one order of magnitude smaller than transparency fluctuations seen in the green light (top) (note that both sets of measurements were made simultaneously with the use of the same 'parallel' beam, and under quite poor atmospheric conditions (hazy sky). Figure 5c shows the time behaviour of the ratio $\Delta' = \delta'/KI_c''$ approximated by the best-fitted parabola Δ_0; at the bottom we plotted the residuals $\Delta = \Delta' - \Delta_0$ average over each 5-min interval. One may conceive that yet under poor atmospheric condition the instrument is capable of detecting fluctuations of the differential signal of the order of $\sim 10^{-3}$–10^{-4}, of solar and/or atmospheric origin. Let us point out that all our measurements were carried out under 'quite good' to 'excellent' atmospheric condition with a clear sky, by contrast the example of record obtained during poor sky conditions is shown in Figure 5. By statistical analysis of records collected during many days, ≈ 30 or more, one can achieve an accuracy of about 1 part in $\sim 10^{-4}$ to 10^{-5} – especially when we are interested in oscillations maintaining a phase over many days like the known 160-min pulsation of the Sun, and when all the records are considered as continuous data series, with gaps caused by a lack of observations.

To do more detailed analysis we investigated further the data obtained in 1977 (11 August – 6 November, in all 32 days) and in 1978 (21 July – 21 October, in all 35 days); a total of 354 hr of observations during 67 days were suitable for a reduction. The results where further reinforced by analysis of the latest data of 11 days in 1981 made, as it was previously noted, without the use of the profiled mask.

4. The Results

A number of the 1977 year records showed fluctuations of the differential intensity signal $\Delta = \Delta' - \Delta_0$ with a characteristic times of 2–3 hr. An example of an original record made on 8 October 1977 and exhibiting a clear ≈ 160-min variations may be seen in Figure 6 where we plotted I_c' values (top), the run of the ratio $K\Delta' = \delta'/I_c''$ (middle) and the residuals $K\Delta = K(\Delta' - \Delta_0)$ (bottom).

One particular feature can be noted: maxima on many records of 1977 tended to concentrate near almost the same UT-moments: $\approx 6^h \approx 9^h \approx 11^h 20^m$ and $\approx 14^h$ with about $2^h 40^m$ mean spacing – despite the fact that all these records were obtained in different months and did show quite various character of diurnal drifts (see, for example, the same Figure 6).

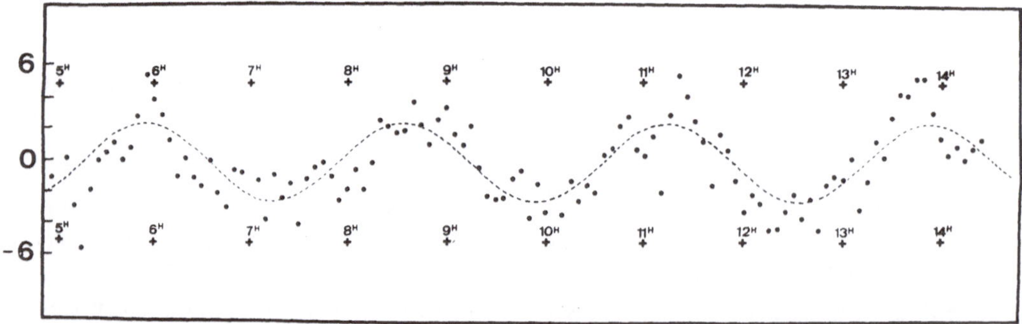

Fig. 7. The averaged measurements of 1977 (a total of 156 hr corresponding to 32 days) arranged using the superposed epoch method. Time is UT and amplitude is in units of 10^{-4} of the 'average Sun'. Dotted line represents the best fitted 160-min sinusoid.

Strong indication of the presence of the 160-min variation in 1977 data can be judged from Figure 7 which is the result of averaging the whole set of 1977 year observations (32 days). Plotting this average result of 1977 we assumed that the 160-min oscillation remained in phase from day to day (as it does from daily velocity observations), so we added the daily sums coherently in time (i.e. keeping in mind that 160 min = 1/9 of a day). One can see that the differential IR intensity signal reveals here a remarkable 160 min variation through a whole sunny day from about 05:00 UT to ≈ 14:00 UT, with a mean amplitude $\approx 2.5 \times 10^{-4}$ in units of the 'average Sun' intensity.

We have computed also the power spectra of these two, 1977 and 1978, data sets, with zeros put in for the time intervals not covered by observations; thus, here the data were also analysed as a coherent time series. Both spectra (Figure 8) showed prominent peak just at the 160 min period; the same signature of the presence of a significant and dominating 160 min oscillation may be seen in the power spectrum averaged for 1977 and 1978 years, see Figure 2 in Koutchmy *et al.* (1980).

It is well-known that reports about an existence of 160 min global pulsation of the Sun were regarded with skepticism by some solar physicists due to the fact that 160 min = $(1/9)^{d}$; it explains why the existence and significance of this 160 min period was subject to controversial discussion during the last 5 years. However, if real, the phenomenon of this sharply tuned and long-period phase-coherent oscillation might have an important issue for the structure of solar interior (Severny *et al.*, 1979; Gough, 1980). To make alternative explanations, it was suggested by several authours (see for instance, Dittmer, 1977; Grec and Fossat, 1979) that 160-min period in velocity measurements might be a pure result of a 1-day regularity in the observations (sampling effect) and daily trends caused mainly by the differential atmospheric extinction. Further, it was noted that the use of data series of about 5.4 hr duration, i.e. $\approx 2 \times 160$ min, can favour 1/9th harmonic of a day.

In our IR data, however, the histogram of data series lengths for 1977 plus 1978 exhibited two maxima, around 3.4 and 7.8 hr, with a minimum, instead of a maximum, near 5.5 hr. To test further this 1/9-day hypothesis, we used again our IR data obtained in 1977–1978, and after numerical analysis of the distribution of the observing 'windows', the conclusion was reached that 160-min periodicity in IR intensity measurements (and, similarly, in velocity measurements, since the latter were made simultaneously with IR records) could not be produced by a 1-day sampling, or by noise and diurnal trends (see extensive discussion of this subject in Koutchmy *et al.*, 1980; Kotov *et al.*, 1982). In essence, we established that there was virtually no evidence of 160-min regularity in the time distribution of 'windows' for IR records at all.

It should be also noted that diurnal trends of the ratio $\delta'(t)/I''(t)$ which are greatly reduced by low-frequency filtering using a 2nd order polynomial approximation do not reveal any systematic behaviour from day to day; these trends are presumably caused by the instruments (mainly by the guiding error slowly changing throughout a day) rather than by the differential atmospheric effect; these trends appear very much like low-frequency noise and certainly cannot be modelled as a differential effect of the atmospheric extinction, as supposed by Grec and Fossat (1979).

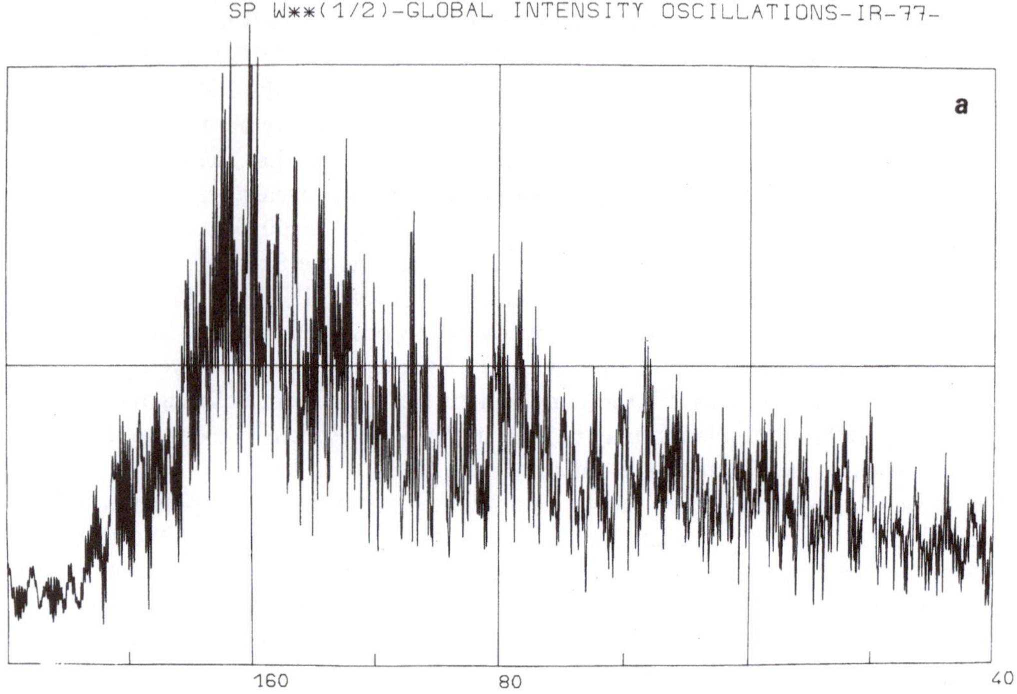

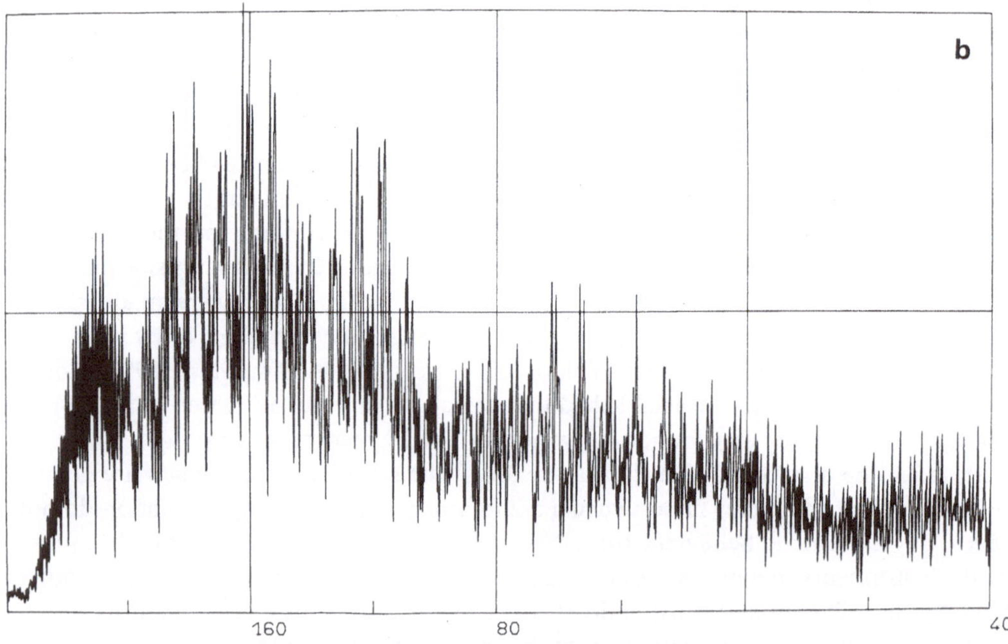

Fig. 8. Power spectra of the IR intensity measurements for 1977 (a) and 1978 (b) years data. A harmonic amplitude $W^{1/2}$, in arbitrary units, is plotted vs period expressed in minutes.

The latest IR data obtained in summer of 1981 also indicated the presence of 160-min intensity variations. Records suitable for reduction were made during 11 clear days in the interval 20 July–15 August, in all about 71 hr of data string. The superposed epoch plot of these data (subjected again to low frequency filtering by 2nd order polynomials, prior to analysis) for the period 160.010 min showed statistically significant ($2A/\sigma \approx 4$) sinusoidal wave with a harmonic amplitude $A \approx 2 \times 10^{-4} I_{\odot}$, see Figure 9, bottom; of great interest is that the phase of the maximal IR signal (i.e. the time of maximum of a sinusoid assigned to the data points), $\approx 01^h45^m$ UT, happened to be in close agreement with the phase of maximum line-of-sight velocity, $\approx 01^h37^m$ UT, inferred from the Crimean velocity observations in 1974–1981 (top); the zero phase everywhere corresponds to the UT moment 00^h00^m on 1 January 1974, and the phase of 160-min modulation at a given epoch is determined by fitting the data with a 16-points numerical (sinusoidal) approximation to the mean 160-min oscillation shape.

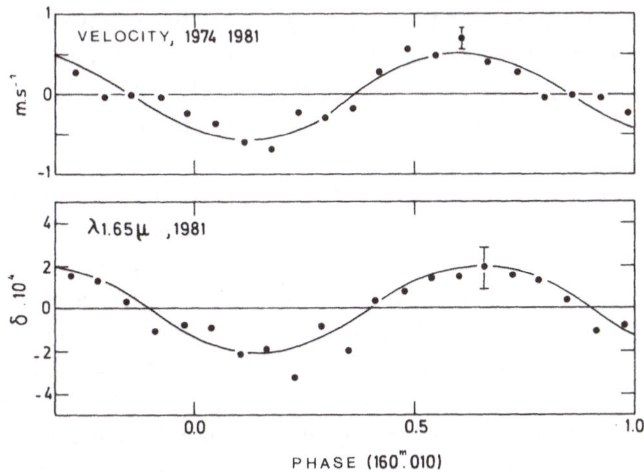

Fig. 9. Superposed epoch plots of the Doppler velocity according to Crimean observations in 1974–1981 (top) and that of the solar IR differential intensity in 1981 (bottom) for the period 160.010 min. The error flags are standard deviations from the mean.

One should note, however, that immediate comparison of phases – at 160.010 min period – of the IR intensity oscillations inferred from data of different years, 1977, 1978, and 1981, could be misleading since the late measurements in 1981 (Figure 9, bottom) were made with the profiled mask removed, in contrast to observations undertaken in previous years. The removal of the mask can inevitably change the weighting function for measurements of the amplitude of intensity variations over the Sun's disk and, hence, to unknown phase shift for the 160.010-min superposed epoch plot.

5. Conclusions

A differential technique for the detection of time variations in the solar limb darkening law has been developped and proved to be quite sensitive, up to 10^{-4} in units of 'the average Sun' intensity, when it is used at the atmospheric transparency window of the near infrared (λ 1.65 μm). When a statistical analysis, based on the power spectrum and superposed epoch routine is applied to a sufficient amount of data, then the oscillatory component and the prevailing phase-coherent oscillations (16.010 min) can be easily extracted from the available data.

On the basis of our IR differential intensity measurements made in 1977, 1978, and 1981 (a total of 78 days) and analysis of the data series we conclude that the limb darkening function in the near IR does undergo 160-min oscillations with a mean amplitude of about $\pm 2 \times 10^{-4}$ in units of the 'average Sun' intensity. The phase of the 160-min oscillations in 1981 data obtained without the profiled mask (see Figure 2), i.e., when measured as almost pure 'center minus limb' difference, is found strongly correlating with the peak of the line-of-sight velocity (see Figure 9).

This 160-min periodic variation cannot be explained by either a 1-day regular sampling of observations or influences of a broad band noise source (see also Koutchmy *et al.*, 1980). Since atmospheric and instrumental explanations have also failed to give a significant effect for the appearance of the 160-min periodicity (see Severny *et al.*, 1979; Kotov *et al.*, 1982), one may conclude that the IR differential intensity observations strongly support the solar origin of the 160-min oscillations. Thus, efforts towards a more elaborate modeling of the adiabatic or non-adiabatic processes and the solar convection zone are now warranted not only by velocity, but also by other types of observational data available in optical, radio, and IR spectral ranges.

However, the precise physical mechanisms (presumably connected with periodic temperature variations in the photosphere) responsible for the IR variations are still not clear; also the relation of the limb darkening variations to the solar constant monitoring (see Deubner, 1977; Woodard and Hudson, 1983) is open to question.

We expect that in the near future the intensity (particularly, IR) observations will be increasingly used to refine our understanding of solar global oscillations and the Sun's structure, in particular, – of the nature of 160-min oscillations for which no completely satisfactory explanation has yet been avanced.

Acknowledgements

This project was initiated thanks to the agreement between the French CNRS and the Soviet Academy of Sciences for cultural and scientific exchanges.

We have also benefited from fruitful discussions with A. B. Severny, H. A. Hill, and P. Delache, to whom we express our gratitude. Also the assistance of T. T. Tsap in obtaining the observations is gratefully acknowledged. Finally, this research was facilitated by a visit of V.A.K. to the Institut d'Astrophysique, Paris, sponsored by the French CNRS (ATP 4272) and by the USSR Academy of Sciences.

References

Brookes, J. R., Isaak, G. R., and van der Raay, H. B.: 1976, *Nature* **259**, 92.
Deubner, F. L.: 1977, *Astron. Astrophys.* **57**, 317.

Dittmer, P. H.: 1977, Ph. D. Thesis, Stanford Univ. IPR Rept. No. 686.
Gough, D. O.: 1980, in H. Hill and W. Dziembowski (eds.), *Lecture Notes in Physics* **125**, 279, Springer, Heidelberg.
Grec, G. and Fossat, E.: 1979, *Astron. Astrophys.* **77**, 351.
Grec, G., Fossat, E., and Pomerantz, M.: 1980, *Nature* **288**, 541.
Hill, H. A. and Caudell, T. P.: 1979, *Monthly Notices Roy. Astron. Soc.* **186**, 327.
Hill, H. A., Rosenwald, R. D., and Caudell, T. P.: 1978, *Astrophys. J.* **225**, 304.
Kotov, V. A., Severny, A. B., and Tsap, T. T.: 1981, *Izv. Krymsk. Astrofiz. Obs.* **66**, 3.
Kotov, V. A., Severny, A. B., Tsap, T. T., Moiseev, I. G., Efanov, V. A., and Nesterov, N. S.: 1983, *Solar Phys.* **82**, 9 (this volume).
Koutchmy, S., Koutchmy, O., and Kotov, V. A.: 1980, *Astron. Astrophys.* **90**, 372.
Scherrer, P. H., Wilcox, J. M., Severny, A. B., Kotov, V. A., and Tsap, T. T.: 1980, *Astrophys. J.* **237**, L97.
Severny,A. B., Kotov, V. A., and Tsap, T. T.: 1976, *Nature* **259**, 87.
Severny,A. B., Kotov, V. A., and Tsap, T. T.: 1979, *Astron. Zh. (USSR)* **56**, 1137.
Woodard, M. and Hudson, H.: 1983, *Solar Phys.* **82**, 67 (this volume).

STRUCTURE OF THE SOLAR OSCILLATION WITH PERIOD NEAR 160 MINUTES*

PHILIP H. SCHERRER and JOHN M. WILCOX

Institute for Plasma Research, Stanford University, Stanford, CA 94305, U.S.A.

Abstract. The solar oscillation with period near 160 min is found to be unique in a spectrum computed over the range of periods from about 71 to 278 min. Our best estimate of the period is 160.0095 ± 0.001 min, which is different from 160 min (1/9 of a day) by a highly significant amount. The width of the peak is approximately equal to the limiting resolution that can be obtained from an observation lasting 6 years, which suggests that the damping time of the oscillations is considerably longer than 6 years. A suggestion that this peak might be the result of a beating phenomenon between the five minute data averages and a solar oscillation with period near five minutes is shown to be incorrect by recomputing a portion of the spectrum using 15 s data averages.

Oscillations of the Sun with a period near 160 min were discovered by Severny *et al.* (1976) using a technique in which the Doppler shift of the central portion of the solar disk was compared with that of an outer annulus, using a Babcock solar magnetograph specially modified for this observation. This discovery was further described by Kotov *et al.* (1978). Similar observations using resonant scattering to measure the mean velocity of the Sun as a whole were reported by Brookes *et al.* (1976).

These reports were met with some skepticism for three reasons: (1) The observed amplitude of less than 1 m s^{-1} is very small and near the limits of observing capability; (2) a period of 160 min is exactly one ninth of a day, and could therefore appear in a power spectrum as a harmonic of the power at 24 hr that is present as an artifact in data obtained from observatories at mid-latitudes, i.e. when the observations must be interrupted every night; and (3) the source of the oscillation is not understood. It could be a *g*-mode, but if so one would expect to observe a number of adjacent *g*-modes, just as there are many *p*-modes with periods near 5 min.

Observations at the Stanford Solar Observatory (Scherrer *et al.*, 1979, 1980) have helped to obviate the first two problems. The third has not yet been solved.

This report is of a more detailed analysis of the observations. We start the analysis by combining the observations obtained at the Stanford Solar Observatory (for the observational arrangement see Scherrer *et al.* (1983)) during the summers of 1977 through 1980 with those obtained at the Crimean Astrophysical Observatory by V. A. Kotov, A. B. Severny, and T. T. Tsap during the summers 1974 through 1979.

For this computation 5 min averages of the observations were used. The observations on each day were fit with a parabola to remove a daily drift that at Stanford is of the

* Proceedings of the 66th IAU Colloquium: *Problems in Solar and Stellar Oscillations*, held at the Crimean Astrophysical Observatory, U.S.S.R., 1–5 September, 1981.

Solar Physics **82** (1983) 37–42. 0038–0938/83/0821–0037$00.90.

order of one m s^{-1} per hour. The residuals were then normalized to have a standard deviation equal to one, so as to give equal weight to each day's observations from each observatory. This procedure was done to allow combining data from both observatories in one analysis. The implicit assumption in such a procedure is not just that the sensitivity of the two instruments differ but also that variations in signal strength from day-to-day or year-to-year within each observatory's dataset are due to instrumental variations. This assumption is probably unwarranted but the method does provide a way to combine datasets with different characteristics with the understanding that all information about signal amplitude is lost in order to gain additional resolution in frequency with a reduced contribution from ghost lines. The main difficulty when combining many days of observations is the effect in the spectrum of the observing time window. An observation made with regularly spaced data gaps produces a spectrum with a well defined set of ghost lines for each real line. The data used in the present analysis does not have evenly spaced gaps, but has a very complex distribution of observing times wity a resulting complex set of ghost lines in the spectrum. By combining the Stanford and Crimean data the ghosts from the diurnal data gaps are greatly reduced since the observatories are situated nearly 180 degrees apart in longitude.

The Stanford and Crimean observations were thus combined as a single data set to compute harmonic amplitude spectra using a simple least-squares method to find the Fourier coefficients. For an observation duration of 6 years, the spectral resolution is about 5 nanoHz so we can compute the spectrum in 2 nanoHz steps to identify all peaks corresponding to oscillations with lifetimes of more than several years. To examine a large range of possible long period oscillations, we computed the spectrum from 60 to 240 microHz. This corresponds to periods of roughly 70 to 280 min.

Figure 1 shows a small part of the resulting spectrum in the range from 159.9 to 160.1 min (104.10 to 104.22 microHz). The peak near 160 min is near the center of this

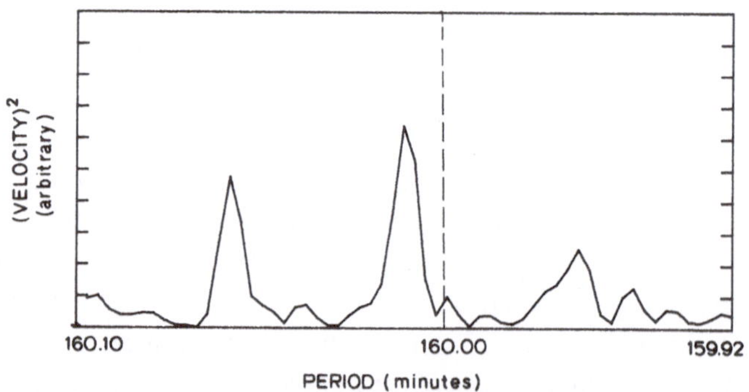

Fig. 1. Harmonic amplitude spectrum of the combined solar velocity observations from Stanford and the Crimea in the range from about 159.9 min to 160.1 min. The vertical scale is arbitrary because the daily residuals at each observatory were normalized to have a standard deviation equal to one. The central peak has a period of 160.0095 min and is surrounded on each side by a peak displaced by 32 nanoHz, which is the splitting associated with an interval of one year (see text). This spectrum was computed with a resolution of 2 nanoHz.

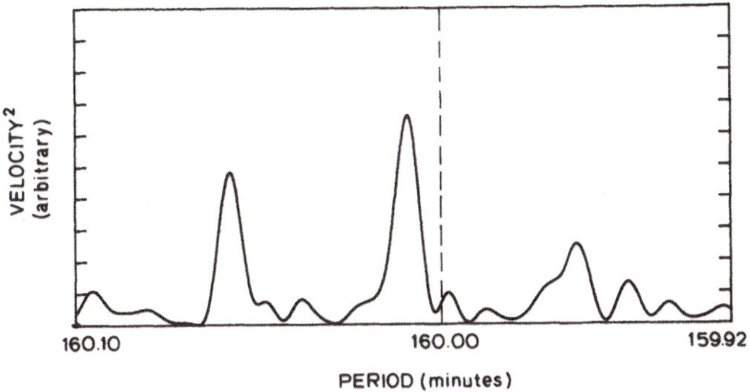

Fig. 2. Same spectrum as Figure 1, but computed with a resolution of 0.1 nanoHz. The vertical line is drawn at a period of 160 min and it is clear that the central peak is significantly removed from this period.

figure. In order to define this peak more clearly, part of the spectrum was recomputed with a resolution of 0.1 nanoHz. This high-resolution spectrum is shown in Figure 2. The centroid of the central peak in Figures 1 and 2 is at 104.1605 microHz (160.0095 min). Thus 160.0095 ± 0.001 min is our best estimate of the period of this solar oscillation, which is different from 160.000 min by a highly significant amount. This analysis is consistent with the agreement in the phase of the oscillation at the two observatories reported by Scherrer *et al.* (1979, 1980) and the 'impressively good' agreement observed at the South Pole by Fossat *et al.* (1981; see also Grec *et al.*, 1980). The first two objections mentioned above are thereby apparently resolved.

The central peak shown in Figures 1 and 2 is surrounded on each side by a peak displaced by 32 nanoHz, which is the splitting associated with an interval of one year. This interval appears in the spectrum because the observations are concentrated during the summer months.

The full width at half maximum of the central peak in Figure 2 is 5.3 nanoHz, which is approximately the limiting resolution to be obtained from an observation lasting 6 years. This suggests that the damping time of the oscillation is considerably longer than 6 years.

Figure 3 shows the full spectrum in the range from 71 to 278 min (60 to 235 microHz) computed in steps of 2 nanoHz. The central peak shown in Figures 1 and 2 now appears in Figure 3 as an isolated line whose amplitude is several times larger than that of any other peak within this frequency range (with the exception of the two yearly satellite peaks explained above). The singular nature of the oscillation at 160.0095 min is apparent in Figure 3. To our knowledge an accepted theoretical explanation of this situation does not yet exist. For a concise review of the theoretical situation see de Jager (1981).

Childress and Spiegel (1981) have proposed that this oscillation may be described as a strange attractor. They suggest that such a signature would include long 'periods'

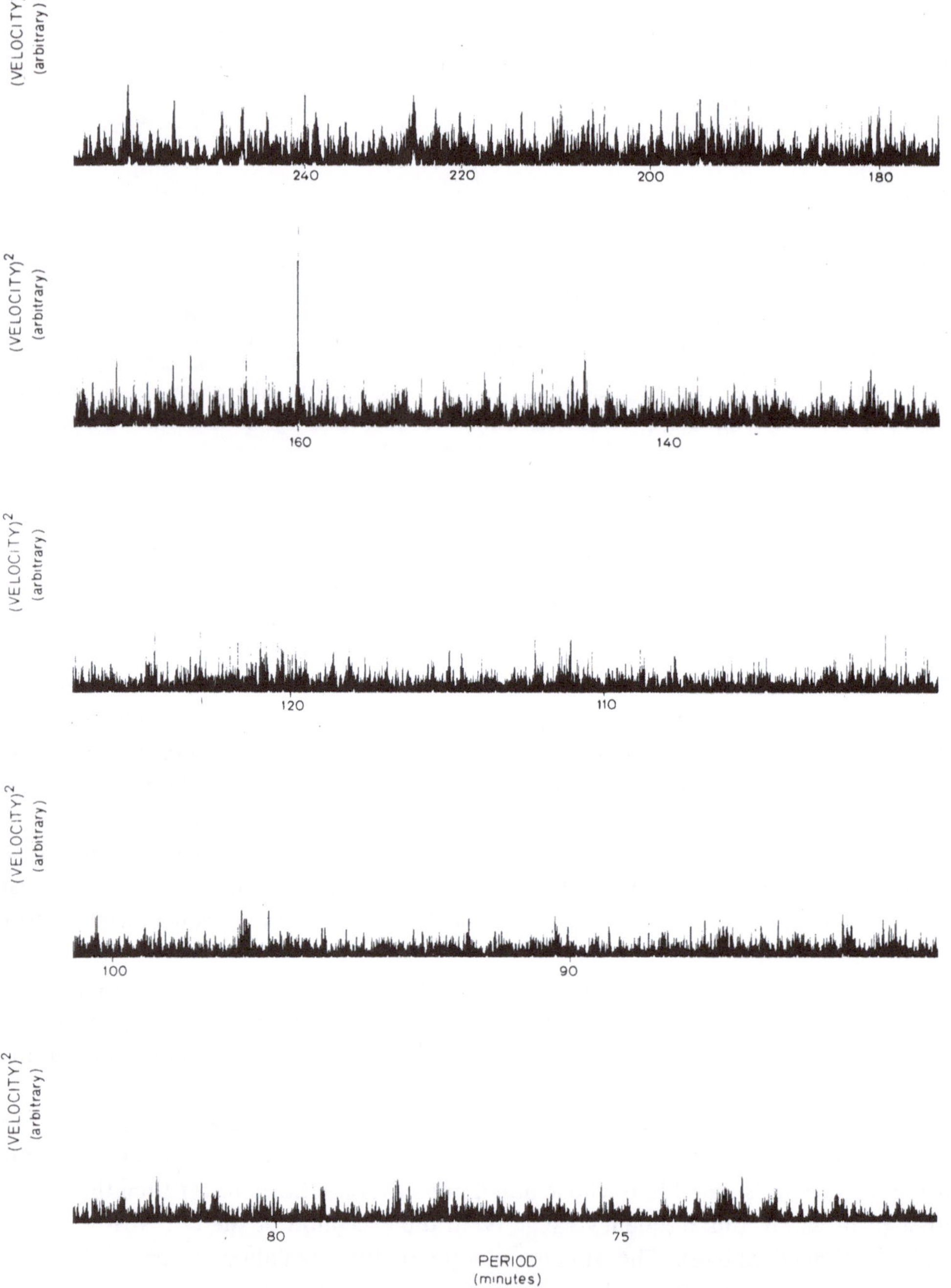

Fig. 3. Spectrum similar to that computed in Figure 1 but over an extended range from about 1.2 hr to 4.6 hr, computed in steps of 2 nanoHz. The central peak shown in Figure 1 at a period of 160.0095 min is now seen to be unique within the period range shown in Figure 3.

(compared to that of the radial fundamental mode), erratic behavior, and intermittency. The physical mechanism may involve instabilities driven in a thin layer.

The amplitude of the oscillation cannot be examined in the spectrum described above since the data was normalized. Figure 4 shows the spectrum computed in the same range as Figures 1 and 2 using only five minute averages of the Stanford data. For this computation the individual day's observations were *not* normalized. This spectrum is similar to the spectra computed from the combined data of Stanford and the Crimea shown in Figures 1 and 2. The amplitude of the central peak in Figure 4 corresponds to a velocity of 17 cm s^{-1}, which is consistent with earlier analyses and with the South Pole observations (Grec *et al.*, 1980; Fossat *et al.*, 1981).

Philippe Delache (1981) has suggested that the solar oscillation with a period near 160 min might be understood as a beating between the five minute intervals in which the data were averaged for the above computations and one of the solar oscillation modes with period near five minutes discovered by Claverie *et al.* (1979). Although some objections to this suggestion could be proposed, the basic test is to compute the spectrum using other than five minute data averages. The spectrum in Figure 5 was

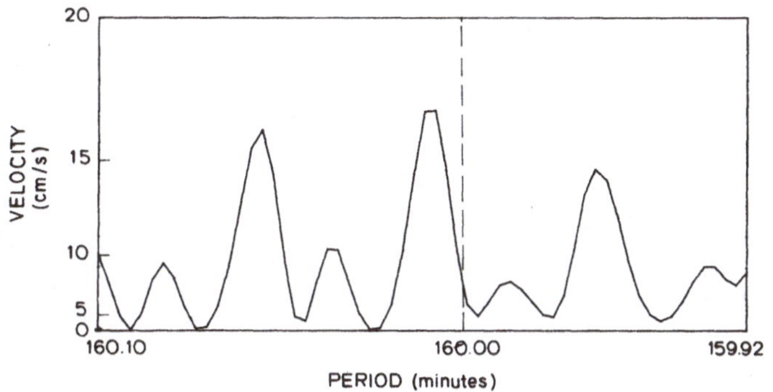

Fig. 4. Spectrum computed in the range of periods from about 159.9 to 160.1 min using only Stanford observations that were not normalized on each day. The maximum of the central peak corresponds to a velocity of 17 cm s^{-1}. This spectrum was computed using 5 min averages.

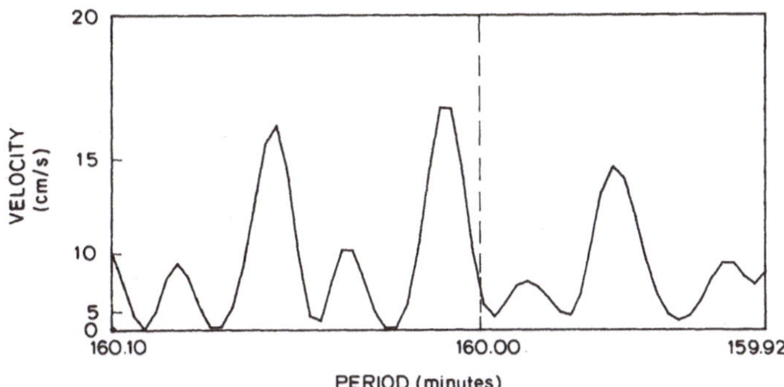

Fig. 5. The same as Figure 4, except that in response to a suggestion by Delache (1981) the spectrum was computed using 15 s averages. The central peaks in Figures 4 and 5 are very similar.

computed using the same data as in Figure 4, but using 15 data averages. These two spectra are essentially identical. It therefore appears that the suggestion of Delache can be excluded.

Acknowledgements

We thank A. B. Severny, V. A. Kotov, and T. T. Tsap for sharing their data with us, and for many interesting discussions at the Crimean Astrophysical Observatory and at the Stanford Solar Observatory.

This work was supported in part by the Office of Naval Research under Contract N00014–76–C–0207, by the National Aeronautics and Space Administration under Grant NGR05–020–559 and Contract NAS5–24420, by the Atmospheric Sciences Section of the National Science Foundation under Grant ATM77–20580 and by the Max C. Fleischmann Foundation.

References

Brookes, J. R., Isaak, G. R., and Van der Raay, H. B.: 1976, *Nature* **259**, 92.

Claverie, A., Isaak, G. R., McLeod, C. P., van der Raay, H. B., and Roca Cortes, T.: 1979, *Nature* **282**, 591.

Childress, S. and Spiegel, E. A.: 1981, 'A Prospectus for a Theory of Variable Variability', preprint No. A14 of Department of Astronomy and Astrophysics, Columbia University.

de Jager, C.: 1981, *Solar Phys.* **74**, 11.

Delache, P.: 1981, *Compt. Rend. Acad. Sci. Paris* **293**, Serie II, 949.

Fossat, E., Grec, G., and Pomerantz, M.: 1981, *Solar Phys.* **74**, 59.

Grec, G., Fossat, E., and Pomerantz, M.: 1980, *Nature* **288**, 541.

Scherrer, P., Wilcox, J. M., Kotov, V. A., Severny, A. B., and Tsap, T. T.: 1979, *Nature* **277**, 635.

Scherrer, P., Wilcox, J. M., Severny, A. B., Kotov, V. A., and Tsap, T. T.: 1980, *Astrophys. J.* **237**, L97.

Scherrer, P., Wilcox, J. M., Christensen-Dalsgaard, J., and Gough, D. O.: 1983, *Solar Phys.* **82**, 75 (this volume).

THE MEASUREMENT OF LONG-PERIOD OSCILLATIONS AT SACRAMENTO PEAK OBSERVATORY AND SOUTH POLE*

ROBIN STEBBINS

*Sacramento Peak Observatory***, *Sunspot, NM 88349, U.S.A.*

and

CHRISTOPHER WILSON***

Department of Physics, University of California at San Diego, La Jolla, CA 92093, U.S.A.

Abstract. A program to measure long-period brightness oscillations at the solar limb has been pursued at Sacramento Peak Observatory for several years. Past improvements in observing technique and data analysis are reviewed. The encouraging results aid in the verification of the reality and the origin of oscillatory signals. However, the main stumbling block to this and other observational programs is the length of observing sequences imposed by the day/night cycle. The South Pole has received considerable attention as a site where extended observations might be possible. Currently, the Sacramento Peak program is developing a South Pole telescope designed for the observing technique and data analysis proven in Sunspot. A review of pertinent South Pole site parameters is given here for other workers who may be considering South Pole observations. Observing sequences longer than 150 hr are possible, though rare. Data sets of this duration are very attractive for solar oscillation studies.

1. Introduction

The history of observations of solar oscillations has been a series of improvements in observational technique, database, and data analysis. Three examples demonstrate this: The first report of global oscillations (Hill and Stebbins, 1975) was a serendipitous byproduct of a novel observing method. The case for the 160 min oscillation rests largely on the vast database compiled by the Crimean and Stanford groups. The analysis leading to the diagnostic diagram revealed the real character of the five minute oscillation (Deubner, 1975). Because of their complex nature and subtle manifestations, future progress in the observational study of solar oscillations will be achieved largely through further improvements in technique, database and analysis.

The program of oscillations study reported here has developed and tested improvements in observing methods and data analysis. The main focus is now on extending the database. While reviewing these efforts, it would be well to keep the three goals of any such study firmly in mind. Initially, the observer will want to demonstrate the existence

* Proceedings of the 66th IAU Colloquium: *Problems in Solar and Stellar Oscillations*, held at the Crimean Astrophysical Observatory, U.S.S.R., 1–5 September, 1981.

** Operated by the Association of Universities for Research in Astronomy, Inc., under contract AST 78–17292 with the National Science Foundation.

*** Summer Research Assistant at Sacramento Peak Observatory.

Solar Physics **82** (1983) 43–54. 0038–0938/83/0821–0043$01.80.

of an oscillatory signal, along with some measure of its robustness. Secondly, the origin of the oscillatory signal must be identified – in particular, whether the signal is of solar rather than terrestrial or instrumental origin and whether the signal is caused by an organized pulsation or some stochastic process. Finally, having established the observation of a solar oscillation, the observer will want to elucidate the properties of the oscillation. While restating these goals may seem trivial, they are frequently overlooked in the excitement of developing new instruments and reporting new results. This paper wil attempt to report progress using these goals for perspective.

2. The Basic Method and Its Generalization

The basic observation senses brightness oscillations in the limb of the Sun. These oscillations are expected to be localized at the extreme limb. The detection is achieved by analyzing the shape of the limb darkening function with the Finite Fourier Transform Definition (FFTD) of the edge of the Sun. This definition of the edge was developed for measuring the solar oblateness (see Hill *et al.*, 1975, for a complete discussion of the FFTD). The FFTD defines the edge of the Sun to be the center of an interval over which a particular Fourier component is missing. For the purposes of this discussion, application of the FFTD involves selecting an interval size, multiplying the limb profile by a prescribed weighting function, and moving the interval until the weighted sum vanishes. The interval size is a free parameter about which more will be said. The weighting function has been chosen to minimize the effects of seeing on the edge location, yet remain sensitive to changes in the limb darkening. Roughly, the weighting function measures a second derivative averaged over the interval. The behavior of the FFTD is documented by a well developed theory and empirical tests.

The FFTD has been previously implemented as part of the telescope requiring the choice of a single interval size. However, the FFTD can be employed more effectively for oscillations studies if multiple definitions with different intervals can be applied simultaneously. A family of edge definitions, with interval size as family parameter, can be used to trace the reality and origin of an oscillatory signal. Since each edge follows an apparent brightness change in the limb profile with a sensitivity dependent on the shape of the brightness perturbation, their relative responses constitute a signature which characterizes the origin of the oscillatory signal. This signature can be examined to test the reality of the power at every frequency without reliance on the broadband statistics of the power spectrum. Further, the signature will also reveal the origins of that power.

The generalized application of the FFTD has two other advantages. First, since the relative motions of the edge are found by differencing all the edges against one chosen as the reference, oscillations can be detected by observing only one edge of the Sun. In previous applications of the FFTD where the edge on one side of the Sun is referenced to the edge diametrically opposed, Fossat *et al.* (1981) have suggested that the observed oscillations were due to atmospheric differential refraction changes. This point has been forcefully contested (Hill, 1979). However, a measurement based on a single limb rather than a diameter renders the objection moot. The single-limb approach does have a disadvantage for deconvolving the actual brightness (Knapp *et al.*, 1980). The second advantage lies in the ease of implementation. By digitizing and recording a portion of the limb, minimal specialized hardware is required to implement the FFTD. The definitions can be chosen and applied numerically as part of the data reduction process. The instantaneous value, the average value, and the power spectrum of the seeing can also be derived from the reduction procedure. This will give average values and power spectra for a data set. This will be useful in identifying the source of an oscillation.

The observation, as described, consists of digitizing and recording a patch of the solar limb at regular intervals. The data reduction consists of computing the locations of each edge, and then subtracting the location each from a chosen reference edge. The difference between two edge locations reflects only brightness changes and hence is called a brightness signal. The Fourier transform and power spectrum of each of these brightness signals are generated in the conventional way (see Stebbins, 1980, for details). For display purposes, the power spectra are converted to amplitude spectra.

3. Results

Sixteen data sets comprising 156 hr were gathered over a span of two years. For each data set, an amplitude spectrum was computed for each brightness signal. These amplitude spectra can be thought of collectively as a single amplitude function of two variables, frequency and interval size. To concentrate on the simplest portion of the solar acoustic spectrum (above the g mode band head and in the least dense region of the p modes), the amplitude function is averaged from 0.45 to 0.6 mHz, and this cut in frequency is averaged for all data sets. The resultant signature, that is amplitude vs interval size, is shown in Figure 1, along with one standard deviation error bars. Clearly, there is a distinctive and statistically significant signature. The amplitude is clearly peaked toward small interval sizes, indicating a brightness fluctuation concentrated at the extreme limb. Further, the phases of all these brightness signals are very nearly the same, -0.94 ± 1.01 degrees. The interpretation of these facts has been given elsewhere (Stebbins, 1980), but will be reviewed and extended here.

Clearly, the signature and phases are not random. What is the origin of this systematic signal? The differential nature of the measurement eliminates most telescope problems. The signature clearly shows an apparent brightening change, and the phase uniformity indicates that it must be concentrated within a few arc seconds, half the smallest interval size, of the intensity onset. Exotic sensitivity variations in the photo-diodes near the limb could produce such a signature. However, discontinuities in the brightness signal should appear when the periodic recalibration is done. No significant discontinuities are seen.

As mentioned previously, atmospheric differential refraction does not affect the brightness signals. Seeing changes will alter the shape of the limb profile, particularly within a few arc seconds of the intensity onset. However, the amplitude of the brightness signal most sensitive to seeing is more than twice that which would have been caused by the actual seeing changes alone. Further, the signature is the wrong shape for seeing.

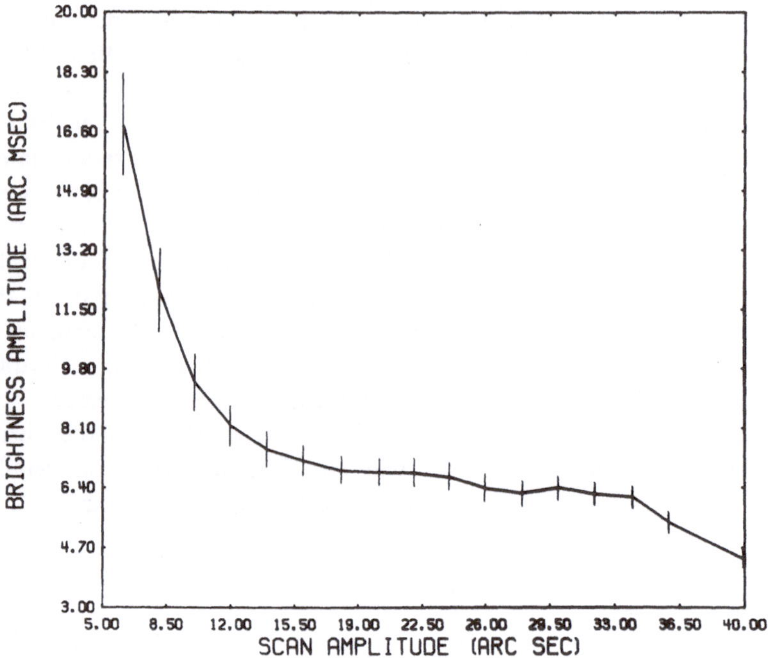

Fig. 1. The reality and origin of a solar oscillation are demonstrated by the signature, the oscillatory amplitude versus interval size. (Scan amplitude is one half of the interval size.) This signature has been averaged over sixteen data sets and 0.45 to 0.60 mHz. Error bars are shown at each value of interval size.

The seeing signature is essentially flat with a minimum at 8 arc sec, and a slight turn up at 6 arc sec.

If the observed signal is not of instrumental or atmospheric origin, could it not be from brightness features on the solar surface moving through the aperture or enduring while in the aperture? Brightness features rotating into the interval of each edge definition would cause the phases to be delayed by 250–720 degrees from one brightness signal to the one with the next largest interval size. So, brightness features rotating into the apertures are clearly inconsistent with the observed phases. Since the contrast of surface features declines toward the limb, evolutionary effects would be most pronounced at the inner boundary of the FFTD interval. This leads again to large phase shifts between brightness signals which are not observed.

Long-period oscillations could explain the observed signature and phases. Knapp *et al.* (1980) show evidence for a limb brightening sharply peaked within 6 arc sec of the intensity onset. Although no widely accepted calculation of the brightening exists, the effect should be most evident at the limb, and all of the brightness signals should respond in phase. Hence, the signature and the phases have established the reality of an oscillatory signal and indicated its origin as being a solar pulsation.

A signature, such as shown in Figure 1, could be drawn for every frequency point of every amplitude function and every data set. To demonstrate this, Figures 2a and 2b

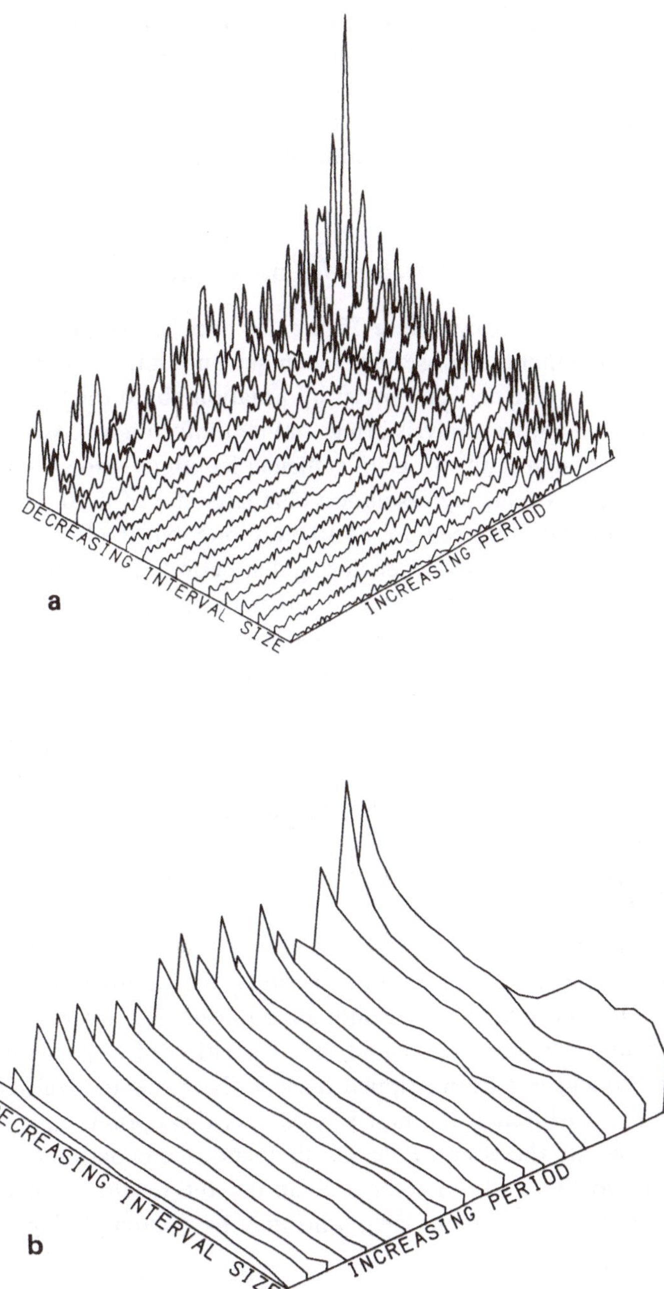

Fig. 2. Isometric plots depict the amplitude function. (a) Cuts along constant interval size are just the amplitude spectrum for each relative edge. (b) Cuts along constant period, at arbitrarily chosen frequencies, show the signature at each frequency. Period increases to the right, from 300 s to infinity. Interval size decreases to the left, from 80 to 12 arc sec.

show isometric plots of the amplitude function for one data set. Figure 2a shows a series of cuts along constant interval size, and Figure 2b shows a series of cuts along constant frequency. The latter demonstrates particularly well the presence of a pronounced signature, peaked toward small interval size. The cuts in Figure 2b were chosen at a fixed frequency spacing rather than for the presence of peaks. The conclusion to be drawn is that there is power from solar pulsations at many frequencies, not just the highest peaks.

4. A Telescope for the Antarctic

For the purposes of this review, these Figures 2a and 2b are presented to show the benefits of the improvements in observing technique and data analysis. The generalization of the FFTD does give new and better information about the reality and the origin of oscillations. The signature analysis brings out that information. What is lacking here is a database with enough time strings of sufficient length to resolve individual modes. One would ideally like frequency resolution characteristic of a 10-day-long observing sequence. The actual data set need not be 10 days long, but the longest gap should not exceed the longest continuous sequence during that period. To put the situation simply, long observing sequences can be acquired either by going to the appropriate space platform, employing a globe girdling network of ground based stations, or going to the polar regions in the right season. The last is the easiest of a difficult lot. Of the two polar regions, the Arctic offers poor weather, no land at the pole, and dismal accessibility; the Antarctic has reputedly good weather, a scientific station at the pole, and adequate assessibility. As a result, a telescope has been built especially for using the above-mentioned techniques at the South Pole (see Stebbins, 1981, for an extensive discussion).

The design goals of this telescope are the formation of a stable, high quality image, the digitization of a portion of the solar limb, assembly and operation in a polar climate, and portability appropriate to the logistical system. A cross-sectional diagram is shown in Figure 3 and a picture of the telescope on site at the South Pole in Figure 4.

The telescope has been tested at South Pole during the austral '80–81 summer. Several instrumental problems prevented the acquisition of any useful data on oscillations. The instrument has been shipped back to Sacramento Peak Observatory for further testing there. Although the first season was disappointing, it did provide an opportunity to evaluate the South Pole site firsthand. Considering the importance of long observations to the study of solar oscillations, the results of this evaluation are presented for other workers in the field who may be contemplating observations at South Pole.

5. South Pole Site Survey

Little data (Pomerantz, 1981a; Wyller, 1970) exists on the suitability of the South Pole for doing astronomy. Available meteorological observations, with one notable excep-

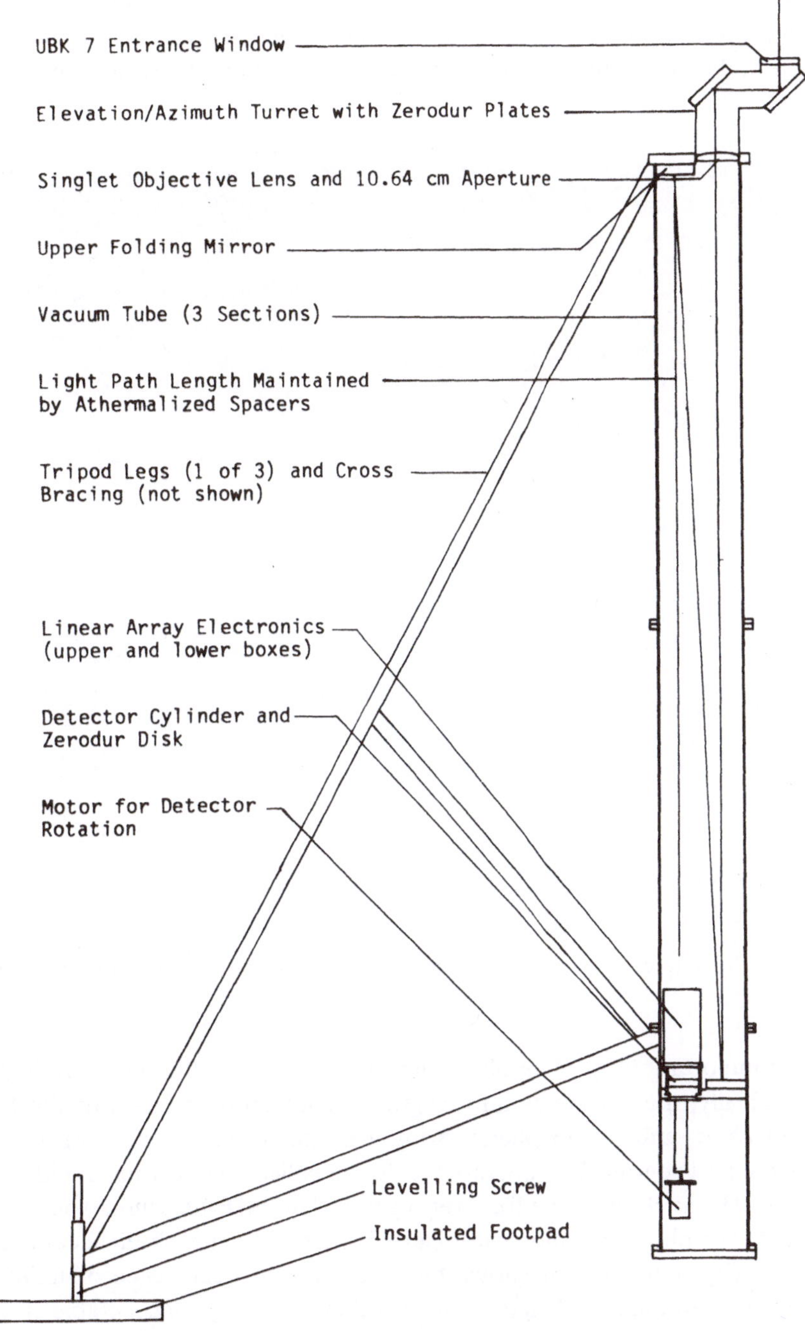

Fig. 3. A schematic cross-section of the Antarctic telescope.

tion, are not terribly useful, and the bulk of the evidence is in the form of casual observation. Since no one has spent the entire season for many years at South Pole, even personal experiences are limited. Herewith is a summary of all the available quantitative information about the suitability of the South Pole as an astronomical site. The quality and quantity of the information about most parameters is woefully lacking.

Fig. 4. The Antarctic telescope on location at the Amundsen–Scott South Pole Station.

The most important facet of an observing site for solar oscillations work is the cloud cover, specifically, the frequency and length of uninterrupted views of the Sun. The National Oceanic and Atmospheric Administration's Geophysical Monitoring for Climatic Change Program has operated a device called the Normal Incidence Pyroheliometer (NIP) at South Pole for over 4 years. The NIP measures solar irradiance every minute. Sample data is shown in Figure 5. These data have been accumulated into a histogram (Figure 6a) which shows the frequency of occurrence of uninterrupted viewing segments of different lengths. The threshold defining unacceptable cloud cover is a drop in intensity of 30% for one minute. There were three segments lasting longer than 90 hr in four year period (November through January, 1977–1981), the longest

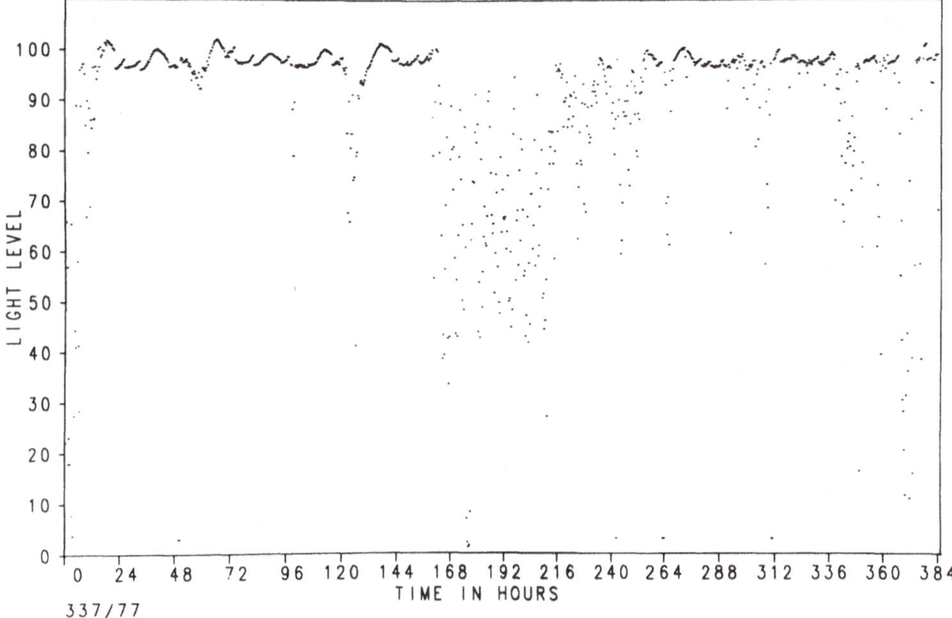

Fig. 5. Sixteen days of sample data from the NIP instrument are shown. The span begins at 00 : 00 GMT, 3 December, 1977. Only every twenty-fourth point is plotted because of the one-minute sampling interval. Brief intensity drops which result from the shadows of nearby instruments can be seen, but are excluded from the analysis. Both clear and cloudy weather can be seen.

being 98 hr. For a threshold of 50%, there are three segments longer than 100 hr, the longest being 133 hr.

A light level drop for one minute is an overly severe criterion for most observations. Figure 6b shows a similar histogram where light level drops lasting 60 min or less are ignored. In this circumstance, there are ten segments lasting longer than 100 hr in a four year period, the longest being 155 hr. If the dropout threshold is reduced to 15 min, there are only six such long spans, the longest being 152 hr.

The lesson from these data is that observing runs lasting from 100 to 150 hr, depending on the selection criteria, are possible at South Pole. It may be possible to link together phases over a much longer time span with correspondingly higher frequency resolution. However, practical considerations, such as start up time and instrument down time, will degrade these uninterrupted viewing periods. While one might like a longer time base for this analysis, these NIP data are a high quality measure of an important site parameter which does not exist for any conventional astronomical site.

The next site parameter to be considered is seeing. While not important for the observations described here, it will be of interest to other workers. Normally, the standard deviation of the atmospheric transfer function could be derived from the FFTD analysis. Lacking the anticipated electronic limb data, substitute photographic data was obtained by placing 35 mm film in the image plane of the telescope. The

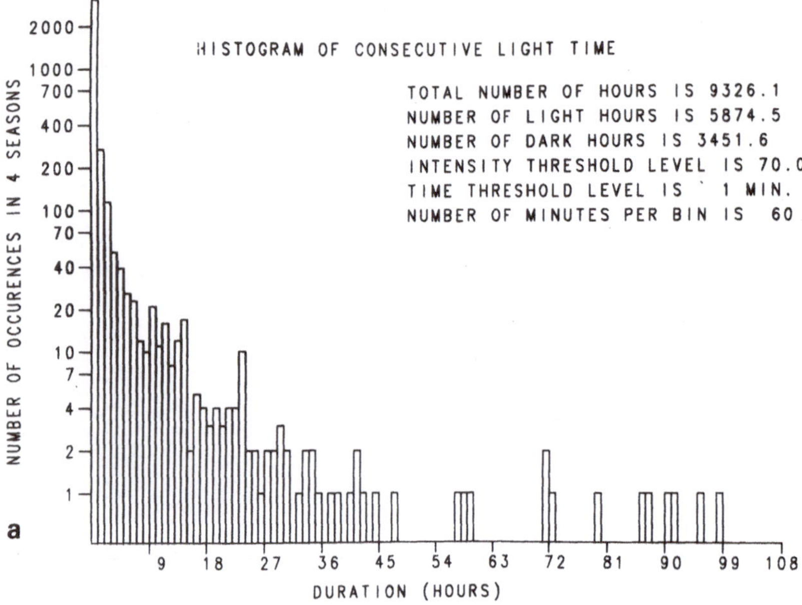

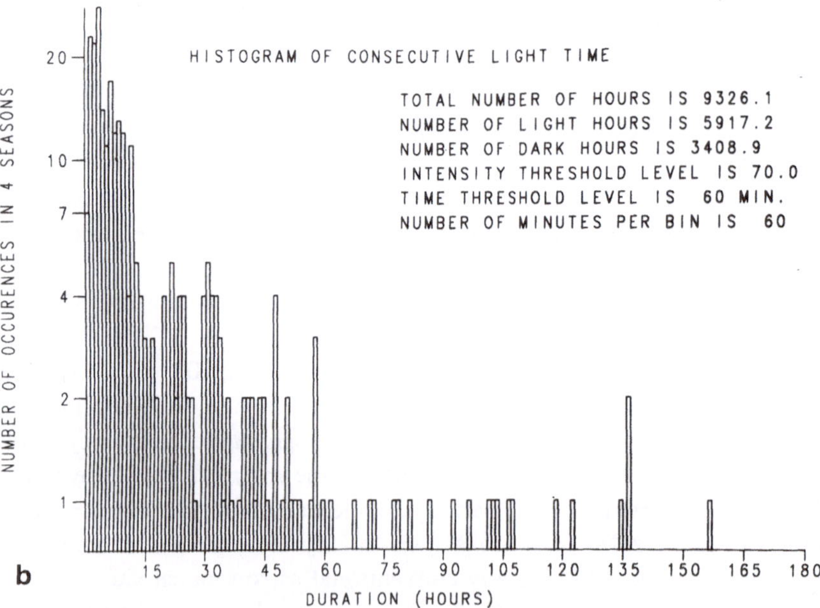

Fig. 6. The frequency of occurrence of clear skies with a given duration are shown in these histograms. The threshold defining clear skies is about 70% of normal light level. (a) An intensity drop below the threshold in any one-minute sample delimits a period of clear skies in this histogram. (b) An intensity drop lasting longer than 60 min delimits clear skies in this histogram. Four Antarctic summers (November, December, and January, 1977–78 to 1980–81) are shown here.

resulting images of the limb were microdensitometered, converted to intensities, and averaged to yield a limb profile. These profiles were then subjected to the FFTD analysis in order to obtain an upper limit on the seeing. (Only an upper limit can be found, owing to the vagaries of the photographic process which contribute to a blurring of the limb profile.) The result is a standard deviation of 2.26 arc sec, a seeing disk with a full width at half maximum of 5.34 arc sec, assuming a Gaussian transfer function. A direct visual check of the solar image at the same time suggested a full width at half maximum equal to 2 arc sec, consistent with the quantitative upper limit. These results stem from samples spanning not more than an hour of time. A much longer monitoring of the seeing is in order. However, the stability of the polar meteorology and the colder-than-air snow surface would lead one to expect long term good seeing.

The next parameter to be considered is the sky brightness, of interest in coronal observations. Using a visual sky photometer (Evans, 1948), the sky brightness was measured to be 20 millionths of the central disk intensity at 1.6 solar radii. For lack of clear weather, this measurement could only be made once. Again, a much longer baseline of observation is desireable. It might also be noted that the clear skies at South Pole are most impressive looking – no visible aureole around the Sun can be seen most of the time. The sky brightness measurement is less impressive. The objective measurement and the naked eye observation could be resolved by the fact that ice crystals in the air are the dominant scattering source (recall that there is no vegetation and essentially no local sources of pollutants). These ice crystals constitute high angle scattering centers, and consequently the whole sky may be brighter without the customary aureole caused by low angle scatterers such as dust and pollution. In support of that contention is the fact that the horizon is oftentimes imperceptible, the white snow graduates into the blue sky.

Other atmospheric phenomena of interest to astronomers are the precipitable water vapor, the ice crystals, and ambient temperature and winds. The precipitable water vapor hovers around 0.45 mm in the summer (Pomerantz, 1981b). The ice crystals, an everpresent phenomenon to greater or lesser extent, give a clear sky a sparkling appearance. Their effect on astronomical observations will probably depend on the particular measurement. The ambient temperature ranges from -20 to -32 degrees centigrade over most of the South Pole summer. Some consideration needs to be given to equipment design and to anticipated human activities (see Stebbins, 1981, for example), but the environment is not insurmountable. The winds are extremely constant, averaging 10 knots along the 20 degree east longitude meridian. The winds are never absent and never exceed about 30 knots, assuring a continuous wash of air over a telescope.

Some miscellaneous points of interest to observers contemplating an Antarctic observation: equipment transport can be extremely hazardous, and every precaution must be taken in packing fragile instrumentation. As befits a site as remote as the South Pole, maintenance and spare parts are hard to acquire. Finally, an Antarctic tour can tax an observing team far more than a conventional site.

In summary, the South Pole appears to be a good site for extended viewing of the

Sun when viewed as a compromise between limitations of a mid-latitude site and the effort needed to construct a grid of stations or a space platform. Providing that the observation is compatible with the weather, most of the other difficulties can be coped with.

6. Conclusions

Advances in the means of collecting and analyzing data have yielded better information about solar oscillations. A novel technique has been developed and tested at Sacramento Peak Observatory. The encouraging results have led to application at the South Pole, where longer data sets can be had. Although no results on oscillations are available yet, a preliminary review of the site indicates that extended observations will produce another advance in the understanding of solar oscillations.

Acknowledgements

We would like to acknowledge the work of G. Richard Mann the South Pole in acquiring some of the data discussed here. Raymond Smartt and George Streander prepared and calibrate the Evans Visual Sky Photometer. We would like to thank Ellsworth Dutton of NOAA's GMCC project for sharing their unpublished NIP data. We also acknowledge the National Science Foundation's support through a logistics grant (DPP80–01469).

References

Deubner, F.-L.: 1975, *Astron. Astrophys.* **44**, 371.

Evans, J. W.: 1948, *J. Opt. Soc. Am.* **38**, 1083.

Fossat, E., Grec, G., and Harvey, J. W.: 1981, *Astron. Astrophys.* **94**, 95.

Hill, H. A.: 1981, in R. B. Dunn (ed.), *Solar Instrumentation: What's Next*, Sacramento Peak National Observatory, Sunspot, New Mexico, Chapter 6, p. 350.

Hill, H. A. and Stebbins, R. T.: 1975, *Ann. N.Y. Acad. Sci.* **262**, 472.

Hill, H. A., Stebbins, R. T., and Oleson, J. R.: 1975, *Astrophys. J.* **200**, 489.

Knapp, J., Hill, H. A., and Candell, T. P.: 1980, in H. A. Hill and W. A. Dziembowski (eds.), *Nonradial and Nonlinear Stellar Pulsation*, Springer-Verlag, Heidelberg, Chapter 3, p. 394.

Pomerantz, M. A.: 1981a, in R. B. Dunn (ed.), *Solar Instrumentation: What's Next*, Sacramento Peak National Observatory, Sunspot, New Mexico, Chapter 6, p. 379.

Pomerantz, M. A.: 1981b, private communication.

Stebbins, R. T.: 1980, in H. A. Hill and W. A. Dziembowski (eds.), *Nonradial and Nonlinear Stellar Pulsation*, Springer-Verlag, Heidelber, Chapter 3, p. 191.

Stebbins, R. T.: 1981, in R. B. Dunn (ed.), *Solar Instrumentation: What's Next*, Sacramento Peak National Observatory, Sunspot, New Mexico, Chapter 6, p. 390.

Wyller, A. A.: 1970, in *Polar Research: A Survey*, National Academy of Sciences, Washington, D.C., p. 170.

FULL-DISK OBSERVATIONS OF SOLAR OSCILLATIONS FROM THE GEOGRAPHIC SOUTH POLE: LATEST RESULTS*

(Invited Review)

GÉRARD GREC

Department D'Astrophysique de l'I.M.S.P., E.R.A. du C.N.R.S.
Université de Nice, Parc Valrose F-06034, Nice Cedex, France

ERIC FOSSAT

Observatoire de Nice, BP252, 06007 Nice Cedex, France

and

MARTIN A. POMERANTZ

Bartol Research Foundation of The Franklin Institute, University of Delaware, Newark, Delaware, U.S.A.

Abstract. This paper presents the latest results obtained from the analysis of the full-disk Doppler shift observations obtained at the geographic South Pole in 1981. About 80 normal modes of oscillation (l = 0–3) have now been identified. Their frequencies range from 1886 μHz ($l = 1$, $n = 12$) to 5074.5 μHz ($l = 2$, $n = 35$), and their amplitudes are as low as 2.5 cm s^{-1}. Amplitude modulation occurs with periods of 1–2 days, and the individual oscillations appear to be excited randomly and independently. In cases where other groups have observed some of the modes identified by us, the agreement in frequency is good.

1. Introduction

The first uninterrupted observations of the Sun significantly longer than one day were made at the geographic South Pole during the first week of 1980. They consisted of Doppler shift measurement integrated over the entire solar disk, using a sodium cell optical resonance spectrometer (Grec *et al.*, 1976). The very high temporal resolution provided by the unusual duration (5 days) of this continuous data set has made it possible, for the first time, to resolve many individual normal modes of the oscillation of the solar sphere. A preliminary analysis of these data has already been published (Grec *et al.*, 1980). The present paper describes the more complete results which have since been obtained. They include the unambiguous identification of about 80 different normal modes. This identification has now made possible the use of the standard inversion technique for a seismological investigation of the solar internal structure (Christensen-Dalsgaard and Gough, 1981; Scuflaire *et al.*, 1982).

* Proceedings of the 66th IAU Colloquium: *Problems in Solar and Stellar Oscillations*, held at the Crimean Astrophysical Observatory, U.S.S.R., 1–5 September, 1981.

Solar Physics **82** (1983) 55–66. 0038–0938/83/0821–0055$01.80.

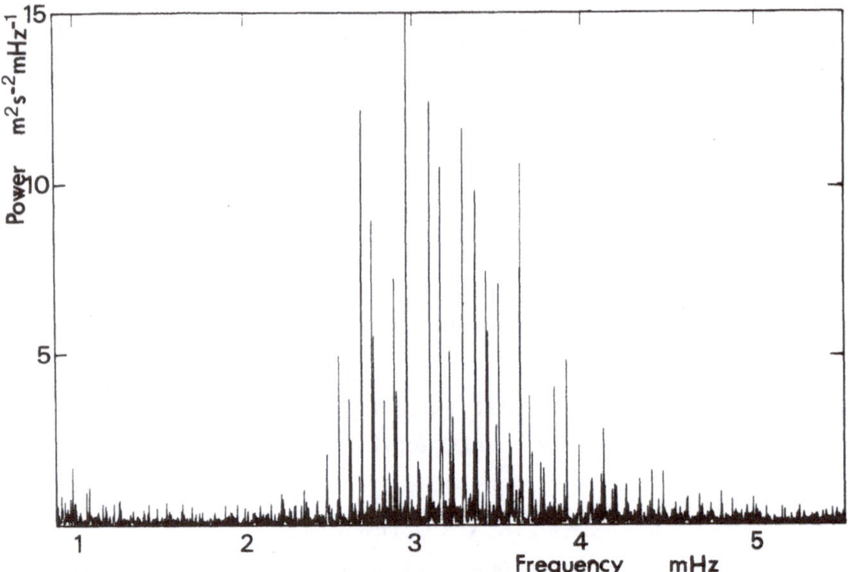

Fig. 1. Power spectrum of the South Pole data, which consists of the full-disk Doppler shift measurements recorded during the first week of January, 1980. The sample analyzed here has a duration of about 6 days, including three gaps of 2 to 3 hr. The frequency resolution is $\Delta\nu = 1.97$ μHz.

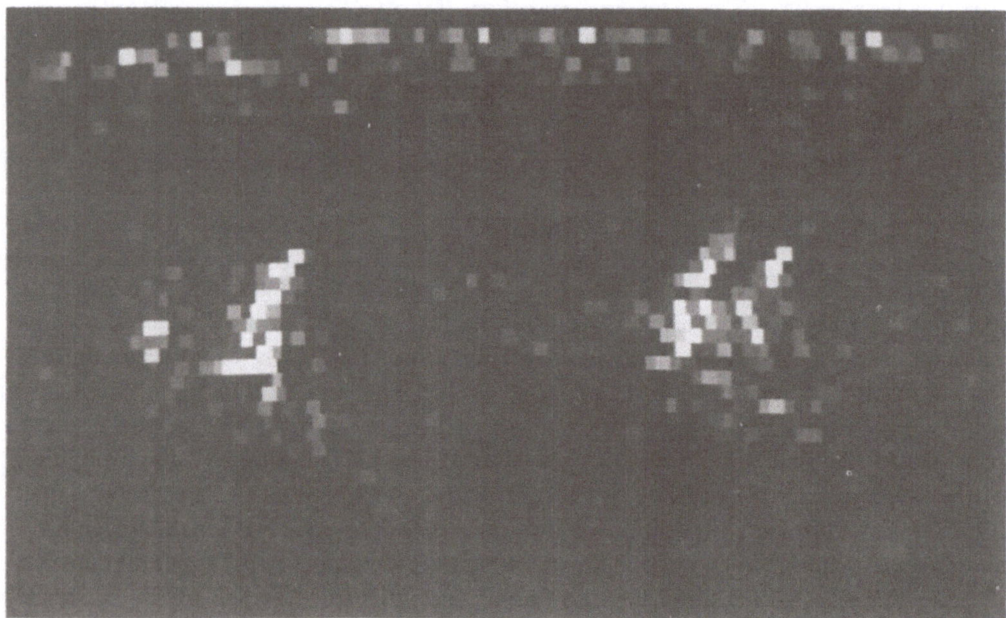

Fig. 2. Same as Figure 1, but the spectrum is now cut into slices of 136 μHz each. Each slice is displayed under the previous one on a video screen. The two lines on the right are the modes of even degrees ($l = 0$ is at the extreme right). The curvature of the four lines displays the departure from equidistance of the four sequence of radial harmonics. This curvature, which is expected from theory, even in the asymptotic approximation, makes possible the identification of each peak with its radial order, leaving room for almost no ambiguity. The first line displayed at the top of the picture starts at the frequency $\nu = 295.7$ μHz.

TABLE I

List of frequencies (in μHz) of normal modes which have been identified in the South Pole data

n \ l	0	1	2	3
12		1886.5	1947.5	
13	1958.5	2021		
14	2093	2157	2217	
15	2229	2294	2352	
16	2363	2427.5	2486	
17		2560.5	2621.5	2679
18	2631	2694	2756	2813
19	2765.5	2829.5	2890	2948.5
20	2900.5	2964.5	3026.5	3087
21	3034	3099	3161	3219.5
22	3169.5	3234	3296	3353.5
23	3303.5	3370	3432	3490
24	3442	3506.5	3568	3628.5
25	3578	3640.5	3701	3765.5
26	3714.5	3779.5		3900
27		3915.5		4035.5
28		4051.5	4111	4173
29	4123	4193	4252	
30	4262	4327	4389	4446
31	4399.5	4464	4526	4583.5
32	4537	4603	4663	
33	4673	4742	4800	
34			4937	
35			5074.5	

N.B.: 1 – The authors apologize for the fact that this list is not identical to the one which had been distributed during the meeting. A scaling error was hidden in the program of analysis and resulted in a decrease of all frequencies by a factor 0.9989.

We also call attention to a typographical error in the caption to Figure 1 in the preliminary power spectrum published in Grec *et al.* (1980). That figure actually comprised data extending over only 3.5 days, in contrast with Figure 1 in the present paper.

However, the other figures in the earlier paper did, in fact, cover the continuous 5-day run.

N.B.: 2 – Two pairs of even modes have not been listed because, as can be seen in Figures 2 and 3, the power spectrum displays for both these pairs a unique peak located in between the interpolated frequencies of $l = 0$ and $l = 2$ modes.

2. Radial Order Identification

In Grec *et al.* (1980), a superposed frequency analysis assuming a uniform spacing of 136 μHz was used to identify the degree l of each individual peak. However, this procedure did not allow the identification of the radial order n because the lowest order modes have too small amplitudes to be measured. This ambiguity, in fact, can be resolved thanks to the fact that the spectral peaks are not uniformly spaced. This is illustrated by Figure 2 which shows the same power spectrum as that in Figure 1, uniformly cut into slices of 136 μHz, with each slice displayed below the previous one.

In such a display, a uniform frequency spacing between consecutive radial harmonics would provide a distribution of power along straight lines (vertical for a spacing of 136 μHz). The curvature observed is then indicative of the departure from this uniform spacing. This general behaviour can be compared with an asymptotic approximation given by Tassoul (1980):

$$\nu(n, l) = \nu_0 \left(n + \frac{l}{2} + \varepsilon \right) + [l(l+1) + \delta] \frac{A \nu_0^2}{\nu}$$

in which ν is the frequency,

$$\nu_0 = \left[2 \int_0^R \left(\frac{\mathrm{d}r}{c} \right) \right]^{-1}, \qquad c = \text{sound speed}\,,$$

$$\varepsilon = \frac{n_e}{2} + \frac{1}{4}\,,$$

and A, δ, and n_e are constant integrals of the solar model.

This comparison has been made by Grec (1981) who found values of $A \approx 0.267$ and $\delta \approx -17$. Being given those two numbers, the value of ε depends on the identification of the order n of the modes. Keeping in mind that n_e has something to do with a polytropic index in Tassoul's calculation, the most reasonable values of n must provide a value of ε of the order of 1. This leads to a possible identification of the order n of each observed normal mode, as summarized in Table I. The frequencies listed in this table have been estimated through the barycentre of each peak, which is in certain cases very broad (10 μHz or more), as illustrated by the mode ($l = 1$, $n = 25$, $\nu = 3.64$ mHz). There are a few pairs of even modes which cannot be separated and, consequently, have not been listed although they are located in a frequency range which contains considerable power.

The same identification has been made independently by Christensen-Dalsgaard and Gough (1981), using no asymptotic approximation, but only least square differences between oscillation frequencies calculated with an almost standard solar model and our observed frequencies.

3. Amplitudes

As we will see in the next section, the amplitude of each individual oscillation is randomly modulated with a characteristic time of about two days. Consequently the use of a sample of data the duration of which is about 6 days does not allow to measure each amplitude with an uncertainty lower than about a factor two. This fact must be kept in mind when using the numbers listed in Table II. These amplitudes have been calculated by integrating the power in a bandwidth of 4 to 8 μHz around each peak frequency and by substracting the mean background noise. The amplitude is not listed if the signal to noise ratio was not larger than 2, in power.

4. Lifetime of Oscillations

Figure 2 clearly shows that the mean width of the peaks is broader than the intrinsic resolution of the data set, which is equal to 1.97 μHz. To estimate the average shape of these peaks, a superposed frequency analysis has been carried out in a more precise way than that originaly used by Grec *et al.* (1980). For this purpose, the curvature of the 'vertical' lines in Figure 2 has been estimated by fitting a polynomial as an additive component to $\Delta\nu$ = constant. The best choice of this polynomial was found to be:

$$\Delta\nu = 134.40 + 0.011\,(n_0 - 13.5)^2\,,$$

where n_0 represents the line taken as the order of the radial (l = 0) normal mode frequency present in this line. $\Delta\nu$ is the spacing between line n_0 and line $(n_0 + 1)$. As is shown in Figure 3, the sliced spectrum now displays power lines which are roughly vertical and can be used for vertical additions.

A vertical sum covering the entire frequency range which contains significant power (from line 10 to line 35) is shown in Figure 4. The following numbers measured on this

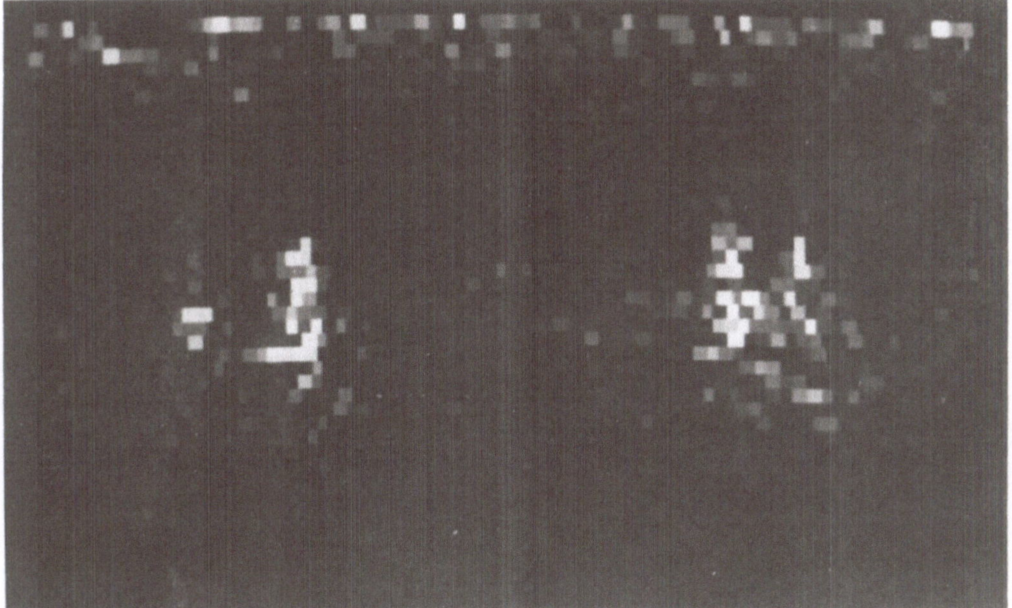

Fig. 3. Same as Figure 2, but each slice of the spectrum has been translated to make the power appear to be distributed in vertical lines. This was accomplished by replacing the constant spacing $\Delta\nu$ = 136 μHz by the polynomial $\Delta\nu = 134.40 + 0.011\,(n_0 - 13.5)^2$ (see text for details).

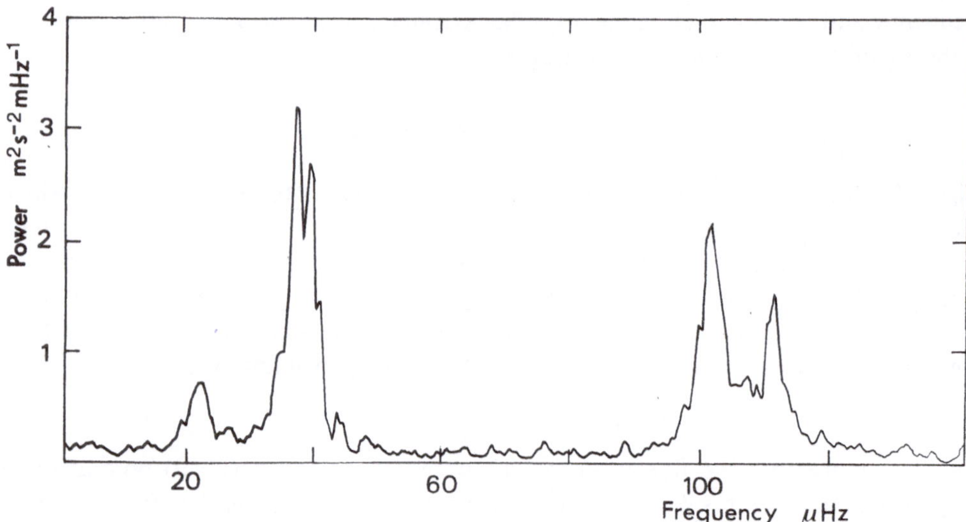

Fig. 4. A vertical summation of Figure 3 (superposed frequency analysis following the curvature) shows the mean shape of the four spectral peaks corresponding to normal modes of degree 3, 1, 2, and 0 (left to right).

figure are of interest (weighted mean values over the whole frequency range):

$$\nu(0, n) - \nu(2, n - 1) = 9.4\ \mu\text{Hz}\,,$$

$$\nu(1, n) - \nu(3, n - 1) = 15.3\ \mu\text{Hz}\,,$$

$$\nu(0, n) - \nu(1, n - 1) = 71.4\ \mu\text{Hz}\,.$$

The mean shape of all four peaks is reasonably well fitted by an exponential function $e^{-|\nu - \nu_0|/\sigma}$ with $\sigma = 2.35\ \mu\text{Hz}$. However, it must be noted that this function is not indicative of the shape of any individual peak, because it is an average of many peaks which have different heights and different widths. A very important point is that the natural width appears to increase very rapidly with the frequency as is demonstrated by the next two figures.

In Figure 5, the sum comprises lines 10 to 20 (roughly, frequencies below 3 mHz). In this case, the shape appears to be consistent with almost monochromatic peaks broadened only by the resolution of the analysis.

In Figure 6, the sum is made over the highest frequency range (above 3 mHz) and clearly proves that the peaks are much broader in this range. This tendency can be seen already on the power spectrum of Figure 1. It is also clearly seen on the power spectrum of brightness oscillations observed by the SMM spacecraft and analyzed by Woodard and Hudson (1983). This is not due only to a random scatter of the centroids of the individual resolved peaks, which could all be intrinsically sharp. There are several intrinsically broad peaks (e.g. $\nu = 3640$ and $\nu = 3432$ which have a width of respectively 9 and 7 μHz). There are also multiple peaks (such as $\nu = 3370$ or $\nu = 3578$) which have

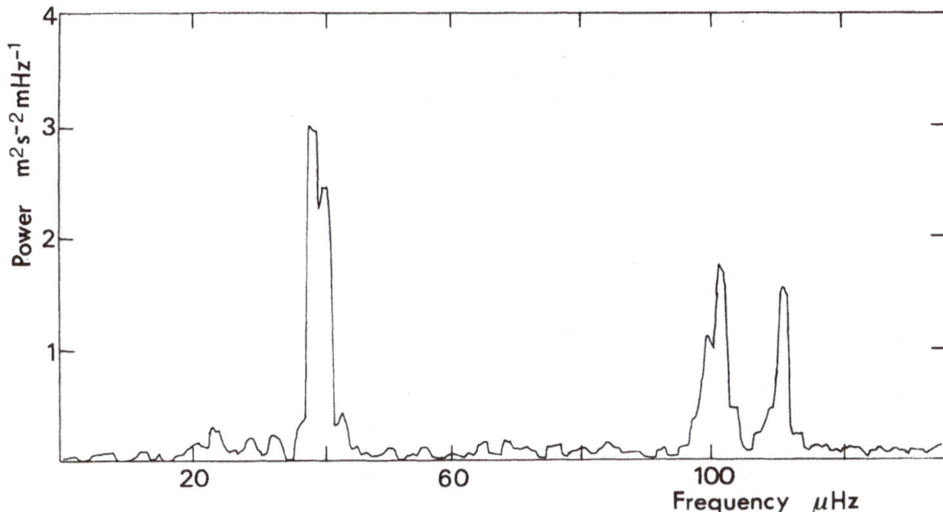

Fig. 5. Same as Figure 4, for frequencies below 3 mHz. The width of the mode $l = 0$ in this case is consistent with the resolution (1.97 μHz) of the data set. The slightly broader shape of the two others is not inconsistent with a small, unresolved, rotational splitting of about 0.5 μHz or less (although a small difference between the curvature of the four lines in Figure 2 can also result in a small difference among the widths of these three peaks).

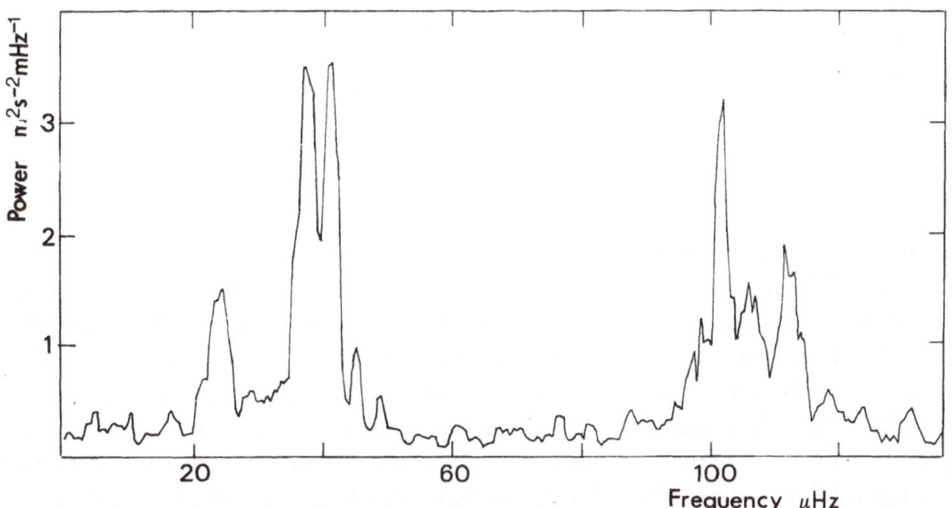

Fig. 6. Same as Figure 4, for frequencies higher than 3 mHz. It appears that the width of the spectral peaks increases dramatically in this frequency range.

to be regarded as a random distribution of power inside a broad envelope, which would be filled if the amount of data available was infinite.

These peaks are broadened, more and more with increasing frequency, by a modulation which can affect the amplitude, the frequency, or both. A direct information concerning the amplitude modulation has been obtained by cutting the data in 12 hr samples, and

TABLE II

List of amplitudes, in cm s^{-1}, of the normal modes identified in Table I. For the two pairs of even modes which cannot be separated in the power spectrum, the amplitude has been arbitrarily equidistributed (number in parenthesis).

n \ l	0	1	2	3
12		2.5	3	
13	3.5	3		
14	3.5	4	5	
15	4	3	4	
16	4	5	9	4
17		11.5	11.5	6.5
18	9	21	14	6.5
19	11.5	13	15	7
20	12	22	7.5	6.5
21	7.5	17.5	21	14
22	10.5	13.5	20	11
23	10	23	15.5	7.5
24	14	15.5	13	6.5
25	9.5	24.5	12	7.5
26	9.5	8.5	(9)	6
27	(9)	11	(10)	6.5
28	(10)	8.5	10	6
29	12.5	9.5	8.5	
30	6	9.5	6	6
31	8	8.5	4	6
32	6	5.5	4	
33	6		4.5	
34			4	
35			3.5	

by doing a Fourier analysis of each. In this case the resolution is not sharp enough to separate $l = 0$ from $l = 2$ or $l = 1$ from $l = 3$, but the odd and even pairs are separated. Figure 7 shows the power of the main pairs plotted as a function of time. For even pairs, the mixing between amplitude modulation and beating phenomena confuses the analysis. However, for odd pairs of peaks, the power is dominated by the contribution of $l = 1$ (except for 3.22 mHz), and the beating is almost negligible. Hence, it is essentially the amplitude modulation of the $l = 1$ modes which shows exponential increases or decreases with a typical time scale of the order or 1–2 days. Furthermore, there is no correlation between two neighboring harmonics.

Is there equipartition of a constant total power among all different normal modes, or is each mode independently and randomly excited and damped? Some information on this topic can be obtained from the time variation of the total power in the five-minute range (Figure 8). The r.m.s. relative fluctuation of the total power with time is 20%. Regarding the fact that there are essentially 30 main peaks which contribute to the total power, this r.m.s. fluctuation should have a value of about $(\sqrt{30})^{-1} \approx 18\%$ if each peak has an independant modulation of 100% in amplitude. This is in quite good agreement

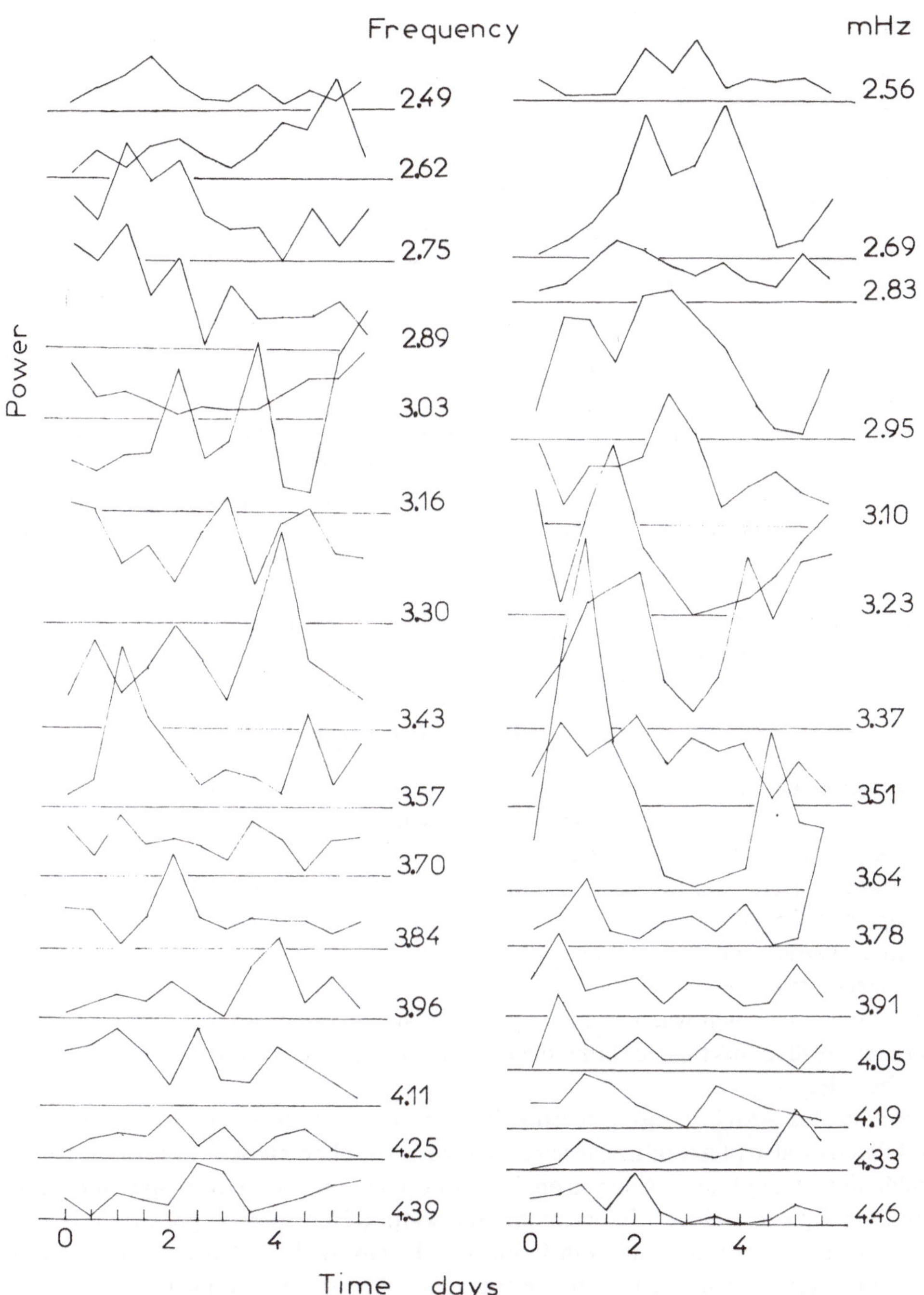

Fig. 7. Power, in $m^2 s^{-2}$, in each pair of even (left) or odd (right) peaks, versus time. The time series has been cut into samples of 12-hr durations for this analysis. The resolution in this case is 23 μHz, sharp enough to separate odd from even pairs, but not sufficient to resolve each individual mode inside a pair.

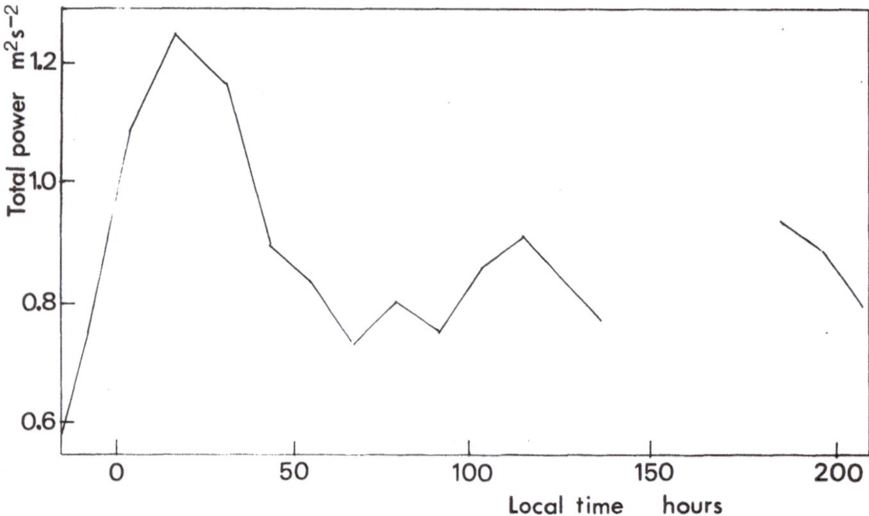

Fig. 8. Total power in the five-minute range (from 2 to 5 mHz) versus time. More data have been utilized in this figure than for the power spectrum, which was limited to the analysis of the almost continuous 6-day run.

with the measured value, and hence favors a random and independant excitation of each individual oscillation.

5. Discussion – Conclusions

(a) *Frequencies* – Table I contains many more frequencies than have been identified by any experiment conducted thus far. Where there is overlap, a comparison can be made with measurements by Claverie *et al.* (1981), Scherrer *et al.* (1983) and by Woodard and Hudson (1983). This comparison is shown in Figure 9, which displays all reported frequencies in a frequency-radial order coordinates diagram similar to Figure 2. The general agreement is quite satisfactory, since the mean difference between two observations never exceeds about 1 μHz. The scatter around the hypothetical continuous line which was fitted by a polynomial is also of the same order for all the different results, and this confirms the conclusion that the natural width of the peaks is a few μHz.

(b) *Lifetime* – An interesting question is to know if the oscillations are only amplitude modulated or also phase or frequency modulated. The characteristic time of the random modulation of amplitude displayed on the Figure 7 does not seem to change significantly with the frequency. Now it happens that the width of the spectral peaks does change by a very significant amount, with frequency. It may well be that the longest period oscillations maintain a long term phase coherence and are only modulated in amplitude. In this case, all the oscillations would have the same kind of amplitude modulation, while the lifetime of the phase would decrease dramatically at frequencies higher than 3 mHz. Further analysis and more data are required to clarify this issue.

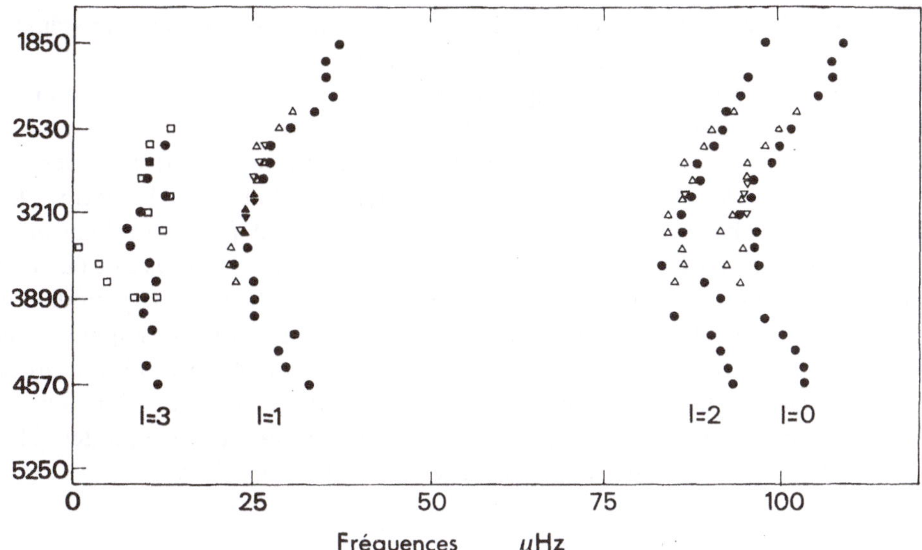

Fig. 9. Comparison of our frequencies determined from the South Pole data (●) with those published by Claverie *et al.* (△), Scherrer *et al.* (□), and Woodard and Hudson (▽). ▲, ■, and ▼ indicate coincidence (better than 0.5 μHz) with our frequencies.

(c) *Rotational splitting* – For frequencies smaller than 3 mHz, Figure 5 shows a peak width almost consistent with the resolution of the analysis in the case $l = 0$. In the case $l = 2$, we must keep in mind that the width contains the resolution of the observation (2 μHz), the unresolved splitting, the residual width of each split component and a possible slight extra-broadening due to the imperfect fitting of the real central frequencies with our polynomial. In these conditions, if the $l = 2$ modes are split in 5 components (or 3 if only the even values of m can be seen), the relatively sharp mean peak on this Figure 5 (about 3.5 μHz) does not seem to be consistent with the relatively large frequency splitting (0.75 μHz) reported by Claverie *et al.* (1981). It may be that all split components have not equally contributed to the mean peak of Figure 5 because of the random amplitude modulation. It may also be that the reported 0.75 μHz splitting is an artifact of analysis. A basic difference between the two respective analysis is that we have made the superposition of Figures 4–6 by vertical summation of lines in Figure 3, while Claverie *et al.* have shifted each line to follow the apparent central component of the $l = 0$ peak. This procedure makes the shape of the $l = 2$ mean peak arbitrarily dependent upon the scatter of the $l = 0$ measured frequencies around their unknown real values (see Figure 9). Following the same procedure but normalizing on the central component of the $l = 2$ peak would make the average shape of $l = 2$ appear as a unique peak, and would split the $l = 0$ mean peak into a few components, separated by about the same amount (close to the resolution of their data). This is contrary to the fact that the $l = 0$ peak is non-degenerate.

This question of the splitting is a very important one which has not yet been satisfactorily answered. A display of the power spectrum on a figure similar to our

Figure 2 or 3 could probably help to avoid a possible confusion between real splitting and random distribution of power in a broad envelope.

(d) *Amplitudes* – Another interesting point to be mentioned is the sharp drop of amplitude with decreasing frequencies around 2.5 mHz. This drop (a factor 4 in power) appears to be limited to a narrow band close to 2.5 mHz, and for lower frequencies, the amplitude decreases slowly to reach the increasing noise level at about 1.9 mHz. This behavior will have to be explained by an adequate excitation – damping mechanism. Is there some coupling between p-modes and surface g-modes (2.5 mHz is not far from the buoyancy frequency of the solar atmosphere, above which no gravity mode can exist)? In any case, in the absence of any accurate prediction of the amplitudes of low-order, low-degree p-modes, the very slow decrease of amplitude can be regarded as very encouraging for the future progress of helioseismology. The next major step will hopefully be the detection of low-degree, low-order p-modes and a few g-modes. This detection may require only a moderate improvement of the sensitivity.

Acknowledgements

This research is supported by the National Science Foundation's Division of Polar Programs under grant DPP 7822467.

References

Christensen-Dalsgaard, J. and Gough, D. O.: 1981, *Astron. Astrophys.* **104**, 173.

Claverie, A., Isaak, G. R., McLeod, C. P., Van Der Raay, H. B., and Roca Cortes, T.: 1981, *Nature* **293**, 443.

Grec, G.: 1981, These de Doctorat, Universite de Nice.

Grec, G., Fossat, E., and Vernin, J.: 1976, *Astron. Astrophys.* **50**, 221.

Grec, G., Fossat, E., and Pomerantz, M.: 1980, *Nature* **288**, 541.

Scherrer, P., Wilcox, J., Christensen-Dalsgaard, J., and Gough, D. O.: 1983, *Solar Phys.* **82**, 75 (this volume).

Scuflaire, R., Gabriel, M., Noels, A., Grec, G., and Fossat, E.: 1982, preprint.

Woodard, M. and Hudson, J.: 1983, *Solar Phys.* **82**, 67 (this volume).

SOLAR OSCILLATIONS OBSERVED IN THE TOTAL IRRADIANCE*

M. WOODARD and H. HUDSON

University of California, San Diego

Abstract. The total solar irradiance measurements obtained by the active-cavity radiometer on board the Solar Maximum Mission have been analyzed for evidence of global oscillations. We find that the most energetic low-degree p-mode oscillations in the five-minute band have amplitudes of a few parts per million of the total irradiance, and we positively detect modes with $l = 0$, 1, and 2. The distribution in l differs from that of the velocity spectrum, with relatively more power at lower l values. The individual modes have narrow line widths, corresponding to values of Q greater than a few thousand, or lifetimes of at least a week. We do not detect the 160-min oscillation in the power spectrum, and place an upper limit of 5 parts per million (99.9% confidence) on its amplitude.

1. Introduction

The new field of 'solar seismology' (e.g. Gough, 1980) is providing a tool to study the solar interior. The observational material consists of the solar global oscillations in three different period ranges: p-modes in the 5-min band (Deubner, 1975; Claverie *et al.*, 1979); the mysterious 160-min oscillation (Severny *et al.*, 1976; Brookes *et al.*, 1976); and the structure at periods of about 50 min or less reported by Hill *et al.* (1976). The mode identifications, amplitudes, frequencies, and general morphology of the oscillations will all contribute to our knowledge of solar interior structure and dynamics, as detailed in other papers from these proceedings.

The main results so far come from velocity measurements, namely the Doppler shifts of Fraunhofer lines. These measurements vary in their degree of spatial resolution; measurements of low-degree modes are more-or-less averages over the whole solar disk. We report here the first observations in integral solar irradiance (flux), using data provided by the Active Cavity Radiometer Irradiance Monitor (ACRIM; see Willson, 1979), on the Solar Maximum Mission (Willson and Hudson, 1981; Willson *et al.*, 1981). These data form a very long time series that will provide unsurpassed frequency resolution.

In this initial report we describe a preliminary time-series analysis of the ACRIM data in two period ranges: 5-min band (Sections 3 and 4) and the 160-min band (Section 5). To summarize briefly, we find that the 5-min modes are detectable and provide interesting information, while the 160-min mode is not but might not have been expected in this data set.

* Proceedings of the 66th IAU Colloquium: *Problems in Solar and Stellar Oscillations*, held at the Crimean Astrophysical Observatory, U.S.S.R., 1–5 September, 1981.

Solar Physics **82** (1983) 67–73. 0038–0938/83/0821–0067$01.05.

2. Data

The ACRIM instrument provides bolometric measurements of the total solar irradiance. The thermal (pyrheliometric) detector responds accurately – at all solar wavelengths of any significance – to variations in the intensity. The instrument is aboard the Solar Maximum Mission spacecraft, launched on February 15, 1980. We have analyzed data from a 137-day interval: March 5 to July 20, 1980. The heart of the detector is a blackened conical cavity, whose rate of heat absorption is kept constant by servo-control (the thermal relaxation time is about 1.5 s). A mechanical shutter exposes the cavity to the full Sun during half of a 131.072-s cycle. When the shutter is closed the cavity is heated by an electrical coil. When the shutter is open, the Sun, of course, supplies part of the power input, hence the solar flux is recorded as the difference of the electrical heating rates on alternate halves of the cycle. The voltage across the heating coil is read (to 13 significant bits) every 1.024 s. Half of the readings cannot be used, because the thermal transients produced by the action of the shutter are much larger than any solar-induced voltage changes. We therefore ignore the 32 readings immediately following the opening or closing of the shutter and average only the succeeding 32 samples. To summarize, we sample the solar flux about once every two minutes, with a solar viewing time of about half a minute. The Nyquist frequency of the basic shutter cycle period is 3.815 mHz, placing it just above the main envelope of the 5-min oscillations (aliasing is apparently not a severe problem in our spectral analysis).

The orbital period of the satellite is close to 96 min (frequency = 0.174 mHz) but changes irregularly by about 0.1% during the observations. Periodic gaps occur in the data when the low-earth orbit SMM satellite is in the earth's shadow. We extrapolate the measured flux to one astronomical unit by making the standard earth-sun distance correction and by compensating for the effects of satellite orbital motion on the perceived irradiance.

Large, slow variations in the data reflect the development of solar active regions (Willson *et al.*, 1981). In searching data for global oscillations, we view these long-term drifts as a source of noise. We describe in the following sections the filtering that we have used to minimize this noise. Slow drifts in the instrument itself are quite small and have negligible effects on the analysis of high frequencies.

3. Observations of Low-Degree *p*-Modes

The basic tool in our analysis of the short-period oscillations is spectral analysis via the fast Fourier transform. Because of the periodic data gaps mentioned above, the spectral power cannot be straightforwardly estimated from the direct transform of the time series, even with some crude estimate of the missing data. However the sunlit portion of the satellite orbit is long enough that a continuous, unbiased auto-covariance function can be calculated and transformed into a spectral estimate. In practice we need some high-pass filtering, so we first calculated the power spectrum of the difference time series (the differences of adjacent two-minute samples), then restored the power spectrum of the original data by the appropriate frequency attenuation factor.

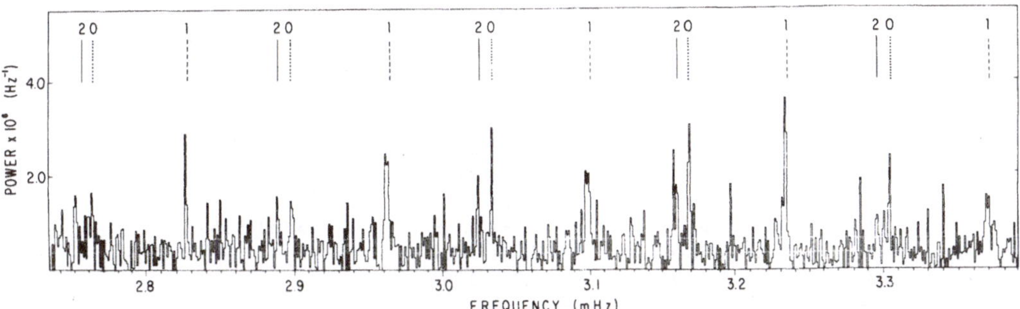

Fig. 1. Portion of the power spectrum of 137 days of total irradiance data from the ACRIM instrument on board the Solar Maximum Mission. The text describes the analysis technique in detail. The modal frequencies measured by Claverie *et al.* (1981) are shown for comparison, along with their mode identifications (l values).

We have analyzed eleven non-overlapping contiguous data intervals, covering the period March 5–July 20, 1980, as described. The resulting periodograms were averaged into a power spectrum of moderate frequency resolution ($\Delta\nu = 0.931$ μHz, corresponding to 2^{13} shutter cycles or ~12 days). Figure 1 shows a relevant high-frequency portion of this spectrum along with a set of comparison frequencies. For each of the 12-day intervals the auto-covariance function was calculated for positive lags and extended symmetrically to negative ones (2^{14} lags total, lag = shutter cycle). A symmetrical triangular envelope was applied to the function to reduce the influence of badly estimated lags near the ends.

To judge the significance of the spectral estimates we have isolated the frequency interval 2.688 to 3.517 mHz, in which oscillations have been detected in velocity. We assume that the individual points are χ^2-distributed and use the mean and scatter of the spectral points to determine the effective number of degrees of freedom. We assume the mean power in the chosen frequency band fits a straight line. We obtain ~4 degrees of freedom by this method; smaller than the nominal value, ~11 (which takes data gaps into account). The discrepancy confirms our impression that the underlying power spectrum is not as smooth as we might desire; a better model must be devised (naturally, the normal modes themselves contribute to an anomalously large scatter). We regard a spectral feature as real if it lies within 2 μHz of an already published frequency, and if its likelihood of occurring by chance (in this small frequency range) is less than 5%. For each significant point or group of points we calculate a mean frequency. We compare our observed frequencies with those of the Birmingham group (Claverie *et al.*, 1981) in Table I, and find no significant discrepancies. Table I adopts the mode classification of Grec *et al.* (1980). The power spectrum of Figure 1 is much noisier than the spectra of the best velocity measurements, but there seem to be additional significant frequencies in this plot besides the ten listed in Table I, probably because the number of degrees of freedom has been conservatively underestimated.

Following Grec *et al.* we have performed a summed-frequency analysis (but with a best folding interval of 135.25 μHz), and display the result in Figure 2.

TABLE I

Observed p-mode frequencies, μHz (Claverie *et al.* (1981)/present results)

n	$l = 0$	$l = 1$	$l = 2$
18		2.692/2.693	
19		2.829/2.828	
20		2.964/2.963	
21	3.033/3.033	3.099/3.099	3.160/3.160
22	3.168/3.169	3.234/3.234	
23	3.303/3.304	3.370/3.369	

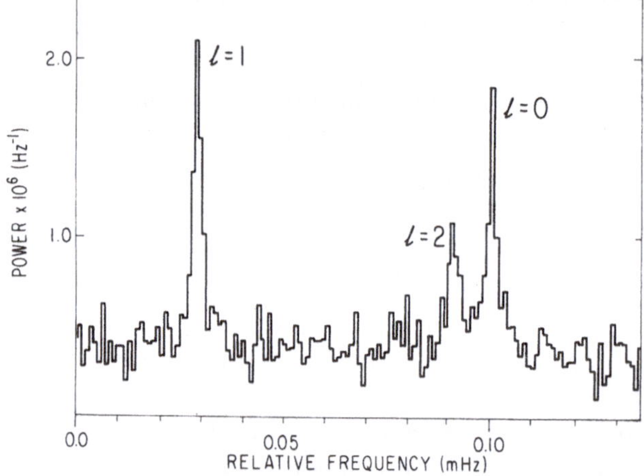

Fig. 2. Summed-frequency analysis of the type performed by Grec *et al.* (1980). The folding interval was 135.25 μHz, and the sum spans the frequency range 2.688 to 3.517 mHz. Note the relative strength of the lower-l modes in comparison with the velocity spectrum.

4. Analysis of *p*-Modes

In Table II we crudely estimate the average power, over the frequency band 2.688 to 3.517 mHz, of the modes $l = 0$, 1, and 2 (a sinusoidal wave train whose amplitude is C, in units of the average solar irradiance, has a total power of $C^2/2$). The amplitudes of the brightest individual modes are about four parts per million (half-amplitude) of the

TABLE II

Mean power in p-modes

l	Power
0	4.4×10^{-12}
1	5.8×10^{-12}
2	2.9×10^{-12}

total irradiance. The ratio of total-irradiance power to velocity power falls rapidly with l, so that at $l = 2$ the ratio is about half that at $l = 0$. We conclude from the rather narrow line widths that the oscillations are phase-coherent over a period of a week and possibly much longer; the $l = 0$ modes in particular appear to be completely unresolved at the resolution of Figure 1. We have not observed variations in the frequencies in different sub-intervals of our 137-day data set.

To test our understanding of the Sun's outer envelope we should compare the observed coupling between geometrical distortions and brightness oscillations with the coupling predicted by theory. Naturally, we cannot infer the vertical displacement of a patch of the solar surface until we know the precise relationship between actual modal amplitude and the observed velocity, which requires not only a theoretical understanding of the solar atmosphere but also an understanding of the instrumental response.

The observed l dependence probably arises in different spatial filtering resulting from the effects of limb darkening and velocity projection. Taking the interval 2.688 mHz–3.517 mHz in the published power spectrum of the Nice group (see Christiansen-Dalsgaard and Gough, 1982), we find that the r.m.s. velocity amplitude of the $l = 0$ mode is ~ 25 cm s^{-1}. Based on this value, we find that the corresponding irradiance amplitude agrees (within uncertainties) with the theoretical predictions of Gough (1980) and is therefore consistent with the notion that photospheric compression and heating are the main cause of brightness fluctuations.

5. Results for the 160-min Oscillation

To study the long-period spectral feature reported in the velocity signal (Severny *et al.*, 1976; Brookes *et al.*, 1976) we have computed the summed-epoch signal for a range of folding periods near 160 min. These composite time signals are used mainly for spectral analysis, but several were examined directly. This method of spectral analysis suffers from periodic data gap problems similar to those which occur when a power spectrum is calculated from the direct Fourier transform of the time series. A particularly annoying difficulty – the appearance of spurious, periodic, summed epoch signals – arises when large, low frequency trends in the data are seen through a slightly aperiodic data window. Because the period changes, the samples which are averaged to obtain a given phase bin of the summed epoch signal are unevenly distributed in time. Therefore, the estimate for a particular bin will be dominated by the values of the long-term trend near the temporal centroid of the distribution. However the distribution of samples is not the same for each bin. Therefore, the summed epoch signal will vary from bin to bin. This results in a non-Gaussian noise that is much larger than the true noise at periods near 160 min and leads to a poor spectral estimate. We emphasize that this leakage of low-frequency power into high frequencies does not occur in data with strictly periodic gaps. To circumvent this effect we have subtracted a running linear fit from the data (a high-pass filter with a cutoff frequency of ~ 1 cycle/day). A spectral estimate was obtained from the filtered data set at 100 equally spaced frequencies centered on 1/160 min (frequency spacing, $\Delta f = 0.0845$ μHz). The spectral power, P, is computed according to

$$P = (A^2 + B^2)/2\Delta f,$$

where A and B are the fundamental Fourier coefficients in the expansion of the summed epoch signal:

$$\delta F/F = A \cos \Phi + B \sin \Phi \quad (+ \text{ higher harmonics}),$$

where $\delta F/F$ = flux variation/mean solar flux, and Φ = phase.

A plot of the resulting power spectrum appears in Figure 3. Within errors the data are consistent with no oscillation of 160 min period. Hence, we compute the maximum amplitude, $C = (A^2 + B^2)^{1/2}$, that a phase-stable oscillation could have. We assume that the Fourier components (A, B) are uncorrelated Gaussian variables with zero mean and determine their variance from the 100 sample frequencies. We find $C < 4.5 \times 10^{-6}$, with greater than 99.9% confidence.

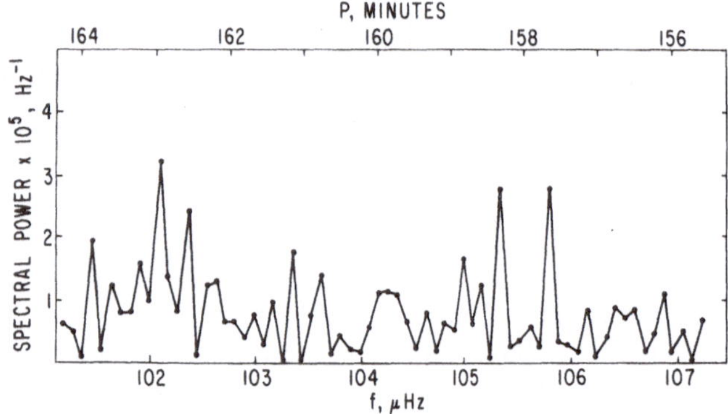

Fig. 3. Power spectrum in the 160-min band. The results set a strong upper limit of 5 parts per million (99.9% confidence) on the amplitude of a sinusoidal variation in solar irradiance at 160 min period.

To compare the ACRIM data with the velocity data for the 160 min oscillation, we select, as the proper measure of physical displacement, $\delta R/R$, the amplitude of the radial displacement in units of the solar radius. (Photospheric heating will probably be an unimportant source of brightness variation, since atmospheric compression seems to be negligible at this low frequency (Whitaker, 1963).) Defining a radiative efficiency, W, according to $\delta F/F = W\delta R/R$, and taking $\delta R \sim 5 \times 10^6$ cm (from the data of Scherrer *et al.*, 1979), we rule out oscillatory modes (or atmospheric models) for which $W < 7$. To put this result into perspective, a purely radial oscillation, in which the photosphere oscillates isothermally, would have $W = 2$ (since $F \sim R^2 T^4$). Thus, we see no conflict between the velocity observations and our apparent null detection. However it seems difficult to reconcile this result with the intensity observations reported by Kotov *et al.* (1978).

6. Conclusions

The total-irradiance observations from the Solar Maximum Mission show that the low-degree *p*-modes affect the bolometric flux from the Sun. We have confirmed the frequencies reported by Claverie *et al.* (1981), suggesting that there are no slow drifts of the modal frequencies. We have demonstrated that the oscillatory modes persist for periods of a week or longer. The amplitudes – a few parts per million – are small, but appear to be consistent with the theory of the oscillatory response of the solar atmosphere (Gough, 1980). We do not find any evidence for the 160-min mode, but this is probably not in conflict with the results of the velocity spectroscopy.

The results reported here are preliminary. In the future we will extend the data base to take full advantage of the high resolution afforded by the long time series, and will also study the intermediate spectral range observed by Hill *et al.* (1976). We would also like to establish the relative phase between the velocity and the irradiance oscillations. The detection of solar flux variations, caused by non-radial oscillations, encourages us to believe that precise photometric observations of other stars can similarly help us to understand their interior structures.

Acknowledgements

We would like to thank Dick Willson for obtaining the beautiful ACRIM data, for designing the instrument in such a way that this unanticipated analysis could be carried out, and for assistance and discussions. We have also benefited greatly from discussions of time-series analysis techniques with Bill Coles, Rod Frehlich, and Charley Lindsey. The National Aeronautics and Space Administration supported this research under grant NSG–5322.

References

Brookes, J. R., Isaak, G. R., and van der Raay, H. B.: 1976, *Nature* **259**, 92.

Christiansen-Dalsgaard, J. and Gough, D. O.: 1982, submitted to *Astron. Astrophys.*

Claverie, A., Isaak, G. R., McLeod, C. P., van der Raay, H. B., and Roca Cortes, T.: 1979, *Nature* **282**, 591.

Claverie, A., Isaak, G. R., McLeod, C. P., van der Raay, H. B., and Roca Cortes, T.: 1981, *Nature* **293**, 443.

Deubner, F.-L.: 1975, *Astron. Astrophys.* **44**, 371.

Gough, D. R.: 1980, in H. A. Hill and W. A. Dziembowski (eds.), *Nonradial and Nonlinear Stellar Pulsation*, Springer, p. 273.

Grec, G., Fossat, E., and Pomerantz, M.: 1980, *Nature* **288**, 541.

Hill, H. A., Stebbins, R. T., and Brown, T. M.: 1976, in J. H. Sanders and A. H. Wapstra (eds.), *Proc. V International Conference on Atomic Masses and Fundamental Constants*, Plenum, New York, p. 622.

Kotov, V. A., Severny, A. B., and Tsap, T. T.: 1978, *Monthly Notices Roy. Astron. Soc.* **183**, 61.

Scherrer, P., Wilcox, J. M., Kotov, V. A., Severny, A. B., and Tsap, T. T.: 1979, *Nature* **277**, 635.

Severny, A. B., Kotov, V. A., and Tsap, T. T.: 1976, *Nature* **259**, 87.

Willson, R. C.: 1979, *J. Appl. Opt.* **18**, 179.

Willson, R. C. and Hudson, H. S., 1981, *Astrophys. J.* **244**, L185.

Willson, R. C., Gulkis, S., Janssen, M., Hudson, H. S., and Chapman, G. A.: 1981, *Science* **211**, 700.

Whitaker, W. A.: 1963, *Astrophys. J.* **137**, 914.

DETECTION OF SOLAR FIVE-MINUTE OSCILLATIONS OF LOW DEGREE*

PHILIP H. SCHERRER and JOHN M. WILCOX

Institute for Plasma Research, Stanford University, Stanford, CA 94305, U.S.A.

J. CHRISTENSEN-DALSGAARD

National Center for Atmospheric Research, Boulder, CO 80307, U.S.A.

and

D. O. GOUGH

Institute of Astronomy, Madingley Road, Cambridge CB3 OHA, U.K.

Abstract. Solar five-minute oscillations of degree $l = 3$, 4, and 5 have been observed at Stanford, in the Doppler shift of the Fe 5124 line. The frequencies and amplitudes are in broad agreement with previous observations of modes with $l \leq 3$, though we note that there are some systematic discrepancies between the results of different observers.

1. Introduction

Whole-disk observations of solar five-minute oscillations have been reported by Claverie *et al.* (1979, 1981a, b), by Grec *et al.* (1980), and by Fossat *et al.* (1981), Grec *et al.* (1983), who measured Doppler shifts, and by Deubner (1981) and Woodard and Hudson (1983), who measured intensity variations. These observations are sensitive mainly to oscillation of degree $l = 0, 1, 2$. In addition, current whole-disk Doppler observations can detect octupole modes, though with low sensitivity.

By reducing the spatial scale over which light from the solar disk is integrated, the maximum sensitivity can be moved towards larger l. Here we report Doppler observations that were made at Stanford, and which are differences between mean spectrum line shifts in light from a central portion of the disk and from an outer annulus. Severny *et al.* (1981) have already reported detecting five-minute oscillations with a similar, though different, filter geometry. On comparing the oscillation frequencies with those reported by Grec *et al.* (1983) we were able to identify modes with $l = 3$. Once that was accomplished, it was then easy to identify modes with $l = 4$ and $l = 5$. Contributions to the signal from modes of higher degree are under investigation.

2. The Observing Technique

The observations made at Stanford use a technique suggested by Severny *et al.* (1976). A Babcock (1953) solar magnetograph, designed for measuring Zeeman splitting, has

* Proceedings of the 66th IAU Colloquium: *Problems in Solar and Stellar Oscillations*, held at the Crimean Astrophysical Observatory, U.S.S.R., 1–5 September, 1981.

Solar Physics **82** (1983) 75–87. 0038–0938/83/0821–0075$01.95.

been converted into an instrument with equivalent sensitivity for measuring Doppler shifts. The measurement is made by observing a Fraunhofer line that has no Zeeman splitting ($g = 0$). In the Stanford instrument a central circular area of the solar disk, of radius 0.50 $R_\odot$, is covered with a right-hand circular polarizer, and the annulus between radii 0.55 $R_\odot$ and 0.80 $R_\odot$ is covered with a left-hand polarizer. The transmitted light-beam enters the magnetograph which, as usual, measures the wavelength difference between right and left circularly polarized light. This is interpreted as the difference in the mean line-of-sight velocity between the central portion of the solar disk and the outer annulus. Details of the observing procedure are described by Dittmer (1978).

The sensitivity of these observations to modes of different degree can be characterized by spatial filters S_l. These are defined as the ratio between the signal amplitude and the root-mean-square amplitude over the entire solar surface of the quantity that is presumed to be measured. Spatial filters S_l^v for Doppler velocity measurements have been tabulated by Christensen-Dalsgaard and Gough (1982) and plotted by Christensen-Dalsgaard (1980), for both whole-disk and Stanford observations. These are mean filters, averaged over all spherical orders m of the surface harmonics that factor from the oscillation eigenfunctions. Thus, granted that there is no strongly preferred orientation of the modes, they represent the expectation value of the actual filter for an observation of any particular modes of degree l. For whole-disk observations S_l^v is

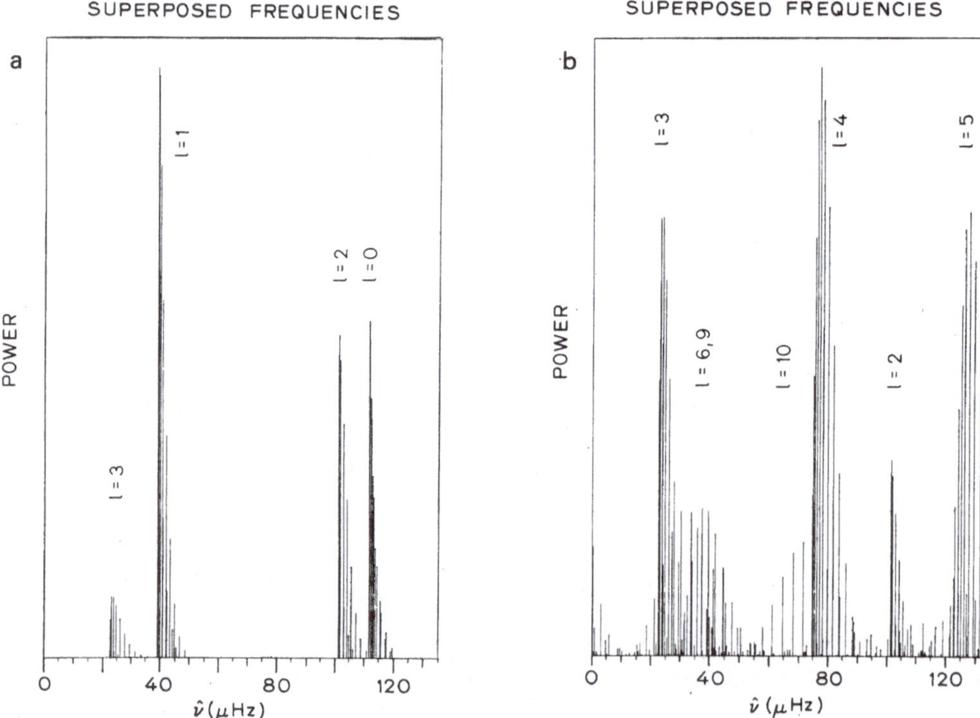

Fig. 1. Superposed frequency diagrams for model 1 with $\Delta\nu = 136.2$ microHz: (a) for whole-disk filters, (b) for Stanford filters.

largest when $l = 1$, and is greater than $\frac{1}{2}S_1^\nu$ only for $l = 0$, 1, and 2. On the other hand, for the Stanford geometry the largest value is for $l = 4$, and $S_l^\nu \gtrsim \frac{1}{2}S_4^\nu$ for $l = 2$–6 and also for $l = 9$ and 10. Thus, compared with whole-disk observations, the Stanford observations are sensitive to modes over a fairly wide range of l. Though this complicates the interpretation, such observations evidently increase the richness of the data.

3. Anticipation of the Results, and Comparison with Whole-Disk Data

The most pronounced difference between the results expected from the whole-disk and the Stanford observations can be illustrated with a superposed frequency diagram (Grec *et al.*, 1980; Christensen-Dalsgaard, 1980). Each frequency ν is reduced to a frequency $\hat{\nu}$ lying in the range $(\nu_0, \nu_0 + \Delta\nu)$ by adding or subtracting an integral multiple of $\Delta\nu$, which is characteristic of the separation in frequency between modes of adjacent order n and of like degree l; the value of ν_0 is arbitrary. The frequencies $\hat{\nu}(\nu)$ are plotted as a bar diagram, each bar having height $(S_l^\nu)^2 \mathrm{A}(\nu)$, where $A(\nu)$ is the expectation of the square of the amplitude of the mode; here we use the function that is defined by the first equation (D12) of Christensen-Dalsgaard and Gough (1982), which is a Gaussian approximation to the envelope of the power observed. The results are shown in Figures 1a and 1b, for whole-disk and Stanford filters, respectively. They represent idealized superposed power spectra, which one would obtain if the mode lifetimes and the observing interval were both infinite. In practice the observing interval is too short to enable the individual bars to be resolved, but it is easy to imagine how Figure 1 should be modified in that case. The striking resemblance between our Figure 1a and Figure 2 of Grec *et al.* (1980), which is a superposed power spectrum of 120^h continuous whole-disk Doppler data obtained at South Pole, is the basis for the identification of the solar modes responsible for that data.

It is evident from Figure 1a that modes with $l = 0$, 1, and 2 dominate the whole-disk data, and that their frequencies are nearly uniformly spaced in n with a spacing that depends only weakly on l. To each value of l there corresponds a narrow range of frequencies $\hat{\nu}$; moreover the frequencies of the groups of modes with $l = 0$ and $l = 2$ (and with $l = 1$ and $l = 3$) are very close together. Thus in a power spectrum computed from observations limited to a single day, those closely spaced peaks merge; the spectrum has two peaks in each frequency interval $\Delta\nu$, as seen in the observations of Claverie *et al.* (1979). Adjacent peaks are expected to contain about the same power, and the peaks produced by modes with odd l lie roughly midway between those with even l. Thus direct identification of the degrees of the modes producing a particular peak in such a spectrum is not possible.

In contrast Figure 1b is dominated by nodes with $l = 3$, 4, and 5, with substantial contributions from modes with $l = 6$ and 9, whose frequencies nearly coincide with those of the $l = 3$ group, and modes with $l = 2$ and 10 which are close to the $l = 4$ group. Although the frequencies of modes with $l \gtrsim 3$ are still approximately uniformly spaced in n, the value of that separation increases more rapidly with l (see Figure 7). Therefore the groups are more weakly confined. In spectra of a single day's observations, or averages of such spectra, one would expect three, rather than two, dominant peaks per frequency interval $\Delta\nu$, corresponding to modes with $l = 3, 4$, and 5 (cf. Christensen-Dalsgaard and Gough, 1982), with perhaps some evidence for smaller peaks from other modes. Therefore the temporal resolution requirements are more stringent than for whole-disk observations. Once again the major peaks should be roughly equally spaced, with no obvious systematic variation in power from modes of different degree. Direct identification of l from such data alone is therefore not possible.

4. The Observations

The observations were made in June and July, 1981, at the Stanford Solar Observatory. We obtained good quality data on 15 days; the intervals during which that data was acquired are listed in Table I. The average duration was 10.3 hr, and the shortest was 8.0 hr. The observations were made with a 0.1 s resolution but were averaged into 15 s intervals during the observations. The 15 s resolution was maintained throughout the subsequent analysis.

TABLE I

Times of observations (UT). The duration refers to the filtered data, and does not include the first and last 20 min

Date at start	Start	Stop	Duration (hr)
17 June 1981	13:50	01:00	10.5
20 June	16:10	02:00	9.3
21 June	18:00	02:00	7.3
22 June	14:00	22:00	7.3
24 June	16:00	01:40	9.0
26 June	16:05	01:40	8.9
27 June	15:00	01:30	9.8
28 June	16:30	01:20	8.2
7 July	14:00	02:15	11.6
8 July	14:00	01:00	10.3
10 July	14:15	02:00	11.1
11 July	14:00	02:00	11.3
12 July	13:40	02:00	11.7
13 July	14:30	22:50	8.3
14 July	15:00	01:40	10.0

A typical day's observation is shown in Figure 2. Figure 2a is a plot of the line-of-sight Doppler velocity observed. It can be seen that a slow drift of about 1 m s^{-1} per hour is present. This was removed before analyzing the data by first applying a 10 min running mean four times, and then subtracting the smoothed data so obtained from the original. The result is shown in Figure 2b. With this kind of filtering the first and last 20 min of each data sequence must be discarded. Hence the average duration of filtered data that we analyzed was 9.6 hr. This is sufficient to resolve the major peaks in the power spectrum, whose mean separation we anticipated to be a little more than

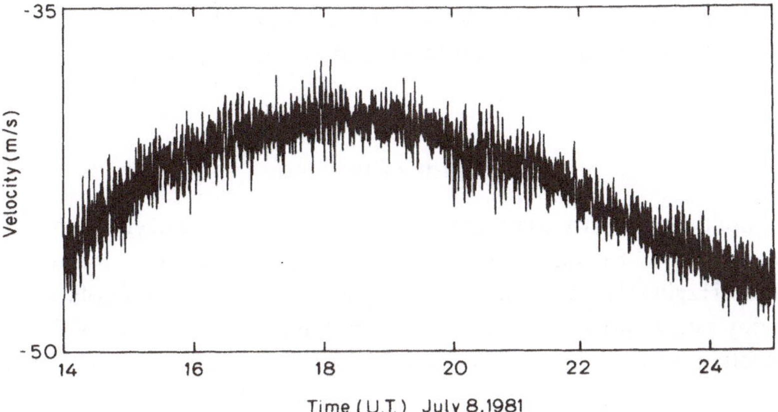

Fig. 2a. Solar velocity signal observed on a typical day, July 8, 1981. The drift through the day is caused by variations in sky transparency and by other unknown causes.

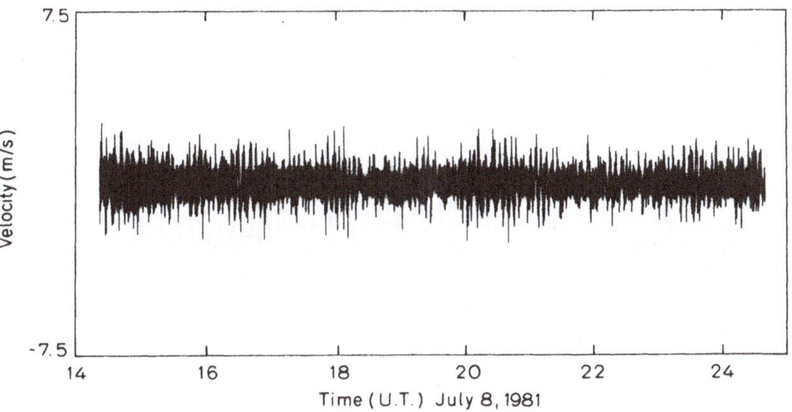

Fig. 2b. The same as Figure 2a after the filtering described in the text has been applied.

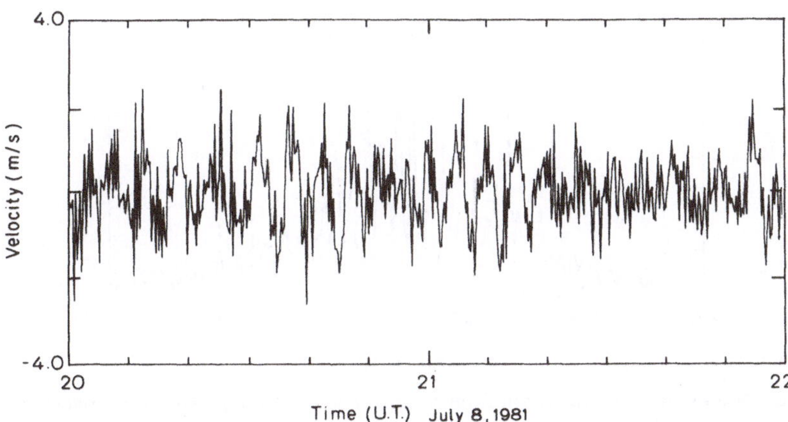

Fig. 2c. A typical two hour interval from Figure 2b. This is included to show that five minute oscillations are visible in the raw data.

45 microHz. Figure 2c shows a 2 hr interval from Figure 2b. Oscillation with periods near 5 min are clearly visible. The amplitude waxes and wanes as the various modes beat with each other.

5. Analysis of the Data

Harmonic amplitude spectra were computed for each day's observation by a direct least-squares method to find the Fourier coefficients in the frequency range 2.0 mHz–4.5 mHz, in steps of 1.0 microHz. Figure 3 is the squared amplitude spectrum of the single day's data shown in Figure 2c. Figure 4 depicts the average of the 15 equally weighted spectra. Also included, for comparison, are the frequencies identified by

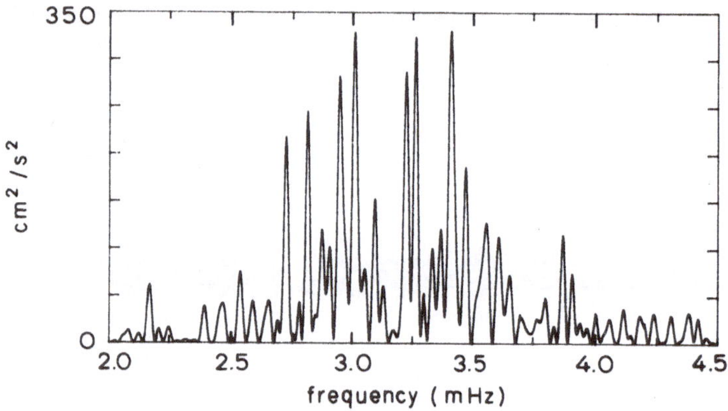

Fig. 3. Power spectrum of data illustrated in Figure 2b. The widths of the peaks are the natural widths corresponding to the 10 hours of data.

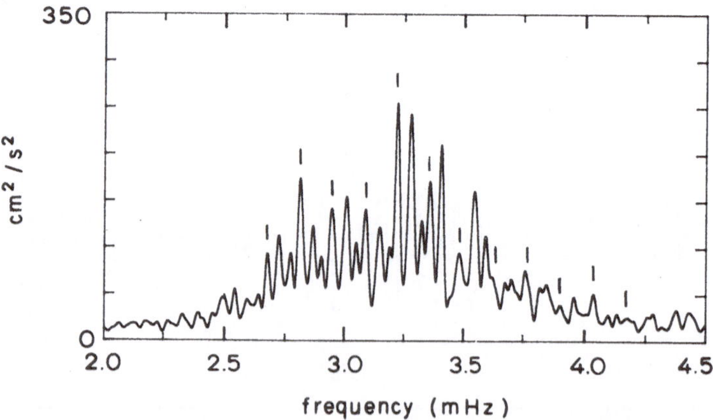

Fig. 4. Average of all power spectra taken from residual data of the type shown in Figure 2b collected during the intervals listed in Table I. The vertical lines drawn above some of the peaks are the peak frequencies for oscillation modes with $l = 3$ observed by Grec *et al.* (1983) at the South Pole.

TABLE II

Frequencies and amplitudes of the peaks in Figure 4. The method of determining the degree l of each mode is described in the text; the order n was then obtained by comparison with eigenfrequencies of the solar model. The last column is the rms surface velocity amplitude per mode, computed under the assumption that the spectrum line shift measured is a pure Doppler shift associated with the line-of-sight velocity of oscillation.

n	l	Frequency (microHz)	Amplitude ($cm^2\ s^{-2}$)	rms velocity ($cm\ s^{-1}$)
16	3	2544	55.4	4.1
16	4	2596	44.2	3.2
16	5	2642	49.3	3.9
17	3	2677	93.3	5.3
17	4	2724	112.8	5.1
17	5	2773	93.6	5.4
18	3	2815	173.9	7.3
18	4	2868	123.2	5.3
18	5	2905	90.0	5.2
19	3	2948	141.8	6.6
19	4	3010	154.4	6.0
19	5	3050	105.0	5.7
20	3	3088	140.9	6.6
20	4	3148	121.7	5.3
20	5	3189	100.3	5.5
21	3	3221	254.0	8.8
21	4	3279	242.5	7.5
21	5	3325	128.9	6.3
22	3	3359	170.8	7.2
22	4	3407	209.9	7.0
23	3	3482	94.3	5.4
23	4	3551	160.4	6.1
23	5	3597	112.3	5.9
24	3	3623	68.9	4.6
24	4	3678	63.1	3.8
25	3	3703	66.8	6.8
25	4	3759	75.9	4.8
25	5	3820	57.5	3.6
26	2	3846	61.0	6.5
26	3	3902	38.6	3.4
26	4	3958	47.5	3.3
26	5	4007	29.5	3.0
27	3	4036	50.8	3.9
27	4	4097	28.5	2.6
27	5	4133	29.6	3.0
28	3	4175	25.5	2.8
28	4	4280	30.3	3.0
29	2	4258	26.8	2.5

Fossat *et al.* (1983) as p modes with $l = 3$. Table II lists the frequencies of the maxima of all the peaks in Figure 4, to a precision of 1 microHz.

For $l = 3$ the rms difference between the peak frequencies of Fossat, Grec, and Pomerantz observed at the South Pole and those reported here observed at Stanford is 3.3 microHz. This gives an estimate of the uncertainty of our tabulated peak frequencies of the $l = 3$ modes.

The overall properties of the mean spectrum can be investigated by computing its auto-correlation. This is shown on Figure 5a. It has fairly sharp peaks centered on lags of 135.5 and 271.0 microHz, corresponding to the uniform spacing of modes of like degree. These are separated by broader asymmetrical features whose maxima are at about 56 and 191 microHz. For comparison we have found auto-correlations of simulated power spectra computed as in Christensen-Dalsgaard and Gough (1982), using all the modes of model 1 of Christensen-Dalsgaard (1982) with l between 0 and 20 and frequencies between 2.2 and 4.2 mHz; the observing intervals listed in Table I were used. Before computing the auto-correlation the average spectra were multiplied by $A(\nu)$. The results are shown in Figure 5b. The continuous curve was obtained using the spatial filters computed by Christensen-Dalsgaard and Gough (1982) for the Stanford observations and assuming that the modes preserve their amplitudes and phases for 5 days at a time; the well-defined maxima at lags of 46, 92, 138 ... microHz reflect the three-peak structure shown in Figure 1b. This is qualitatively different from the auto-correlation of the spectrum of the observations.

To try to account for this difference we have made two additional simulations. The dashed curve was obtained by assuming that the modes preserve their phases and amplitudes over the entire observing interval; this is consistent with the estimates of the coherence times quoted by Claverie *et al.* (1981a, b). Values of the auto-correlation are generally lower than in the previous case, and there are fewer peaks. The dominant feature is a peak at 139 microHz. The reason for the changed appearance of the curve is that now there is no averaging over different sets of random amplitudes and phases, and so the spectrum is less regular. The second simulation used filters computed as in Christensen-Dalsgaard and Gough (1982), but assumed a central disk of radius 0.6 $R_\odot$ and an outer annulus between 0.65 and 0.9 $R_\odot$. The spectrum is then dominated by modes with lower l: principally $l = 3$ and the closely spaced pair $l = 2, 4$. A coherence time of 5 days was assumed. The resulting auto-correlation, shown as a dot-dashed line in Figure 5b, reflects the basic double-peaked structure of the spectrum. Evidently a similar result would be obtained with our Stanford filters if modes with one of the principal values of l were selectively excited to a lower amplitude than the other. A glance at Figure 4 or Table II reveals that where the amplitude is high there is a tendency for every third peak (which we infer is associated with $l = 5$) to be weak. It is evident that an increase in coherence time and a change in the sensitivity that leads to the domination of fewer modes both give better agreement with the observed autocorrelation, the latter probably being the more successful. In fact, although the dimensions of the observing apertures are known accurately, the spatial filters computed are rather uncertain, owing to the neglect of variations in the line profile over the disk of the Sun and the effects

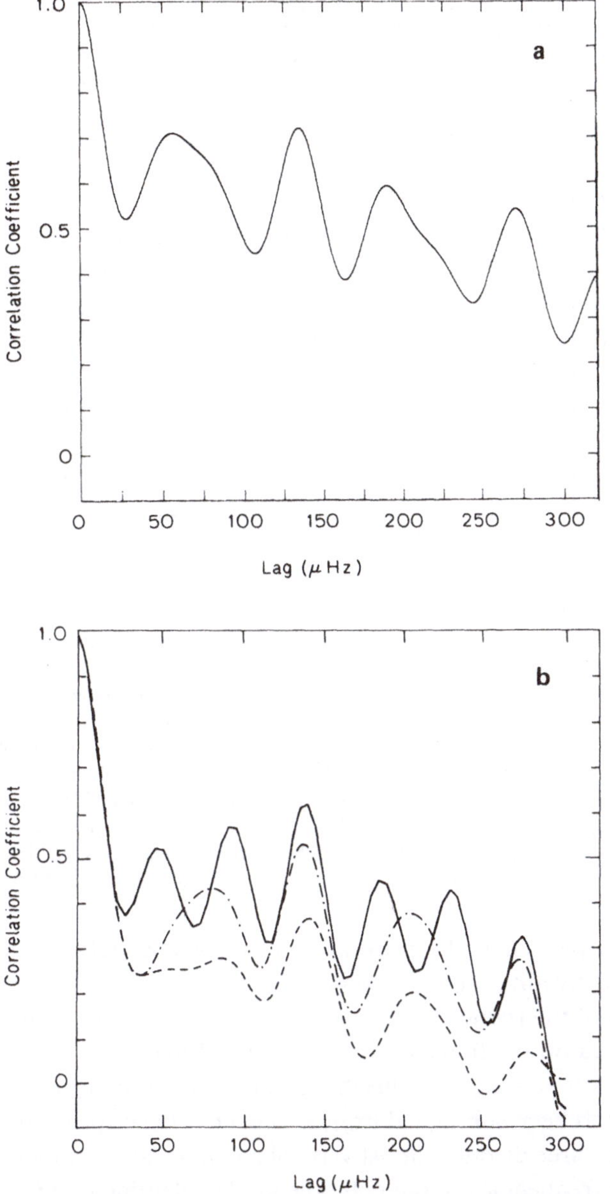

Fig. 5. Autocorrelation of power spectra: (a) of the mean spectrum shown in Figure 4, (b) of simulated mean spectra. In (b), the continuous curve represents simulated data with coherence times of 5 days, using Stanford filters; the dashed curve is similar, but with perfect phase coherence throughout the duration of the simulated observations; the dot-dashed curve represents data obtained with filters that give more weight to the modes of low degree, as described in the text.

of rotation; changes to the spatial filters would also occur if the observed line shift were not caused solely by Doppler displacements. Thus further investigations of the sensitivity of the observations are needed. Nevertheless, the autocorrelations of the spectrum of

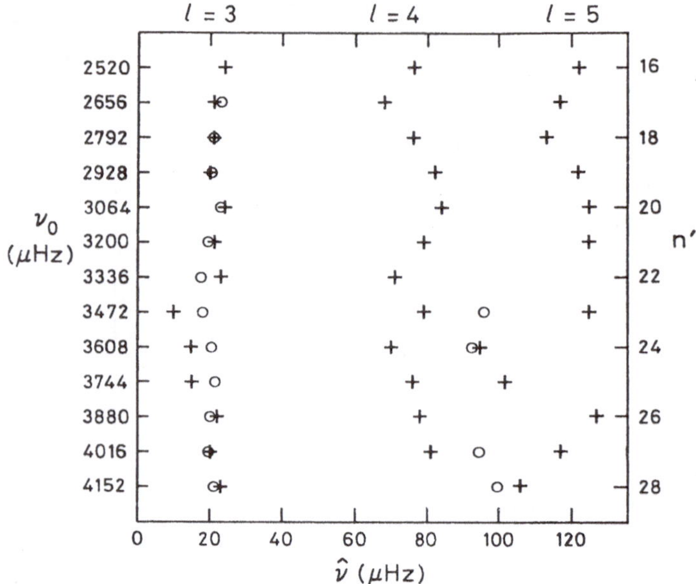

Fig. 6. Echelle diagram of the frequencies between 2.5 and 4.3 mHz of low-degree modes, constructed with $\Delta\nu$ = 136.0 microHz. The frequencies ν of the modes represented in the diagram are given by $\nu = \nu_0 + \hat{\nu}$. Crosses indicate the frequencies of the peaks in Figure 4. Circles represent octupole modes (with $\hat{\nu}$ near 20 microHz) and some dipole modes (with $\hat{\nu}$ near 100 microHz) observed from the South Pole by Grec *et al.* (1983). Thus by comparing our results with the South-Pole data and with theory we identify the degrees of the modes responsible for the three principal columns; these are indicated at the top of the diagram. We presume the frequencies 3703, 3846, and 4258 microHz to correspond to dipole modes. The ordinate scale n' on the right is the order n of modes with $l \geq 3$, and is $n - 1$ when $l = 2$. This identification was obtained by the least-squares comparison with theory described in the text. Fitting regression lines $\alpha + \beta\,(n - n_0)$, with $n_0 = 22 - \frac{1}{2}l$, to the three principal columns of Stanford frequencies ν yields $(\alpha, \beta) = (3152 \pm 1, 135.7 \pm 0.3)$, $(3140 \pm 1, 136.2 \pm 0.4)$, and $(3117 \pm 1, 136.5 \pm 0.3)$ for l = 3, 4, and 5, respectively.

the observations, in particular the peak at 135.5 microHz, demonstrates clearly the presence of resolved five-minute oscillations.

Figure 6 is a plot of the reduced frequencies $\hat{\nu}$ associated with the frequencies listed in Table II, computed on the frequency interval $\Delta\nu$ = 136.0 microHz. Following Grec (1981), $\hat{\nu}$ has been plotted at different heights on the diagram in decreasing order of the multiple of $\Delta\nu$ that had been subtracted from ν, as in an echelle spectrum. On the whole, the frequencies form three distinct almost vertical columns, as we anticipated. Included in the figure are the frequencies of the octupole modes identified by Grec *et al.* (1983) in the Antarctic data, and also some of the dipole modes. The former almost coincide with our first column of modes, which we therefore identify with l = 3. On consulting Figure 1b, we infer that the remaining two columns are produced by modes with l = 4 and l = 5. We also have evidence for quadrupole modes lying between the l = 4 and l = 5 columns.

Our columns are not as smooth as those of Grec *et al.* The reason is probably that our results are more prone to interference from other modes. The presence of a mode of lower amplitude whose frequency is too close to one of the principal modes for the separation to be resolved has the effect of shifting the line in the power spectrum associated with the principal mode. We conjecture that the break in the octupole column at 3482 microHz, for example, is due to the intersection of the locus of the octupole modes with the locus of modes with $l = 9$.

6. Comparison with Previous Observations

In almost all cases our octupole frequencies are in good agreement with those reported by Grec *et al.* (1983). We can make little direct comparison with the other observations of five-minute modes because, aside from our few quadrupole modes, we have not observed modes of the same degree. However, an indirect comparison is possible.

It is a property of the modes under consideration that $n \gg l$. When this condition is satisfied, the frequencies ν are roughly proportional to $n + \frac{1}{2}l$. The asymptotic analysis by Tassoul (1980) shows that to a higher approximation

$$\nu \sim (n + \tfrac{1}{2}l + \varepsilon)\nu_0 - [l(l+1) + \delta]A\nu_0^2/\nu\,, \tag{6.1}$$

where

$$\nu_0 = \left(2\int_0^R C^{-1}\,\mathrm{d}r\right)^{-1}, \tag{6.2}$$

R is the radius of the Sun, $c(r)$ is the sound speed and the quantities ε, A and δ are other constants of the equilibrium model. Thus the deviation from the simple linear approximation to ν is a linear function of $y \equiv l(l+1)$. Introducing $x = n + \frac{1}{2}l$ one can approximate (6.1) in the neighbourhood of $x = x_0$, where x_0 is constant, by

$$\nu \simeq \alpha + \beta(x - x_0)\,. \tag{6.3}$$

Then

$$\alpha \simeq \nu_0(X + A\delta X^{-1} + AX^{-1}y) \tag{6.4}$$

and

$$\beta \simeq \nu_0(1 - A\delta X^{-2} - AX^{-2}y)\,, \tag{6.5}$$

where $X = x_0 + \varepsilon$. Thus, to the order of accuracy of (6.1), α and β should vary linearly with y, and indeed we have found that to be approximately true of numerically computed eigenfrequencies of realistic solar models. For Model 1 of Christensen-Dalsgaard (1982), for example, $\mathrm{d}\alpha/\mathrm{d}y \simeq -1.76$ μHz and $\mathrm{d}\beta/\mathrm{d}y \simeq 0.050$ μHz when $x_0 = 22$.

We have computed estimates of α and β for the Sun for each value of l separately by linear regression with our own data and all the other published observations that have succeeded in distinguishing modes of different degree. The results are shown in Figure 7. To aid comparison, lines with slopes -1.76 μHz and 0.050 μHz are included. The absolute positions of these lines have been chosen to pass between the data; they do not agree precisely with the numerical results, partly because the solar model used to

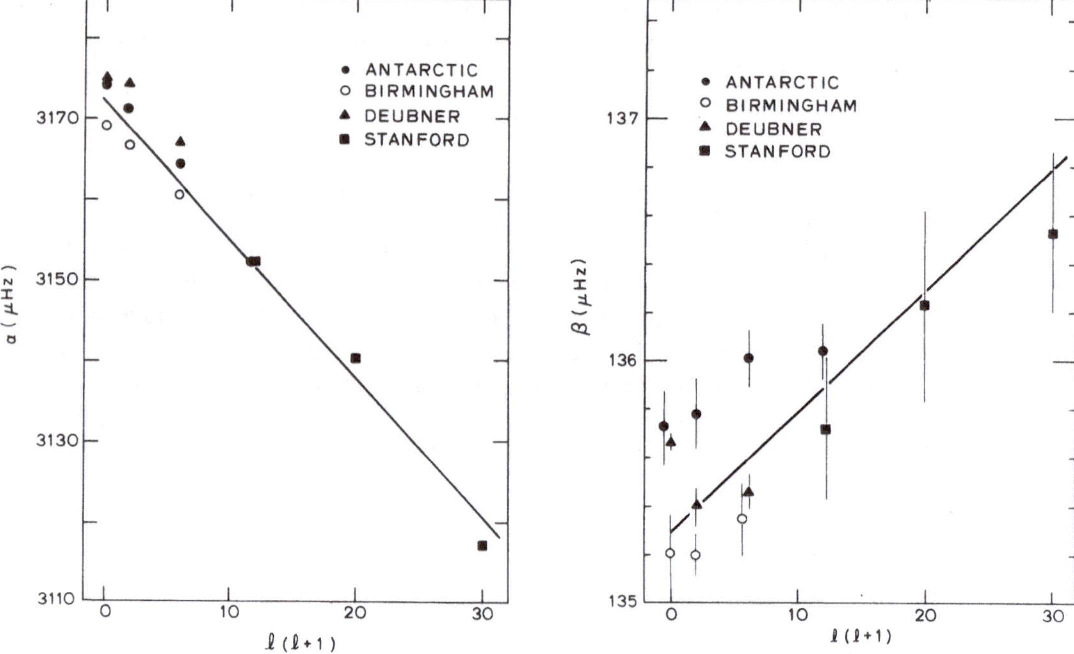

Fig. 7. Regression coefficients: (a) α, (b) β computed with $x_0 = 22$. Squares represent the Stanford results reported in this paper, filled circles the Antarctic data of Grec *et al.* (1983), open circles the Birmingham data of Claverie *et al.* (1981a, b) and triangles the data of Deubner (1981). The recent SMM–ACRIM data of Woodard and Hudson (1983) agree very well with those of Claverie *et al.* (1981a, b). The standard errors in α are roughly the size of the symbols; those of β are indicated by vertical lines. Where two data overlap, they have been shifted slightly to the left or right. The diagonal lines in (a) and (b) have slopes −1.75 microHz and 0.050 microHz, respectively.

compute the eigenfrequencies appears to be somewhat helium deficient (cf. Christensen-Dalsgaard and Gough, 1980, 1981). It is evident that our results are roughly in accord with the others, though the differences between the observations are greater than the standard errors. Part, though not all, of these differences results from having analysed data over different ranges of frequency.

7. Conclusion

We have observed and identified solar oscillations with $l = 3, 4, 5$ in the frequency range 2.5–4.3 mHz. The frequencies of the octupole modes are in good agreement with those found by Grec *et al.* (1983), and with the help of a theoretical extrapolation we have demonstrated fair agreement with other observations that have been published. There are, however, systematic discrepancies between the observations of modes with $l \leq 2$. These discrepancies are not insubstantial, if interpretated in terms of the differences in the internal structure of the sun that they seem to imply.

Acknowledgements

We thank E. Fossat, G. Grec, and M. Pomerantz for communicating to us their observations prior to publication, and E. Fossat for suggesting the format of Figure 6.

This work was supported in part by the Office of Naval Research under Contract N00014–76–C–0207, by the National Aeronautics and Space Administration under Grant NGR 05–020–559 and Contract NAS5–24420, by the Division of Atmospheric Sciences, Solar Terrestrial Research Program of the National Science Foundation under Grant ATM77–20580, by the Max C. Fleischmann Foundation, by the National Center for Atmospheric Research, which is sponsored by the National Science Foundation, and by the Science and Engineering Research Council.

References

Babcock, H. W.: 1953, *Astrophys. J.* **118**, 387.

Christensen-Dalsgaard, J.: 1980, in P. Ledoux (ed.), *Proc. 5th European Meeting in Astronomy*, Institute d'Astrophysique, Liège, p. A.1.1.

Christensen-Dalsgaard, J.: 1982, *Monthly Notices Roy. Astron. Soc.* **199**, 735.

Christensen-Dalsgaard, J. and Gough, D. O.: 1980, *Nature* **288**, 544.

Christensen-Dalsgaard, J. and Gough, D. O.: 1981, *Astron. Astrophys.* **104**, 173.

Christensen-Dalsgaard, J. and Gough, D. O.: 1982, *Monthly Notices Roy. Astron. Soc.* **198**, 141.

Claverie, A., Isaak, G. R., McLeod, C. P., van der Raay, H. B., and Roca Cortes, T.: 1979, *Nature* **282**, 591.

Claverie, A., Isaak, G. R., McLeod, C. P., van der Raay, H. B., and Roca Cortes, T.: 1980, *Astron. Astrophys.* **91**, L9.

Claverie, A., Isaak, G. R., McLeod, C. P., van der Raay, H. B., and Roca Cortes, T.: 1981a, *Nature* **293**, 443.

Claverie, A., Isaak, G. R., McLeod, C. P., van der Raay, H. B., and Roca Cortes, T.: 1981b, *Solar Phys.* **74**, 51.

Deubner, F.-L.: 1981, *Nature* **290**, 682.

Dittmer, P. H.: 1978, *Astrophys. J.* **224**, 265.

Fossat, E., Grec, G., and Pomerantz, M.: 1981, *Solar Phys.* **74**, 59.

Grec, G.: 1981, These de Dr Sci. Phys., Université de Nice.

Grec, G., Fossat, E., and Pomerantz, M.: 1980, *Nature* **288**, 541.

Grec, G., Fossat, E., and Pomerantz, M.: 1983, *Solar Phys.* **82**, 55 (this volume).

Severny, A. B., Kotov, V. A., and Tsap, T. T.: 1976, *Nature* **259**, 87.

Severny, A. B., Kotov, V. A., and Tsap, T. T.: 1981, *Solar Phys.* **74**, 65.

Tassoul, M.: 1980, *Astrophys. J. Suppl.* **43**, 469.

Woodard, M. and Hudson, H.: 1983, *Solar Phys.* **82**, 67 (this volume).

DETECTION OF INDIVIDUAL NORMAL MODES OF OSCILLATION OF THE SUN IN THE PERIOD RANGE FROM 2 HR TO 10 MIN IN SOLAR DIAMETER STUDIES*

RANDALL J. BOS and HENRY A. HILL

Department of Physics and Arizona Research Laboratories, University of Arizona, Tucson, AZ 85721, U.S.A.

Abstract. New observations of solar oscillations are reported. Power density spectra derived from these observations reveal narrow-band oscillations that are spatially global, have spatial symmetry properties that are either symmetric or antisymmetric for reflection about the center of the solar disk and also about the solar equator, and have coherence times $\gtrsim 41$ days. Large-scale differential refraction effects have been reduced by a factor of 10^5 over that found in previous solar diameter studies by the design of the experiment; thus, these effects are eliminated as a possible source of the oscillations. A discussion is presented of this reduction as well as other features of the observing and analysis program. It is concluded that the probability is very high that individual normal modes of oscillation of the Sun have been detected in the period range from 2 hr to 10 min.

1. Introduction

Evidence of solar oscillations has been obtained from meaurements of solar diameters made in 1973, 1975, and 1978 (Brown *et al.*, 1978; Hill and Caudell, 1979; Caudell and Hill, 1980; Caudell *et al.*, 1980). This paper reports the preliminary analysis of observations obtained in 1979.

The 1979 observations are yielding much more information than the previous three sets of observations because of changes in the solar detector which lead to more detailed spatial information and better signal-to-noise ratio and because of a longer time span for the observations. The improved spatial information permits identification of certain spatial symmetry properties of the eigenfunctions for reflections about the center of the solar disk and about the solar equator. These properties are invaluable in mode classification. They may be equally useful in evaluating the viability of such solar models as the oblique magnetic rotator model proposed by Dicke (1978) and recently brought forward by Isaak (1982). From an observational point of view, the additional spatial information makes it possible to significantly reduce the effects of differential refraction in the Earth's atmosphere, because their spatial characteristics can be readily distinguished from those of solar oscillations. An improved signal-to-noise ratio is obtained because of the simultaneous observation of three solar diameters instead of one as in previous work. The time span for the 1979 data set is 41 days, leading to a

* Proceedings of the 66th IAU Colloquium: *Problems in Solar and Stellar Oscillations*, held at the Crimean Astrophysical Observatory, U.S.S.R., 1–5 September, 1981.

Solar Physics **82** (1983) 89–102. 0038–0938/83/0821–0089$02.10.

frequency resolution of 0.28 μHz. For oscillations which are stable for periods of time $\gtrsim 41$ days, this increased time span also contributes to an improved signal-to-noise ratio.

Power density spectra derived from this data set spanning 41 days offer a frequency resolution of 0.28 μHz. However, with such analyses, one is often left with the difficult task of separating the signal from the noise. One way to address this is to subdivide the data set into two or more subsets and compare the power density spectra of these smaller sets. This technique, used in some variation in all of the previous analyses of solar diameter measurements, was successfully employed in the analysis of the 1979 data. However, the variation used in this latter analysis allowed for the identification of statistically significant oscillatory signals while maintaining frequency resolution which is adequate to distinguish individual modes of oscillation.

It has been recognized for some time that the oscillatory signals obtained in solar diameter measurements at SCLERA* are the direct manifestation of fluctuations in the solar limb darkening function. In the previous work, the data were analyzed in various ways to demonstrate this property. In the analysis of the 1979 observations, full advantage of this property is taken by using it as a means to significantly reduce atmospheric differential refraction effects and to separate the signal from the noise arising from a broad class of noise sources. As a consequence, individual eigenstates of the Sun have been clearly seen for the first time in the frequency range 125 μHz to 1.5 mHz (periods between ≈ 2 hr and 10 min). This statement, based on confidence levels typically $> 99.9\%$, represents significant progress in the development of the field of helioseismology.

2. Observations

The solar diameter observations that are used here to identify individual eigenstates of the Sun were obtained in June and July of 1979 at SCLERA. These observations are of the same general class as those made at SCLERA starting in 1972 with Clayton (1973) and Patz (1975), and followed by Hill and Stebbins (1975), Brown *et al.* (1978), and Caudell *et al.* (1980). These preceding works defined positions on diametrically opposite limbs and then observed the change in the separation of these positions as a function of time. The positions on the limbs are determined by the Finite Fourier Transform Definition (FFTD) (see Hill *et al.*, 1975; Hill, 1978) and the relative changes in the edge separation recorded interferometrically. A detailed description of the data-taking procedure is found in these earlier works and a description of the facility is provided by Oleson *et al.* (1974).

However, there are two important differences between the earlier work and that done in 1979. The most recent work utilized information on six limb positions rather than on the separation of only two diametrically opposite edges. These six limb positions form

* SCLERA is an acronym for the Santa Catalina Laboratory for Experimental Relativity by Astrometry, a facility jointly operated by the University of Arizona and Wesleyan University.

three diametrically opposite pairs with the center diameter coinciding with the equator and the other two being rotated $\frac{1}{8}$ of a radian either side of the equator. The slit configuration is shown in Figure 1.

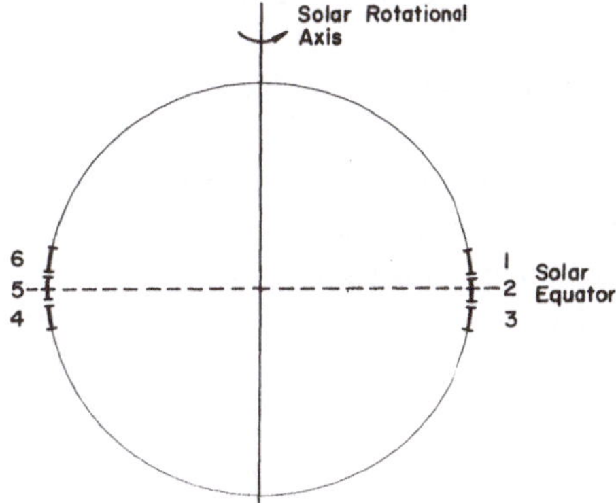

Fig. 1. Slit configuration relative to the solar disk.

The second important difference is that at each of the six limb positions, two different edges, $r_1(\theta, \phi)$ and $r_2(\theta, \phi)$, were defined by applying the FFTD first with a scan amplitude a_1 of 6.8 arc sec and then with a scan amplitude a_2 of 27.2 arc sec. The angles θ and ϕ are the spherical coordinates of the slit position on the limb (see Figure 2). The separation of these two edges is typically 3.5 arc sec (see Hill *et al.*, 1975). It is the separation of these two edges, S_i with $i = 1$–6, that is recorded for each of the six limbs and combinations of the six sets of data that are subsequently Fourier analyzed for evidence of solar oscillations. The S_i are defined such that a positive value of S_i indicates net displacement in the positive radial direction.

It has been well-documented that the oscillations are detected primarily through changes in the radiation intensity, i.e., perturbations in temperature, with spatial frequencies on the order of one (arc sec)$^{-1}$ near the limb (Hill and Caudell, 1979; Knapp *et al.*, 1980). Because of this, the sensitivity of the S_i to oscillations is quite similar to that obtained in diameter measurements, $r_1(\theta, \phi) + r_1(\pi - \theta, \pi + \phi)$. However, the

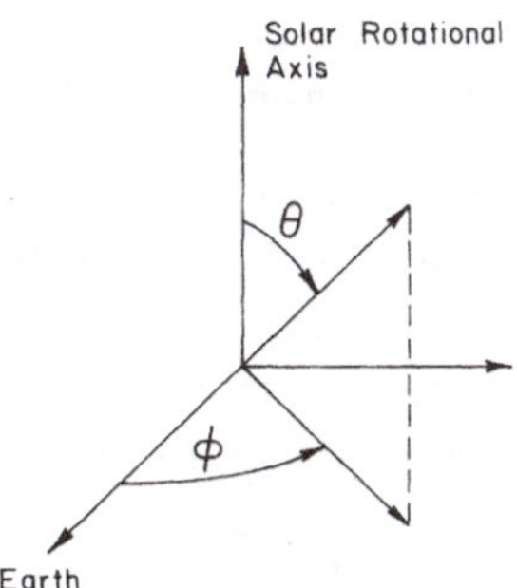

Fig. 2. Spherical coordinate system defined relative to the solar rotational axis.

sensitivity of the S_i relative to that of the diameter measurements is significantly reduced for all sources of noise which change the scale in the telescope, such as differential refraction in the Earth's atmosphere. This clearly discriminates against certain problems that may otherwise be presented by the Earth's atmosphere. Section 8 evaluates this important point in more detail.

Power density spectra for various combinations of the signals, S_i, were calculated. In this, the mean of S_i for each day was removed and a 10% cosine bell applied to the residual, $S_i - \overline{S}_i$, for each day. Information from these various combinations is used in the following sections to ascertain various properties of the data and oscillations. An example of the power density spectrum is shown in Figure 3 for a 30 μHz frequency range near 450 μHz. This power density function is for the combination of signals:

$$\left(\sum_{i=1}^{6} S_i\right)\Big/6\,. \tag{1}$$

Because of the sign convention on the individual S_i, this combination of S_i is sensitive to eigenfunctions that are symmetric about the equator of the Sun and symmetric for reflection through the center of the solar disk ($\theta \to \pi - \theta$; $\phi \to \pi + \phi$).

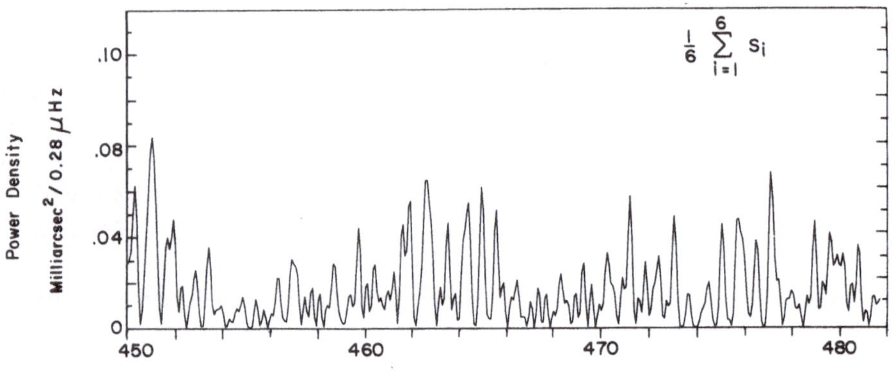

Fig. 3. Power density spectrum of the slit combination ($S_1 + S_2 + S_3 + S_4 + S_5 + S_6$) for the frequency range 450 to 480 μHz.

3. Fractional Power Due to Temporally and Spatially Noncoherent Signals

Noncoherent signal in the power density spectrum can be classified as such by either of two criteria: signals having spatial coherence lengths which are less than a solar diameter or those with temporal coherence times less than 2 days. In principle, signals can exist that meet only one of these criteria, but signals arising from global oscillations should not meet either. The 1979 observations can be analyzed using both criteria.

A. Spatially noncoherent signals

Six independent signals, S_i for i = 1–6, are acquired from the six separate limb positions.

From a given combination of the S_i, the quantity of interest is the fraction of power density which is due to spatially coherent signals. The following model is adopted for defining a measure of coherence: a spatially coherent signal will have a coherence length $>$ a solar diameter, while a noncoherent signal will have a coherence length $\lesssim 100$ arc sec; in addition, the coherent signals are symmetric or antisymmetric for reflections about the disk center and also for reflections about the equator. Should the measure of coherence turn out to be less than 100% by a statistically significant amount, then the assumptions in the above model may not be valid although a certain degree of coherence may be established. If, on the other hand, the measure of coherence approaches 100%, then this result is also a measure of the validity of the above model. The combination of S_i that has been examined for spatial coherence is

$$\left(\sum_i S_i\right)\Big/6\,,$$

the combination used for the power density reported in Figure 3. For this combination, only the spatially coherent signals that are symmetric for reflections both about the disk center and the equator survive. For this combination of S_i and using the model outlined in the preceding paragraph, the measure of coherence, γ, can be written as

$$\gamma = \frac{3}{2}\left(1 - \frac{2y}{x}\right), \tag{2}$$

where

$$x = \frac{1}{6}\left|F\left(\sum_i S_i\right)\right|^2, \tag{3}$$

$$y = \frac{1}{6}\sum_{i=1}^{6}\left|F(S_i)\right|^2 - \left|F\left(\frac{S_1 + S_2 + S_3 - S_4 - S_5 - S_6}{6}\right)\right|^2 - $$

$$- \frac{2}{3}\left|F\left(\frac{S_1 + S_4 - S_3 - S_6}{4}\right)\right|^2 - \frac{2}{3}\left|F\left(\frac{S_1 - S_4 - S_3 + S_6}{4}\right)\right|^2 \tag{4}$$

and $F(f)$ is the Fourier transform of function f. This coherence function is normalized such that $\gamma = 1$ for a 100%-coherent signal and $\gamma = 0$ when there is no spatially coherent signal present. Thus, any statistically significant departure of γ from zero is a direct measure of the component in the S_i that is spatially coherent in the combination $\sum S_i$.

Using the observational data in the frequency range from 400 μHz to 450 μHz, the calculated coherence function γ implies that 98% ± 5% of the power density is spatially coherent and that 2% ± 5% is spatially noncoherent. The power density of the spatially noncoherent type, $P_{s,nc}$, is thus

$$P_{s,nc} = (0.5 \pm 1.3) \times 10^{-3}\ (\text{milli arc sec})^2/0.28\ \mu\text{Hz}\,. \tag{5}$$

As previously noted, the fact that γ is found to be quite close to one serves as a check on the validity of the assumptions that went into the derivation of Equation (2). The validity of these assumptions is also supported by the results from Section 5 on symmetry properties.

B. Temporally noncoherent signals

In a procedure which is an extension of that considered by Groth (1975), temporal coherence was examined by numbering the days of the data set sequentially from one to 18 and then dividing the days into even-numbered and odd-numbered sets. The frequency resolution for power density spectra from these two data sets is essentially the same as that for the total 18-day set spanning 41 days since the mean length of both data sets is ≈ 38 days. Any variation in power density in the peaks from the even- to odd-numbered spectra must result from signals with coherence times < 2 days, the mean spacing of the 18 days of data. If signals from any eigenstates are present, they will very likely be resolved and cannot give rise to any interference effects to complicate the interpretation.

The decision to set the boundary distinguishing coherent and noncoherent signals at approximately 2 days was dictated in part by signal-to-noise considerations in the design of the study. It was also recognized that a 2-day limit is sufficiently long to discriminate effectively against most noise sources such as the Earth's atmosphere and the observing instrument.

A comparison of the even-and odd-numbered power density spectra was made for the frequency range 400 to 500 μHz. From the average of the square of the observed scatter in the peak power densities, 98.9% of the power density was found to be temporally coherent and 1.1% temporally noncoherent. The power density of the temporally noncoherent signal, $P_{t,nc}$, is thus

$$P_{t,nc} = 4.1 \times 10^{-4}\,(\text{milli arc sec})^2/0.28\,\mu\text{Hz}\,. \tag{6}$$

The high proportion of temporally coherent vs noncoherent power density should enhance the statistical significance of any results that may be drawn from the analysis of these observations.

The determination of the temporally noncoherent power density level is approximately ten times more accurate than that for the spatially noncoherent power density level. It is interesting to note that the two independently determined power levels are consistent with the hypothesis that temporal and spatial coherence must both be associated with an oscillation. This conclusion is further supported by the observed symmetry properties discussed in Section 5.

4. Phase Coherence Time > 41 Days

After classifying the observed power density as coherent or noncoherent, the next question to be addressed concerns the actual length of the phase coherence. The question arises in part because the future of the field of helioseismology depends on the ability to study the spatial properties of individual solar eigenstates and to measure very accurately the eigenfrequencies themselves (Gough, 1978). Both of these can only be achieved if the coherence times are $\gg$ days.

An initial approach to this problem might be to extend the technique used to establish the 1.1%–98.9% distribution of noncoherent to coherent power from a limiting coherence time of, for example, 2 days to 4 days or more. However, this procedure brings about a rapid decrease in the signal-to-noise ratio, making some other analysis possibly more attractive.

A concept of correlation time (Stratonovich, 1963) that gives some idea of the size of the time interval over which correlation extends between values of a process is defined by the relation

$$\tau_{\text{cor}} = \frac{1}{K(0)} \int_0^\infty |K(\tau)| \, d\tau , \tag{7}$$

where

$$K(\tau) = \int_{\nu_1}^{\nu_2} P e^{-i2\pi\nu\tau} \, d\nu \tag{8}$$

and P is the power density spectrum. This correlation time is appropriate for the oscillatory signals with frequency between ν_2 and ν_1 as long as $\tau_{\text{cor}} \gg 1/(\nu_2 - \nu_1)$.

The τ_{cor} was calculated for the power density of $\sum S_i/6$ over the frequency range 450 to 480 μHz and compared to that expected for oscillations with autocorrelation functions

$$K(\tau) = K(0) e^{-\beta|\tau|} \cos \omega_0 \tau . \tag{9}$$

The results of this analysis yield

$$\beta^{-1} = 51 \text{ days} . \tag{10}$$

Therefore, the coherence time of these oscillations must be > 41 days.

5. Spatial Symmetries of Observed Eigenfunctions

The identification of spatial symmetries in the observed eigenfunctions can be valuable for several reasons. Should the phase and amplitude of an oscillation at one solar limb have a significant and well-defined relation to that of an oscillation at a diametrically opposite limb, the global character of the oscillation is demonstrated. The classification of the symmetry properties also helps to identify any preferred direction within the Sun, a preferred direction which might arise, for example, as a result of a Sun that is not spherically symmetrical as seen by the oscillations. In addition, symmetry properties might provide a means for ascertaining whether the oscillations originate in the Sun or the Earth's atmosphere. For example, symmetries that clearly reflect an axis on the Sun would be quite difficult to ascribe to effects arising in the Earth's atmosphere while, on the other hand, symmetries that reflect the local elevation and azimuth system of the observatory would be hard to interpret in terms of a solar phenomenon.

The observations reported here were made in such a way so as to detect changes in the solar limb darkening function at six different places on the solar perimeter. These locations are shown in Figure 1, with the details of the geometry given in Section 2. The more interesting combinations of the S_i are

$$\begin{aligned}&(S_1 + S_2 + S_3) \pm (S_4 + S_5 + S_6)\,,\\ &(S_1 + S_4) \pm (S_3 + S_6)\,,\\ &(S_1 - S_4) \pm (S_3 - S_6)\,.\end{aligned} \tag{11}$$

Analysis of these combinations shows that an oscillating signal appears predominantly in only one of the six combinations; the secondary signals appearing in any one of the remaining combinations are < 0.2 of the prominent signal.

These results can be expressed more formally: the eigenfunctions are global and, to an accuracy of better than 20%, are either symmetric or antisymmetric for reflection on the solar disk through the Sun's center ($\theta \to \pi - \theta$, $\phi \to \phi + \pi$) and for reflection about the solar equator where θ and ϕ are defined in Figure 2. So for each observed oscillation, there exists at this time a set of symmetry properties.

These symmetry properties show that the Sun's observed axis of surface rotation is strongly reflected in the observed eigenfunctions. It is well-known that the spherical harmonic $Y_l^m(\theta, \phi)$ for the coordinate system defined in Figure 2 is also either symmetric or antisymmetric, depending upon the values of l and m for the two types of reflections discussed above. Therefore, these observations support the description of the horizontal spatial properties of the eigenfunctions by Y_l^m, and the l and m value for each oscillation can be classified as either even or odd.

Claverie *et al.* (1981) recently reported detecting a series of oscillating modes that they classified as high order with $l = 1$, $m = 0$. Whole disk velocity measurements such as these should have very low sensitivity to states of odd $l + m$ if the (θ, ϕ) dependence of the eigenfunction is given by Y_l^m. Therefore, they should not see the $l = 1$, $m = 0$ or the $l = 2$, $m = \pm 1$ states. However, they obtain strong signals for all three cases.

One way out of this dilemma is to propose a modification of the Sun that would remove the previously expected symmetry. Isaak (1982) has made such a proposal, suggesting that the Sun is an oblique magnetic rotator. But the symmetry property that this model removes, i.e., the existence of antisymmetric wavefunctions for reflection about the equator, is in fact one of the properties found in the observations reported here. Thus, there does appear to be a fundamental discrepancy between the two sets of observations. The results reported here indicate that solar rotation is the primary spherical symmetry-breaking phenomenon on the Sun.

As mentioned in the introductory remarks of this section, the observed symmetry properties may be quite restrictive in the search for the origin of the oscillations. First, let us consider the symmetry property for reflection through the center of the solar disk.

The observed existence of a strong correlation between the motions of the limbs over a distance of 2000 arc sec clearly shows that the oscillations are spatially global, a result consistent with that found in Section 3. Second, consider the fact that the observed oscillations have symmetry properties that track the solar axis of rotation. These properties plainly point to the Sun as the origin of the observed oscillations and, in particular, rule out differential refraction effects in the Earth's atmosphere.

6. Reproducibility of Power Density Spectra

Inspection of Figure 3 shows that the spectrum is quite dense, a characteristic which is typical of the entire band from 125 μHz to ≈ 1.5 mHz. This is encouraging with respect to future work on the inverse problem, i.e., determining the internal structure of the Sun from the eigenfrequency spectrum. However, this complexity makes it quite difficult to make any meaningful comparison with results previously reported from SCLERA. In all of these preceding works, the frequency resolution that was used in the analysis was ≈ 30 μHz. The peaks in these earlier spectra probably represented the superposition of several of the peaks in the work reported here.

However, the data set from the 1979 observations, in which the peaks are probably associated with individual eigenstates, can be divided into two independent subsets and the reproducibility of peaks in the power density spectra verified. This procedure employed the power density spectra used to determine the mean of the temporally noncoherent power level, i.e., the 18-day data set was divided into even- and odd-numbered subsets and their power density spectra intercompared. In this manner, two independent data sets were generated, both with a frequency resolution of approximately 0.3 μHz. In the frequency range from 400 to 500 μHz and counting all peaks with a power density $> 1.25 \times 10^{-3}$ (milli arc sec)^2, there are 206 peaks in one spectrum and 230 in the second. Superimposing these two spectra, there are 175 peaks which coincide, to within 0.3 μHz, from one spectrum to the other. No attempt has been made to remove the sidebands, i.e., the extra peaks, associated with an oscillation and introduced by the window function. However, a preliminary analysis indicates that about $\frac{2}{3}$ of the peaks counted in either power spectrum are sidebands. Taking this into account, one finds that there are approximately 70 independent peaks in the range from 400 to 500 μHz and that 80% of these are coincident.

This observed coincidence rate is eleven standard deviations away from that expected on the hypothesis that the peaks result from noise. However, this result should come as no surprise since it was found that 98.9% of the power density in the spectrum had coherence times > 2 days.

7. Amplitudes of Resolved Modes of Oscillation

The mean power of the observed oscillatory signals in the 1979 observations is 2.5×10^{-2} (milli arc sec)^2/0.28 μHz at 450 μHz. The root mean square amplitude, $\langle a^2 \rangle^{1/2}$, can be obtained by integrating the power density function of a single state with respect to frequency. This yields a mean square amplitude of

$$\langle a^2 \rangle^{1/2} = 0.4 \text{ milli arc sec/state} . \tag{12}$$

This will be a very useful number in future work on these oscillations.

A comparison of the power density observed here with that of Brown *et al.* (1978) is of interest. To do this, several factors must be considered: the difference in sensitivity of the two measurements (S_l vs r); the reporting of diameter vs radius amplitude; different sensitivities to spatially noncoherent signals; and different frequency resolution. Taking these factors into account and converting to a representation that permits direct comparison of these results with those of Brown *et al.*, the power density of 2.5×10^{-2} (milli arc sec)^2/0.28 μHz is equivalent to a power density of

$$24 \text{ (milli arc sec)}^2/30\ \mu\text{Hz} \tag{13}$$

at 450 μHz. This is to be compared to 28 (milli arc sec)^2/30 μHz, the value actually obtained by Brown *et al.* (1978).

8. Atmospheric Differential Refraction and Seeing Effects Reduced to the Submilliarcsecond Level

The mean power of the observed oscillating signals in the 1979 data is 2.5×10^{-2} (milli arc sec)^2/0.28 μHz. As noted in Section 7, this value compares quite well with the previously reported power density of Brown *et al.* (1978). Clearly, atmospheric differential refraction and seeing effects must be below this level if they are not to present any problem.

In the 1979 observations, large-scale differential refraction and seeing effects are estimated to be $\lesssim 3 \times 10^{-6}$ (m arc sec)^2/0.28 μHz and $\lesssim 1 \times 10^{-5}$ (m arc sec)^2/0.28 μHz, respectively. These upper limits are considerably less than the observed oscillatory power given above. These low power levels for atmospheric effects were achieved as the result of two features of the observational method pertaining to the technique used to detect motion of the solar limb and to the number of positions on the solar limb which were examined. These features are discussed in this section.

As previously described, the oscillations were detected as changes in the location of the solar limb as defined by the application of the FFTD using a 6.8-arc sec scan amplitude relative to that defined by the FFTD using a 27.2-arc sec scan amplitude. This procedure reduces the amplitude of differential refraction effects by a factor of 300 or the corresponding power density by a factor of 300^2 relative to those found in simple diameter measurements.

The manner in which this reduction comes about can be easily seen. The magnitude of the correlation, C, between column density fluctuations in the Earth's atmosphere measured along the two lines of sight to opposite solar limbs was found by Fossat *et al.* (1981) to be very close to 1, i.e.,

$$1 - C = 0.5 \times 10^{-2} . \tag{14}$$

This measurement puts constraints on the characteristic horizontal scale of fluctuations in the atmospheric column density. In particular, this length in angular measure must be greater than the diameter of the Sun, a point also noted by Fossat *et al.* (1981). Therefore, to a good approximation, a Taylor series expansion can be made of the atmospheric refraction R about the center of the Sun in the direction of the diameter being measured. Keeping terms to second order, we have

$$R(r) = R(0) + R'r + \tfrac{1}{2}R''r^2 \tag{15}$$

or

$$\Delta R(r) = R'r + \tfrac{1}{2}R''r^2\,, \tag{16}$$

where

$$\Delta R(r) = R(r) - R(0)\,.$$

Thus, in the combination of S_i, i.e.

$$\left(\sum_{i=1}^{6} S_i\right)\Big/ 6\,,$$

used in Figure 3, the residual effect, $\Delta R_{\Sigma S_i}$, is

$$\Delta R_{\Sigma S_i} = 2R'(r_2 - r_1)\,, \tag{17}$$

where r_1 and r_2 correspond to the radii defined by the FFTD 6.8- and 27.2-arc sec scan amplitudes, respectively. Note that second order effects drop out. For a diameter measurement defined by the FFTD 27.2-arc sec scan alone, the refraction effect, ΔR_d, is

$$\Delta R_d = 2R'r_2\,. \tag{18}$$

Thus we have

$$\frac{\Delta R_{\Sigma S_i}}{\Delta R_d} = \frac{r_2 - r_1}{r_1}\,. \tag{19}$$

The difference $r_1 - r_2$ is 3.5″ (see Section 2) so that

$$\frac{\Delta R_{\Sigma S_i}}{\Delta R_d} = -3.5 \times 10^{-3}\,. \tag{20}$$

Thus, power density for large-scale differential refraction effects in the 1979 observations was reduced by a factor of 10^5 below that present in the earlier diameter measurements.

Previous attempts, such as the work of Fossat *et al.* (1981), to infer the size of atmospheric differential refraction assumed that the correlation C is very close to one, as supported by their observations, and considered only large-scale effects. Since large-scale differential refraction effects have clearly been reduced to a negligible level, differential refraction with horizontal length $\lambda_a < 2R$ may also receive attention. In the range $\lambda_a > 100$ arc sec, the sensitivity to this noise source is also reduced by a factor similar to that given by Equation (19). The factor in this case is

$$\frac{\Delta R_{\Sigma S_i}}{\Delta R_d} = \frac{4\pi(r_2 - r_1)}{\lambda_a}. \tag{21}$$

For $\lambda_a < 100$ arc sec and a coherence length also < 100 arc sec, the resulting power density is additionally reduced by a factor of 6 because the signal from six limb positions is averaged together. Thus the average power density from atmospheric differential refraction in the 1979 data has been reduced by a rather large factor over most of the spatial frequency range.

Atmospheric seeing is at the low level of $< 1 \times 10^{-5}$ (milli arc sec)^2/0.28 μHz because of the utilization of the FFTD in defining the position of the solar limb. The low sensitivity of the FFTD to seeing is one of its strong features (see Hill *et al.*, 1975). The net seeing effect in these observations is also reduced by a factor of 3 below that present in the earlier diameter measurements because the data contain the average of the signals from six limb positions rather than only two.

Should any questions remain regarding the elimination of the Earth's atmospheric effects in the analysis of the 1979 observations, there is yet another consideration. The spatial symmetry properties of the observed signals as classified in Section 5 are indeed difficult to understand if the oscillations are the result of atmospheric effects. These symmetry properties comprise one of the five sufficient condition tests discussed by Hill *et al.* (1983). The symmetry of the signals for reflection about the center of the Sun establishes the global spatial character of the phenomenon. The contribution to a global signal from atmospheric seeing with an aplanatic patch size of the order of 10 arc sec must be reduced by many orders of magnitude below the seeing contribution to noise in the location of a single limb of 1×10^{-5} (milli arc sec)^2/0.28 μHz given above. Actually, for reflections through the center of the Sun, the oscillating signals were observed to be either symmetric or antisymmetric with the magnitude of the amplitudes of the two sides being equal to $\pm 20\%$. Such properties would certainly make difficult an interpretation of the oscillations in terms of atmospheric differential refraction.

Finally, some additional statements can be made which are quite relevant. In Section 3, it was determined that the power of the temporally noncoherent signals was 1.1% of the mean peak power of the coherent component. Atmospheric effects which are temporally noncoherent must be less than or equal to this upper limit. Similarly, the spatially noncoherent atmospheric component must be less than $2 \pm 5\%$ of the mean peak power. Diurnal terms, a potentially serious source of coherent signal arising from the atmosphere, are also not present, as can be seen by examining Figure 3.

Atmospheric effects have received much attention in work on solar diameter measurements. However, the analysis of the 1979 observations shows that these effects have been reduced to a nonsignificant level while not seriously degrading the sensitivity of the observations to the stable, oscillatory phenomena which are of interest.

9. Discussion

This section brings together evidence for the interpretation of the individual spectral peaks as the resolved frequencies of individual normal modes of solar oscillation. Observational results are presented which are sufficient to demonstrate: (1) that the oscillatory phenomena are spatially global; (2) that the frequencies of the oscillations are stable to better than one part in 1000 over a period of 41 days; and (3) that the horizontal spatial characteristics of the oscillations have symmetry properties that identify an axis in the Sun coinciding with the solar rotational axis. Further, the observational program has been designed to reduce the effects of the Earth's atmosphere to a level well below the observed power density levels of the oscillations.

In order to develop these points, let us examine the postulate that the observed structure in Figure 3 is the result of noise. In this examination, the effects of the Earth's atmosphere can be quickly dismissed because of the extremely low sensitivity of the present observations to atmospheric effects, a feature built into the program (see Section 8). The oscillatory signals might also be arising within the observing instrument. However, this is not a viable explanation in light of the horizontal symmetry properties of the observed signals. The oscillations can be classified as either symmetric or antisymmetric for reflection on the solar disk through the center of the Sun and for reflection about the solar equator, properties which cannot be easily ascribed to instrumental effects but which are natural if the oscillations are solar in origin.

The noise postulate is thus reduced to a statement that the observed structure in Figure 3 is the result of noise, i.e., fluctuations, in the solar limb darkening function. Presumably, the use of the term noise in this case must be used to signify a local perturbation. It is quite clear that this is not the proper interpretation, as the oscillations are observed simultaneously at diametrically opposite limbs of the Sun with the magnitude of these diametrically derived signals equal to $\pm 20\%$ and with the signals either in phase or out of phase with each other (see Section 5). These results are a direct demonstration of the global properties of the oscillations.

The interpretation of the signal as noise in lieu of global oscillations also presumbly implies certain temporal properties. That is, a noise source should not lead to oscillating features which are very stable in time. But only 1.1% of the observed signal was found to be temporally noncoherent for coherence times > 2 days and a significant number of the oscillations was found to have coherence times > 41 days (see Sections 3 and 4).

Thus, the postulate that the observed structure in Figure 3 is the result of noise fails on all accounts: the observed spatial and temporal properties are only consistent with the postulate that individual normal modes of oscillation of the Sun have been detected.

10. Summary

Evidence has been given which supports the conclusion that individual eigenstates of the Sun have been detected in the period range from ≈ 2 hr to ≈ 10 min. If the Sun's interior is represented fairly well by one of the standard solar models, then we have clearly observed both individual acoustic and gravity mode eigenstates of the Sun. The implications of this are obvious. We are at the threshold of an exciting era in the field of helioseismology in which it will be possible to probe the solar interior at an extremely quantitative level. If, on the other hand, individual gravity mode eigenstates have not been detected, the state of our knowledge of the solar interior is not as advanced as is generally believed, a prospect which is equally exciting.

Acknowledgements

The authors wish to acknowledge helpful discussions with Thomas Caudell and critical reading of the manuscript by Caudell and Philip Goode. This work was partially supported by the Astronomy Division of the National Science Foundation and the Air Force Office of Scientific Research.

References

Brown, T. M., Stebbins, R. T., and Hill, H. A.: 1978, *Astrophys. J.* **223**, 324.

Caudell, T. P. and Hill, H. A.: 1980, *Monthly Notices Roy. Astron. Soc.* **193**, 381.

Caudell, T. P., Knapp, J., Hill, H. A., and Logan, J. D.: 1980, in H. A. Hill and W. A. Dziembowski (eds.), *Nonradial and Nonlinear Stellar Pulsation*, Lecture Notes in Physics, No. 125, Springer-Verlag, Berlin, p. 206.

Claverie, A., Isaak, G. R., McLeod, C. P., van der Raay, H. B., and Roca Cortes, T.: 1981, *Nature* **293**, 443.

Clayton, P. D.: 1973, 'A Precise Measurement of the Sun's Visual Oblateness', University of Arizona (Ph. D. Thesis).

Dicke, R. H.: 1978, in G. Friedlander (ed.), *Proceedings Informal Conference on Status and Future of Solar Neutrino Research*, BNL Report 50879, 2, Brookhaven National Laboratories, Upton, New York, p. 109.

Fossat, E., Grec, G., and Harvey, J. W.: 1981, *Astron. Astrophys.* **94**, 95.

Gough, D. O.: 1978, *Proceedings 2nd European Solar Meeting*, Toulouse, France, March 8–10.

Groth, E. J.: 1975, *Astrophys. J. Suppl.* **29**, 286.

Hill, H. A.: 1978, in J. A. Eddy (ed.), *The New Solar Physics*, Westview Press, Boulder, Colorado, Chapter 5.

Hill, H. A. and Caudell, T. P.: 1979, *Monthly Notices Roy. Astron. Soc.* **186**, 327.

Hill, H. A. and Stebbins, R. T.: 1975, *Astrophys. J.* **200**, 471.

Hill, H. A., Stebbins, R. T., and Oleson, J. R.: 1975, *Astrophys. J.* **200**, 484.

Hill, H. A., Bos, R. J., and Caudell, T. P.: 1983, *Solar Phys.* **82**, 129 (this volume).

Isaak, G. R.: 1982, *Nature* **296**, 130.

Knapp, J., Hill, H. A., and Caudell, T. P.: 1980, in H. A. Hill and W. A. Dziembowski (eds.), *Nonradial and Nonlinear Stellar Pulsation*, Lecture Notes in Physics, No. 125, Springer-Verlag, Berlin, p. 394.

Oleson, J. R., Zanoni, C. A., Hill, H. A., Healy, A. W., Clayton, P. D., and Patz, D. L.: 1974, *Appl. Opt.* **13**, 206.

Patz, D. L.: 1975, 'An Experimental Method to Determine Small Differences Between the Polar and Equatorial Solar Limb Profiles', University of Arizona (Ph. D. thesis).

Stratonovich, R. L.: 1963, *Topics in the Theory of Random Noise*, transl. from the Russian by Richard A. Silverman; Vol. 3 in the series: Mathematics and Its Applications, Gordon and Breach Science Publishers, Inc., New York.

HELIOSEISMOLOGY WITH HIGH DEGREE p-MODES*

(Invited Review)

F.-L. DEUBNER

Institut für Astronomie und Astrophysik, D-8700 Würzburg, F.R.G.

Abstract. The value of p-modes of high degree l as a diagnostic for the structure and dynamics of the solar envelope is reviewed.

1. Definition of the subject

For the purpose of this brief overview we shall define 'high degree' modes pragmatically as coherent non-radial oscillations which are not (well) observed in full disk observations of the Sun. Accordingly the spatial scale of these motions ranges typically from a few arc sec (3×10^3) km) to about $1000''$ ($\sim 10^6$ km). Incidentally, we cannot really quote any 'representative size' of the oscillations nor can it be defined in a meaningful way. It depends entirely on the spatial filter functions implied by the observational setup, since the full spectrum of p-modes extends from the radial modes ($l = 0$) into the range of granular sizes ($l \simeq 2000$) with monotonically decreasing amplitude (cf. Section 5). The contributions from various parts of the solar disk of such a small scale periodic wave field most effectively cancel in full disk observations.

Recognizing the global nature of the low degree p-modes reported earlier in this meeting, we wonder whether eventually we shall be able to trace somewhere within the range of sizes defined above a deviding line between truly globally coherent modes and only locally coherent oscillations. This might shed some light on the relative importance of various excitation mechanisms for p-modes presently under discussion.

2. Brief History

Most of the early observational work on the 5-min oscillations suffered severely from insufficient coverage of the large range of sizes and periods characterizing this phenomenon. The classic work of Leighton *et al.* (1962) – these authors were using two-dimensional photographic records of the Doppler shift – had the potential of revealing the large scale coherent properties of the oscillations. However, the observations were too short. On the other hand, subsequent investigations using photoelectric

* Proceedings of the 66th IAU Colloquium: *Problems in Solar and Stellar Oscillations*, held at the Crimean Astrophysical Observatory, U.S.S.R., 1–5 September, 1981.

Solar Physics **82** (1983) 103–109. 0038–0938/83/0821–0103$01.05.

equipment, by Howard (1967), Deubner (1972), White and Cha (1973), and many others who took data from small areas on the solar surface for many hours at a time, completely failed to recognize the important spatial coherence of the modes.

Nevertheless, in 1970, Ulrich followed by Leibacher and Stein (1971) gave a basically correct description of the evanescent waves which are seen to cover the solar globe all the time and everywhere, in terms of resonant oscillations of the solar interior leaking to the surface.

Only later in the last decade, observers were ready to understand the requirements for really unseful observations of the 5-min oscillations, which consisted of spatially extended records repeated as many times as daylight permitted.

3. Observing techniques

An ideal observation of solar p-modes obviously would cover the entire surface with sufficient spatial resolution (2″) at roughly 100 s intervals for at least one solar revolution, i.e. about $(3 \times 10^6) \times (2.5 \times 10^4) = 7.5 \times 10^{10}$ pixels per observation. Although this requirement appears somewhat unrealistic, the importance of uninterrupted observations from space assisted by ground based monitoring looking at the Sun from a different angle has been hardly enough emphasized yet, especially if transient events (flarequakes) as conceived some time ago by Wolff (1972) are to be evaluated.

These days observations are taken with single-channel receivers mostly by scanning the solar surface in one direction and taking account of the isotropic character of the wave field either by chosing the appropriate transformation of the spatial coordinate afterwards in the data analysis, or by spatially averaging in the other direction (long slit technique) during observation. (A brief summary of the various ways of sampling two-dimensional isotropic phenomena, and the relations among the pertinent data transforms is given in Aime and Ricort (1975).) Multi-channel receivers (linear and areal diode arrays) permit to obtain – at the cost of a largely increased data flow – two-dimensional images of the velocity (or intensity) distribution, which are of particular value if the assumption of angular isotropy has to be dismissed, e.g. in the investigation of latitude dependent effects such as differential rotation and other large-scale circulation systems (Hill *et al.*, 1983), and perhaps of low degree modes.

4. Representation of Data

The k, ω or diagnostic diagram has become widely used as the most compact and to theory most directly applicable representation of the properties of oscillations. An explanation of its significance is contained in many standard texts (cf. Leibacher and Stein, 1981), and doesn't need to be repeated here. The discrete eigenfrequencies clearly separate the various modes of different radial index n in the k, ω plane as shown in

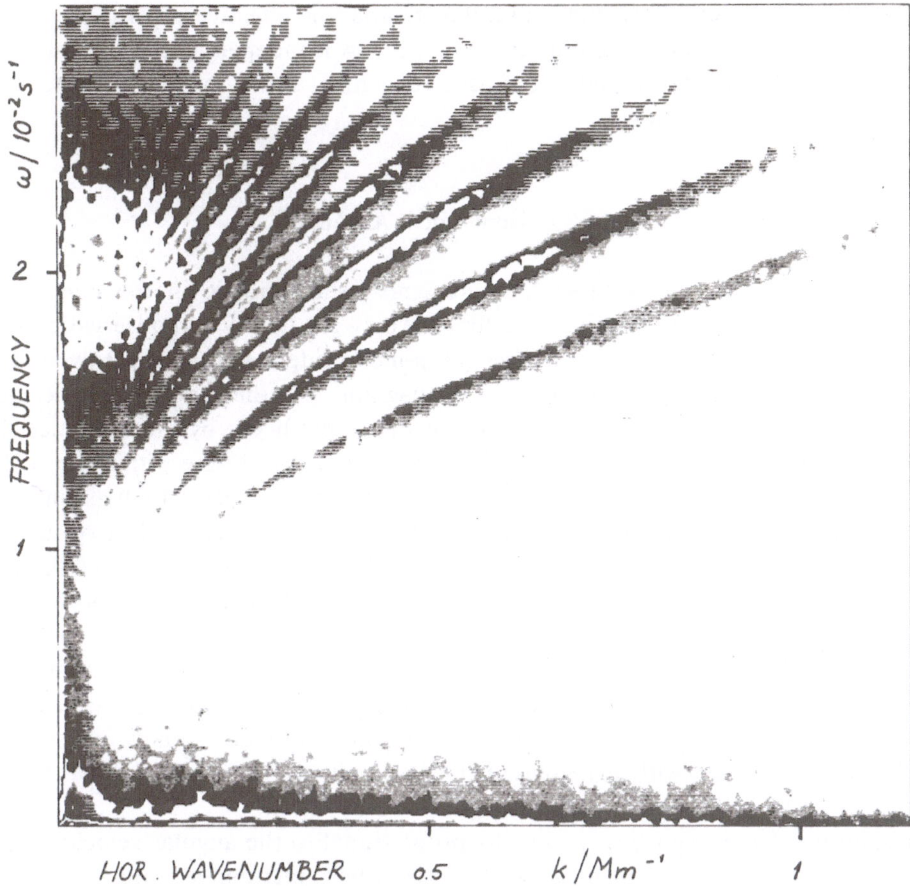

Fig. 1. k, ω diagram of high degree solar p-modes obtained with the domeless Coudé refractor of the Anacapri station of the Kiepenheuer-Institut. Observed spectral line: Fe I 5576.099. Total duration of the observation: 4 × 7.54 hr.

Figure 1, and the dependence of these frequencies on horizontal wavenumber k (or surface harmonic index l) can be directly seen. Also, in this diagram, ranges of 'effective depth' of a mode of given order and degree may be delineated (Ulrich *et al.*, 1979).

For specific diagnostic purposes a number of modifications of the k, ω diagram have been suggested: since asymptotically the p-mode ridges (see Figure 1) approach a parabolic shape with increasing wavenumber, a plot of ω^2 vs k immediately displays the deviations from the asymptotic law which is now represented by straight lines (Christensen-Dalsgaard, 1980). Division of ω^2 by k makes the straight lines all parallel and thus reveals in even more detail the structure of the asymtotic dependence of the eigenfrequencies on l and n (Gough, 1981).

Similar to these modified diagnostic diagrams is a plot of ω/k vs k, which depicts horizontal phase velocity as function of spatial scale. This relates directly to the thermal structure of the solar envelope and is useful in comparing 'effective depths' of various modes. Also, the amount of rotational splitting $\Delta\omega = vk$ is independent of k in this kind of diagram which, therefore, lends itself as a conveniant indicator of any systematic depth-dependence of rotation.

5. Helioseismological Aspects

We now turn to the use of these measurements as diagnostic in the context of helioseismology. The most conspicuous difference between theory and observation was pointed out by Deubner (1975) soon after the p-mode ridges had been resolved. The position of the observed ridges in the k, ω diagram, i.e. the eigenfrequencies are systematically lower by a few percent than the values predicted by theory. Since the p-mode eigenfrequencies decrease with increasing sound travel time within the stellar envelope, this can only mean that the solar convection zone is deeper than assumed in the model or, in terms of mixing length theory, that convection is more effective and the mixing length parameter $\alpha = l/H$ is larger than 1 (Ulrich and Rhodes, 1977). Raising the value of α to 2.5 brings theory and observation more closely together, however, a further increase of α would create a convection zone reaching considerably deeper than 200 000 km, which is then in conflict with the observed surface abundance of lithium.

Another possibility of lowering the eigenfrequencies is by reducing the pressure, e.g. by taking into account electrostatic forces in the equation of state. Lubow *et al.* (1980) have shown that together with an increased mixing length parameter of $\alpha = 2.5$ this leeds to almost perfect agreement between theory and observation.

The splitting of p-mode ridges is directly proportional to the angular velocity of the standing wave pattern of a given mode. It has already been mentioned in the last section as a means to infer rotational velocities not only at the surfaces but also as a function of depth within the solar envelope (Rhodes *et al.*, 1979). First results published by Deubner *et al.* (1979) indicated an increase of the rotation rate with depth. This result has recently been challenged by the Kitt Peak observers (Rhodes *et al.*, 1983) who find no variation whatsoever within their limits of accuracy. If this difference is real, it may indicate a change of the dynamics of the convection zone at different phases of the solar cycle, since the latter observations were obtained during solar maximum, the first ones at the time of the minimum.

Close to the abscissa, in the diagnostic diagram of photospheric motions we find concentration of power due to quasistationary flow patterns, such as granules and supergranules, with time scales long compared to the characteristic period of p-modes. These are surface features, and the splitting, or rather inclination (which was suppressed in Figure 1) of this ridge of power at the abscissa reflects the surface velocity, which we can measure in this way with absolute accuracy comparable to the accuracy of line shift measurements, however without the susceptibility of the latter for systematic degradation of the results, as e.g. caused by straylight.

Surface rotation rates determined in this way in recent years with the Capri instrument come very close to the rotation rates of sunspots and, maybe, indicate a slight decrease of the rotation rate during the rising phase of the present solar cycle:

	rad day^{-1} (sidereal)
1978	0.2563
1980	0.2527 ± 22

Rotation and stellar structure below, say, 50 000 km in the convection zone can only be determined from observation of modes with sufficiently low l-value (cf. Figure 2 in

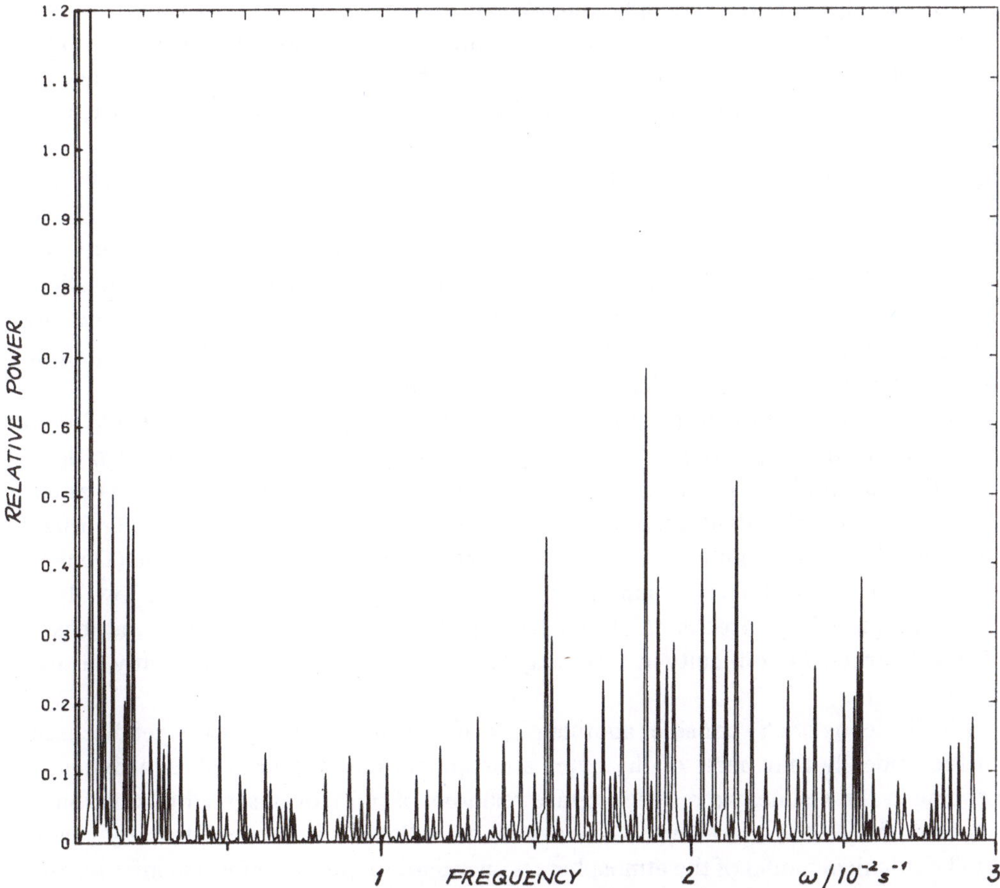

Fig. 2. Power spectrum of the slope of the velocity signal as a function of solar longitude, as measured at 106 s intervals in an area 1000″ × 200″ centered on the solar disk. The spectrum was obtained by sine-wave fitting of a data set from two contiguous days.

Ulrich *et al.*, 1979) which are extremely difficult to resolve in the k, ω diagram with the current techniques of observation of high degree p-modes. Obviously the full disk observations presented earlier in this meeting by Isaak and Fossat are far superior in this important domain.

Nevertheless, we made an attempt to utilize the temporal variation of the slope of the Doppler signal across the solar disk observed during p-mode observations to detect the intermediate range of l-values from about 4 to 10. The resulting power spectrum in Figure 2, obtained by iterative sine-wave fitting looks similar to the spectra of low degree modes. It contains significant power in the 5-min range. Its diagnostic value is presently under study.

Whereas the position of the ridges, i.e. the eigenfrequencies can be determined with little ambiguity ($\pm$ 10 μHz) the distribution of power along these ridges, and across the diagram as a function of ω and k is more difficult to obtain, more subject to observational parameters (spatial and temporal window limitations, pixel size) and observing conditions (seeing and spectrograph stability, etc.).

The distribution of power contains information on excitation, damping, and propagation of the resonant modes, properties which are rather harder to understand quantitatively than the resonant cavity. In the spectrum (Figures 3–9 in Deubner, 1981), which displays oscillatory power per unit wavenumber, there is little variation of power in the low k range up to $k \simeq 0.2\ \mathrm{Mm}^{-1}$ (or $l \simeq 150$), i.e. the typical scale of supergranulation, followed by a monotonic decrease at higher l-values. Does this finding imply that convective motions of this smaller scale interfere with oscillations of corresponding l-values? Does it imply that oscillations of this size are only stochastically excited by turbulent motions of a congruent scale? Or does dissipation become more effective at this size because it fits better to the chromospheric network scale, or simply because the resonant cavity becomes rougher there?

Accurate observation of the intrinsic 'line width' of the p-mode ridges will help to find an answer to these questions. Until today, the intrinsic ridge width of the high degree p-modes has not been resolved. Therefore the coherence or damping time τ is at least equal to the available continuous observing time of the order of 8 to 10 h, and probably much larger. There is an indication that τ increases with decreasing l-value (Deubner, unpublished). The difference among the values of the coherence time of low degree p-modes quoted by Grec *et al.* (1980, $\simeq$ 2 days) and Claverie *et al.* (1983, at least 30 days) illustrates the difficulty in assessing correct numbers from insufficiently resolved observations.

Finally, let us briefly consider another piece of information we have not yet discussed. In the optically thin parts of the outer atmosphere, directly accessible to dynamical probing in the visible, we may use phase relations of pairs of observed fluctuations as a tool to analyse the character of the waves and oscillations in situ, and thereby – taking a full dynamical model of the atmosphere with radiation into account – to infer its mean vertical structure. Extensive work in this vein is being done by a group of French workers (Schmieder, 1978; Mein, 1977; Gouttebroze and Leibacher, 1980; Leibacher *et al.*, 1982). Linear adiabatic theory predicts that in the regime of evanescent waves brightness should precede upward velocity by about 90°, and vertical phase velocity should be very large for frequencies less than the acoustic cut-off frequency $N_{ac} \sim 3 \times 10^{-2}\ s^{-1}$. This is approximately true in the Na D lines, however $\Delta\phi_{I\text{-}V}$ is more like 120° in the medium strong Fe I 5383 line, and it is 45° in the C I 5380 line. At higher frequencies where waves become progressive and transport energy, brightness and velocity are expected to propagate in phase, and the phase velocity should approach the velocity of sound. Convincing evidence of the compliance of nature with these ideas is still missing; however, observation of these waves with periods shorter than ~ 150 s is exceedingly difficult due to their strongly filtered amplitudes and comparatively high horizontal wavenumbers. For succesful interpretation of the data a two-dimensional treatment of radiation transfer is likely to become important. k, ω diagrams of short period oscillations may prove useful to distinguish between genuine solar and terrestrial atmospheric disturbances in the signal.

References

Aime, C. and Ricort, G.: 1975, *Astron. Astrophys.* **39**, 319.

Christensen-Dalsgaard, J.: 1980, private communication.

Claverie, A., Isaak, G. R., McLeod, C. P., van der Raay, H. B., and Roca Cortes, T.: 1983, *Solar Phys.* **82**, 233 (this volume).

Deubner, F.-L.: 1972, *Solar Phys.* **22**, 263.

Deubner, F.-L.: 1975, *Astron. Astrophys.* **44**, 371.

Deubner, F.-L.: 1981, in S. Jordan (ed.), *The Sun as a Star*, NASA/CNRS monograph series on Nonthermal Phenomena in Stellar-Atmospheres, Washington, Paris, p. 65.

Deubner, F.-L., Ulrich, R. K., and Rhodes, Jr., E. J.: 1979, *Astron. Astrophys.* **72**, 177.

Gough, D.: 1981, private communication.

Gouttebroze, P. and Leibacher, J. W.: 1980, *Astrophys. J.* **238**, 1134.

Grec. G., Fossat, E., and Pomeranz, M.: 1980, *Nature* **288**, 541.

Hill, F., Toomre, J., and November, L. J.: 1983, *Solar Phys.* **82**, 411 (this volume).

Howard, R.: 1967, *Solar Phys*, **2**, 3.

Leibacher, J. W. and Stein, R. F.: 1971, *Astrophys. Letters* **7**, 191.

Leibacher, J. W. and Stein, R. F.: 1981, in S. Jordan (ed.), *The Sun as a Star*, NASA/CNRS monograph series on Nonthermal Phenomena in Stellar Atmospheres, Washington, Paris, p. 263.

Leibacher, J. W., Gouttebroze, P., and Stein, R. F.: 1982, *Astrophys. J.* **258**, 393.

Leighton, R. B., Noyes, R. W., and Simon, G. W.: 1962, *Astrophys. J.* **135**, 474.

Lubow, S. H., Rhodes, Jr., E. J., and Ulrich, R. J.: 1980, in H. Hill and W. Dziembowski (eds.), *Non-Radial and Non-Linear Stellar Oscillation*, Springer Verlag, Berlin, p. 300.

Mein, N.: 1977, *Solar Phys.* **52**, 283.

Rhodes, Jr., E. J., Harvey, J. W., and Duvall, Jr., T. L.: *Solar Phys.* **82**, 111 (this volume).

Rhodes, Jr., E. J. Deubner, F.-L., and Ulrich, R. K.: 1979, *Astrophys. J.* **227**, 629.

Schmieder, B.: 1978, *Solar Phys.* **57**, 245.

Ulrich, R. K.: 1970, *Astrophys. J.* **162**, 993.

Ulrich, R. K. and Rhodes, Jr., E. J.: 1977, *Astrophys. J.* **218**, 521.

Ulrich, R. K., Rhodes, Jr., E. J., and Deubner, F.-L.: 1979, *Astrophys. J.* **227**, 638.

White, O. R. and Cha, M. Y.: 1973, *Solar Phys.* **31**, 23.

Wolff, C. L.: 1972, *Astrophys. J.* **176**, 833.

RECENT OBSERVATIONS OF HIGH-DEGREE SOLAR *p*-MODE OSCILLATIONS AT THE KITT PEAK NATIONAL OBSERVATORY*

(Invited Review, Abstract)

EDWARD J. RHODES, Jr.**

Department of Astronomy and Earth and Space Sciences Institute, University of Southern California, Los Angeles, CA 90007, U.S.A.

and

Space Physics Section, Jet Propulsion Laboratory, California Institute of Technology, Pasadena, CA 91109, U.S.A.

JOHN W. HARVEY

Kitt Peak National Observatory, Tucson, Ariz. 85717, U.S.A.

and

THOMAS L. DUVALL, Jr.

NASA Southwestern Station, Kitt Peak National Observatory, Tucson, Ariz. 85717, U.S.A.

Abstract. A brief summary is given of a program which is currently being carried out with the McMath telescope of the Kitt Peak National Observatory in order to study high-degree ($l \gtrsim 150$) solar *p*-mode oscillations. This program uses a 244×248 pixel CID camera and the main spectrograph of the McMath telescope to obtain velocity-time maps of the oscillations which can be converted into two-dimensional ($k_h - \omega$) power spectra of the oscillations. Several different regions of the solar spectrum have been used in order to study the oscillations at different elevations in the solar atmosphere. The program concentrates on eastward- and westward-propagating sectoral harmonic waves so that measurements can be made of the absolute rotational velocities of the solar photospheric and shallow sub-photospheric layers. Some preliminary results from this program are now available. First, we have been unable to confirm the existence of a radial gradient in the equatorial rotational velocity as was previously suggested. Second, we have indeed been able to confirm the presence of *p*-mode waves in the solar chromosphere as was first suggested by Rhodes *et al.* (1977). Third, we have been able to demonstrate differences in photospheric and chromospheric power spectra.

Reference

Rhodes, Jr., E. J., Ulrich, R. K., and Simon, G. W.: 1977, in *Proceedings of the November 7–10, 1977 OSO-8 Workshop*, Univ. of Colorado, Boulder, Colo., p. 365.

* Proceedings of the 66th IAU Colloquium: *Problems in Solar and Stellar Oscillations*, held at the Crimean Astrophysical Observatory, U.S.S.R., 1–5 September, 1981.

** Also,visiting astronomer KPNO.

Solar Physics **82** (1983) 111. 0038–0938/83/0821–0111$00.15.

STATISTICAL CONSIDERATIONS IN THE ANALYSIS OF SOLAR OSCILLATION DATA BY THE SUPERPOSED EPOCH METHOD*

S. E. FORBUSH, M. A. POMERANTZ, S. P. DUGGAL†, and C. H. TSAO

Bartol Research Foundation of The Franklin Institute, University of Delaware, Newark, Delaware 19711, U.S.A.

Abstract. Although the method of superposed epochs (Chree analysis) has been utilized for seven decades, a procedure to determine the statistical significance of the results has not been available heretofore. Consequently, various subjective methods have been utilized in the interpretation of Chree analysis results in several fields. The major problem in the statistical treatment of Chree analysis results arises from the fact that in most studies of natural phenomena, data are neither random nor sequentially independent. In this paper, a statistical procedure which takes this factor into account is developed.

1. Introduction

For investigating the possible relationship between two sets of geophysical observations, Chree (1912, 1913) introduced a procedure for analyzing one set of measurements during epochs which were selected on the basis of a specific type of feature in the second set of measurements. The method of superposed epochs can also be used for investigating basic periodicities in time series of data (see e.g., Chapman and Bartels, 1940). This version of the superposed epoch technique is germane in the present context, since it has been utilized by several groups for investigating certain aspects of global solar oscillations (Severny *et al.*, 1976; Scherrer *et al.*, 1979; Grec *et al.*, 1980).

However, unfortunately, despite its long history, a proper statistical test for evaluating the significance level of the results obtained by superposed epoch analysis has not been available heretofore. This lack of a quantitative 'figure of merit' of the results of applications of the Chree procedure has led to controversial situations arising from different interpretations of the reality of an apparent signal. The fact that proper statistical methods have generally not been available for assessing Chree analysis results arises from a fundamental problem: Data representing natural phenomena are neither random nor sequentially independent. Consequently, the basic criterion for the application of standard statistical procedures is, in fact, violated.

The pitfalls of ignoring the non-randomness of data representing observations of natural phenomena were first demonstrated by Bartels (1935; see also Chapman and Bartels, 1940). He introduced the concept of quasi-persistency and developed a procedure for calculating the standard error by evaluating the extent of its effect. In this

* Proceedings of the 66th IAU Colloquium: *Problems in Solar and Stellar Oscillations*, held at the Crimean Astrophysical Observatory, U.S.S.R., 1–5 September, 1981.

† Shakti P. Duggal died, July 11, 1982.

Solar Physics **82** (1983) 113–122. 0038–0938/83/0821–0113$01.50.

paper we will describe a statistical procedure based on analysis of variance that takes into account the quasi-persistency and is suitable for testing the significance of Chree analysis results. (For a complete review, see Forbush *et al.*, 1982.)

It should be emphasized that the purpose here is not to discuss previous analyses of solar oscillation data by superposed epoch analysis. Rather, we wish to issue a caveat to the solar-physics community that conclusions based upon superposed epoch analysis must be viewed with extreme skepticism unless it is unambiguously demonstrated that both the nature of the statistical tests that are applied, and the assignment of error bars or other indices of the probable reality of a signal are strictly legal.

2. Chree Analysis

Let us assume that on the basis of some observational criterion (e.g., an unusually high value of a particular geomagnetic index) N key-days are associated with some variation in the data under investigation (e.g., the cosmic ray intensity). In the method of superposed epoch analysis, each key-day is designated as the center of an epoch (day zero), the length of which is selected on the basis of a physically plausible period (e.g., the 27-day solar rotation period). We then list the data in the form of a matrix in which the rows r_j represent the epochs, and the columns c_i represent days before and after the individual key-days c_{13} as in Table I.

The column averages of this matrix, which will invariably show some variations, represent the Chree analysis result. We will refer to this result as the signal. The objective of the procedures described in this paper is to determine its significance level (i.e. the probability that it did not occur by chance).

Classical statistical procedures may be (and ordinarily are) followed for evaluating the apparent significance level of the variance attributable to the signal. However, as will become clear later, this is grossly erroneous because the data for any epoch are not sequentially independent. In general, there are real effects, in addition to the one under study, which can cause the measured phenomenon to vary over different time scales. This is the nub of the problem.

3. Statistical Test

A. Analysis of Variance

Table I represents the data matrix in a typical Chree analysis. In this example, we assume that there are 150 epochs ($r = \sum r_j = 150$) each comprising 27 days ($c = \sum c_i = 27$). The statistical test of the resulting signal can be performed as follows:

(1) Remove the linear trend, if any, from each row r_j.

(2) Calculate the variance of the population S_c^2 from the column means $\bar{c}_i$:

$$S_c^2 = r \sum_{i=1}^{c} \frac{(\Delta \bar{c}_i)^2}{c - 1}, \tag{1}$$

where

r = total number of rows,

c = total number of columns.

(3) Calculate the variance S_r^2 of the population from the row means $\bar{r}_j$:

$$S_r^2 = c \sum_{j=1}^{r} \frac{(\Delta \bar{r}_j)^2}{r-1} \, . \tag{2}$$

(4) Calculate the total variance S_T^2 from individual data points (x_{ij}):

$$S_T^2 = \sum_{i=1}^{c} \sum_{j=1}^{r} \frac{(\Delta x_{ij})^2}{rc-1} \, . \tag{3}$$

(5) Calculate the residual variance of the population S_R^2:

$$S_R^2 = \frac{[(rc-1)S_T^2] - [(c-1)S_c^2] - [(r-1)S_r^2]}{(c-1)(r-1)} \, . \tag{4}$$

(6) Test whether the variances of single rows and columns are homogeneous. One possible test is described in Appendix I.

(7) At this stage, let us first assume for simplicity that there is no quasi-persistency in the data. Under this assumption (which in most cases is not valid) the signal variance which is represented by S_c^2 can be compared with the residual variance S_R^2 of the data by using the F test with $(c-1)$, $(c-1)(r-1)$ degrees of freedom, df. If this test reveals

TABLE I

Data matrix that is generally used in Chree analysis. x_{ij} represents a single data point for column i and row j. $\bar{c}_i$ and $\bar{r}_j$ represent the averages for columns and rows respectively. Day 0 is termed key day.

CHREE MATRIX – $\alpha_{ij}(1)$

DAY →					
EPOCH ↓	−13	−12	------------ 0 ------------	+12	+13
	1	2	------------ 13 ------------	26	27
1					$\bar{r}_1$
2					$\bar{r}_2$
, , , , , ,			x_{ij}		$\bar{r}_j$
150					
	$\bar{c}_1$	$\bar{c}_2$	$\bar{c}_i$	$\bar{c}_{26}$	$\bar{c}_{27}$

that the signal is not significant, there is no need to proceed further, since the determination of quasi-persistency leads only to an increase in the residual variance.

B. Quasi-persistency

In order to provide a physical picture, let us assume that the signal in each row can be represented by a sine wave. In this case, following Bartels (1935), we define quasi-persistence as a periodicity which repeats for a certain number of epochs with approximately the same phase and amplitude; each such sequence ending more or less abruptly without any relation to other sequences. An example of quasi-persistent vectors derived from simulated data (see Appendix 2) is shown in Figure 1. To determine the

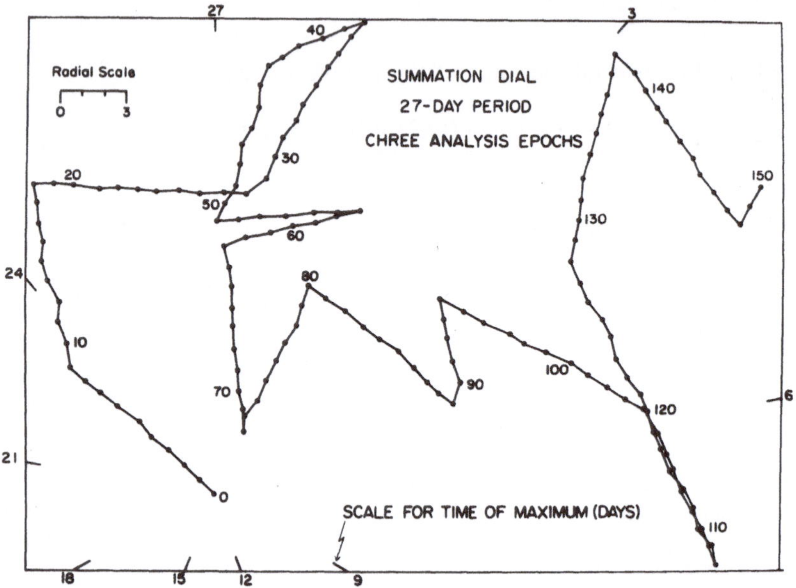

Fig. 1. Summation dial for the 27-day period. Vectors represent the 150 epochs of the simulated data. Every 10th epoch is marked at the end of the corresponding vector.

realistic residual variance, S_R^2 for quasi-persistent data, we transform the original Chree analysis matrix α_{ij} (1) into a new matrix in which the first row represents the mean of the first two rows of the original matrix, i.e.:

$$X'_{i1} \text{ (new matrix)} = \frac{(X_{i1} + X_{i2})}{2},$$

where the columns i extend from 1 to 27.

Similarly, the 2nd row of the new matrix α_{ij} (2), represents the average of the 3rd and 4th rows of α_{ij} (1). Following the same procedure, we construct h matrices:

α_{ij} (1) – Original data matrix with r rows.

α_{ij} (2) – Each row represents average of 2 consecutive rows of matrix α_{ij} (1). Total number of rows = $r/2$.

$\alpha_{ij}(3)$ – Each row represents average of 3 consecutive rows. Total number of rows = $r/3$.

$\alpha_{ij}(h)$ – Each row represents average of h rows. Total number of rows = r/h.

For each matrix, we repeat the first five steps described in the last section to obtain the residual variances $S_R^2(1)$, $S_R^2(2)$, $S_R^2(3) \ldots S_R^2(h)$ corresponding to the aforementioned h matrices. It is evident from Appendix III that if the data contain quasi-persistency, and if it is assumed that there is no persistent signal in the epochs, the ratio

$$\frac{S_r(h)h^{1/2}}{S_r(1)} = \frac{\zeta(h)}{\zeta(1)} \tag{5}$$

will exhibit some relationship with $h^{1/2}$. Note that in the analysis of variance, the signal is eliminated from the residual variance (4). However, this relationship will break down at some limiting value: $\zeta(h)/\zeta(1) = \zeta(\infty)/\zeta(1)$. In fact the quasi-persistency is negligible beyond $\sigma = [\zeta(\infty)/\zeta(1)]^2$ rows.

In the above discussion, it has been assumed that the chronological order of the epochs has been maintained in all the matrices $\alpha_{ij}(1)$, $\alpha_{ij}(2) \ldots \alpha_{ij}(h)$.

C. Data analysis

To clarify the procedure described thus far, and to demonstrate the pitfalls of using ordinary statistical tests for evaluating the significance level of Chree analysis results, let us consider the vectors shown in Figure 1. The data corresponding to these vectors (Appendix II) consist of a matrix with $c = 27$ and $r = 150$. The Chree analysis results derived from these data, i.e., the column means of this matrix, are plotted as a function of the day (column number c_j) in Figure 2. At first sight, Figure 2 reveals an impressive trend. In general, there appears to be a significant difference between the column means before and after the key-day (0-day). In fact, the plot in Figure 2 has the distinct appearance of a sine wave. Now let us examine, by using an ordinary statistical test, whether the variation evident in Figure 2 is actually significant. In other words, let us ignore the quasi-persistency in the data and perform the calculations enumerated in the seven tests outlined in Section A above. The final results are shown in Table II (random data). Application of the F-test reveals that the probability that the signal in Figure 2 has appeared by chance is very low (3 in 10^{15}). On the basis of this result, the 'signal' would be accepted as real.

An examination of Figure 1 suggests that this result cannot be valid because the vectors show abrupt change in direction after each sequence. Let us now apply the new statistical test described in the preceding section. To evaluate the quasi-persistency, we plot $\zeta(h)/\zeta(1)$ vs $h^{1/2}$ in Figure 3. It is clear from this figure that $\zeta(h)/\zeta(1)$ shows a definitive relationship with $h^{1/2}$ up to a point ($h \approx 4$) where the relationship breaks down. The break occurs at $\zeta(h)/\zeta(1) = \zeta(\infty)/\zeta(1) \approx 2.4 = \sqrt{\sigma}$. Thus the equivalent length of sequences is $[\zeta(\infty)/\zeta(1)]^2 = 5.76$. In other words, the quasi-persistency lasts for about six rows, which is consistent with the appearance of the summation dial in Figure 1. On the basis of this derived equivalent length of sequences, the results listed in Table II are

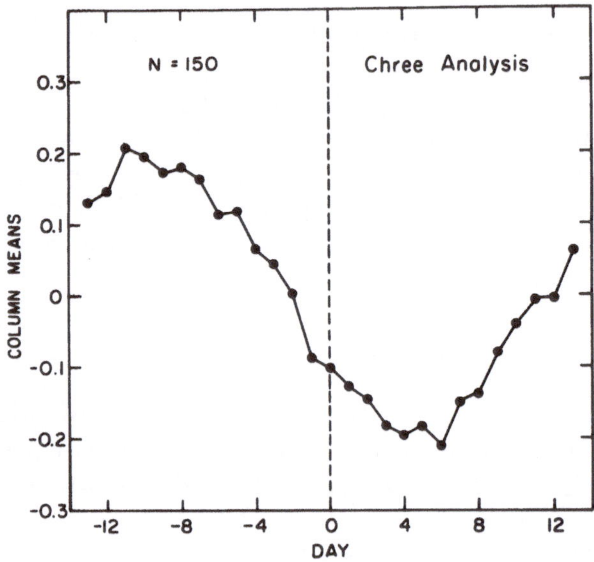

Fig. 2. Column means ($\bar{c}_i$) representing Chree analysis results are plotted as a function of days (Table I). Total number of epochs for this analysis is 150.

TABLE II

Comparison of the signal variance in the Chree analysis result with the residual variance in order to evaluate the probability that the signal has appeared by chance.

Analysis of variance	Random data: quasi-persistency ignored	Non-random data: quasi-persistency included
Signal variance, df* = 26	2.94	2.94
Residual Variance, df = 3874	0.59	3.40 (0.59 × 5.76)
F(26, 3874)	4.96	1.16**
Probability that the signal has appeared by chance $(1 - P)$	3×10^{-15}	3×10^{-1}

* df = degrees of freedom.

** Note that for this case, residual variance is larger than the signal variance, hence F(3874, 26) = (residual variance/signal variance) > 1, in accordance with standard practice (Hald, 1952).

obtained. The probability that the 'signal' has appeared by chance has increased to 0.33; hence, the apparent effect displayed in Figure 2 is *not* significant.

4. Discussion and Conclusion

For obtaining information concerning periodicities, or for understanding the relationship between two phenomena, superposed epoch analysis is unquestionably a useful procedure. However, this method of analysis yields meaningful results only if the

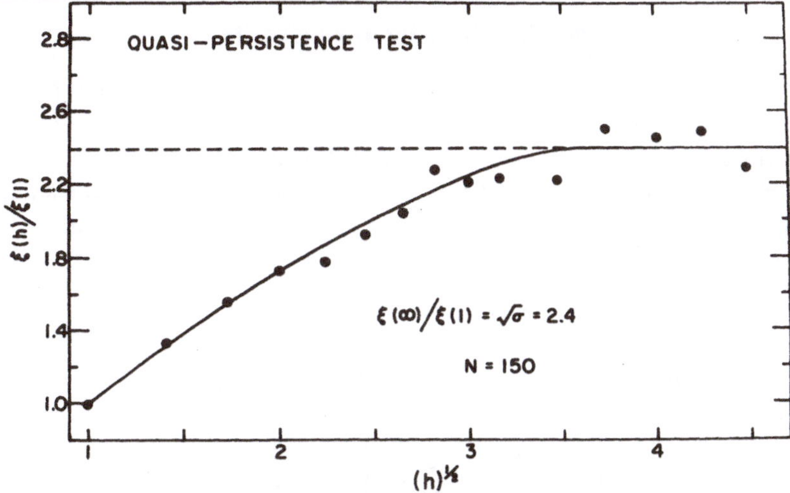

Fig. 3. Plot of the ratio $\zeta(h)/\zeta(1)$ vs $h^{1/2}$ derived from 150 epochs of Table I. The increase in the ratio for low values of $h^{1/2}$ indicates quasi-persistency in the data.

inherent quasi-persistency of natural phenomena is properly taken into account in the evaluation of the result. In fact, as was originally emphasized by Bartels (1935), the proper evaluation of quasi-persistency is of utmost importance in all types of analysis of problems in geophysics (and astrophysics). We have demonstrated here that the standard error can be grossly underestimated by ignoring the almost inevitable quasi-persistency in the data. In a statistical analysis of simulated data, we showed that ordinary (textbook) statistical tests led to the incorrect conclusion that the signal in Figure 2 is highly significant, whereas in reality it is not.

A new method, based on two-way classification analysis of variance, has been developed to determine the quasi-persistency (equivalent length of sequences) in the data. This analysis revealed that the effective (independent) number of epochs (rows) in the Chree matrix is much less than that determined under the invalid assumption that the data are strictly random. Thus, the standard error is modified, and the new effective standard error is found to be *larger* than the signal. It should be pointed out that an alternative method based on vectorial representation can be used to test the Chree analysis result. A study is in progress to determine the relative merits of the two procedures in various cases. It is hoped that application of the procedure developed here for including the effects of quasi-persistency in evaluating the statistical uncertainty of superposed epoch results will lead to more objective conclusions in future studies utilizing this powerful analytical tool.

Acknowledgements

This research is supported by the National Science Foundation's Division of Polar Programs under grants DPP–7923218–01 and DPP–7822467 and Atmospheric Research Section under grant ATM–8005866.

Appendix I: Test for Homogeneity

The hypothesis to be tested is that the variances of k normally distributed populations are equal. If there is no quasi-persistency in the data, Bartlett's test (see e.g., Hald, 1952; Dixon and Massey, 1957) can be utilized to test this hypothesis.

Let the variance of the ith sample of size n_i be given by S_i^2. Note that the sample size will take care of the fact that data in some rows or columns are missing.

Let

$$\eta = (N-k)\ln S_p^2 - \sum (n_i - 1)\ln S_i^2 ,$$

$$S_p^2 = \sum (n_i - 1) S_i^2/(N-k) ,$$

$$A = \frac{1}{3(k-1)}\left[\sum \frac{1}{n_i - 1} - \frac{1}{N-K}\right] ,$$

$$\nu_1 = k - 1 ,$$

$$\nu_2 = \frac{k+1}{A^2} ,$$

$$b = \frac{\nu_2}{1 - A + (2/\nu_2)} ,$$

$$N = \sum n_i .$$

Then the sampling distribution of $F = \nu_2 \eta / \nu_1 (b - \eta)$ is approximately $F(\nu_1, \nu_2)$. It should be emphasized that this test is not valid for non-independent data. In case there is quasi-persistency, the equivalent length of sequences (σ) can be evaluated. Bartlett's test can then be applied to sets of k/σ independent samples.

Appendix II: Data Simulation

The simulated data $D(t)$ for each epoch, for the tests described in this paper, are generated from:

$$D(t) = R_q \sin\left[\omega t + \phi_q(t)\right] + \zeta(t) + \beta t ,$$

where R_q, $\phi_q(t)$ represent amplitude and phase of a quasi-persistent signal, $\omega = 2\pi/27$ and $\zeta(t)$ and βt represent random and linear effects in each epoch.

Harmonic analysis of these simulated data, after linear term corrections, yields the 27-day period vectors in the summation dial in Figure 1. Note that each vector represents a single epoch row in Table I.

Appendix III: Evaluation of Quasi-Persistency

Let

$M_i(h)$ = ith among N/h means of h consecutive means,
N = total number of $M_i(1)$,
$N(h)$ = total number of $M_i(h)$,
$r_i(1)$ = contribution of random effects to $M_i(1)$,
$r_i(h)$ = contribution of random effects to $M_i(h)$,
$q_i(1)$ = the quasi-persistent contribution to $M_i(1)$,
$q_i(h)$ = the quasi-persistent contribution to $M_i(h)$,
m = the contribution of the persistent wave to $M_i(1)$, constant for all i from 1 to N,
$c^2(1)$ = variance of $M_i(1)$,
$c^2(h)$ = variance of the means of h successive sequential means,

$$M_i(1) = [m + q_i(1) + r_i(1)]\,,$$

$$c^2(1) = \frac{1}{N}\sum_1^N [m + q_i(1) + r_i(1)]^2$$

$$= \frac{1}{N}\sum \{[(m + q_1(1)]^2 + r_i^2(1)\}\,.$$

Since for large N, $\sum m r_i(1) = 0$ and $\sum r_i(1) q_i(1) = 0$,

$$\begin{aligned} c^2(1) &= \frac{1}{N}\sum [m^2 + 2mq_i(1) + q_i^2(1) + r_i^2(1)] \\ &= m^2 + 2m\bar{q}(1) + S_q^2(1) + S_r^2(1)\,. \end{aligned} \tag{1A}$$

Since

$$\frac{2m \sum q_i(1)}{N} = 2m\bar{q}(1)\,,$$

$$\bar{q}(1) = \text{mean of all quasi-persistent steps}$$

and

$$\sum r_i^2(1)/N = S_r^2(1)\,,$$

$$\sum q_i^2(1)/N = S_q^2(1).$$

Similarly

$$\begin{aligned} c^2(h) &= \frac{1}{N/h}\sum_1^{N/h} [m^2 + 2mq_i(h) + q_i^2(h) + r_i^2(h)] \\ &= m^2 + 2m\bar{q} + S_q^2(h) + S_r^2(h)\,. \end{aligned} \tag{2A}$$

Since

$$2m \sum_{1}^{N/h} q_i(h) = 2m\bar{q}\,.$$

Then

$$\begin{aligned} c^2(h)h &= m^2h + 2mh\bar{q} + S_q^2(h)h + S_r^2(h)h\,, \\ c^2(h)h &= h(m^2 + 2m\bar{q}) + S_q^2(h)h + S_r^2(1)\,. \end{aligned} \tag{3A}$$

As a special case, assume that the data contain no persistent wave, i.e. $m = 0$, then

$$c^2(h)h = hS_q^2(h) + S_r^2(1)\,. \tag{4A}$$

For large values of h, the right-hand side becomes constant, i.e.

$$c^2(h)h = \text{const.} = c^2(1)\sigma\,, \tag{5A}$$

where σ is defined as 'equivalent length of sequences' (Bartels, 1935).

Equation (5A) can be written as

$$c(h)h^{1/2}/c(1) = \zeta(h)/\zeta(1) \approx \zeta(\infty)/\zeta(1) = \sigma^{1/2}\,.$$

References

Bartels, J.: 1935, *Terrs. Magnetism Atmospheric Electricity* **40**, 1.

Chapman, S. and Bartels, J.: 1940, *Geomagnetism*, Vol. II, Oxford University Press.

Chree, C.: 1912, *Phil. Trans. London* **A212**, 75.

Chree, C.: 1913, *Phil. Trans. London* **A213**, 245.

Dixon, W. J. and Massey, F. J.: 1957, *Introduction to Statistical Analysis*, McGraw-Hill Book Co.

Forbush, S. E., Duggal, S. P., Pomerantz, M. A., and Tsao, C. H.: 1982, *Rev. Geophys. Space Phys.*, in press.

Grec, G., Fossat, E., and Pomerantz, M.: 1980, *Nature* **288**, 541.

Hald, A.: 1952, *Statistical Theory with Engineering Applications*, John Wiley and Sons.

Scherrer, P. M., Wilcox, J. J., Kotov, V. A., Severny, A. B., and Tsap, T. T.: 1979, *Nature* **277**, 635.

Severny, A. B., Kotov, V. A., and Tsap, T. T.: 1976, *Nature* **259**, 8.

ON THE ACCURACY OF FREQUENCY DETERMINATION BY AN AUTOREGRESSIVE SPECTRAL ESTIMATOR*

G. KOVÁCS

Konkoly Observatory, Hungary

Abstract. The accuracy of frequency determination by a least squares technique for an autoregressive spectral estimator is studied and compared with the Fourier method. Using numerical tests the probability distribution function of the peak location is calculated. The autoregressive filter order is optimized in the sense of minimum variance of the peak location. Simple sinusoidal signals with additive Gaussian noise are considered and the effect of other components is only indicated. Generally, a filter order between 1/3 and 1/2 of the total data number and a not very dense data sampling, gives the most stable spectrum. The results are numerical.

1. Introduction

Since the introduction of the maximum entropy method (MEM) of spectral analysis, it has become widely used and studied in various fields dealing with problems of data processing. Though Burg's recursive scheme (Burg, 1975) has proved to be far superior to conventional methods, especially for short data records, there are two limitations in practice:

(1) Splitting and shifting of spectral peaks (Chen and Stegen, 1974; Fougere, 1977).

(2) Difficulty in obtaining general analytical expressions for its statistical appearance since MEM is a nonlinear spectral estimator.

The anomalies listed first do not occur in the nonlinear method as proposed by Fougere (1977) nor in the least squares (LS) spectral estimate suggested by Ulrych and Clayton (1976). The advantage of the LS estimate over that of Burg has been demonstrated by Ulrych and Clayton (1976) and Swingler (1979). Therefore, we limit our considerations here to the LS method.

In the case of large filter order and data number the statistical properties of the MEM spectral estimator have been determined and compared with those of the Fourier method by Kromer (1970) and Berk (1974). Confidence limits for a MEM spectral estimator using a one-way (predicting in one direction only) autoregressive (AR) model have been described by Reid (1979).

No work is available relating to the statistical properties of the general two-way LS spectral estimator. Because the frequency structure of the light variation of a variable star is the most important quantity for comparison with theory, our considerations are focused only on the accuracy of the frequency determination when the LS spectral estimator as proposed by Ulrych and Clayton (1976) is used.

* Proceedings of the 66th IAU Colloquium: *Problems in Solar and Stellar Oscillations*, held at the Crimean Astrophysical Observatory, U.S.S.R., 1–5 September, 1981.

Solar Physics **82** (1983) 123–128. 0038–0938/83/0821–0123$00.90.

2. Tests by Artificial Data

First the role of sampling time in the stabilization of the spectrum is studied. Using a test signal of the form

$$x_j = \sum_{k=1}^{3} A_k \sin(P\Delta t f_k(j-1)) + B_k \cos(P\Delta t f_k(j-1)),$$

$$A_1 = 1, \quad B_1 = 0, \quad A_2 = 0.8, \quad B_2 = -0.5, \quad A_3 = 0.2, \quad B_3 = 0.1,$$
$$P = 6.283\,185,$$

$$f_1 = 0.07 \text{ Hz}, \quad f_2 = 0.09 \text{ Hz}, \quad f_3 = 0.12 \text{ Hz}, \quad j = 1, 2, \ldots, 45,$$
$$\Delta t = 0.5 \text{ s},$$

LS spectra were calculated with different sampling rates. Once the filter order (M) and an integer number (L) of Δt is chosen, L spectra were calculated using data corresponding to the sequences: $j = 1, L+1, 2L+1, \ldots$; $j = 2, L+2, 2L+2, \ldots$; $j = L, 2L, 3L, \ldots$. Taking the average of these spectra the plot shown in Figure 1 was obtained. Numbers on the curves denote the various M, L combinations. The spectra were normalized to unity at the highest peak in each case. It is clearly seen that a denser data sampling makes the spectra unstable and it is possible to choose a proper M, L combination to deduce the exact frequencies.

A fairly crucial point in estimating MEM spectra is the determination of the order of the AR process. Several objective criteria may be used for the proper choice of the optimal value of filter order. For harmonic processes with noise, these criteria underestimate the AR order and an empirical rule can be adopted. Percy (1977) has found that the best values of M are such that $M\Delta t$ is of the same order of magnitude as the period present. Ulrych and Ooe (1979) have adopted the rule which constrains M to $N/3 \leq M \leq N/2$ (N is the number of data points).

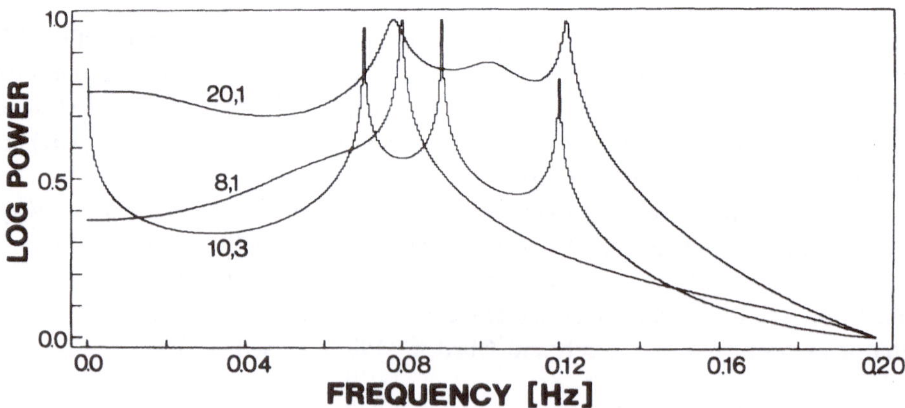

Fig. 1. LS AR spectra of a test signal with three sinusoidal components of different frequencies. The various filter order and averaging numbers are indicated near each curve (for details, see text).

As regards the stability of the frequencies obtained by MEM, a natural criterion for the optimal filter order is that choice of M at which the variance of the peak location is minimal. Because it is almost impossible to follow this problem analytically, numerical tests are required.

The logarithm of the normalized standard deviation of peak location as a function of M/N is plotted in Figure 2. Each curve has been shifted vertically somewhat, but the scale is the same, as indicated and the zero points correspond to the first points of the curves. The same hundred realizations of signals in the form *deterministic signal* + *Gaussian component* were used to calculate the variance of the peak location. The results did not change significantly on using more realizations. For the curve at the bottom, the deterministic component consisted of two sinusoids of the form

$$\sin(P(j-1)\Delta t 0.07) + 0.5\sin(P(j-1)\Delta t 0.09)\,, \qquad P = 6.283\,185\,,$$

and for the other curves one sinusoid only, i.e.

$$\sin(P(j-1)\Delta t 0.07)\,, \qquad P = 6.283\,185\,.$$

The location of the peak at ~ 0.07 Hz was used to calculate the variance in the two sinusoids case. Numbers near the curves indicate the data number, the sampling time (Δt) and the standard deviation of the noise. Vertical arrows show the minima of each curve. It is seen that in the case of a signal with a single sinusoidal component the best result is expected when $\frac{1}{3} \leq M/N \leq \frac{1}{2}$. The shift of the optimal M above half of the data

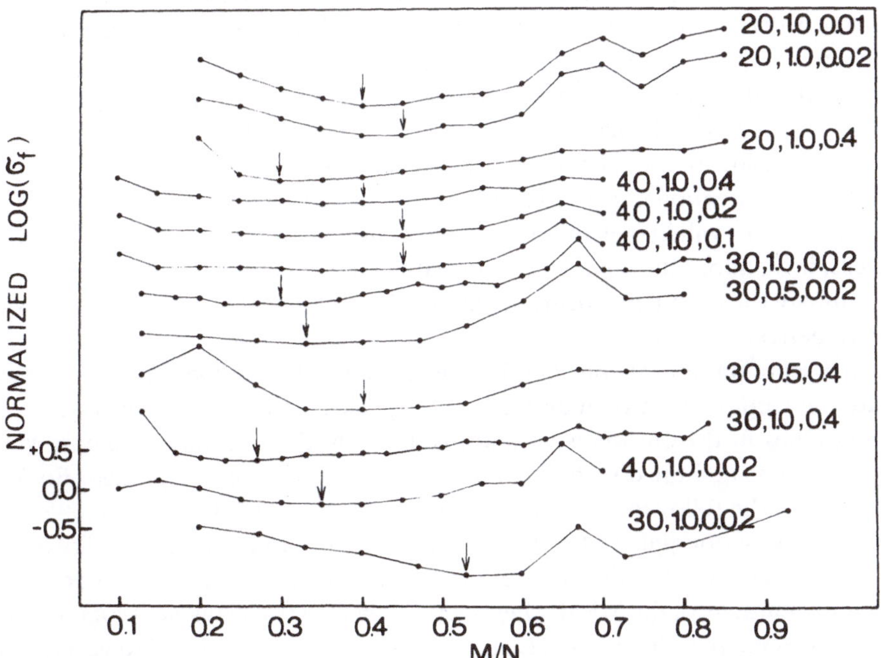

Fig. 2. Dependence of the standard deviation of peak location on the relative filter length (M/N) for different signals (for details, see text).

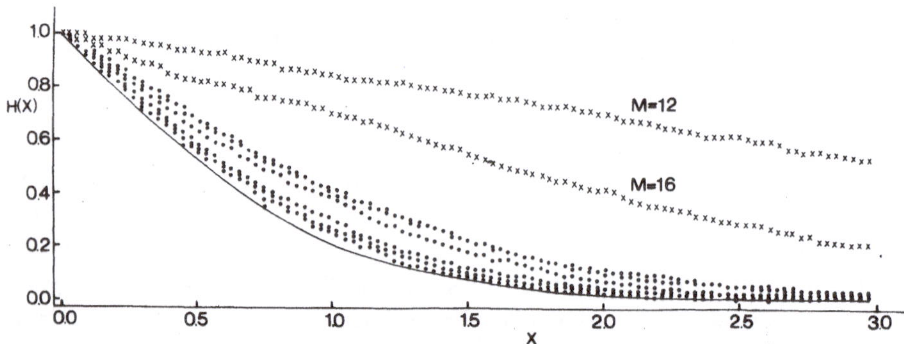

Fig. 3. Empirical probability distribution functions of the frequency shift when using the Fourier method (continuous line) and the LS AR spectral estimator (other symbols). $H(x)$ means the probability that the value of $\sqrt{N}\, N\Delta t\, |\Delta f|\, A/\sigma$ exceeds the value of x (for details see text).

number in the two sinusoids case is obvious. We conclude that Ulrych and Ooe's (1979) empirical rule adopted for optimal M generally holds and there is no hope of improving it.

As a further step in studying the statistical properties of peak location, we have calculated the empirical distribution function of peak location and compared it with that of the Fourier method. Denoting the frequency shift by Δf, the signal amplitude by A, and the standard deviation of noise by σ, the probality $H(x)$ that the value of $\sqrt{N}\, N\Delta t\, |\Delta f|\, A/\sigma$ is greater than x is plotted in Figure 3 for the Fourier method (continuous line) for the optimal filter order in the one sinusoid case (dots) and for two different filter orders in the two sinusoids case (crosses). (In the Fourier method the distribution functions were not sensitive to the number of sinusoidal components, therefore only the curve related to the one sinusoid case is plotted.) The test signals were the same as used for the calculation of the curves in Figure 2. Because the probability distribution of peak location in the Fourier method depends only on the signal parameters in the combination mentioned above (Kovács, 1981), the same combination was used here too, though the validity of this statement is not proved in the case of MEM. Nevertheless, if we adopt the combination of parameters mentioned above, it makes it easier to compare the probability distribution functions of peak location of signals with various parameters.

For a given set of parameters one hundred realizations of a random series were used to calculate the empirical distribution function of peak location. The calculation was repeated ten times in the one sinusoid and five times in the two sinusoids case with different realizations and the distribution functions were averaged and plotted. For the one sinusoid case the following parameters were used: $N = 20, 30, 40$; $\sigma = 0.005, 0.1, 0.2, 0.4, 0.8$ and $M =$ optimal M, $A = 1$, $\Delta t = 1$ s. It can be seen that the stability of the MEM against noise is weaker than that of the Fourier method. The situation becomes even more unfavourable for the MEM when a test signal with two sinusoidal components is used (parameters: $N = 30$, $\sigma = 0.02$, $\Delta t = 1$ s, $A = 1$, $M = 12, 16$). However, it is important to remark that whereas MEM was able to give the correct average frequency tested (i.e. 0.07009 Hz for $M = 12$ and 0.06999 Hz for $M = 16$), the Fourier method

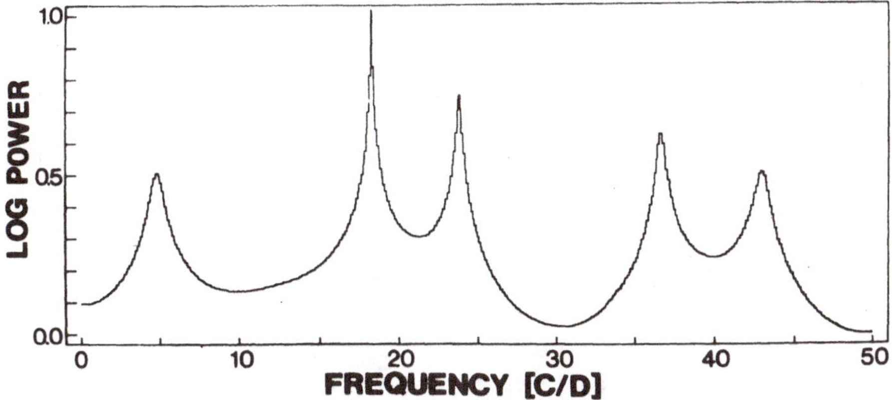

Fig. 4. LS AR spectrum of *SX Phe* (data from Stock *et al.*, 1972).

suffered a serious frequency shift (i.e. the average frequency was 0.07500 Hz). As a conclusion we may say that numerical tests only are unable to give confidence intervals in a general case for the frequency stability of the two-way LS spectrum.

3. Tests by Real Astronomical Data

Finally, we show the efficiency of the two-way LS frequency spectrum estimator by using real astronomical data.

The LS frequency spectrum of *SX Phe* is shown in Figure 4. The data published by Stock *et al.* (1972) were used with $M = 24$, $L = 4$ (the data were made equidistant by quadratic interpolation and the data number was left unchanged). The power was normalized to unity at the highest peak. The primary (0.05491 day) and the secondary (0.04202 day) periods show very good agreement with their previously known values (0.0549648 and 0.042773 day, respectively). For comparison, Percy (1977) obtained the values 0.0557 and 0.0423 day respectively using Burg's technique. The nonlinear interaction frequencies are also clearly seen and are very close to their predicted values.

As a second example a part of the photoelectric data of Θ *Tuc* published by Stobie and Shobbrook (1976) was analysed. The data were made equidistant in the same way as for *SX Phe*. The frequency spectra of the individual nights of observation are shown in Figure 5, where each spectrum has been normalized to unity at its highest peak. The letters *a*, *b*, *c*, *d* stand for JD 2441000 + 597, 611, 612, 634, respectively. $M = 14$, $L = 3$ in each case except in case *d*, where $M = 14$, $L = 5$. Except for *d* (where it is indicated only), the spectra show (similarly to Percy's (1977) result) a more or less stable double peak structure in the vicinity of 17 and 20.4 cycle/day (however, by using $M = 30$, $L = 2$ in case *d*, the peak at 19.8 cycle/day splits into peaks at 17.5 and 19.8 cycle/day, respectively). This result is consistent with the Fourier method based frequency spectra obtained by Kurtz (1980) and Pelt (1980) and the small instability in our spectra may partly be accounted for by the frequencies grouped around 20, 18, and 16 cycle/day as was claimed by those studies.

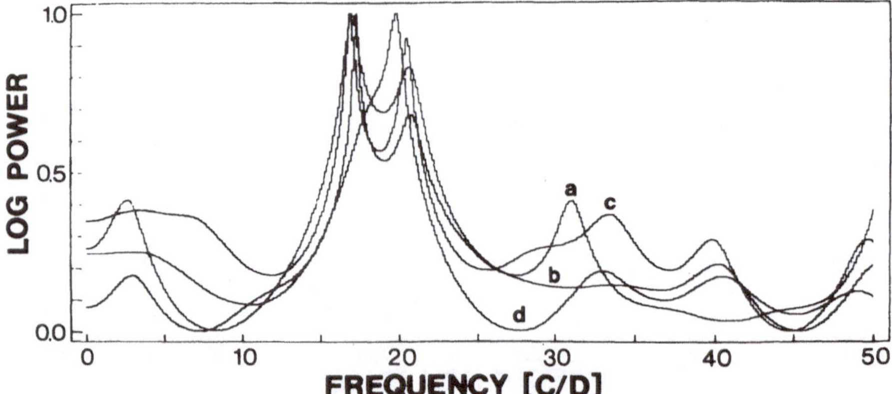

Fig. 5. LS AR spectra of the individual nights of observation of Θ *Tuc* (data from Stobie and Shobbrook, 1976). The letters a, b, c, d stand for JD 2441000 + 597, 611, 612, 634, respectively.

Acknowledgements

It is a pleasant duty to acknowledge that a considerable part of the computation for this note was performed with the PDP 11/45 computer donated to the N. Copernicus Astronomical Center by the U.S. Academy of Sciences following the initiative and through the action of Dr C.R. O'Dell and Mr A. M. Baer.

The author sincerely thanks the Director of the Center, Prof. W. Dziembowski, for hospitality during his stay in Warsaw; Director of the Konkoly Observatory, Dr B. Szeidl, for supporting the author's initiatives to work there; and Dr M. Kozłowski for aspects related to computer handling.

The financial support of the Hungarian and of the Polish Academy of Sciences is also acknowledged.

References

Berk, K. N.: 1974, *Ann. Statistics* **2**, 489.

Burg, J. P.: 1975, 'Maximum Entropy Spectral Analysis', Ph.D. Thesis, Stanford University, Palo Alto, Calif.

Chen, W. Y. and Stegen, G. R.: 1974, *J. Geophys. Res.* **79**, 3019.

Fougere, P. F.: 1977, *J. Geophys. Res.* **82**, 1051.

Kovács, G.: 1981, *Astrophys. Space Sci.* **78**, 175.

Kromer, R. E.: 1970, 'Asymptotic Properties of the Autoregressive Spectral Estimator'; Ph.D. Thesis, Dept. of Statistics Tech. Rpt. No. 13, Stanford University, Stanford, Calif.

Kurtz, D. W.: 1980, *Monthly Notices Roy. Astron. Soc.* **193**, 61.

Pelt, J.: 1980, 'Fequency Analysis of Astronomical Time Sequences', Publication of Tartu Astrophysical Observatory.

Percy, J. R.: 1977, *Monthly Notices Roy. Astron. Soc.* **181**, 647.

Reid, J. S.: 1979, *J. Geophys. Res.* **84**, 5289.

Stock, J., Kunkel, W. E., Hesser, J. E., and Lasker, B. M.: 1972, *Astron. Astrophys.* **21**, 249.

Stobie, R. S. and Shobbrook, R. R.: 1976, *Monthly Notices Roy. Astron. Soc.* **174**, 401.

Swingler, D. N.: 1979, *J. Geophys. Res.* **84**, 679.

Ulrych, T. J. and Clayton, R. W.: 1976, *Phys. Earth Planet. Inter.* **12**, 188.

Ulrych, T. J. and Ooe, M.: 1979, in S. Haykin (ed.), *Nonlinear Methods of Spectral Analysis*, Topics in Applied Physics, Springer-Verlag, Berlin, Heidelberg, New York, 1979, Vol. 34, p. 81.

ON THE ORIGIN OF OSCILLATIONS IN A SOLAR DIAMETER OBSERVED THROUGH THE EARTH'S ATMOSPHERE: A TERRESTRIAL ATMOSPHERIC OR A SOLAR PHENOMENON*

HENRY A. HILL, RANDALL J. BOS, and THOMAS P. CAUDELL

Department of Physics and Arizona Research Laboratories, University of Arizona, Tucson, AZ 85721, U.S.A.

Abstract. Interpretations of current and past results from ground-based solar diameter measurements, as well as the planning of scientific programs for the 1980's, are strongly dependent on the perceived level of the degrading effects of the Earth's atmosphere. One of the more effective approaches has been to design the observing program and the subsequent data analysis such that the solar diameter measurements themselves could provide an evaluation of atmospheric effects. Many important results have been obtained in studies of this type and these results are collected here to help in appraising the current situation. This evidence all points in one direction: the Earth's atmosphere, while complicating the design of observational programs, is not the source of the oscillations observed in solar diameter measurements. Further, this same evidence indicates that the Earth's atmosphere will not pose any serious limitations in ground-based solar diameter studies during the 1980's.

1. Introduction

The signals which have been interpreted as representing solar oscillations are indeed quite small in magnitude when measured against more typical astrophysical signals. For the highest frequency resolution that has been reported to date in the observations, the whole disk velocity amplitudes are of the order of 10 cm per sec (Claverie *et al.*, 1981), while the amplitudes of the diameter variations are of the order of a few milli arc sec (Bos and Hill, 1983). When the small size of these amplitudes is put into perspective, it is reasonable to expect that the presence of the Earth's atmosphere may pose problems. Various practices and techniques have been employed to reduce the effects of such problems in the measurements while numerous efforts have been undertaken to identify the residual effects in the reported results. Efforts to identify residual atmospheric effects usually may be classified as belonging to one of two classes. The first attempts to study the Earth's atmosphere through selected observations and, from this, to extrapolate to what might be expected in other observational programs. In many instances, this approach may be the only option available. However, it must be recognized that uncertainty exists in this extrapolation. Consequently, the search for these extrapolated properties in the observations may serve either to identify residual atmospheric effects in the observations, as projected, or possibly only serve to test the

* Proceedings of the 66th IAU Colloquium: *Problems in Solar and Stellar Oscillations*, held at the Crimean Astrophysical Observatory, U.S.S.R., 1–5 September, 1981.

Solar Physics **82** (1983) 129–138. 0038–0938/83/0821–0129$01.50.

atmospheric model used in the extrapolation process. The actual identification of effects in the observations which match those that have been projected in the extrapolation process satisfies a necessary condition for the accuracy of the extrapolation process. A demonstration, on the other hand, that the process is correct must entail a test of the sufficient condition class and it is at this level that the following discussion of existence tests will proceed.

The second of these two classes of studies to detect residual terrestrial atmospheric effects works to identify properties present in the observations which are not consistent with those of an atmospheric source. This type of property is often difficult if not impossible to identify, but, if actually found in the observations, implies that the observed phenomenon is very probably not due to the Earth's atmosphere. Should more than one such property be identified and detected, then a considerably stronger statement can be made regarding the origin of the effect. This second class of study furnishes, to a degree, tests of a *sufficient* condition class. Stated more emphatically, it is sufficient to observe such a property to demonstrate that the phenomenon does not originate in the terrestrial atmosphere.

In the field of solar oscillations, observations and analyses continue to be done which fall into the necessary condition class. Since the relevance of this work depends very strongly on the degree to which results of the more powerful sufficient condition class of studies are taken into account, a summary of the currently available material in the latter category is given here. This summary, in conjunction with the various reports in the literature which primarily treat work of the necessary condition class, should provide a resource for those interpreting results, analyzing data and designing future experiments; in addition, it will serve to point out the tight constraints imposed by the results of studies that fall into the sufficient condition class.

2. Tests of Sufficient Condition Class

Five properties of the reported solar oscillations exist which are not consistent with the properties expected for the Earth's atmosphere and which therefore furnish the bases for a series of tests in the sufficient condition class. In each of the five cases, the placement of the origin of the oscillations in the terrestrial atmosphere would require a radical modification in our understanding of the Earth's atmosphere. These five properties of the oscillations are: a high degree of phase stability over extended periods of time; reproducibility of observed frequencies of oscillation from one data set to another; the radial dependence of the observed oscillatory signal; a direct measurement of the effect of differential refraction in the Earth's atmosphere; and spatial symmetry properties and spatial coherence of the observed oscillatory signal. These properties are also listed in Table I and have been reported in a number of papers spread over the last ~ six years. These references and the salient points from these works will be brought together in the following paragraphs, culminating with the most recent results of Bos and Hill (1983).

3. Results Satisfying a Sufficient Condition

Four different sets of observations made on the SCLERA† telescope have yielded information on global oscillations of the Sun. These were obtained in 1973 (Hill and Stebbins, 1975), in 1975 (Brown *et al.*, 1978), in 1978 (Caudell *et al.*, 1980), and in 1979 (Bos and Hill, 1983). Each of these sets of observations of solar oscillations has, upon analysis, exhibited properties which fall into two or more of the five categories which are listed in Table I. Table II summarizes the correspondence between the original data sets and the five tests of the sufficient condition class which have been examined. Table II should also be quite useful in pointing the reader to the results of the rather extensive list of works which fall into this class.

Very important points in regard to Table II are that all work on solar diameter measurements has been included which belongs to the sufficient condition class and that tests for two or more of the five conditions have been made in each of the four sets of data. It should also be noted that a statistically significant positive result was obtained, i.e., the corresponding sufficient condition was satisfied, for all of the entries in Table II by the observations made at SCLERA.

TABLE I

List of sufficient condition tests

1. High degree of phase stability over extended periods of time.
2. Reproducibility of observed frequencies of oscillation from one data set to another.
3. The radial dependence of the observed oscillatory signal.
4. A direct measurement of the effect of differential refraction in the Earth's atmosphere.
5. Spatial symmetry properties and spatial coherence of the observed oscillatory signal.

TABLE II

Sufficient condition-type properties*

Observations	1	2	3	4	5
1973 (a)	e, f	e	e		
1975 (b)	b		b, c		
1978 (c)	c	g	c	g	
1979 (d)	d	d	d		d

* The sufficient condition tests identified by the heading numbers correspond to those in Table I.

a. Hill and Stebbins (1975).
b. Brown *et al.* (1978).
c. Caudell *et al.* (1980).
d. Hill and Bos (1983).
e. Hill and Caudell (1979).
f. Caudell and Hill (1980).
g. Knapp *et al.* (1980).

† SCLERA is an acronym for the Santa Catalina Laboratory for Experimental Relativity by Astrometry, a facility jointly operated by the University of Arizona and Wesleyan University.

A. Temporal phase stability

Theoretical considerations indicate that low-order global oscillations of the Sun should have coherence times that are much longer than a day (Wolff, 1972). Terrestrial atmospheric waves, on the other hand, have rarely been observed to have coherence times greater than 2 or 3 hr. Exceptions to this are found in atmospheric tidal waves, which are diurnal and may be coherent over longer periods of time, and in atmospheric waves associated with storms and/or topographically special cases. The latter waves can be generated only when the atmospheric parameters are within certain very limited boundaries (Gossard and Hooke, 1975). One may argue that the SCLERA site has some sort of topography which produces coherent atmospheric waves at all times. However, it is well-documented that the weather pattern changes daily over the Santa Catalina Mountains where the SCLERA telescope is located.

There is yet another important factor. Even if a type of atmospheric wave existed which had a long coherence time, it very probably could not produce a signal in the solar diameter which also had a long coherence time. The changing observing conditions through the day are responsible for this. During an 8- or 9-hr observing day, there is a large systematic variation in the zenith angle, a change of approximately 180° in the azimuth and a change of approximately 120° in the paralactic angle (see Smart, 1965). A stable atmospheric wave could give rise to a temporally coherent signal in the solar diameter measurements only if each of these three angles, particularly the latter two, were constant in time, which is clearly not the case.

It should be noted that it was just this property of a long coherence time that led to the discovery of the global oscillations with periods around an hour in the 1973 observations. Hill and Stebbins (1975), in analyzing measurements of solar oblateness, averaged results from 10 days of data which covered a 17-day period. The average oblateness results were then plotted as a function of the paralactic angle in order to facilitate the identification of systematic errors introduced by the mirrors of the elevation and azimuth tracking system. The paralactic angle, η, increases monotonically during the day but the rate of increase, $d\eta/dt$, varies by a factor of about 2, reaching a maximum at noon. Thus, by averaging in η and plotting as a function of η, a procedure was performed which was equivalent to putting the data through a tuned filter whose pass band frequency scanned a range from approximately 1/hr to approximately 1/30 min. In this way, a statistically significant oscillatory signal was detected in this frequency range which could be seen in the plot of oblateness vs η. Figure 10 in the paper by Richard (1975) shows a plot of oblateness vs η that has not been processed to eliminate the oscillations.

It should thus come as no surprise that, when Hill and Caudell (1979) reexamined the 1973 observations, they found six oscillations that exhibited a high degree of temporal phase coherence over the 17 days; this phase coherence was found to be statistically significant (Caudell and Hill, 1980). Since the oscillations in the solar oblateness were seen in the 10 days of data analyzed by Hill and Stebbins, and since that is a number typical of the work which followed at SCLERA, the statistical significance of the oscillating oblateness signal vs η should not be unlike that found for the later works.

A test for temporal phase stability was made in the same way for the 1973, 1975, and 1978 observations. For each day that observations were available, the Fourier transform was made of the data and the phase of a peak in the power density spectrum noted. These phases were then plotted as a function of time. If the coherence time of the oscillation was greater than the total length of the data set, then the plot of phases vs time would be in a straight line. In practice, the situation is not quite this simple since the actual number of oscillations between one phase determination and the next is not known. This additional complexity enters the picture in evaluating the statistical significance of the interpretation of the phase vs time data. However, this is a well defined problem and can be easily treated.

TABLE III

Temporal phase coherence*

Observations	Number of oscillations	Probability of noncoherent source**	Probability of noncoherent source***	Analysis
1973 (a)	6	0.37	2.6×10^{-3}	f
1975 (b)	2	0.30	0.09	b
1978 (c)	12	0.8×10^{-2}	$(0.8 \times 10^{-2})^{12}$	c
1979 (d)	$\approx$ 200	(average peak power extremely small)		d

* The references a–f are given in Table II.

** For individual oscillations.

*** For set of oscillations.

Table III gives average probabilities, for the first three sets of observations, that the oscillations result from noise. Since the oscillations are not coupled in any obvious way, i.e., they are not diurnal nor are they harmonics of some other frequency, they are very likely independent. The probability, then, that the complete set of oscillations for a particular data set results from noise is simply the average probability found for the individual oscillations raised to a power equal to the number of peaks. The results are given in column 4.

Sufficient additional constraints were imposed and information obtained in the 1979 observations to clearly eliminate atmospheric effects (see Bos and Hill, 1983). A phase coherence test may nevertheless be applied, and the level of noncoherent signal ascertained. The complete data set was analyzed with a frequency resolution of 1/41 days or 0.28 μHz, a resolution sufficient to distinguish individual eigenstates. However, in achieving this resolution, the previous tests for phase coherence are no longer viable as only one phase/oscillation is available. This problem was remedied by Bos and Hill (1983), who found that 1.1% of the power density was due to signals with coherence times $\lesssim 2$ days.

This test shows that the coherence time for the majority of the oscillations in the 1979 observations is $\gtrsim 2$ days. By additional analysis, a coherence time > 41 days was established by Bos and Hill (1983) for the frequency range 450–480 μHz.

It would appear that temporal phase stability over periods of time numbering tens of days is a well-documented property of the oscillations in solar diameter measurements. This singular characteristic of the observations is a very important 'road sign' in the interpretation of the oscillations.

B. Reproducibility of observed frequencies of oscillation from one data set to another

Global oscillations of the Sun should have well-defined frequencies since they are expected to have long coherence times (Wolff, 1972), while atmospheric phenomena are not expected to exhibit such a property. Thus, an obvious test for the origin of these oscillations is to compare one set of observations with another, examining the two spectra for peaks which are common to both. This can be done by comparing the results from one year to those of another or by breaking up one set into subsets and making a comparison. These two types of comparisons are referred to as external and internal comparisons respectively.

Both external and internal comparisons have been made involving each of the four sets of observations, the results of which are listed in Table IV. In three of the comparisons, a probability has been estimated that the number of peaks common to the

TABLE IV

Repeatability of observations*

Observations	Internal comparison		External comparison			References
	Number of peaks coincident	Probability of noncoherent source	Two data sets	Number of peaks coincident	Probability of noncoherent source	
1973 (a)			1975	5/6**	no estimates made	e
1975 (b)	10/15					b
1978 (c)			1975	20/(29, 36)***	9.4×10^{-4}	c
1979 (d)	80% of peaks coincident between 400 μHz and 500 μHz	11σ				d

* The references a–e are given in Table II.

** In the frequency range considered, there were six peaks in the 1973 observations and five from the 1978 observations.

*** In the frequency range considered, there were 29 peaks in the 1975 observations and 36 from the 1978 oscillations.

spectra under consideration was the result of random noise. The new results shown in Table IV for the 1979 observations are quite important because of the statistical statements that can be derived from them, and also because, for the first time in this frequency range, probable individual eigenstates are being compared.

An interesting question, motivated by the results in Table IV, concerns why the coincidence rate is not higher than observed. In the 1973, 1975, and 1978 observations, the frequency resolution was 30 μHz. It is quite apparent that this resolution is not adequate to resolve the individual eigenstates. Over time intervals of one or more years, changes in the levels of excitation of the individual eigenstates, a quite feasible occurrence, could lead to considerable changes in the structure of the power density spectra. This leads to reduced coincidence rates in external comparisons. In the case of the 1979 observations, where individual modes of oscillation are resolved, the coincidence rate is 80% for peaks in the two independent power density spectra. This is not inconsistent with what would be expected when the level of the observed noncoherent signal is taken into consideration. A similar statement applies also to the internal comparison for the 1975 observations, where approximately 67% of the peaks were found to coincide.

The results shown in Table IV do not prove that the observed oscillatory signals are global solar oscillations. However, these results are necessary if the signals are to be classified as global oscillations, and they are sufficient, to a statistically high level, to rule out random noise and, in particular, effects in the Earth's atmosphere as the origin of the signals.

C. Radial dependence of the observed oscillatory signal

The observed radial dependence of the oscillating component of the limb darkening function has proven to be one of the more important properties uncovered in solar diameter studies. The amplitude of the oscillatory signal, $I'(r)$, increases rather rapidly in the last few arc sec as the limb is approached. This property, while important in discriminating between different interpretations of the observations, will also aid in improving detector design for future studies of solar oscillations and in identifying properties of the solar atmosphere not anticipated.

The manner in which the 1975 observations were performed did not permit a test for the enhancement of the oscillatory signal at the extreme limb. However, this property has been found in the 1973, 1978, and 1979 observations by Hill and Caudell (1979), Knapp *et al.* (1980), and Bos and Hill (1983), respectively. The existence of this property is sufficient to rule out differential refraction effects in the Earth's atmosphere.

The edge definition of the solar limb which is employed in all of these observations is extremely insensitive to seeing (Hill *et al.*, 1975). However, the spatial properties of I' can also be used to test for this type of atmospheric effect by comparing them with those which should arise from seeing. The seeing effect is expected to produce a term that is proportional to $\mathrm{d}^2I/\mathrm{d}r^2$, where $I(r)$ is the static limb profile. The fact that the observed properties of I' are quite different from $(\mathrm{d}^2I/\mathrm{d}r^2)$ is sufficient to rule out seeing as the source of the oscillatory signals (Knapp *et al.*, 1980).

D. Direct measurement of the differential refraction effects in solar diameter observations

As discussed in the previous section, differential refraction, atmospheric seeing and solar oscillations produce different forms of radial dependence for the perturbations in $I(r)$. The signatures of these phenomena are sufficiently different that it is possible to measure directly the contribution of differential refraction to the solar diameter measurements.

A data analysis program was designed and implemented by Knapp *et al.* (1980) to identify the power density of the differential refraction effects in the solar diameter measurements. The program, based on the different radial properties described above, was applied to the 1978 observations; the power density, P_{ar}, in a radius measurement due to differential refraction was found to be:

$$P_{ar} = 0.25 \pm 0.09 \text{ (milli arc sec)}^2/30 \ \mu\text{Hz} \tag{1}$$

in the frequency range of 670 μHz (periods ≈ 25 min). This is significantly below the power density levels found to be due to oscillations in the 1978 and the 1979 observations.

Other projections of the contribution of differential refraction to solar diameter measurements differ significantly from the results given in Equation (1). However, these projections, which fall into the class of necessary conditions, give information primarily about the extrapolation process used in making the projections. Because of this, this type of work actually falls under the study of atmospheric physics.

E. Spatial symmetry properties and spatial coherence of observed eigenfunctions

The observations obtained by Bos and Hill (1983) were made in such a way as to yield information on changes in the limb darkening function at six different locations on the perimeter of the solar disk. These six independent signals have been analyzed for certain spatial properties of the oscillatory signals. It was observed that the eigenfunctions are spatially global and, to an accuracy of better than 20%, either symmetric or antisymmetric for reflection on the solar disk through the Sun's center and for reflection about the solar equator.

So, for the majority of the oscillations observed in 1979, a distinct set of symmetry properties may now be defined. With respect to identifying the origin of the oscillations, an equally important result is that their symmetry axes coincide, within the observational error, with the rotational axis of the Sun.

The observed spatial properties place very strong constraints on the types of random atmospheric effects which may contribute to the observations. It would be quite difficult for atmospheric seeing, with a typical aplanatic patch size of ≈ 10 arc sec, to produce oscillatory signals at diametrically opposite edges of the Sun that are highly correlated. Even more discriminating are the observed spatial symmetries, which are either symmetric or antisymmetric for reflections through the center of the solar disk and reflections about the solar equator. Only an extremely complicated epicycle model of the Earth's atmosphere could give rise to such properties.

It should also be noted that if the atmosphere were the origin of the oscillations, then any observatory near to SCLERA must observe a different set of spatial properties and, in particular, not the symmetric and antisymmetric ones observed at SCLERA. In other words, to insist that the atmosphere is the source requires that the SCLERA telescope be at a unique spot on the Earth.

4. Summary

Investigations framed as sufficient tests for atmospheric effects in solar diameter measurements have been reviewed in this paper. It is important to note that no filtering has been made in gathering this material. These investigations were designed such that the demonstrated presence of a specific property or combination of properties in the observations would be sufficient to rule out terrestrial atmospheric effects as a source of the oscillatory signals in the solar diameter measurements. In each instance where a test of this nature was performed, a statistically significant result was obtained which does not support the interpretation of the observed oscillatory signals as arising from atmospheric effects and/or random noise. These results furnish very important clues as to the nature of the oscillations. Furthermore, any proposal put forward for the interpretation of the oscillations in solar diameter measurements should be subjected to nothing less than the tests made possible by these research results.

Acknowledgements

We wish to acknowledge valuable discussions with Jerry Logan and to thank Philip Goode for a critical reading of the manuscript. This work was supported in part by the Astronomy Division of the National Science Foundation and by the Air Force Office of Scientific Research.

References

Bos, R. J. and Hill, H. A.: 1983, *Solar Phys.* **82**, 89 (this volume).

Bown, T. M., Stebbins, R. T., and Hill, H. A.: 1978, *Astrophys. J.* **223**, 324.

Caudell, T. P. and Hill, H. A.: 1980, *Monthly Notices Roy. Astron. Soc.* **193**, 381.

Caudell, T. P., Knapp, J., Hill, H. A., and Logan, J. D.: 1980, in H. A. Hill and W. A. Dziembowski (eds.), *Nonradial and Nonlinear Stellar Pulsation*, Lecture Notes in Physics, No. 125, Springer-Verlag, Berlin, p. 206.

Claverie, A., Isaak, G. R., McLeod, C. P., van der Raay, H. B., and Roca Cortes, T.: 1981, *Nature* **293**, 443.

Gossard, E. E. and Hooke, W. H.: 1975, *Waves in the Atmosphere*, Developments in Atmospheric Sciences, No. 2, Elsevier Scientific Publishing Co., Amsterdam, Oxford, New York.

Groth, E. J.: 1975, *Astrophys. J. Suppl.* **29**, 286.

Hill, H. A. and Caudell, T. P.: 1979, *Monthly Notices Roy. Astron. Soc.* **186**, 327.

Hill, H. A. and Stebbins, R. T.: 1975, *Astrophys. J.* **200**, 471.

Hill, H. A., Stebbins, R. T., and Oleson, J. R.: 1975, *Astrophys. J.* **200**, 484.

Knapp, J., Hill, H. A., and Caudell, T. P.: 1980, in H. A. Hill and W. A. Dziembowski (eds.), *Nonradial and Nonlinear Stellar Pulsation*, Lecture Notes in Physics, No. 125, Springer-Verlag, Berlin, p. 394.

Richard, J.-P.: 1975, in G. Shaviv and J. Rosen (eds.), *General Relativity and Gravitation*, John Wiley and Sons, New York, Toronto, and Israel Universities Press, Jerusalem, p. 169.

Smart, W. M.: 1965, *Textbook on Spherical Astronomy*, 5th edition, Cambridge University Press, Cambridge, p. 49.

Wolff, C. L.: 1972, *Astrophys. J.* **176**, 833.

SOLAR DIAMETER(S)*

J. RÖSCH

Université Pierre et Marie Currie, Paris

and

Observatoire du Pic-du-Midi et de Toulouse, LA No. 285 du CNRS

and

R. YERLE

Observatoire du Pic-du-Midi et de Toulouse, LA No. 285 du CNRS

Abstract. Because of the renewed attention now paid to the solar diameter, its variations from equator to pole, or its secular or long-period changes, the question: *what is a solar diameter?* is not meaningless. Two kinds of definitions may be given: either astrophysical, each one relating to a specific physical parameter, or observational, relating to a given quantity to be measured. Only the second kind is directly accessible, and astrophysical definitions should be linked to these quantities, once they are determined with the highest possible accuracy. In practice, all the programs under way refer to the point of the limb where the brightness gradient is maximum, or to a higher order approximation of the shape of the profile. Two of them are compared: the Pic-du-Midi experiment, using fast scans of the limb to define the inflection point after a correction for the blurring effect of the atmosphere, and the SCLERA experiment, using the algorithm called FFTD to eliminate this correction. The advantage of a fast scan is emphasized, and the remark is formulated that, once the signal is digitized and stored, FFTD or any processing of it can be performed. In collecting day-long one-limb scans to calibrate the blurring correction, the authors have found fluctuations of the maximum brightness gradient which provide a new entry to the field of solar oscillations.

1. Introduction

The idea was born already ten years ago, while the controversy around the solar oblateness announced by Dicke and Goldenberg (1967) was very active, on both observational and theoretical sides, to undertake solar diameter measurements at Pic-du-Midi. The good observing conditions already proven, together with a straightforward method, could be expected, indeed, to afford some valuable data in the matter. For a number of cumulative reasons, although the principles had been laid down from the beginning, the program progressed very slowly, and it is not before 1978 that it really began to develop. As a counterpart, several facts occured in the meantime which did not make it obsolete, but rather enhanced its interest, and even caused a branching towards two different goals, as will be seen.

These facts are:

– the publication in 1975 by the SCLERA Group (Hill *et al.*, 1975) of results contradicting Dicke's ones but indicating pulsations of the solar diameter, later on interpreted as photometric fluctuations in the limb profile;

– the extensive development of theoretical and observational work on solar oscillations;

* Proceedings of the 66th IAU Colloquium: *Problems in Solar and Stellar Oscillations*, held at the Crimean Astrophysical Observatory, U.S.S.R., 1–5 September, 1981.

Solar Physics **82** (1983) 139–150. 0038–0938/83/0821–0139$01.80.

– the attention recently paid to possible secular variations of the solar diameter (Eddy and Boornazian, 1979), which could represent long period oscillations, and lead to the need for absolute measurements of the angular diameter of the Sun (Parkinson *et al.*, 1980);

– and finally, from our own side, the observation, on daily measurements of the maximum brightness gradient of the solar limb, of oscillations ressembling those found in Doppler or brightness measurements.

This last point caused the splitting of our program into diameter measurements on one hand, and limb oscillation studies on the other one. The present paper will deal mostly with the first part, starting with a preliminary question.

2. What Does Solar Diameter Mean?

Perhaps some thinking on this point could help in clarifying the controversies of the last decade.

The Sun is not a stainless steel ball like those in the bearings. Nobody argues about the significance of the diameter of such a ball, because it is physically defined with an accuracy of the order of the dimension of iron atoms, and we are far from being able to measure it, in practice, with such an accuracy.

The situation has been the same for the Sun as long as everybody believed it to appear with a 'very sharp' edge (something like the surface of the ocean as seen at the horizon), and as long as the observational uncertainties were largely predominant; various authors found various values of the 'diameter', but, admitedly, all of them were measuring the same thing.

Now the scene is completely different. It has been theoretically and observationally proved that nowhere, from the center to the outer corona, does a zero exist in the density and emissivity of the solar atmosphere, so that no infinite brightness gradient is to be observed on the limb. On another hand, sensitivity and accuracy of the observational techniques have gained orders of magnitude, and the effects of the terrestrial atmosphere, which were, if not ignored, at least unexplored as late as fifty years ago, are now seriously taken into account.

Clearly, as soon as the non-existence of a 'vertical' edge is established, a definition of what is to be called 'solar diameter' is required.

3. Two Kinds of Definitions of the Solar Diameter

It would be nice to define the radius of the Sun by the level in its atmosphere where a given *physical* parameter has a specified value or a particular property. It could be, say, the level of the temperature minimum. That would probably lead to make use of different definitions, according to the astrophysical problem under consideration. To change numerically from one kind of radius to another would need a model of the atmosphere. For instance, what is the relationship between the locus of temperature minimum and an equipotential surface of the gravitational field of the Sun, the oblateness of which was the primary aim of Dicke's experiment? But much more severe is the fact that such physical quantities are not directly observable, and that, consequently, a model, with its uncertainties, will be needed to relate them to measurable ones.

Therefore, it appears logical and simpler to take the opposite way: choose a definition by an *observable* parameter, establish it as firmly as possible upon careful measurements, and deliver the concluded numerical value of the diameter as a boundary condition to any model to be computed later on for whatever astrophysical problem. In fact, only definitions of this second type are practicable. But then, the drawback is that, as *observable* parameters result from a number of local physical parameters which may vary in time (oscillations) or from one place to another (e.g. pole/equator), one should compute – unavoidably through a model again – differential coefficients giving the variation of the observable as a function of the variation of the physical parameters, provide they are known from elsewhere. However, if the definition is properly chosen, the variations should be small, and consequently not very sensitive to the uncertainties in the model.

4. The Limb-Profile Observations

Dicke's experiment, based upon what can be called 'integrated photometry', was essentially designed for pole-equator differential measurements. Not surprisingly, all those being developed at the moment, or at least the five of them presented at the Sacramento Peak Meeting in 1980 (Dunn, 1981) make use, in some way, of the limb profile, in modern improvements of the old visual observations. Indeed, the 'steepest part of the profile' still appears as the most evident feature to be observed, provided that more insight is looked for as regards its physical meaning and its dependence from various parameters.

First of all these parameters is the effect produced by the spreading function of the instrument *plus* the terrestrial atmosphere. It is a classical result that the convolution of the edge of a uniformly bright half-plane by a spreading function having a center of symetry gives a smooth symetrical profile having its inflection point exactly on the edge of the object, whereas if the object is limb-darkened, this inflection point is shifted inside increasingly with the width of the transfer function.

Among the five papers mentioned above, two only (SCLERA and Pic-du-Midi) refer to this effect, and consequently the discussion will be limited to these two.

Both experiments operate by radial photoelectric scans of the limb, but they differ in the procedure of these scans, and in the way in which the effect of the atmospheric blurring is treated.

For the clarity of the discussion, it may be useful to recall the parameters which characterize the atmospheric effects on the image of a point-source.

These parameters clearly appear on Figure 1, recovered from old records (Rösch, 1962) of the integrated flux during the occultation of a stellar image by a moving knife-edge (time scale about 2 s for a complete tracing): *scintillation* makes the fluctuations out of occultation; *blurring* (or *spreading*) governs the ratio of the maximum

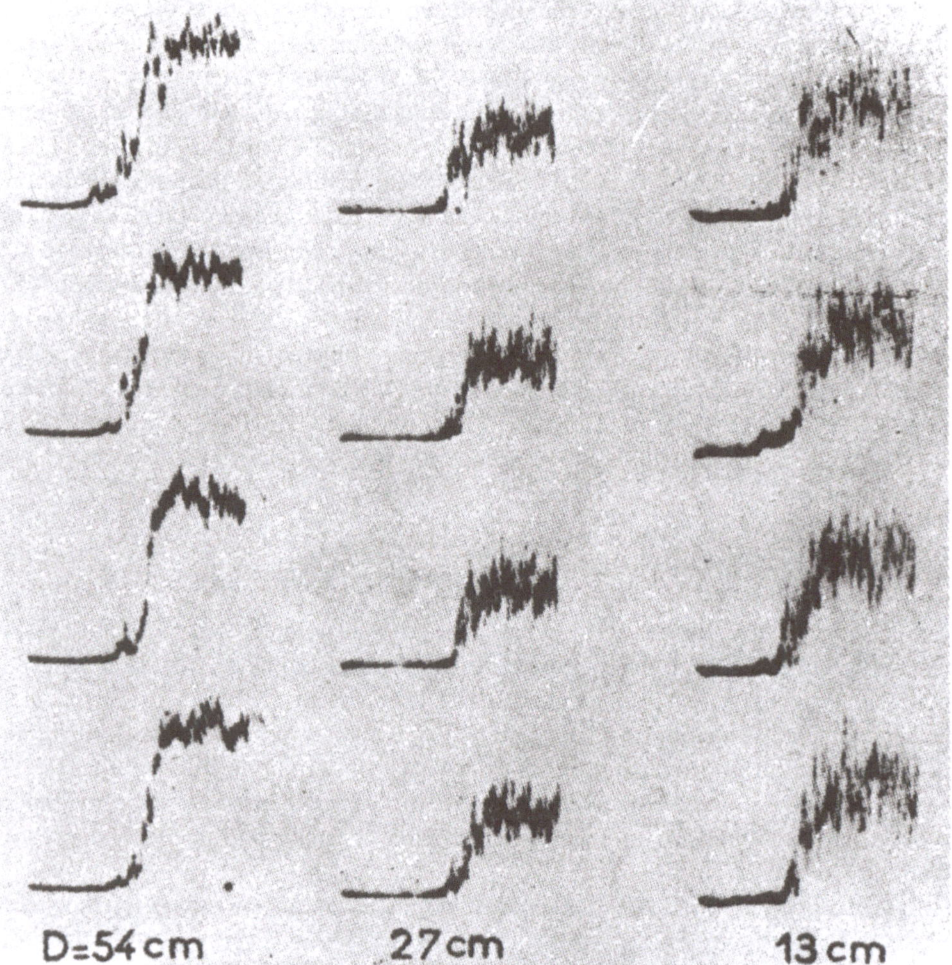

Fig. 1. Occultation of a stellar image by a kinfe-edge, showing the effects of *scintillation*, *blurring* and *image motion*. Duration of a scan ~2 s. D: diameter of the objective.

slope of the tracing to the maximum in the case of a diffraction-limited image (this ratio is equivalent to the reciprocal of the factor D/r_0 defined by Fried, 1966); random *image-motion*, algebraïcally added to the motion of the knife-edge, widens the outer enveloppe of the tracing in the direction of the motion.

How do these parameters influence a scan of the solar limb?

– *Absorption*, if uniform over the field observed, would not affect the brightness ratio between two given points;

– *scintillation* is not detrimental within the extent of the aplanetic patch (i.e. ~ 10″) or somewhat more; the tracings show how it decreases for larger and larger objectives;

– image *blurring* causes the 'shift' of the inflection point, as mentioned above; of course, if the outer conditions are such that a larger objective gains towards its theoretical resolution, as seen from the slopes of the tracings, the shift will be smaller;

– *image motion* will have a dramatic effect, again larger for a small objective than for a large one; first, of course, it will cause a random fluctuation of the measured diameter, but at least this fluctuation is purely additive, and can be treated directly by an arithmetic average; but, second, as it is clear on these pictures and conspicuous too on limb profiles, if the scan is not fast enough as compared to the image motion, the tracing is completely hashed and does not resemble anymore something like a solar limb profile.

Any device intended for solar diameter measurements should take these remarks into account.

5. The Pic-du-Midi Experiment

We deliberately adopted the definition of the solar diameter as the distance between inflection points on opposite limbs, as being the most directly observable parameter; we ought therefore to explore, on one hand, the calibration of the shift *versus* the spreading function which should be known at any time, and on the other hand the stability of the inflection point towards the physical parameters in the solar atmosphere. Figure 2a shows the principle of the observation. The focal image of the Sun (60 mm) is projected onto a CERVIT rod cut with sharp edges, somewhat shorter than a solar diameter. The beams from the opposite limbs are transported by two rhombohedra so as to project both limbs (after a magnification) onto one and the same slit and detector. In front of the slit, a rotating cube, as used by Brandt (1970) for the JOSO Site Survey, produces a scan of the limbs in succession, as seen on Figure 2b; in between the limbs is the scan of a reference patch sampled from the center of the solar disk, which, measured by the same detector, gives at any time the darkening at any point of the limb. Forty scans per second are produced, resulting in a rate of 3000″ per time second; therefore, the 'hashing effect' described above is practically eliminated and the profile is frozen.

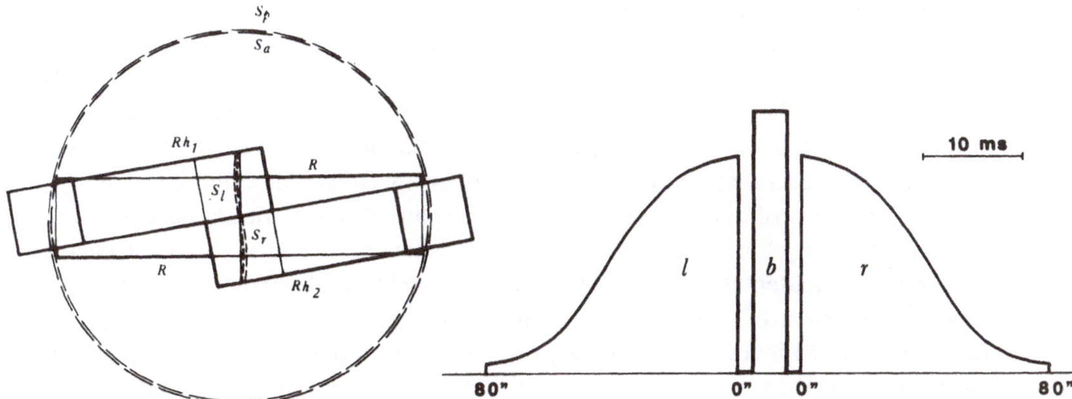

Fig. 2. *Left*: the solar image, as seen looking to the sky. S_p and S_a are the solar images at perihelion and aphelion; R is the CERVIT rod; Rh_1 and Rh_2 are the rhombohedra; S_l and S_r are the portions of the limb as seen through the rhombohedra. *Right*: the profile of a complete scan, on left limb l, calibration beam b, and right limb r, with scales in arc and time seconds.

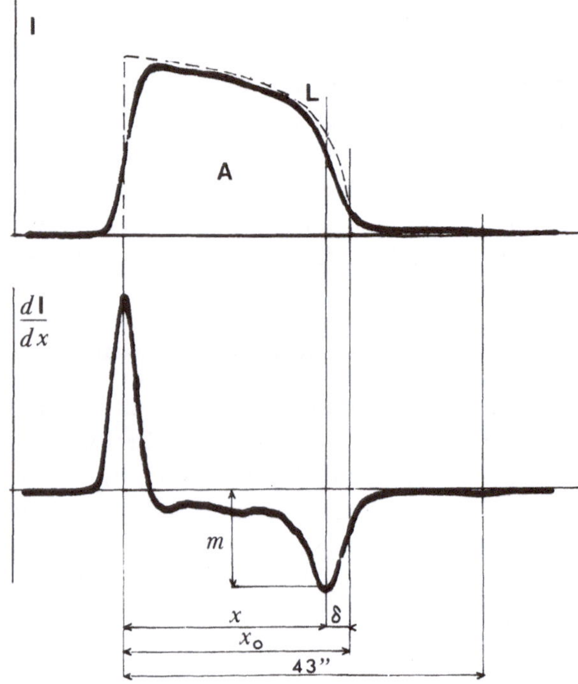

Fig. 3. Oscilloscope tracings of the signal (top) and of its first derivative (bottom), showing the definitions of the abscissae of points of maximum gradient on the true limb L (x_0) and on the blurred limb (x) (δ is the shift), of the maximum gradient itself m, and of the integral flux $\mathcal{A}$.

The upper part of Figure 3 shows a real CRO tracing on one limb. As the length of the slit, parallel to the limb, does not exceed 1″, local brightness fluctuations at the granulation scale can be detected; some are visible on the tracing presented, and consecutive tracings do confirm their solar nature.

Of course, we thought about using diode arrays instead of scanning. But we gave up the idea, at least for the moment, for two reasons: first, we considered that a classical photometric device where a Fabry lens forms a fixed pupil onto an extended cathode whatever the point treated on the image was more reliable than a detector containing a large number of discrete elements which should be carefully calibrated individually; and second, to take profit of the optimal angular resolution, given the actual size of the diodes, calls for a very long equivalent focal length, which in turn raises some problems in optical design. Incidentally, it should be noted that Stebbins (1980) used a diode array the individual element of which covered 1 arc-second.

The lower part of Figure 3 refers to the problem of the shift of the inflection point by the image blurring. It shows, added to the CRO display of the signal, the simultaneous tracing of its first derivative, produced by a differentiating circuit (for the effective work, the signal is sampled and the derivative computed). The dashed line superimposed represents what the *true* profile would be; x and x_0 are the abscissae of the true and apparent inflection point (origin at the limit of the scan) and $\delta = x_0 - x$ is the *shift*. For *each* scan are computed m, the maximum of $dI/I\,dx$, the abscissa x of this maximum, and the flux $\mathscr{A}$ integrated over the scan. The basic idea is that if the blurring is sufficiently defined by one parameter, both m and δ are functions of this unique parameter, and therefore, through a calibration to be established, the knowledge of m for each scan would give the correction δ for the same one. By a procedure (using $\mathscr{A}$ as an auxiliary quantity) which has already been described (Rösch and Yerle, 1980), we could make daily plots of δ (with arbitrary origin) versus $1/m$, so as to extrapolate towards a non-blurred profile. Figure 4 is an example of such a plot; the abscissae ε are calibrated in FWHM of the Gaussian spreading function which would give the apparent slope m to a classical limb profile. The increase (in absolute value) of δ with increasing ε is evident. The correction computed by the SCLERA group would show, at the same scale, a very small slope, and leave most of the points below it for the large values of ε; this is not surprising, since the wings of a real spreading function are known to decrease much more slowly than a Gaussian, and they should contribute to increase δ when the atmospheric blurring is important.

We know several reasons which contribute to the dispersion of the points around an average curve. The first one is that the sampling interval of the signal is much too coarse; operation with adequate value is just to start now; next to this improvement an iteration process should be applied for a more accurate plotting; and finally, real oscillations of

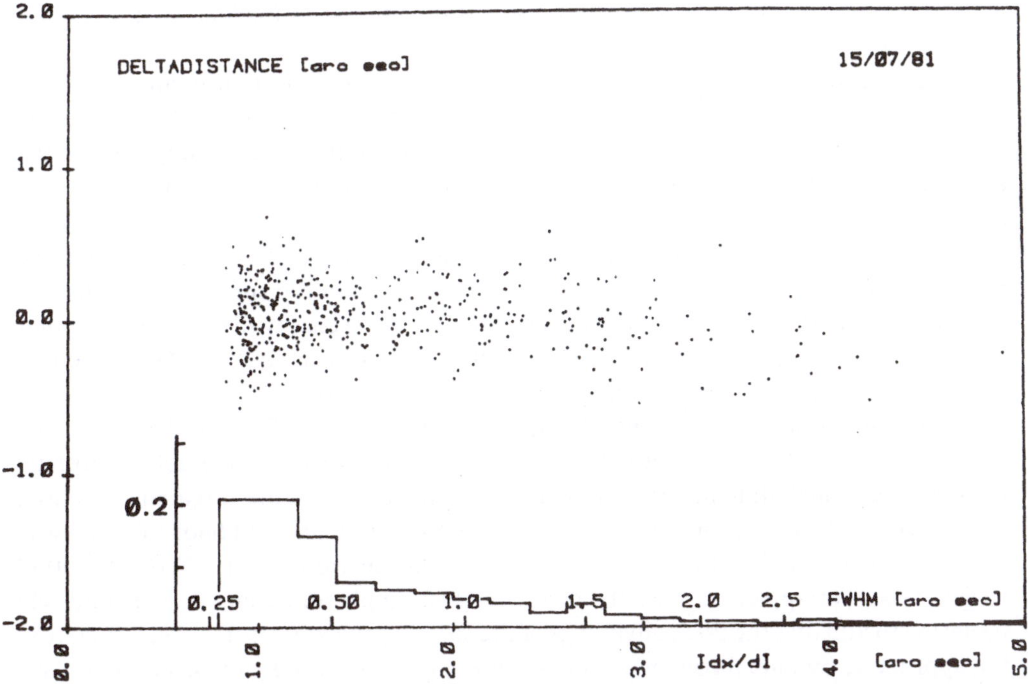

Fig. 4. Shift of the inflection point (arc seconds, arbitrary origin) versus the reciprotical of the maximum brightness gradient. The histogram shows the frequency (in decimal fraction of the total) of the equivalent FWHM in arc-seconds.

the gradient would influence the diagram, but there are solutions to take that into account. Also some incidental causes of dispersion do exist, as seen on Figure 4, which is a diagram for what we consider as a 'very good day'. The crowding of the points towards low values of ε is conspicuous: as the maximum gradient on the solar limb is not infinite, no points can be observed for $\varepsilon < 0\overset{''}{.}30$. We have other cases where 60% of the points fall below $\varepsilon = 0\overset{''}{.}50$. Anyhow, some features need to be explained, like the increased and asymetrical dispersion for small ε's, or the apparent slight maximum around $\varepsilon = 1''$; we have some reasons to suspect technical and perhaps atmospheric origins, which are being explored.

To summarize this point, work remains to be done for ascerting quantitatively the δ versus m calibration curve; but because of the advantages, which will be developed, of using individual fast scans, we are definitely to pursue it.

6. A Comparison with the SCLERA Experiment

The choice of the SCLERA group is based, on one side, upon their conclusion that the shift of the inflection point could not be corrected, in practice, an on the other side, upon the establishment of a particular algorithm, the Finite Fourier Transform, leading to the definition of the solar edge as the point where it vanishes (FFTD). As said above, any definition of the limb is conventional and permitted: it remains, afterwards, to look after its connections with other definitions, its observational behaviour and its solar significance.

The connection of the FFTD with the definition by the inflection point is clear. A Taylor expansion of the Finite FT integral readily shows that all the coefficients af the odd derivatives vanish, and therefore that both definitions are equivalent at the approximation of the second derivative, and differ only at the level of the *fourth* and higher even ones.

The peculiar property of the FFTD demonstrated by the SCLERA Group is that the zero defined in that way is stationary for the small values of the spreading function: in other words, no 'shift' correction has to be applied, as is the case with the inflection point definition. Indeed, the value published by the SCLERA Group for the solar oblateness found by this method is given with a very small r.m.s. error.

One should notice that the Finite FT is obtained by a scan of period 0.6 s, and that the adjustment to the zero takes much more time. Therefore, a considerable smoothing in time occurs, including image-motion and the fluctuations of the spreading function. It may work. However, one may have a different methodological philosophy, as was once nicely expressed by Lallemand, the father of the electron camera: 'You can either weight a sack of apples as a whole, or weight each apple and compute the sum'. He neglected to add that the great advantage of the second procedure is that you can choose the apples one by one, throw away the bad ones, and weight only the good ones. This applies very easily to our experiment, since for *each* scan, the quality is measured by the maximum slope, and discarding the scans with low slope is fully permitted because they are known to be the bad ones.

Still another great advantage of fast scans with short slit is that any local brightness accident on the Sun appears on the signal. Dicke and Goldenberg have been criticized because their observations integrated the flux over a 30° sector of the limb, so that faculae could induce apparent pole-equator differences. The FFTD method as used at SCLERA integrates over the total scanning amplitude ($\pm 6''.8$ to $\pm 27''.2$) and over the length of the slit (100″). A fast scan with short slit allows for discarding not only the tracings having a low maximum gradient, but also those showing a definite brightness excess, like a small facula, at a given position angle of the diameter under measurement.

In fact, the FFTD method and ours can overlap. The drawback of the long time-constant does not result from the FFT definition, but from the technical constraints of its application. By using diode-arrays, the problems of the time-constant and of the length of the slit should vanish. We have explained above why we sticked to direct fast scans, but we do not mean that diode-arrays could not serve in such measurements.

Conversely, we have undertaken to compute FFTDs' from the signals we have sampled and stored, since the analysis of how the FFTD works on a frozen fast scan, and how it compares with the inflection point definition, would be extremely instructive. But it could be still more useful: in his presentation at the Sacramento-Peak Meeting, Hill (1980) mentioned, on the basis of the latest observations of the SCLERA Group (Knapp *et al.*, 1980), that the second derivative d^2I/dr^2 at the limb is modified in the course of oscillations. Indirectly, we are reaching the same conclusion, since we have found – see below – oscillations in the value of the maximum gradient, which imply changes of the second derivative on both sides of its zero. Again, processing a pure fast scan of the limb could yield more directly the same information as a series of FFTD of varying amplitude.

7. The Solar Stability of These Definitions

As the solar problems we are dealing with imply small variations of the local physical parameters, either in time (oscillations) or from one region to another (e.g. pole/equator) one must question about the individual or combined influence of these parameters upon what is being measured.

It is interesting, at this point, to quote a section of the above mentioned paper by Hill (1980):

> 'It is important to note that these results (modifications of d^2I/dr^2) are based on observations rather than theoretical modeling, which has encountered some difficulty (Hill *et al.*, 1978). The above observational based value for the inflection point edge definition is to be contrasted with the much smaller results from theoretical modeling such as those from Rösch and Yerle (1980). This only serves to accent the vulnerability that exists when results from a theoretical analysis are used where theory itself does not properly describe the relevant observational phenomena.'

We can agree completely with the last sentence, and we cannot be surprised that the theoretical modeling by Hill *et al.* has encountered some difficulty. Precisely, what we did was to start from the HSRA model, which is semi-empirically built to 'properly describe' a large bulk of 'relevant observational' data, and then to introduce *small departures* from this model and see what happens to the limb.

Up to now, we have only modified the temperature distribution with height; thus, we found the differential coefficient $\Delta r/\Delta T$ (the drift of the inflection point with temperature) which appears to be quite small ($\simeq 0''.006$ per $100°$). We have undertaken to compute the differential coefficient $\Delta r/\Delta \rho$ (the drift with density) which may well be larger and reconcile our results with the SCLERA ones. Of course, differential coefficients are computed for the brightness at any part of the limb, and will serve in the interpretation of the limb profile variations in terms of changes in the local physical parameters.

8. Fluctuations of the Maximum Brightness Gradient

During preliminary observations on *one limb* to calibrate the δ versus m correlation, we currently obtained daily series of several thousands of scans, over up to ten consecutive hours, showing oscillatory variations of the maximum brightness gradient (Yerle, 1981). Obviously, such oscillations could be due as well to fluctuations of the atmospheric spreading function as to real solar oscillations. Anyhow, the fact that part of the effect, at least, should be solar is to be expected since it converges with the results obtained by the SCLERA Group and by the global Doppler experiments. From this point of view, these observations provide a new approach to the general problem of solar oscillations, in detecting them through a directly observable parameter which had not yet been taken profit of, and which could give a new insight into the fluctuations of the physical conditions in the solar layers which contribute to the limb profile. On another hand, the discrimination between solar and terrestrial contributions, on this specific parameter, could be very useful in the ever lasting discussion of the subject.

A peculiar feature in these observations, as compared to others, is that, as they refer to a very limited length along the limb (less than 1″, as compared to 100″ for the

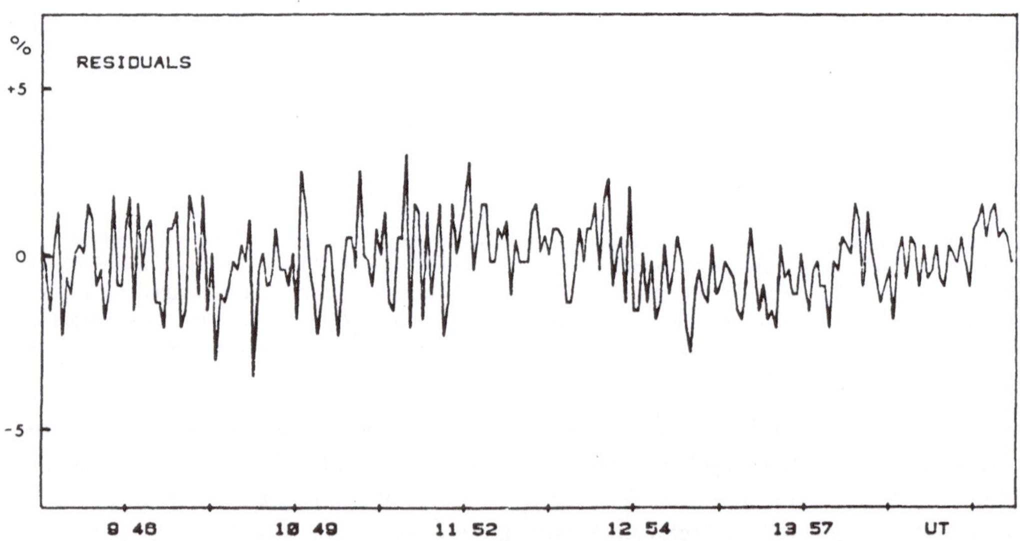

Fig. 5. Residuals of the maximum gradient (after compensation for the daily trend).

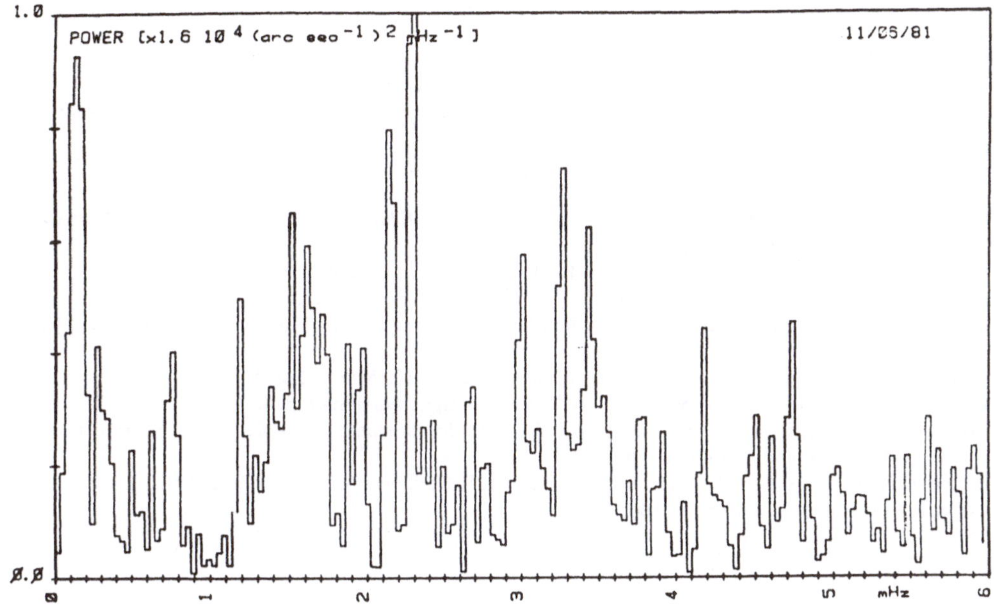

Fig. 6. Power spectrum of fluctuations of the maximum brightness gradient (observations over 8 hr).

SCLERA instrument) they practically integrate all the oscillating modes without filtering.

Figure 5 shows the residuals in m after correction for the trend due to the variation of the spreading function both with zenith distance and with the increasing image deterioration in day-time. A period of the order of the well-known 160 min oscillation, and which seems to be in phase with it, is readily visible, as well as fluctuations of much shorter period. Power spectra computed from data of that type show a number of peaks, as seen, for example, on Figure 6, particularly in the 2–4 mHz region. Because of the already mentioned integration of all modes, conclusions will be possible only on the basis of a large number of spectra.

9. Secular Oscillations and Angular Solar Diameter

The experiment we have undertaken as well as the SCLERA one offers a possibility for detection of secular changes of the solar diameter under the condition of an accurate *angular* scaling. Hill (1980) suggested to use as a reference the diffraction angle of a sharply monochromatic radiation by a grating the groove spacing of which has been controlled by interferences of the same radiation. One may express as a more general principle that, an angle being defined by the ratio of two lengths, it should be accurately known if both are measured in units of the same wave-length. Long ago we have been considering several designs following such a principle. The choice has to be done according to various practical problems arising for each of them.

References

Brandt, P. N.: 1970, *Solar Phys.* **13**, 243.

Dicke, R. H. and Goldenberg, H. M.: 1967, *Phys. Rev. Letters* **18**, 313.

Dunn, R. B. (ed.): 1981, *Solar Instrumentation – What's Next*, Sacramento Peak Observatory, September 1980.

Eddy, J. A. and Boornazian, A. A.: 1979, *Phys. Today* **32**, 17.

Fried, D. L.: 1966, *J. Opt. Soc. Am.* **56**, 1427.

Hill, H. A.: 1981, in R. B. Dunn (ed.), *Solar Instrumentation – What's Next*, Sacramento Peak Observatory, September 1980, p. 300.

Hill, H. A., Stebbins, R. T., and Oleson, J. R.: 1975, *Astrophys. J.* **200**, 484.

Hill, H. A., Rosenwald, R. D., and Caudell, T. P.: 1978, *Astrophys. J.* **225**, 304.

Knapp, J., Hill, H. A., and Caudell, T. P.: 1980, *Lecture Notes in Physics*, No. 125, Springer-Verlag, Berlin, p. 394.

Parkinson, J. H., Morrison, L. V., and Stephenson, F. R.: 1980, *Nature* **288**, 548.

Rösch, J.: 1962, *Symposium on Solar Seeing*, Roma, Consiglio Nazionale delle Ricerche, p. 38.

Rösch, J. and Yerle, R.: 1981, in R. B. Dunn (ed.), *Solar Instrumentation – What's Next*, Sacramento Peak Observatory, September 1980, p. 366.

Stebbins, R. T.: 1980, *Lecture Notes on Physics*, No. 125, Springer-Verlag, Berlin, p. 191.

Yerle, R.: 1981, *Astron. Astrophys.* **100**, L23.

SHORT PERIOD OSCILLATIONS*

F.-L. DEUBNER and J. LAUFER

Institut für Astronomie und Astrophysik, D-8700 Würzburg, F.R.G.

Abstract. Short period oscillations (with periods less than 150 s) are shown to be non-uniformly distributed on the solar surface, and in time. Rather, they appear concentrated in short bursts which preferentially occur in regions with strong instantaneous downflow motion throughout the observed extent of the solar atmosphere.

From a spatio-temporal analysis of extended series of spatial high resolution spectra, obtained with the Sacramento Peak Vacuum Tower, Deubner (1976), henceforth called Paper I, has inferred the presence of short period waves in the upper solar photosphere which contribute observable power to the velocity fluctuations at periods as short as 20 s. In the high frequency tail of these power spectra, power descends to the noise level in a conspicuous series of cascading maxima and minima (particularly in the Na D_1 line data) which in Paper I were interpreted as being due to a filtering effect of the spectral line contribution function. In a forthcoming paper (Deubner and Durrant, 1982) an alternative interpretation of the characteristic shape of the power spectra in terms of non-linear processes in the transfer of the Doppler signal will be discussed.

The present investigation is an attempt to localize *spatially* bursts of short period waves on the solar surface, i.e. with respect to other well recognizable phenomena such as granules and/or photospheric and chromospheric long period oscillations. The aim is to gain further insight into the physical processes responsible for generation of this short period power.

The observational material is the same as in Paper I and Deubner (1974) to which we refer for detailed description. In the series of spectra of September 6 (disk center) short period bursts were defined by the following procedure: The velocity fluctuations measured with the Lambdameter in the Na D_1 line were passed through a Fourier filter with a frequency passband $0.1 < \omega < 0.3\ \mathrm{s}^{-1}$. The rms fluctuations transmitted through this passband were determined as function of position x (along the slit) and time t within a 3-min-times-1-arc-sec-window run across the filtered data. 132 positions with outstanding power in the frequency range just given were selected for further analysis. Chromospheric network areas were explicitly excluded from the sample.

Centered on the selected positions, windows 6 arc sec wide and 7.5 min long were defined in the original unfiltered data arrays and, by superposition, the average variations of intensity and velocity within these windows were determined for each spectral line. Of these distributions the ones for C I 5380 and Na D_1 are displayed in Figure 1.

Most obviously these average distributions are spatially symmetric with respect to the center of the selected window areas. This we would not expect, of course, from a

* Proceedings of the 66th IAU Colloquium: *Problems in Solar and Stellar Oscillations*, held at the Crimean Astrophysical Observatory, U.S.S.R., 1–5 September, 1981.

Solar Physics **82** (1983) 151–155. 0038–0938/83/0821–0151$00.75.

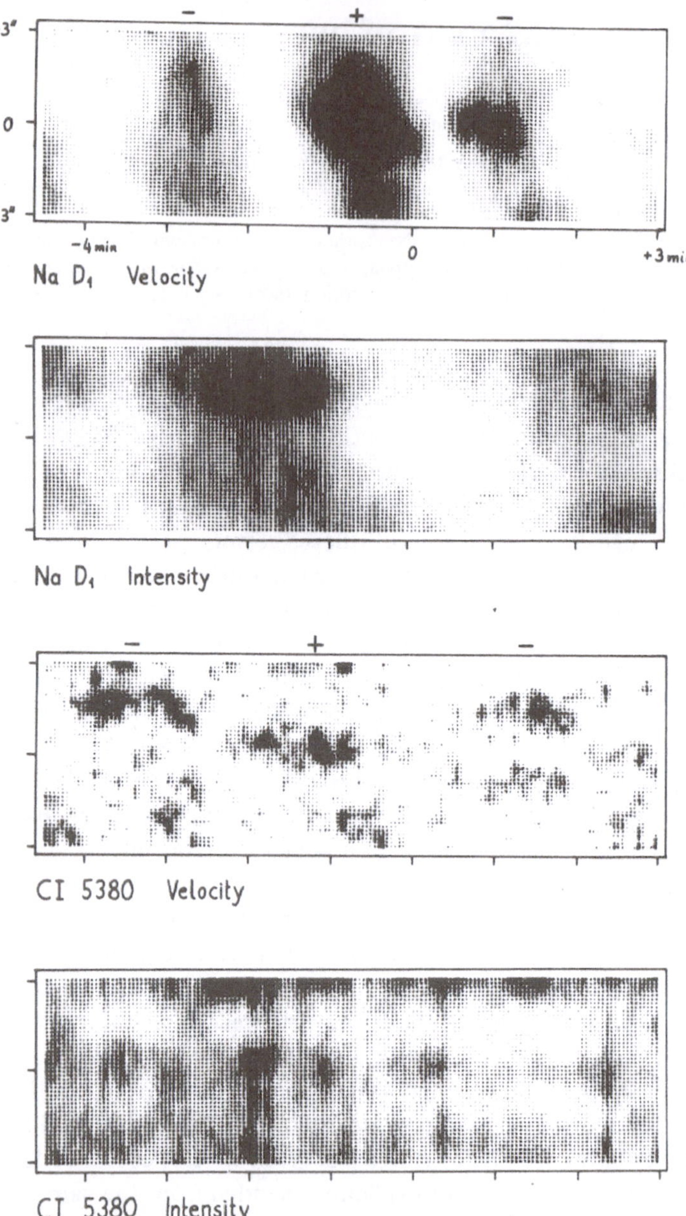

Fig. 1. Composite representation of simultaneous dynamical events in two layers of the solar atmosphere, as measured in the Na D_1 and C I 5380 lines. At time $t = 0$ a short burst of high frequency oscillations (not visible in this figure because of its low amplitude) is observed in Na D_1. 132 'burst events' ocurring in the x, t-area covered by the observation are superimposed in each frame to reveal the average brightness and velocity distribution in the vicinity of these events. + and − signs down- and up-flow. In the C I line, superposition of the oscillatory velocity field (similar to that in the Na D_1 line) and persistent flow patterns (downflow in the middle of the frame) create a velocity distribution which renders the spatial scale of the oscillations much smaller than their actual scale. Note the chromospheric and photospheric oscillations having large amplitudes and being closely in phase at $t = 0$.

randomly chosen sample. However, we would not have easily predicted it for the present sample, either.

In particular, close to the short period event, we observe in the chromosphere (Na D_1) a strong coherent oscillation with a period of ~ 220 s. Downward velocity precedes brightness by about $\pi/2$, as we expect from an evanescent wave. At the photospheric level a persistent downflow with reduced brightness (i.e. an intergranular region) is

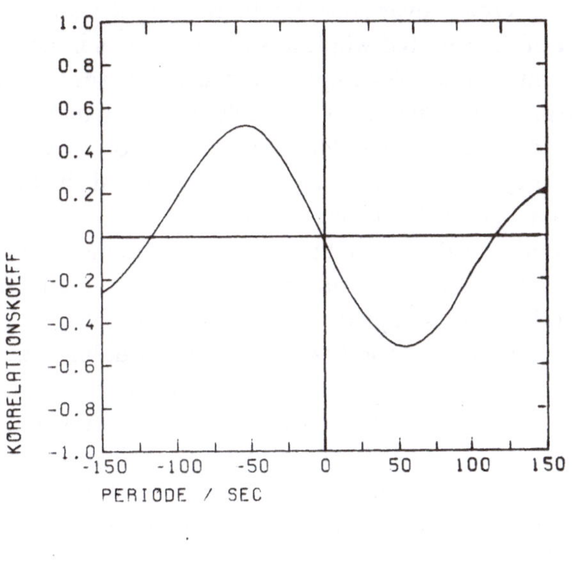

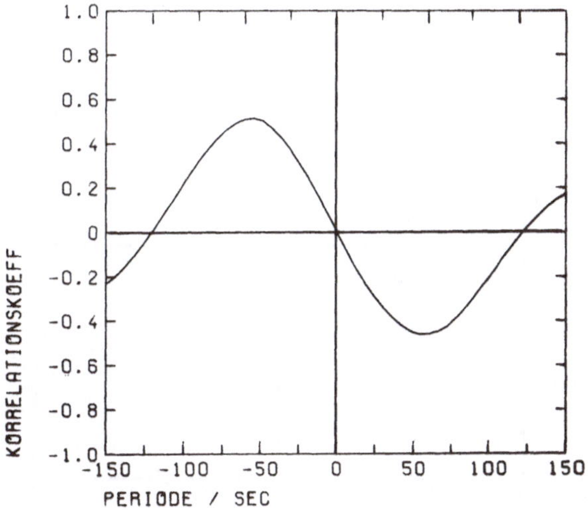

Fig. 2a. Meam cross-covariance functions of velocity and intensity fluctuations observed in Na D_1, brightness leading upward motion by approximately 90°. The upper curve is derived from the temporal variations in the selected 132 areas. The lower curve with a slightly longer period of 240 s represents a similar number of randomly selected areas.

found near the spatial center of the window, and a gradual increase of brightness (emerging granules?) is observed on either side of the central region. Superimposed on these comparatively slow changes a 300 s oscillation is also centered on the same area.

Given the presence of long period velocity fields with high amplitudes in the selected areas, we first need to check the hypothesis that the short period power is merely a spurious consequence of image motion coupled to the long period velocities. (Such effects are certainly present (e.g. see Howard and Livingston, 1968).)

Figure 2 shows mean cross-covariance functions derived from the Na D_1 intensity and velocity data out of the selected window areas as well as in an equivalent number of randomly chosen control areas. A comparison of these curves shows that the average period of the chromospheric oscillations is markedly shortened during the high frequency events, whereas we find the familiar 240 s in the control areas. We propose that additional high frequency power generated, or concentrated, in the selected regions causes the observed change of the mean period.

In conclusion, we find that three well known solar phenomeny, previously regarded as not being directly related to each other, namely: granulation, photospheric and chromospheric p-modes, are coupled together spatially under certain additonal constraints with regard to temporal phase. This coupling produces a fourth phenomenon which may be a short period burst or just a rapid pulse.

The average velocity variations in the chromosphere also show indications of asymmetric wave forms and phase acceleration which we shall discuss in a forthcoming paper. The whole sequence of events in the various distributions shown is strongly reminiscent of earlier observations by Evans and Michard (1962), who suggested a

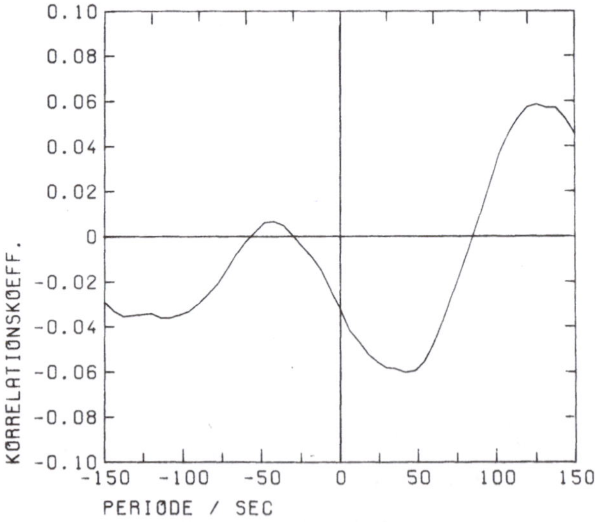

Fig. 2b. The difference of the two cross-covariance functions shown in Figure 2a brings out the presence of enhanced short period power in the selected burst events. The difference signal has a period of roughly 170 s.

causal relation between rising granules and the excitation of both photospheric and chromospheric oscillations. In the light of today's improved knowledge of nonradial resonant solar oscillations we interpret the present observations a being indeed strongly indicative of some kind of coupling between convection and the p-modes.

Since we are inclined to believe that short period bursts, or pulses, are either generated by rising (hot) granules, or are propagated more or less uniformly from lower layers in the convection zone, we may wonder how the observed short period flux gets concentrated in the intergranular lanes. Considering the geometry of the regions involved, it seems to us that wavefront retardation would be the most natural mechanism. By forming a concave wavefront on top of the intergranular lane, short period waves would most efficiently be directed to converge where enhanced short period flux is observed. Sufficient wave front retardation is easily achieved by the concomitant effects of reduced sound speed and stationary downflow in the intergranular region. Further intermittent enhancement may occur during phases of simultaneous oscillatory downflow.

Finally, we wish to draw attention to the result that the acoustic energy appears to be concentrated in the solar atmosphere at those locations which according to common belief are the footpoints of the magnetic network. It is from these places that the acoustic energy flux (in the form of magneto-acoustic and Alfvén waves) is preferentially carried higher up into the chromosphere and corona. It appears that the transfer of energy to the upper layers of the atmosphere is substantially facilitated by processes directing an enhanced flux of short period waves to these locations before the wave energy is dissipated in shocks.

References

Deubner, F.-L.: 1974, *Solar Phys.* **39**, 31.
Deubner, F.-L.: 1976, *Astron. Astrophys.* **51**, 189.
Deubner, F.-L. and Durrant, C. J.: 1982, *Astron. Astrophys.*, in press.
Evans, J. W. and Michard, R.: 1962, *Astrophys. J.* **136**, 493.
Howard, R. and Livingston, W. C.: 1968, *Solar Phys.* **3**, 434.

SPECTRAL-SPATIAL ANALYSIS OF WAVE MOTIONS IN THE REGION OF TEMPERATURE MINIMUM OF THE SUN'S ATMOSPHERE*

V. E. MERKULENKO, V. I. POLYAKOV, L. E. PALAMARCHUK, and N. V. LARIONOV

SibIZMIR, Siberian Department, USSR Academy of Sciences, Irkutsk 33, P.O. Box 4, U.S.S.R.

Abstract. A spectral-spatial analysis is made in the region of temperature minimum of the Sun's atmosphere. Filtergrams in the Ba II 4554 + 0.05 Å line have been used. The wavelengths corresponding to the sequence of peaks on the spectrum are in close agreement with those theoretically calculated by Ulrich and Rhodes (1977) for five-minute oscillations. We thus have good cause to surmize that the observed oscillations in the region of temperature minimum pertain to the *p*-mode class of general oscillations of the Sun, the main fraction of their energy falling on the period of 300 s.

1. Introduction

The problem of five-minute oscillations has long lacked general agreement about the scale of oscillations. Thus, according to Leighton *et al.* (1962) it is 2″.5, while Evans and Michard (1962) report a value of 3″ to 4″, Deubner (1967, 1969, 1974), 6″ to 15″, Howard (1967), 5″ to 10″. The finding of Musman and Rust (1970) concerning the phase coherence of oscillations seems to confirm the possible existence of the waves with λ equal to 22–44 × 10^3 km.

Thus, the scale size of oscillating elements on the Sun embraces a wide range from 2″.5 to 60″. Wolff (1972) suggested that the Sun must be pulsating integrally with high-order nonradial modes (*p*-modes). As follows from numerical calculations of Ando and Osaki (1975), many acoustic modes have a positive growth rate, in which case oscillations with a period of about five minutes are driven most effectively. A two-dimensional spectral analysis (ω, k) of sufficiently large spatial and temporal realizations, made by Deubner (1975) and Rhodes *et al.* (1977), allowed them to reveal fine structure of the power spectrum as separate narrow bands. These bands in the range of large wave numbers agree consistently with solving branches of the dispersion equation for *p*-modes of oscillations of the Sun, with the ratio of the convection zone mass to the Sun's mass being 0.087.

It should be noted that the technique of fast Fourier transform (FFT) employed in the references cited, provides low spectral resolution for small wave numbers k. In order to check to what extent the spatial scale of oscillations on this portion of the spectrum agrees with theory, we shall invoke the fact that the main part of oscillation energy falls

* Proceedings of the 66th IAU Colloquium: *Problems in Solar and Stellar Oscillations*, held at the Crimean Astrophysical Observatory, U.S.S.R., 1–5 September, 1981.

Solar Physics **82** (1983) 157–161. 0038–0938/83/0821–0157$00.75.

on the period of 300 s. Consequently, the spectrum in space for a fixed time must display peaks which correspond to wave numbers of p-modes of oscillations for this period.

2. The Spatial Spectrum of Oscillations at the Formation Level of the Ba II 4554 + 0.05 Å Line

The core of the Ba II 4554 Å line is produced at a height of about 600 km, i.e. near the temperature minimum (Shilova *et al.*, 1966). The ± 0.05 Å line wings exhibit a clear-cut granulation, and therefore they are produced at the upper photospheric level. Since the atomic weight of Ba is rather large (137.4), the profile of this line is insensitive to thermal motions (Rutten, 1978). As a consequence, the brightness variation in its wing implies mainly a drift due to macroscopic motions.

The calculation of the spatial spectrum of oscillations used a filtergram of an area of the quiescent atmosphere at disk center, taken with a birefringent filter in the wing of Ba II 4554 + 0.05 Å. The transmission band of the filter is 0.08 Å. A two-dimensional spectral analysis (k_x, k_y) was made for the intensity matrix on the area 128″ × 128″ with a 1″ step. The spectrum was calculated by the method of two-dimensional correloperiodogramanalysis (CPGA) which consists in the calculation of the multiple correlation coefficient $G(\lambda_x, \lambda_y)$ between the original brightness distribution function $R(x, y)$ and the harmonic function $\exp(i(k_x x + k_y y)$, where k_x and k_y are the wave vector projections on the X and Y axes, running through a discrete series of given values.

Figure 1 shows a two-dimensional periodogram of the distribution of the correlation coefficient $G(\lambda_x, \lambda_y)$. The abscissa axis indicates the wavelength λ_x while the ordinate axis indicates λ_y. The degree of shading is proportional to the value of the function G.

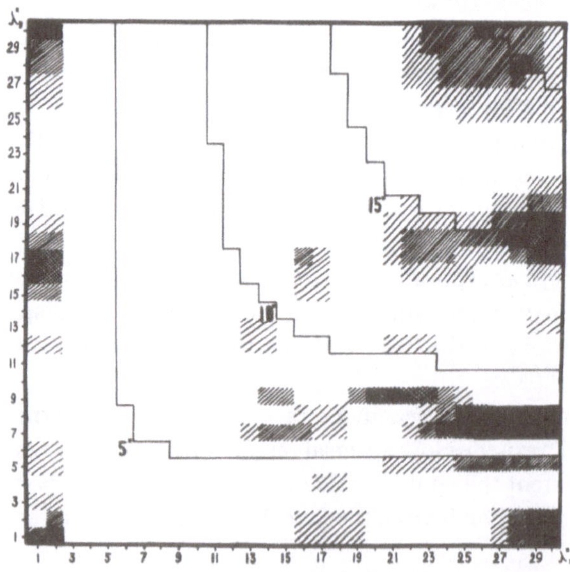

Fig. 1. Two-dimensional correloperiodogram from the intensity distribution on the 128″ × 128″ area obtained using a filtergram in the Ba II 4554 + 0.05 Å line.

Solid lines on the periodogram indicate wavelength values, determined from the condition of geometrical addition for the vector **k**, i.e.

$$|\mathbf{k}| = \sqrt{k_x^2 + k_y^2} \quad \text{or} \quad \lambda = \frac{\lambda_x \lambda_y}{\sqrt{\lambda_x^2 + \lambda_y^2}} \,. \tag{1}$$

It can be seen that maxima of the two-dimensional spectrum correspond to the wavelengths: 7″, 8″.5, 13″.5, 19″.5, and 26–30″. It should also be noted that for waves with $\lambda < 20''$ the condition $\lambda_x > \lambda_y$ is, as a rule, fulfilled. Hence one may suppose the existence of anisotropy in the direction of the vector **k** for waves with $\lambda < 20''$. To investigate the spatial spectrum of waves with $\lambda > 20''$ by the CPGA method, a one-dimensional spectral analysis has been made using another filtergram. The area of study was sized 256″ × 200″. Microphotometric scans of 256″ length consisted of 512 points and were separated by 4″. The microphotometer slit was sized 0″.5 × 1″0.

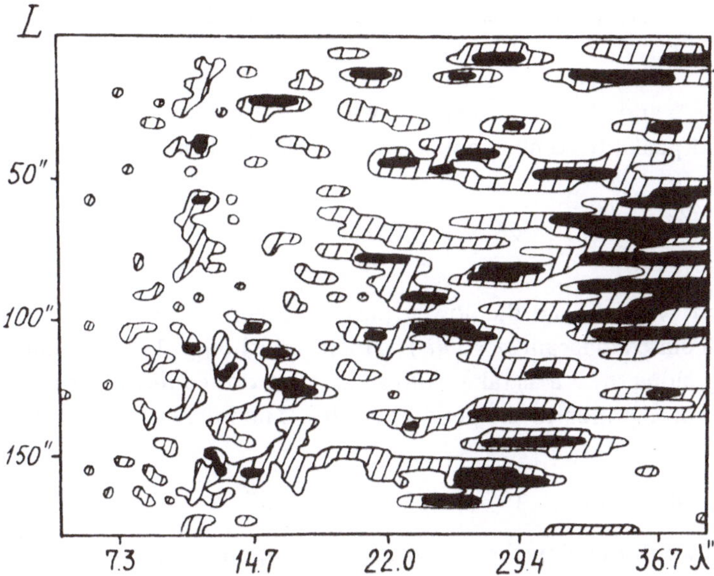

Fig. 2. Diagram representation of one-dimensional correloperiodograms for a sequence of photometric scans on the 256″ × 200″ area.

The spectra of these 50 scans are combined as a diagram, shown in Figure 2. The wave length λ_x is plotted on the abscissa, the distance in arc sec normal to photometric scans is laid off as ordinate. The degree of shading corresponds to two consecutive values of the correlation coefficient in excess of 90% confidence level. It can be seen that the spectrum changes from one scan to the other. A periodogram averaged over the 50 scans is shown in Figure 3. Plotted on the ordinate is the mean value of product of the correlation coefficient by the related value of probability. For $\lambda_x < 20''$ the spectrum is substantially smoothed but in the region of large values of λ_x there are three pronounced

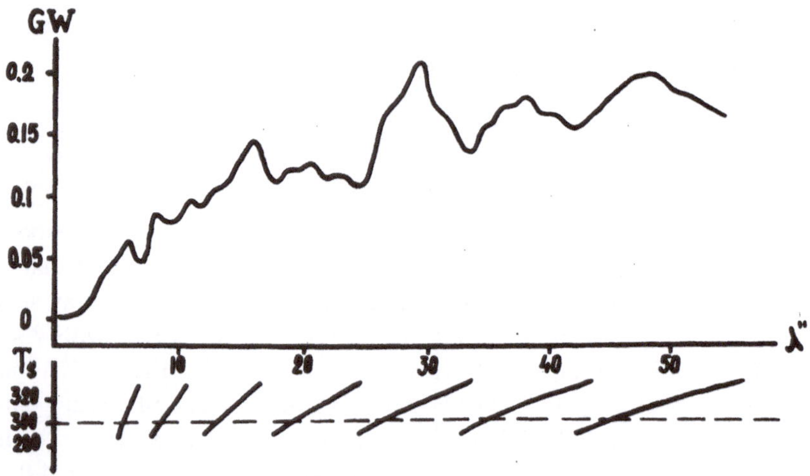

Fig. 3. Periodogram averaged over 50 photometric scans. The lower part of the figure shows a fragment of the diagram $\omega = \omega(k)$ reported by Ulrich and Rhodes (1977).

maxima: 29″, 38″, and 48″. Assuming that when spectra are averaged the values of λ_x for the identified peaks become commensurable with λ_y, from Equation (1) we find:

$$\lambda = \frac{\lambda_x}{\sqrt{2}} = 20\rlap{.}''1\,, \quad 26\rlap{.}''9\,, \quad 34\rlap{.}''0\,.$$

For comparison with theory the lower part of Figure 3 shows a fragment of a theoretically calculated diagram $\omega = \omega(k)$ (Ulrich and Rhodes, 1977), modified with the scale uniformity along the λ axis taken into account. A dotted line crossing the brances of the diagram corresponds to the period of 300 s. The intersection points fall on the wavelengths:

$$5'' , \quad 8'' , \quad 13'' , \quad 19'' , \quad 26'' , \quad \text{and} \quad 35'' .$$

For comparison we show below the wavelengths corresponding to the sequence of peaks on the spectra in space, shown in Figures 1 and 3.

$$7'' , \quad 8\rlap{.}''5 , \quad 13\rlap{.}''5 , \quad 19\rlap{.}''5\text{–}20'' , \quad 27'' , \quad \text{and} \quad 34'' .$$

3. Conclusion

(a) The fairly close agreement of wavelengths for peaks on the spectra in space with wavelengths corresponding to the period of 300 s on the theoretical diagram of Ulrich and Rhodes (1977) enables one to suppose that in the region of temperature minimum there are oscillations which are attributable to the *p*-mode class of general oscillations of the Sun, the main part of the energy of which falls on the period of 300 s.

(b) As follows from the diagram shown in Figure 1, for wave vectors **k** corresponding to wavelengths with $\lambda < 20''$ there seem to exist an anisotropy in directions.

Acknowledgement

We are grateful to Dr G. V. Kuklin for assistance with mathematical handling of the data.

References

Ando, H. and Osaki, Y.: 1975, *Publ. Astron. Soc. Japan* **27**, 581.
Deubner, F.: 1967, *Solar Phys.* **2**, 133.
Deubner, F.: 1969, *Solar Phys.* **9**, 343.
Deubner, F.: 1974, *Solar Phys.* **39**, 31.
Deubner, F.: 1975, *Astron. Astrophys.* **44**, 371.
Evans, J. M. and Michard, R.: 1962, *Astrophys. J.* **136**, 493.
Howard, R.: 1967, *Solar Phys.* **2**, 3.
Leighton, R. R., Noyes, R., and Simon, G.: 1962, *Astrophys. J.* **135**, 474.
Rhodes, E. L., Ulrich, R. K., and Simon, G. M.: 1977, *Astrophys. J.* **218**, 901.
Rutten, R.: 1978, *Solar Phys.* **56**, 237.
Shilova, N. S. and Obridko, V. N.: 1966, *Solnechnye Dannye*, No. 8, 73.
Ulrich, R. K. and Rhodes, E. J.: 1977, *Astrophys. J.* **218**, 521.
Wolff, C. L.: 1972, *Astrophys. J.* **177**, 287.

OBSERVATION OF FIVE-MINUTE-PERIOD GRAVITY WAVES IN THE SOLAR PHOTOSPHERE*

(Invited Review, Abstract)

R. T. STEBBINS

Sacramento Peak Observatory[1], *Sunspot, NM 88349, U.S.A.*

PHILIP R. GOODE

Arizona Research Laboratories[2,3], *University of Arizona, Tucson, AZ 85721, U.S.A.*

and

HENRY A. HILL

Department of Physics and Arizona Research Laboratories[2,3], *University of Arizona, Tucson, AZ 85721, U.S.A.*

Abstract. Vertically propagating traveling waves have been observed in the solar photosphere. These waves have a period of 278 ± 41 seconds and a vertical phase velocity of about 2 km s^{-1}. It is noted that these waves also have approximately the same period as the well-studied five-minute-period acoustic mode, which is evanescent in the photosphere. The only consistent interpretation of the traveling waves implies that they are gravity waves. About half the time the gravity waves are outgoing, while the remainder of the time they are ingoing.

The data were collected by Stebbins *et al.* (1980) to study the vertical structure of the photosphere. They examined velocity pertubations at nine altitudes in the photosphere using a Doppler shift technique. The current work represents a reanalysis of that data which uncovered the five-minute-period traveling waves.

The mean velocity amplitude of disturbances at a given altitude, as registered in the observed Doppler shift, was found to be directly proportional to the mean velocity amplitude at the base of the photosphere. This was the expected relationship between the velocity amplitudes. It was not expected that the standard deviation of the velocity amplitude at a particular altitude would be independent of the velocity amplitude at the

* Proceedings of the 66th IAU Colloquium: *Problems in Solar and Stellar Oscillations*, held at the Crimean Astrophysical Observatory, U.S.S.R., 1–5 September, 1981.

[1] Operated by the Association of Universities for Research on Astronomy, Inc., under contrast AST 78-17292 with the National Science Foundation.

[2] Research supported in part by the U.S. Department of Energy and the National Science Foundation, Astronomy Division.

[3] Arizona Research Laboratories is an interdisciplinary research program at the University of Arizona. The research activities of Hill and Goode are conducted at the Santa Catalina Laboratory for Experimental Relativity by Astrometry (SCLERA), which is administered through Arizona Research Laboratories. SCLERA is jointly operated by Wesleyan University and the University of Arizona.

Solar Physics **82** (1983) 163–164. 0038–0938/83/0821–0163$00.30.

base of the photosphere. In addition, the phase difference between velocities at different altitudes is, unexpectedly, inversely proportional to the velocity amplitude at any altitude. These two traits of the data are consistent with a five-minute-period nonacoustic traveling wave being superimposed on the five-minute-period acoustic mode. This supposition is borne out by a detailed examination of the data in the complex plane of amplitude and phase and by a calculation of the power spectrum of the traveling waves as a function of their vertical wavenumber (see Hill *et al.*, 1982).

References

Hill, H. A., Goode, P. R., and Stebbins, R. T.: 1982, *Astrophys. J. Letters* **256**, L17.

Stebbins, R. T., Hill, H. A., Zanoni, R., and Davis, R. E.: 1980, in H. A. Hill and W. A. Dziembowski (eds.), *Nonradial and Nonlinear Stellar Pulsation*, Lecture Notes in Physics, No. 125, Springer-Verlag, Berlin, p. 381.

RADIATIVE TRANSFER AND SOLAR OSCILLATIONS*

(Invited Review)

JØRGEN CHRISTENSEN-DALSGAARD

Institute d'Astrophysique, Liège, Belgium

and

*Advanced Study Program, National Center for Atmospheric Research***

and

SØREN FRANDSEN

*High Altitude Observatory, National Center for Atmospheric Research**, Boulder, Colo., U.S.A.*

and

Astronomisk Institut†*, Aarhus Universitet, DK-8000 Aarhus C, Denmark*

Abstract. We consider the atmospheric behaviour of solar oscillations in a model including a detailed, semi-empirical atmosphere. Equations of radial and non-radial oscillation, with consistent treatment of the radiation field, are derived. These equations are solved in the radial, grey case; the results show that departure from the Eddington approximation has little effect on the properties of the oscillations. Preliminary results are presented for the non-radial case, indicating substantial deviations from the Eddington approximation when the optical thickness across a horizontal wavelength is of the order of or less than unity.

1. Introduction

A consistent calculation of solar (or indeed stellar) oscillations must take into proper account the interaction between the hydrodynamics of the motion and the radiative energy transfer: the radiation contributes to the heating or cooling of the gas and so affects the dynamics of the oscillation. In addition observations of solar oscillations are made through their effect on the solar spectrum of radiation and so the diagnostics of such oscillations depends on an understanding of how they affect the radiation field.

The dynamical effect of the radiation occurs through the energy equation which relates two of the thermodynamic fluctuations, e.g. of pressure and density, through the fluctuation in the divergence of the flux of energy. The radiative flux has traditionally been treated in the diffusion approximation, although more recently the Eddington approximation (Unno and Spiegel, 1966) has been commonly applied. These approximations both become exact in the limit of large optical depth, and so they are probably adequate for calculations of global properties of the oscillations. This is true in particular for low order modes that reside predominantly in the solar interior. On the other hand

* Proceedings of the 66th IAU Colloquium: *Problems in Solar and Stellar Oscillations*, held at the Crimean Astrophysical Observatory, U.S.S.R., 1–5 September, 1981.

** NCAR is sponsored by the National Science Foundation.

† Permanent address.

Solar Physics **82** (1983) 165–204. 0038–0938/83/0821–0165$04.50.

the solar atmosphere makes a significant contribution to the damping of the solar 5 min oscillations (e.g. Ando and Osaki, 1975, 1977; see also Section 4 below). Furthermore the very high accuracy now attained in the observational determination of the frequencies of the 5 min modes of low degree (Grec *et al.*, 1980; Claverie *et al.*, 1981) suggests that effects of the atmosphere on the computation of oscillation frequencies, although certainly small, may eventually become significant; thus it was shown by Christensen-Dalsgaard and Gough (1980) that for modes close to the upper end of the observed frequency range the atmospheric boundary condition has some influence on the eigenfrequencies. Finally it seems likely that the dynamics of the possible chromospheric mode (Ando and Osaki, 1977; Ulrich and Rhodes, 1977), trapped between the temperature minimum and the transition region, is affected by the treatment of radiative transfer.

The interpretation of observations of solar oscillations in terms of their behaviour in the atmosphere requires theoretical calculations of the response of the features in the solar spectrum to the oscillations. The simplest, and least reliable, approach is to assume that an observed quantity, e.g. a lineshift in a given part of a spectral line, can be ascribed an effective height of formation, determined as an average over a contribution or response function for the observed quantity, such that the observation characterizes the oscillation at this height (e.g. Fossat and Ricort, 1975; Schmieder, 1976, 1978). Thus a lineshift would be interpreted as measuring the oscillation velocity at a given height in the atmosphere. This procedure, however, clearly ignores the variation of the oscillation through the region where the spectral line is formed. To go beyond it one must make a detailed calculation of the perturbations in the atmosphere associated with the oscillation and of the effects of these perturbations on the observed quantities. The latter part of the process has been considered by, e.g., Mein (1971), Canfield (1976), Cram *et al.* (1979), Gouttebroze and Leibacher (1980), and Keil (1980), and clearly involves the interaction between the radiation field, typically in a spectral line, and the oscillation. However radiative effects may also be important in the calculation of the variation of the oscillations with height in the atmosphere.

The diagnostic part of the theory of oscillations in the solar atmosphere has become increasingly important as observations relating to the atmospheric behaviour of the oscillations have become more detailed. These observations include well-resolved two-dimensional power spectra ($k - \omega$ diagrams) in chromospheric spectral lines in addition to the commonly observed photospheric lines (Rhodes *et al.*, 1983) as well as a resolved $k - \omega$ diagram in continuum intensity (Brown and Harrison, 1980), and the whole-disk measurements of intensity oscillations by Woodard and Hudson (1983). In addition there is a large volume of observations with less resolution in space and time, but with simultaneous measurements of line shifts and intensities in several different spectral lines, or in different parts of the same line (e.g. Cram, 1978; Lites and Chipman, 1979; Stebbins *et al.*, 1980). Such observations give potentially detailed information about the variation in oscillation amplitude and phase with the height in the atmosphere, and are thus of great interest as tests of the theory of oscillation. Furthermore, once the theoretical treatment of the oscillations has been sufficiently developed, observations of phase and amplitude variations might possible be used to infer properties of the mean structure of the atmosphere, and thus supplement traditional atmospheric diagnostics.

A special problem is presented by the SCLERA observations of oscillations in the limb darkening function close to the solar limb (e.g. Brown *et al.*, 1978). When interpreted as straight displacement of the solar limb these observations yield amplitudes apparently in contradiction to other observations in the same period range. Calculations by Hill *et al.* (1978) which treated the radiation field in the grey Eddington approximation showed that this discrepancy could be explained by taking into account the variation in the intensity perturbation with the distance from the limb; but this required modifications to the outer mechanical boundary condition that so far have received little additional support. Further, more detailed, oscillation calculations are needed to throw light on this problem.

Apart from calculations using the grey Eddington approximation or the even simpler Newton's law of cooling (e.g. Noyes and Leighton, 1963) relatively little work has been done on the atmospheric behaviour of oscillations in the Sun or other stars. Davis (1971) described a technique for including the radiation field in non-linear computations of radial pulsations of Cepheids. Similar calculations have been made by Karp (1975). Kalkofen and Ulmschneider (1977) presented a method for including radiative transfer in the calculation of radially propagating, non-linear waves in the solar atmosphere. Apparently the only attempt at a better treatment of radiative transfer in linear theory of solar oscillations was made by Schmieder (1977) who considered radially propagating waves in the photosphere. Her calculations essentially used the Eddington approximation, but she went beyond the grey case by considering the radiation field at a number (4 or 6) of different wavelengths. It might also be pointed out that the problem of convective stability, taking into account a consistent, grey radiation field, was studied by Legait (1982a, b), and that detailed dynamical models of the solar granulation with non-grey radiation have been developed by, e.g., Levy (1974) and Nordlund (1982).

There is little doubt that non-linear effects are important for the behaviour of waves in the solar atmosphere, especially in the chromosphere where the combined velocity amplitude approaches the sound speed. On the other hand linear calculations are computationally far less costly than the corresponding non-linear calculations, and so permit a more detailed exploration of the dependence of the oscillations on the parameters of the problem. This, together with the relative simplicity of linear oscillations, make such calculations well suited for attempts to understand the basic physics of the oscillations and perhaps to delimit the regions of validity of commonly used approximations. Furthermore the wave amplitudes are probably sufficiently small in the lower parts of the atmosphere that the results of linear theory provides a reasonably realistic basis for comparisons with observations.

We present initial results of a project aimed at a consistent calculation of linear oscillations in the solar interior and atmosphere, non-radial as well as radial, with as detailed a treatment of the radiation field as permitted by limitations in the available computational resources. It is hoped eventually to include the radiation field at a sufficient number of frequency points to give a realistic description of the variation in the continuum, and to study the effect of spectral lines on the oscillations; the formalism we have developed is sufficiently general to cover these cases. However in this initial paper results are only given for the grey case, where the absorption coefficients are assumed to be frequency independent, and a major concern is to test the accuracy of the Eddington approximation. Section 2 presents the equations of radial oscillations in the solar atmosphere, and discusses the use of variable Eddington factors (e.g. Auer and Mihalas, 1970) to treat the directional dependence of the radiation. In Section 3 we describe the setting up of the equilibrium solar model which consists of a semi-empirical atmosphere model matched to an envelope model. Section 4 presents results on the radial oscillations of this model. In Section 5 is discussed the generalization of the variable Eddington factor technique to non-radial oscillations, and preliminary results are given which indicate that departures from the standard Eddington approximation may become quite important at large horizontal wavenumbers. Finally Section 6 contains a discussion of the results and considers future directions for our work.

2. Equations and Boundary Conditions for Radial Oscillations

We consider small-amplitude radial oscillations about an equilibrium state and linearize the equations in the perturbations. For simplicity only the plane-parallel case with constant gravity is treated here; the full set of spherical equations are presented in the Appendix.

As usual (e.g. Cox, 1980) the perturbations must satisfy the continuity equation which in the plane-parallel case may be written

$$\frac{\mathrm{d}\delta r}{\mathrm{d}r} = -\frac{\delta\rho}{\rho}\,; \tag{2.1}$$

here r is a vertical coordinate increasing outwards, δr is the displacement, ρ the density and δ denotes the amplitude of the Lagrangian perturbation (i.e. the perturbation following the motion; e.g. Cox, 1980).

The equation of motion requires a little more care. The atmospheric model includes a turbulent contribution p_{turb} to the total pressure which may thus be written $p_{\mathrm{tot}} = p + p_{\mathrm{turb}}$, where p is the gas pressure; p_{turb} is determined from the turbulent velocity (the so-called microturbulence) introduced to account for the observed line spectrum. The equation of hydrostatic support is supposed to involve p_{turb}, so that

$$\frac{\mathrm{d}p}{\mathrm{d}r} = -g\rho - \frac{\mathrm{d}p_{\mathrm{turb}}}{\mathrm{d}r} = -g\rho + f_{\mathrm{turb}}\rho\,, \tag{2.2}$$

which defines the turbulent body force f_{turb} (e.g. Mihalas and Toomre, 1981); here g is gravity. Thus the perturbed equation of motion is

$$\rho\frac{\partial^2\delta\mathbf{r}}{\partial t^2} = -\nabla p' - \rho'(g - f_{\mathrm{turb}})\mathbf{a}_r - \rho(\mathbf{g}' - \mathbf{f}'_{\mathrm{turb}})\,, \tag{2.3}$$

where the prime denotes the Eulerian perturbation and $\mathbf{a}_r$ is a unit vector directed vertically upwards. (We have neglected the effects of radiation pressure.)

In the absence of a definite physical model for f_{turb} the specification of $\mathbf{f}'_{\text{turb}}$ is largely arbitrary. Here we shall assume that the Lagrangian perturbation $\delta\mathbf{f}_{\text{turb}}$ vanishes; in this we differ from Mihalas and Toomre (1981) who took instead $\mathbf{f}'_{\text{turb}} = 0$. Taking $\delta\mathbf{f}_{\text{turb}} = 0$ ensures that an infinitely slow uniform vertical displacement of the layer is a solution to the resulting oscillation equations, as it certainly should be; on the other hand this clearly does not guarantee the validity of the approximation at higher frequencies. Then

$$\mathbf{f}'_{\text{turb}} = -\mathbf{a}_r \frac{\mathrm{d}\mathbf{f}_{\text{turb}}}{\mathrm{d}r} \delta r ; \tag{2.4}$$

if the perturbation is assumed to depend on time as $\exp(-i\omega t)$ and we return to the plane-parallel, constant-gravity case it is easy to show that

$$\frac{\mathrm{d}}{\mathrm{d}r}\left(\frac{\delta p}{p}\right) = \frac{\rho}{p}\left(\tilde{g}\frac{\delta p}{p} + \omega^2 \delta r\right) , \tag{2.5}$$

where $\tilde{g} \equiv g - f_{\text{turb}}$ is an effective gravity. We neglect the turbulent pressure in the convection zone.

The energy equation may be written in the general form

$$i\omega\left(\frac{\delta T}{T} - \nabla_{\text{ad}}\frac{\delta p}{p}\right) = \frac{1}{\rho c_p T}\,\delta(\operatorname{div}\mathscr{F}_{\text{tot}}) , \tag{2.6}$$

where T is temperature, $\nabla_{\text{ad}} = (\partial \ln T/\partial \ln p)_s$, s being specific entropy and c_p is the specific heat at constant pressure. $\mathscr{F}_{\text{tot}}$ is the total energy flux, which may be written as

$$\mathscr{F}_{\text{tot}} = \mathscr{F}_R + \mathscr{F}_C + \mathscr{F}_M , \tag{2.7}$$

where $\mathscr{F}_R$ is the radiative flux, integrated over frequency, $\mathscr{F}_C$ is the convective flux, and $\mathscr{F}_M$ is a 'mechanical' flux which must be invoked to explain the heating of the chromosphere. The perturbation of the radiative flux is the major subject of the present paper, and we return to it shortly. The perturbation of $\mathscr{F}_C$ causes problems because no definitive theory of time-dependent convection exists; furthermore even the physical mechanism responsible for $\mathscr{F}_M$ is uncertain, and so here the difficulties are even greater. Although expressions for the perturbation in the convective flux have been derived in the mixing-length formalism (e.g. Unno, 1967; Gough, 1977) we have chosen here to neglect the contributions to Equation (2.6) from $\mathscr{F}_C$ as well as $\mathscr{F}_M$, i.e. to assume

$$\delta(\operatorname{div}\mathscr{F}_C) = \delta(\operatorname{div}\mathscr{F}_M) = 0 . \tag{2.8}$$

This represents a major uncertainty in the present work, which must be kept in mind when evaluating the results. Nevertheless one may hope that it does not invalidate at least the overall features of the behaviour of the oscillations in the solar atmosphere.

To determine the perturbation in the radiative flux we must consider the equation of transfer. The radiation field in the equilibrium state satisfies

$$\mu \frac{\partial I_\nu(r, \mu)}{\partial r} = \rho \kappa_{s,\nu} J_\nu + \rho \kappa_{a,\nu} B_\nu - \rho(\kappa_{s,\nu} + \kappa_{a,\nu}) I_\nu \tag{2.9}$$

(e.g. Mihalas, 1978); for simplicity we have assumed isotropic scattering and local thermodynamical equilibrium (LTE). Here $I_\nu(r, \mu)$ is the specific intensity at position r and frequency ν at the angle $\hat{\vartheta} = \cos^{-1}\mu$ from the vertical; $\kappa_{s,\nu}$ and $\kappa_{a,\nu}$ are the monochromatic scattering and absorption coefficients, respectively, B_ν the Planck function and

$$J_\nu = \frac{1}{4\pi} \oint I_\nu \, \mathrm{d}\Omega \tag{2.10}$$

is the mean intensity, the integration being over solid angle Ω. We also introduce the higher moments

$$H_\nu = \frac{1}{4\pi} \oint \mu I_\nu \, \mathrm{d}\Omega , \tag{2.11}$$

$$K_\nu = \frac{1}{4\pi} \oint \mu^2 I_\nu \, \mathrm{d}\Omega ; \tag{2.12}$$

here H_ν is related to the radiative flux by

$$\mathcal{F}_R = 4\pi \int_0^\infty H_\nu \, \mathrm{d}\nu \, \mathbf{a}_r , \tag{2.13}$$

and K_ν is related to the radiation pressure. By integrating Equation (2.9) over μ and ν we obtain

$$\operatorname{div} \mathcal{F}_R = \frac{\mathrm{d}\mathcal{F}_{R,r}}{\mathrm{d}r} = 4\pi\rho \int_0^\infty \kappa_{a,\nu}(B_\nu - J_\nu) \, \mathrm{d}\nu , \tag{2.14}$$

where $\mathcal{F}_{R,r}$ is the radial component of $\mathcal{F}_R$.

We shall assume that the motion is non-relativistic. Then the time derivative in the complete equation of transfer (e.g. Mihalas, 1980) may be neglected. If we furthermore neglect the effect of lines and other regions with sharp gradients in I_ν as a function of ν (where otherwise Doppler-shift effects caused by velocity gradients might become important) Equation (2.14) may be perturbed in a straightforward manner; substituting the result into the energy equation (2.6), using Equations (2.7) and (2.8), we finally get

$$i\omega\left(\frac{\delta T}{T} - \nabla_{\mathrm{ad}}\frac{\delta p}{p}\right) = \frac{1}{\rho c_p T}\,\mathrm{div}\,\mathscr{F}_R\frac{\delta\rho}{\rho} + $$

$$+ \frac{4\pi}{c_p T}\int_0^\infty [\delta\kappa_{a,\nu}(B_\nu - J_\nu) + \kappa_{a,\nu}(\delta B_\nu - \delta J_\nu)]\,\mathrm{d}\nu\,. \tag{2.15}$$

Here $\delta\rho/\rho$ and $\delta\kappa_{a,\nu}$ can be expanded in terms of $\delta T/T$ and $\delta p/p$ as

$$\frac{\delta\rho}{\rho} = \rho_T\frac{\delta T}{T} + \rho_p\frac{\delta p}{p}\,, \qquad \frac{\delta\kappa_{a,\nu}}{\kappa_{a,\nu}} = (\kappa_{a,\nu})_T\frac{\delta T}{T} + (\kappa_{a,\nu})_p\frac{\delta p}{p}\,, \tag{2.16}$$

where e.g., $\rho_T = (\partial\ln\rho/\partial\ln T)_p$ and may be determined from the equation of state. Furthermore B_ν is an explicit function of T and so δB_ν can immediately be found as

$$\delta B = \frac{\mathrm{d}B_\nu}{\mathrm{d}T}\,\delta T\,. \tag{2.17}$$

Thus Equation (2.15) can be written as

$$\left(i\omega - \frac{4\pi}{c_p T}\int_0^\infty \left\{[(\kappa_{a,\nu})_T + \rho_T](B_\nu - J_\nu) + T\frac{\mathrm{d}B_\nu}{\mathrm{d}T}\right\}\kappa_{a,\nu}\,\mathrm{d}\nu\right)\frac{\delta T}{T} =$$

$$= \left\{i\omega\nabla_{\mathrm{ad}} + \frac{4\pi}{c_p T}\int_0^\infty [(\kappa_{a,\nu})_p + \rho_p](B_\nu - J_\nu)\,\kappa_{a,\nu}\,\mathrm{d}\nu\right\}\frac{\delta p}{p} -$$

$$- \frac{4\pi}{c_p T}\int_0^\infty \kappa_{a,\nu}\,\delta J_\nu\,\mathrm{d}\nu\,. \tag{2.18}$$

This equation can clearly be used to eliminate $\delta T/T$ in terms of $\delta p/p$ and the δJ_ν.

To close the system of equations we therefore need equations for the δJ_ν. By perturbing Equation (2.9) it follows that

$$\mu\frac{\partial\delta I_\nu}{\partial r} = \rho[\delta\kappa_{s,\nu}J_\nu + \delta\kappa_{a,\nu}B_\nu - (\delta\kappa_{s,\nu} + \delta\kappa_{a,\nu})I_\nu +$$

$$+ \kappa_{s,\nu}\delta J_\nu + \kappa_{a,\nu}\delta B_\nu - (\kappa_{s,\nu} + \kappa_{a,\nu})\,\delta I_\nu]\,; \tag{2.19}$$

here δI_ν and δJ_ν are related by

$$\delta J_\nu(r) = \frac{1}{4\pi} \oint \delta I_\nu(r, \mu)\, \mathrm{d}\Omega. \tag{2.20}$$

Thus one might in principle include $\delta I_\nu(r, \mu)$, at suitable discrete points in ν and μ, as variables, and solve the corresponding Equation (2.19) together with Equations (2.1) and (2.5). It is evident that reasonable resolution in ν and μ may result in a very large set of equations.

As is common in the theory of stellar atmospheres, it is more efficient to work in terms of moments of the radiation field. We introduce δH_ν and δK_ν as

$$\delta H_\nu(r) = \frac{1}{4\pi} \oint \mu \delta I_\nu(r, \nu)\, \mathrm{d}\Omega, \tag{2.21}$$

$$\delta K_\nu(r) = \frac{1}{4\pi} \oint \mu^2 \delta I_\nu(r, \nu)\, \mathrm{d}\Omega. \tag{2.22}$$

The zeroth and first moments of Equation (2.19) then yield

$$\frac{\mathrm{d}\delta H_\nu}{\mathrm{d}r} = \rho[\delta\kappa_{a,\nu}(B_\nu - J_\nu) + \kappa_{a,\nu}(\delta B_\nu - \delta J_\nu)], \tag{2.23}$$

$$\frac{\mathrm{d}\delta K_\nu}{\mathrm{d}r} = -\rho[\delta\kappa_\nu H_\nu + \kappa_\nu \delta H_\nu], \tag{2.24}$$

where $\kappa_\nu = \kappa_{s,\nu} + \kappa_{a,\nu}$ is the total opacity. To close the system of equations we now use the variable Eddington-factor technique (e.g. Auer and Mihalas, 1970), by introducing the Eddington-factors

$$f_{\mathrm{osc},\nu} = \delta K_\nu(r)/\delta J_\nu(r). \tag{2.25}$$

Had $f_{\mathrm{osc},\nu}$ been known the equations would clearly form a closed system. This is not the case, but we may determine $f_{\mathrm{osc},\nu}$ by iteration. As an initial gues we take $f_{\mathrm{osc},\nu} = f_{\mathrm{eq}} \equiv K_\nu/J_\nu$ which is known from the equilibrium solution. Equations (2.1), (2.5), (2.23), and (2.24) are then solved, to give a first estimate of the perturbations. This solution is substituted into the right-hand side of Equation (2.19) which may then be integrated for δI_ν; this integration can be done by quadrature, independently at each point in ν and μ, and so it is extremely fast. Finally a new estimate of $f_{\mathrm{osc},\nu}$ is obtained from Equations (2.20), (2.22), and (2.25). This iteration may be performed simultaneously with the iteration for ω which is in general an eigenvalue of the problem, and so it is quite efficient. In this way the number of radiation variables is reduced by a factor corresponding to the number of μ-points, at the expense of a few additional iterations. It might be noticed that the standard Eddington approximation corresponds to taking $f_{\mathrm{osc},\nu} = \frac{1}{3}$.

We still have to specify the boundary conditions. The outer surface of the model, at $r = r_s$ say, is assumed to correspond to the lower boundary of an isothermal corona; by making its temperature different from that of the last point in the model we can simulate the rapid temperature rise in the outer part of the transition region. A condition on δr and $\delta p/p$ at this point is then obtained by requiring the solution to match continuously to an outward propagating, adiabatic wave in the corona (e.g. Cox, 1980). The conditions on the perturbations in the radiation field are that there be no incoming radiation at the surface, i.e.

$$\delta I_\nu(r_s, \mu) = 0 \quad \text{for} \quad \mu < 0 . \tag{2.26}$$

When using the variable Eddington factors these are replaced by the conditions

$$\delta H_\nu(r_s) = g_{\text{osc},\,\nu}\,\delta J_\nu(r_s) , \tag{2.27}$$

where the $g_{\text{osc},\,\nu}$ form another set of variable Eddington factors that may be determined in the iteration for $f_{\text{osc},\,\nu}$; the conditions (2.26) are then applied in the quadrature to determine δI_ν from Equation (2.19).

In a complete stellar model (where one must clearly use the spherical equations) the solution has to satisfy regularity conditions at the centre and these, together with the surface conditions, determine ω as an eigenvalue of the problem. Here we shall be concerned with models truncated at a finite distance from the centre, and so we need a separate set of conditions. From the energy Equation (2.15) follows that the solution becomes increasingly adiabatic with increasing depth, and we shall assume that at the bottom boundary, $r = r_b$, the solution is exactly adiabatic, i.e.

$$\frac{\delta T}{T} = \nabla_{\text{ad}} \frac{\delta p}{p} \quad \text{at} \quad r = r_b . \tag{2.28}$$

A more realistic condition that takes into account departures from adiabaticity has been formulated by Dziembowski (1977), but if the lower boundary is at a sufficient depth the condition (2.28) is adequate. It may also be shown (as in the equilibrium case, e.g. Mihalas, 1978) that the radiation field tends to the diffusion approximation at great depths. This determines the variation of δH_ν with ν, leaving the ν-integrated flux perturbation

$$\delta H = \int_0^\infty \delta H_\nu \, \mathrm{d}\nu \tag{2.29}$$

undetermined, and thus provides an additional $N_f - 1$ conditions, where N_f is the number of points in ν.

The conditions specified so far permit a solution for any value of ω, because no mechanical condition has been invoked at the lower boundary. However it is of interest to look for standing wave solutions in the truncated model, and to do so we have to impose an, essentially artificial, mechanical condition. Here we shall take the condition that the displacement vanish, i.e.

$$\delta r = 0 \quad \text{at} \quad r = r_b . \tag{2.30}$$

Alternatively one might look for solutions at given period (or real part of ω), regarding the growth or damping rate (i.e. the imaginary part of ω) as a real eigenvalue. This requires one real condition at the lower boundary which we take to be that the part of the model interior to r_b perform no work on the part exterior to r_b, i.e.

$$\operatorname{Im}(\delta r^* \delta p) = 0 \quad \text{at} \quad r_b . \tag{2.31}$$

Finally let us consider the grey case, where $\kappa_{a,\nu} = \kappa_a$ and $\kappa_{s,\nu} = \kappa_s$ are assumed to be independent of ν. Then Equation (2.18) becomes

$$\{i\omega - \omega_R[4 - \Delta_c(\kappa_{a,T} + \rho_T)]\} \frac{\delta T}{T} = $$

$$= [i\omega\nabla_{\rm ad} - \omega_R \Delta_c(\kappa_{a,p} + \rho_p)] \frac{\delta p}{p} - \omega_R \frac{\delta J}{B} , \tag{2.32}$$

where B is the frequency-integrated Planck function, $\delta J = \int_0^\infty \delta J_\nu \, d\nu$ and we have introduced

$$\Delta_c = J/B - 1 = - \frac{\operatorname{div} \mathscr{F}_R}{4\pi\rho\kappa_a B} \tag{2.33}$$

as a measure of the departure from radiative equilibrium, J being the integrated mean intensity; furthermore

$$\omega_R = \frac{4\pi B \kappa_a}{c_p T} \tag{2.34}$$

is a characteristic radiative relaxation rate in the optically thin limit related to the one introduced by Spiegel (1957). Equation (2.32) is similar to Equation (12) of Ando and Osaki (1975), but as they assumed that $J = B$ everywhere they did not have the terms in Δ_c. This is equivalent to assuming that the equilibrium model is in radiative equilibrium everywhere, an assumption that is invalid not only in a mechanically heated atmosphere but also in the upper part of the convection zone where there is a transition from radiative to convective energy transport (deeper down in the solar convection zone energy is carried almost exclusively by convection and so once again Δ_c is negligible). As shown in Section 4 the inclusion of the terms in Δ_c appears to have a fairly strong stabilizing effect on acoustic modes in the Sun.

After integration over ν Equations (2.19) and (2.24) are unchanged, apart from the removal of the subscripts ν, whereas Equation (2.23) may be written, using

Equation (2.32), as

$$\frac{\mathrm{d}\delta H}{\mathrm{d}r} = \rho B \kappa_a \left\{ \frac{4[i\omega\nabla_{\mathrm{ad}} - \omega_R \Delta_c(\kappa_{a,p} + \rho_p)]\dfrac{\delta p}{p} - [i\omega + \omega_R \Delta_c(\kappa_{a,T} + \rho_T)]\dfrac{\delta J}{B}}{i\omega - \omega_R[4 - \Delta_c(\kappa_{a,T} + \rho_T)]} - \right.$$

$$\left. - \Delta_c \frac{\delta\kappa_a}{\kappa_a} \right\}. \tag{2.35}$$

(By writing the equation in this apparently more complicated form we avoid difficulties that might otherwise be caused by the near cancellation of δJ and δB in the interior of the model.) The only inner thermal boundary condition is now Equation (2.28). The variable Eddington factor method may be used as before, to iterate for $f_{\mathrm{osc}}(r) = \delta K(r)/\delta J(r)$ and $g_{\mathrm{osc}} = \delta H(r_s)/\delta J(r_s)$. Notice that there are now only four perturbation quantities (e.g. δr, $\delta p/p$, δH, and δK) and so the order of the system is the same as when the Eddington approximation, or the diffusion approximation, is used.

3. The Equilibrium Model

The solar model we have considered consists of a deep envelope matched smoothly to a semi-empirical atmospheric model. The envelope model is constructed by integrating the equations of hydrostatic support, continuity and radiative or convective energy transport inwards, assuming constant luminosity, from boundary conditions determined at the bottom of the atmosphere. The physics of the envelope is, with a few exceptions discussed below, as in the solar evolution calculation of Christensen-Dalsgaard (1982): the equation of state of Eggleton *et al.* (1973), opacity tables from Cox and Tabor (1976) interpolated with stretched splines (Cline, 1974) and convection treated with mixing length theory in the form of Baker and Temesváry (1966).

The atmospheric model is based on a model similar, but not identical, to model C of Vernazza *et al.* (1981) and was kindly supplied by A. Skumanich. This model, however, had to be somewhat modified to make it suitable for oscillation calculations (Mihalas and Toomre, 1981). In particular the turbulent body force f_{turb}, defined by Equation (2.2), had erratic variations in some parts of the original model above the temperature minimum. These were removed by resetting f_{turb} smoothly where needed and redetermining p by integrating Equation (2.2), keeping p fixed at the top and bottom of the atmosphere. All number densities were scaled with the ratio β between the new and the old p, thus keeping all degrees of ionization the same; the value of β was always between 0.9 and 1.1.

The mesh in the original model had only about 50 points from the bottom of the atmosphere to the base of the corona. This is insufficient to resolve adequately the variations in the oscillation quantities, and so the mesh had to be reset by interpolation. The distribution of the new mesh was based largely on the variation in density and mass, but with additional points in the region of rapid temperature rise from 10 000 K to

24 000 K and at the temperature plateau at 24 000 K. To keep f_{turb} smooth p was reset from the interpolated f_{turb}; this required changes in p of less than 1.2% from the directly interpolated value. The final mesh had 300 points in the atmosphere.

It might be argued that the fluctuations in f_{turb} represent a physical effect and that accordingly they should not be eliminated. However the fluctuations are inadequately resolved in the original model and so they cannot be reliably interpolated. We have therefore preferred to remove them. Future, more detailed, atmospheric models are needed to demonstrate their reality.

In the final, interpolated model the maximum deviation in pressure from the original model caused by the smoothing of f_{turb} is about 9% and occurs at the 24 000 K temperature plateau. For temperatures less than 9000 K the deviation is less than 5% everywhere. Thus the smoothing has a relatively small effect on the thermodynamics of the atmosphere.

A number of thermodynamic derivatives, not provided in the original atmosphere, are needed in the oscillation calculation. Of course the notion of an equation of state is not well-defined in the upper part of the atmosphere where LTE is no longer a good approximation. Here, ideally, one should perturb the statistical equilibrium equations (e.g. Mihalas, 1978) which provide a direct coupling between the radiation field and the thermodynamic state of the gas. However this would greatly increase the complexity of the calculation, and in the present work we have avoided such complications. Following Mihalas (1979) we have taken some account of the effect of NLTE by introducing departure coefficients, found from the equilibrium model, in the Saha and Boltzman's equations; these were assumed to be unaffected by the oscillation. The equation of state was otherwise as in Christensen-Dalsgaard (1982), with complete calculation of the ionization of C, N, and O; however here we also include the ionization of the first electron of the 7 most abundant among the remaining heavy elements.

The departure coefficient d_H for hydrogen ionization was taken from the original model. The ionization of He and heavy elements was calculated with a ficticious mean departure coefficient d_Z, assumed to be the same for all elements and all levels of ionization; d_Z was found by iteration, to make the density found from the equation of state calculation, at given temperature and electron density, the same as the density in the original atmosphere. In a limited region around a temperature of 10^4 K d_Z was smoothed to remove fairly drastic fluctuations in the opacity derivatives. This smoothing caused a maximum deviation in density of about 0.4%.

The composition of the atmosphere model assumed a ratio α_{He} between the number densities of He and H of 0.1. A direct, spectroscopic determination of α_{He} is of course difficult; it may be that evolution calculations, where α_{He} is regarded as a free parameter which is adjusted until the luminosity of the model of the present Sun agrees with the observed value, offer a more reliable determination of α_{He}. The values obtained are generally somewhat smaller than 0.1. In the present calculation we have preferred to use the value 0.0858 determined from the evolution calculation of Christensen-Dalsgaard (1982), to obtain an envelope that is as closely as possible consistent with the complete solar model. The atmosphere therefore had to be adjusted for the change in composition.

This was done by regarding the relation between temperature and the total hydrogen number density n_H as fixed; d_H and d_Z were determined as before, using the original α_{He}, and the electron number density was then found by iteration, keeping d_H and d_Z fixed, to get the density corresponding to the new composition. By assuming that p_{turb}/ρ had the same dependence on n_H as in the original model we could then determine the total pressure and hence the height and mass scales, using the equations of hydrostatic support and continuity. This procedure is clearly not strictly correct, but in view of the relatively small change in α_{He} it is probably adequate.

The opacity in the atmosphere was calculated using a programme described by Auer *et al.* (1972), taking into account contributions from H, H^-, H_2^+, C, Si, and Mg. In the present, grey, calculation the value $\kappa_{A,5000}$ at 5000 Å was used. This is fairly close to, and much faster to compute than, the Rosseland mean; furthermore it agrees with the Rosseland mean κ_{CT} found from the tables of Cox and Tabor (1976) at a temperature of about 6300 K, over a wide range in density. In the photosphere a smooth matching between $\kappa_{A,5000}$ and κ_{CT} was therefore accomplished by computing

$$\log \kappa^* = \phi(y) \log \kappa_{CT} + [1 - \phi(y)] \log \kappa_{A,5000}, \tag{3.1}$$

where $y = \Delta_m^{-1} \log(T/T_m)$,

$$\phi(y) = \begin{cases} 0 & \text{for } y < -1 \\ \frac{1}{2} + \frac{3}{4}y(1 - \frac{1}{3}y^2) & \text{for } |y| \le 1 \\ 1 & \text{for } y > 1, \end{cases} \tag{3.2}$$

and T_m and Δ_m were 6300 K and 0.1, respectively. This ensures that κ^* and its first derivatives with respect to T and ρ are continuous. Above the temperature minimum κ^* was set to $\kappa_{A,5000}$. To compensate for the fact that $\kappa_{A,5000}$ is different from the opacity used in constructing the atmospheric model the final atmospheric opacity κ was obtained as $\kappa = \alpha_{op} \kappa^*$, where α_{op} in the atmosphere is a constant determined such that the optical depth at the bottom of the atmosphere, at 5000 Å, was the same as in the original model.

In the integration in the interior we neglect the turbulent pressure, so that $f_{turb} = 0$. We furthermore assume strict LTE, by setting all departure coefficients to unity, use the opacity as given by Equation (3.1) (i.e. take $\alpha_{op} = 1$) and assume no mechanical (as distinct from convective) energy flux, i.e. $\mathscr{F}_M = 0$. To get a smooth transition from the atmosphere to the interior we determine, rather arbitrarily, the values of f_{turb}, d_H, d_Z, α_{op}, and $\mathscr{F}_M$ in the upper part of the interior as

$$A(\tau) = [1 - \phi(\hat{y})] A_{bot} + \phi(\hat{y}) A_{int}, \tag{3.3}$$

where $A_{bot} = A(\tau_{bot})$ is the value of any of these quantities at the bottom of the atmosphere, A_{int} is its value in the interior and τ is optical depth; ϕ is defined in Equation (3.2) and $\hat{y} = (\tau - \tau_{bot})/\Delta\tau - 1$. The bottom of the atmosphere was chosen to be at $\tau_{bot} = 0.44$ (corresponding to $T = 5840$ K), and $\Delta\tau = 1$ was used in the smoothing.

As the envelope integration starts at a fairly small optical depth we cannot treat the radiative flux in the diffusion approximation. Instead we use the grey moment equation

$$\frac{\mathrm{d}K}{\mathrm{d}r} - \frac{1}{r}\left(\frac{1}{f_{\mathrm{eq}}} - 3\right)K = -\kappa\rho H\,, \tag{3.4}$$

together with

$$B = \frac{K}{f_{\mathrm{eq}}(1 + \Delta_c)} \tag{3.5}$$

which follows from Equation (2.32) and the definition of f_{eq}. Thus the temperature gradient is

$$\nabla \equiv \frac{\mathrm{d}\ln T}{\mathrm{d}\ln p} =$$

$$= \frac{1}{4}\frac{P}{\rho\tilde{g}}\left[\frac{\kappa\rho H}{K} - \frac{1}{r}\left(\frac{1}{f_{\mathrm{eq}}} - 3\right) + \frac{\mathrm{d}\ln f_{\mathrm{eq}}}{\mathrm{d}r} + (1 + \Delta_c)^{-1}\frac{\mathrm{d}\Delta_c}{\mathrm{d}r}\right], \tag{3.6}$$

where Equation (2.2) was used. To make a smooth transition in ∇ between the atmosphere and the interior we replaced H in Equations (3.4) and (3.6) by $\Phi(\tau)H$, where $\Phi(\tau_{\mathrm{bot}})$ was determined from the condition that ∇ be continuous at the bottom of the atmosphere, and $\Phi(\tau)$ was found as in Equation (3.3), with $\Phi_{\mathrm{int}} = 1$; $\Phi(\tau_{\mathrm{bot}})$ was always close to 1. H is found from Equation (2.7) as

$$H = \frac{1}{4\pi}(\mathscr{F}_{\mathrm{tot}} - \mathscr{F}_M - \mathscr{F}_C)\,. \tag{3.7}$$

Finally the mixing length treatment of convection was modified appropriately.

The equations for the structure of the envelope thus depend on f_{eq} and Δ_c, and hence on the radiation field. Rather than making a full model atmosphere calculation we have determined the radiation quantities iteratively. Given f_{eq} and Δ_c as functions of τ and the value of $\mathscr{F}_{M,\mathrm{bot}}$, we can determine the envelope structure and hence integrate the equation of transfer in the envelope and atmosphere, to get new $f_{\mathrm{eq}}(\tau)$, $\Delta_c(\tau)$ and $\mathscr{F}_{M,\mathrm{bot}}$. As initial values were used $f_{\mathrm{eq}} = \frac{1}{3}$, $\Delta_c(\tau) = \mathscr{F}_{M,\mathrm{bot}} = 0$. The convergence of this iteration was rather slow; in the calculation we stopped after three iterations, where the mean change in f_{eq} and Δ_c was less than 0.3%.

The ratio α of the mixing length to the pressure scale height was determined such that the depth of the convection zone was the same as in model 1 of Christensen-Dalsgaard (1982); this required $\alpha = 1.8556$. The difference between this value and the value of 1.6364 used by Christensen-Dalsgaard reflects the change in the surface boundary condition. With this calibration the differences between the envelope model and model 1 in p, ρ, and T, compared at fixed mass, was less than 1% at the base of the convection zone. The remaining difference is probably still an effect of the difference in the surface boundary condition, and could presumably be removed by suitably modifying the boundary condition in the evolution calculation. In this way it should be possible to construct a complete model of the present Sun, matched smoothly to a detailed atmospheric model.

4. Results for Radial Oscillations with Grey Radiation

The equilibrium model described in the preceding section was truncated at a fractional radius $x = r/r_{\text{phot}}$ of 0.24, where r_{phot} is the photospheric radius, here taken to correspond to the point where $T = T_{\text{eff}}$, the effective temperature. The temperature at the bottom of the envelope was 8.2×10^6 K; at this temperature the luminosity in a complete solar model is about 0.97 times the surface value, and hence to compute a deeper envelope one would have to take into account the variation in luminosity caused by nuclear energy generation. The outer boundary of the model was taken at a temperature of 49000 K in the transition region, and a coronal temperature of 1.6×10^6 K was assumed in the outer mechanical boundary condition. The complete equilibrium model, including the atmosphere, had 992 mesh points.

The spherical oscillation equations (cf. the Appendix) were solved in the grey case using a second order centred difference scheme similar to the one used by Baker and Kippenhahn (1965) (see also Baker *et al.*, 1971). The Eddington factors were calculated using 10 point Gaussian quadrature in μ. We iterated for ω or its imaginary part until the mechanical boundary conditions (2.30) or (2.31) were satisfied.

Some indication of the possible eigenmodes of the model can be obtained by writing the equations of adiabatic oscillation in the form

$$\frac{d^2\psi}{d\tau_a^2} + \Omega^2\psi = 0 \tag{4.1}$$

(Christensen-Dalsgaard *et al.*, 1983), where $\psi = r(\rho c)^{1/2}\,\delta r$, c being the sound speed, and

$$\tau_a = \int_r^{r_s} \frac{dr}{c} \tag{4.2}$$

is the 'acoustical depth'. Ω^2 is given by Equation (4.4) of Christensen-Dalsgaard *et al.* and may be written as

$$\Omega^2 = 4\pi^2(\nu^2 - \mathscr{V}), \tag{4.3}$$

where $\nu = \omega/2\pi$ is the cyclic frequency; this equation defines an effective 'potential' $\mathscr{V}$ for radial oscillations, such that the mode is oscillatory when $\nu^2 > \mathscr{V}$ and evanescent when $\nu^2 < \mathscr{V}$.

The behaviour of $\mathscr{V}$ in the upper part of the model, as a function of the height $h = r - r_{\text{phot}}$ above the photosphere, is illustrated on Figure 1. $\mathscr{V}$ increases with decreasing temperature in the convection zone until close to its upper boundary. Here the rapid change in the temperature gradient associated with the strong super-adiabaticity causes a dip in $\mathscr{V}$, followed by a gradual increase until near the temperature minimum, where $\mathscr{V}$ has a maximum of height $\mathscr{V}_{\text{max}} = \nu_c^2$, $\nu_c \approx 5.9$ mHz corresponding roughly to Lamb's (1909) acoustical cut-off frequency at the temperature minimum. Further out $\mathscr{V}$ decreases again due to the chromospheric temperature rise until the base of the transition region where the rapid temperature increase causes a rapid rise in $\mathscr{V}$. The sharp peak in $\mathscr{V}$ at zero height is probably an artifact of the matching between the atmosphere and the interior.

There is a possibility of modes with $\nu < \nu_c$ predominantly trapped in the interior of the model; these are the subphotospheric acoustic modes responsible for the 5 min oscillations. In addition to this subphotospheric cavity, however, there is also a cavity between the temperature minimum and the transition region; as first shown by Ando and Osaki (1977) and Ulrich and Rhodes (1977) in the non-radial, and R. Scuflaire (private communication) in the radial case, it is possible to trap a single chromospheric mode in this cavity. Modes with $\nu > \nu_c$ are essentially free to propagate in the entire atmosphere, and so they might be expected to experience stronger damping than modes

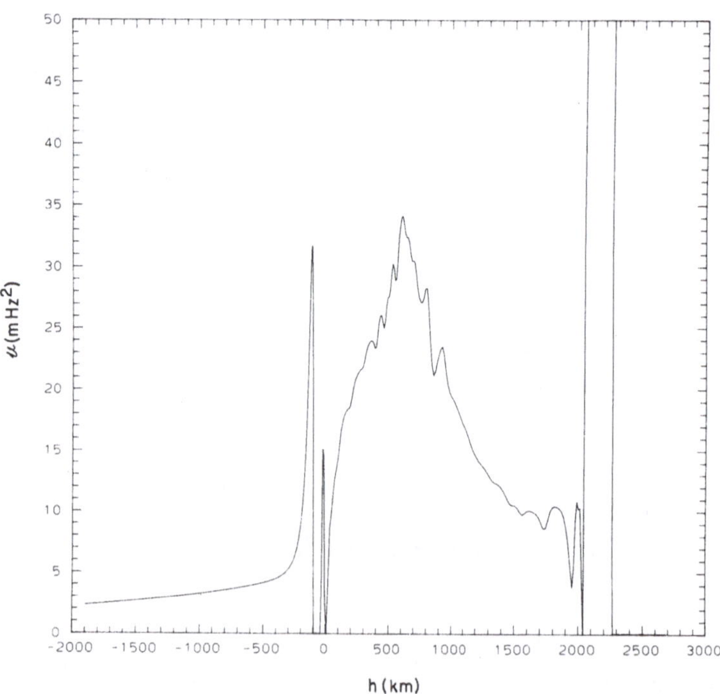

Fig. 1. The effective 'potential' $\mathscr{V}$ for radial adiabatic oscillations (cf. Equations (4.1) and (4.3)), as a function of the height h above the point where $T = T_{\text{eff}}$.

TABLE I

Results for selected eigenmodes of the complete model

(I)		(II)		(III)		(IV)		(V)	
n	ν	ν	η	ν	η	ν	η	ν	η
3	0.586508	0.586508	-5.75×10^{-9}	0.586508	-5.79×10^{-9}	0.586508	-5.69×10^{-9}	0.586508	1.80×10^{-9}
7	1.211290	1.211289	-8.45×10^{-7}	1.211289	-8.43×10^{-7}	1.211289	-8.35×10^{-7}	1.211289	1.56×10^{-7}
11	1.812284	1.812263	-3.08×10^{-5}	1.812264	-3.05×10^{-5}	1.812263	-3.04×10^{-5}	1.812261	3.06×10^{-6}
15	2.399463	2.399172	-2.87×10^{-4}	2.399169	-2.82×10^{-4}	2.399158	-2.83×10^{-4}	2.399126	2.72×10^{-6}
19	2.987472	2.986223	-9.06×10^{-4}	2.986160	-8.91×10^{-4}	2.986128	-8.96×10^{-4}	2.985939	-8.41×10^{-5}
23	3.583508	3.580353	-1.73×10^{-3}	3.580003	-1.70×10^{-3}	3.579969	-1.71×10^{-3}	3.579587	-3.36×10^{-4}
28	4.187176	4.181323	-2.73×10^{-3}	4.180675	-2.80×10^{-3}	4.180649	-2.82×10^{-3}	4.179942	-1.27×10^{-3}
32	4.793326	4.782402	-4.34×10^{-3}	4.780914	-4.55×10^{-3}	4.781012	-4.57×10^{-3}	4.782005	-3.14×10^{-3}
37	5.422156	5.405070	-7.02×10^{-3}	5.405052	-6.89×10^{-3}	5.405001	-6.89×10^{-3}	5.415151	-3.13×10^{-3}
41	6.010853	5.977390	-5.83×10^{-3}	5.973994	-5.50×10^{-3}	5.974366	-5.52×10^{-3}	5.984504	-6.25×10^{-3}
45	6.564577	6.550810	-3.49×10^{-3}	6.550168	-3.41×10^{-3}	6.552798	-3.69×10^{-3}	6.541162	-3.85×10^{-3}
49	7.124885	7.116820	-7.31×10^{-3}	7.113316	-8.19×10^{-3}	7.116736	-8.01×10^{-3}	7.114371	-5.18×10^{-3}
C	3.802524	3.772163	-4.27×10^{-2}	3.774059	-4.19×10^{-2}	3.774702	-4.18×10^{-2}	3.800963	-2.15×10^{-2}

(I) is for the adiabatic case, **n** being the number of modes in the eigenfunction and ν the cyclic frequency. (II) was obtained using consistently iterated Eddington factors; here η is the relative growth rate (a negative value indicating that the mode is damped). (III) used the equilibrium Eddington factors. (IV) assumed the Eddington approximation, as did (V) where in addition the terms in Δ_c were neglected (corresponding to the calculations of Ando and Osaki, 1975, 1977). The line marked '*C*' gives results for the chromospheric mode. To display the fairly small differences between the different cases the results are presented with greater precision than warranted by their absolute accuracy.

with $\nu < \nu_c$; but there is still an appreciable degree of reflection in the transition region (see also Bahng and Schwarzschild, 1963).

Some results on ν and the relative damping (or growth) rate η for selected modes of the complete envelope are shown in Table I; here $\eta = \omega_i/\omega_r$ where $\omega_r = \mathrm{Re}(\omega)$ and $\omega_i = \mathrm{Im}(\omega)$. The modes were computed by imposing the boundary condition (2.30). For comparison we have included adiabatic results obtained by solving just Equations (A.1) and (A.2) and relating $\delta\rho/\rho$ and $\delta p/p$ adiabatically; these are shown in the first column of the table. The second set of values were obtained with consistently iterated Eddington factors. The remaining results show the effects of various approximations; the third set was obtained by replacing f_{osc} and g_{osc} by f_{eq} and g_{eq}, the equilibrium Eddington factors, the fourth by using the standard Eddington approximation, i.e. $f_{\mathrm{osc}} = \frac{1}{3}$, $g_{\mathrm{osc}} = \frac{1}{2}$, and the final set used the Eddington approximation and in addition assumed $J = B$ as did Ando and Osaki (1975, 1977). In addition to the internal acoustic modes, which have been labelled by the number of nodes in the adiabatic case, results are also shown for the chromospheric mode.

It is evident from the table that the treatment of radiation has a very small effect on the cyclic frequencies of oscillation, at most about 0.05%; thus until the observational accuracy has been improved and the uncertainty caused by other approximations in the theory is reduced the Eddington approximation is certainly adequate for the computation of eigenfrequencies. On the other hand the differences caused by assuming adiabatic oscillations, up to about 0.5%, are quite significant. The effects on the growth rates are considerably larger, up to a few per cent, but here the effects of other theoretical uncertainties are correspondingly larger, and so once again the Eddington approximation is probably adequate.

The Ando and Osaki approximation has a strong effect on the damping rates. In contrast to when this approximation is not made modes with frequencies less than about 2.4 mHz are found to be unstable, and for the remaining modes the damping rate is generally considerably smaller. There is also a significant effect on ν. We return to this question in more detail at the end of the section.

The behaviour of the consistent Eddington factors f_{osc} is of some interest. Figure 2 shows the difference $f_{\mathrm{osc}} - f_{\mathrm{eq}}$ for the modes of Table I. Had our treatment of the convective flux perturbation been consistent f_{osc} should have tended to f_{eq} as ω tended to zero; although this is not exactly true, $f_{\mathrm{osc}} - f_{\mathrm{eq}}$ is nevertheless small at low frequencies. With increasing ν f_{osc} departs increasingly from f_{eq}, although the departure is fairly small for frequencies less than 4.5 mHz, where most of the power in the 5 min oscillations is concentrated. The rapid variation in f_{osc} at $\nu = 6.55$ mHz is caused by δJ nearly vanishing here. In fact points where $\delta J = 0$ and $\delta K \neq 0$ are clearly singularities of f_{osc}, and so here the variable Eddington factor method formally breaks down; in practice, however, this has not caused convergence problems. More serious are zeros in δK at points where $\delta J \neq 0$; here $f_{\mathrm{osc}} = 0$, giving rise to singularities in the moment equations as they have been formulated here. In the present calculation such behaviour caused convergence problems in narrow frequency ranges around $\nu = 6.25$ and 7.6 mHz.

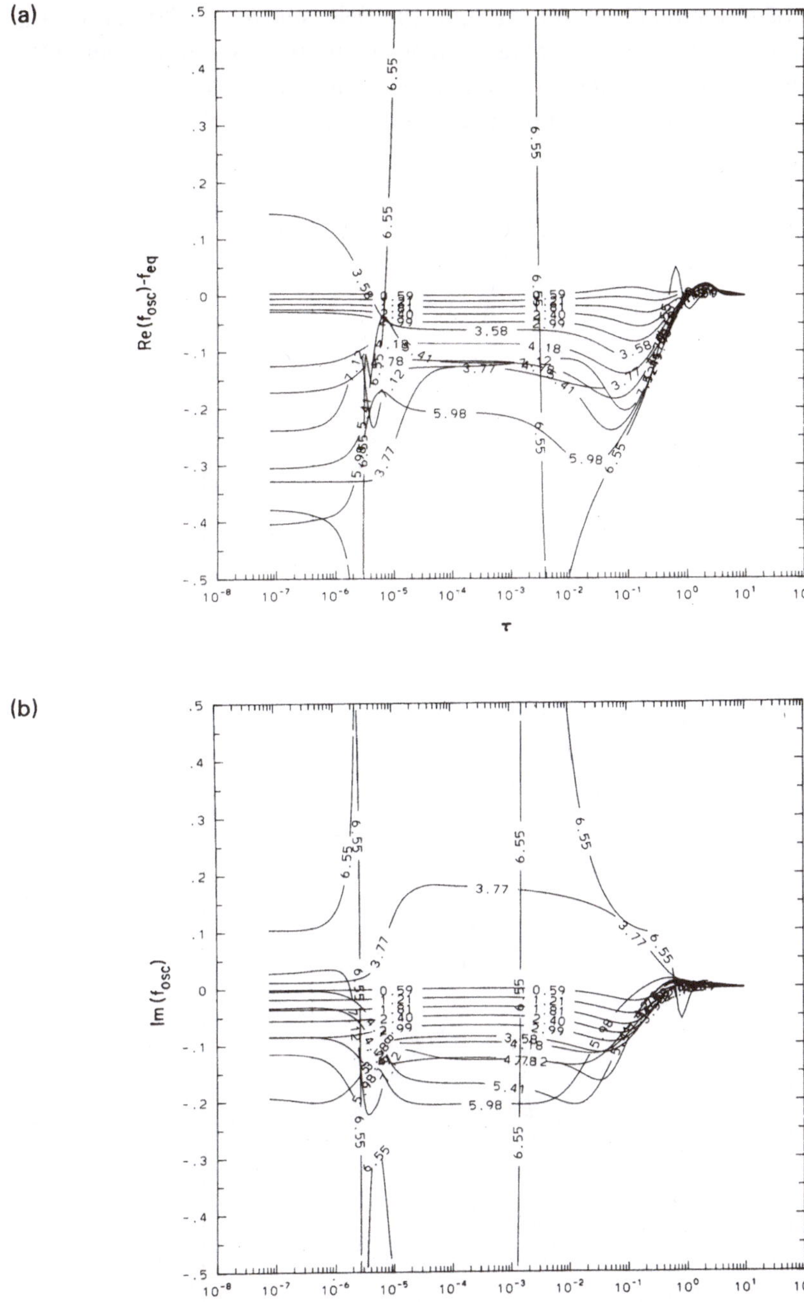

Fig. 2. Real part of $f_{osc} - f_{eq}$ (a) and imaginary part of f_{osc} (b), where f_{osc} is the consistently iterated Eddington factor for the oscillation and f_{eq} the equilibrium Eddington factor, as functions of the optical depth τ, for the modes of Table I. The curves are labelled with the cyclic frequencies ν of the modes, in mHz. For $\nu = 6.55$ mHz f_{osc} is almost singular at $\tau = 2 \times 10^{-3}$ and at $\tau = 3 \times 10^{-6}$, and $\mathrm{Re}(f_{osc}) - f_{eq}$ has a plateau at a value of about 1 between $\tau \approx 10^{-5}$ and $\tau \approx 5 \times 10^{-4}$.

The integration of Equation (2.19) for the radiation field and the evaluation of f_{osc} occupied less than 10% of the total computation time. Thus consistent treatment of the radiation field only requires a modest increase in the numerical work.

Some effects on the oscillation eigenfunctions of the treatment of the radiation field

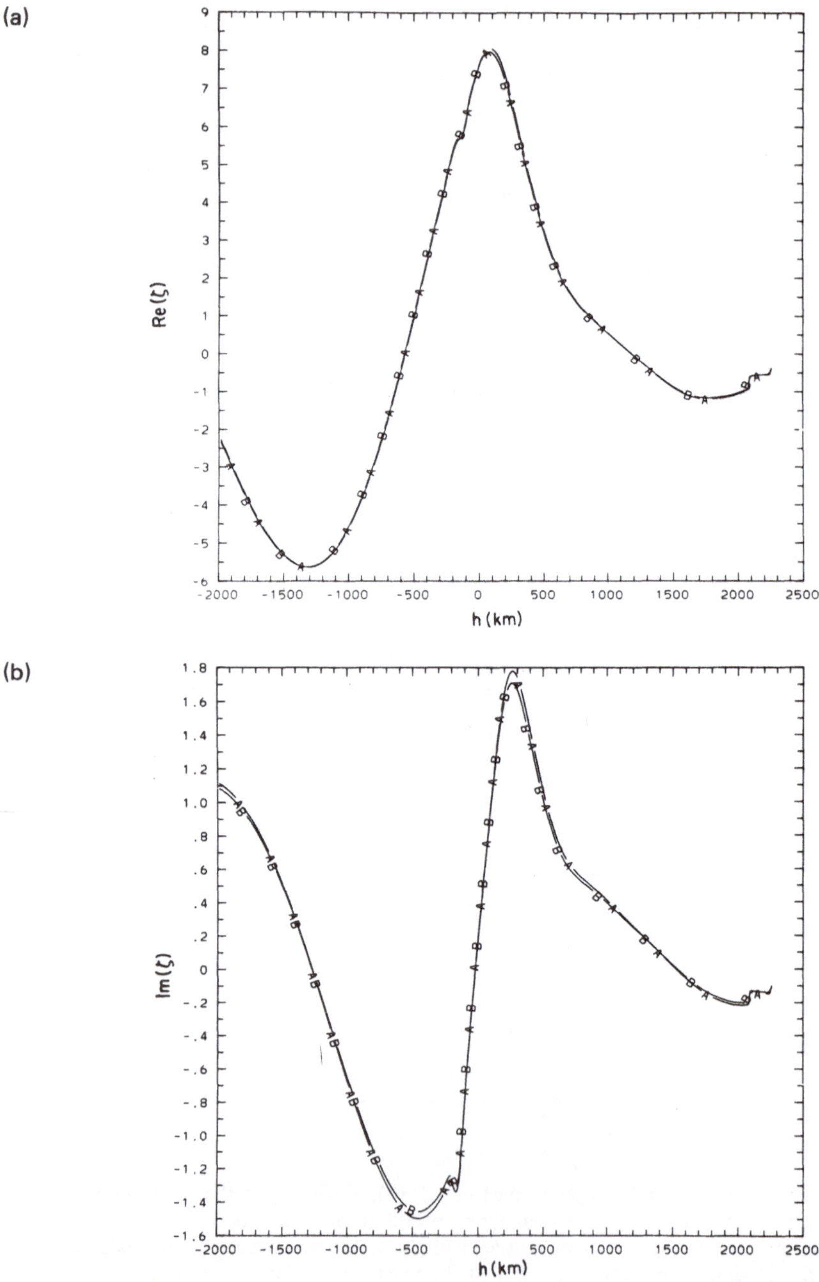

Fig. 3a–b.

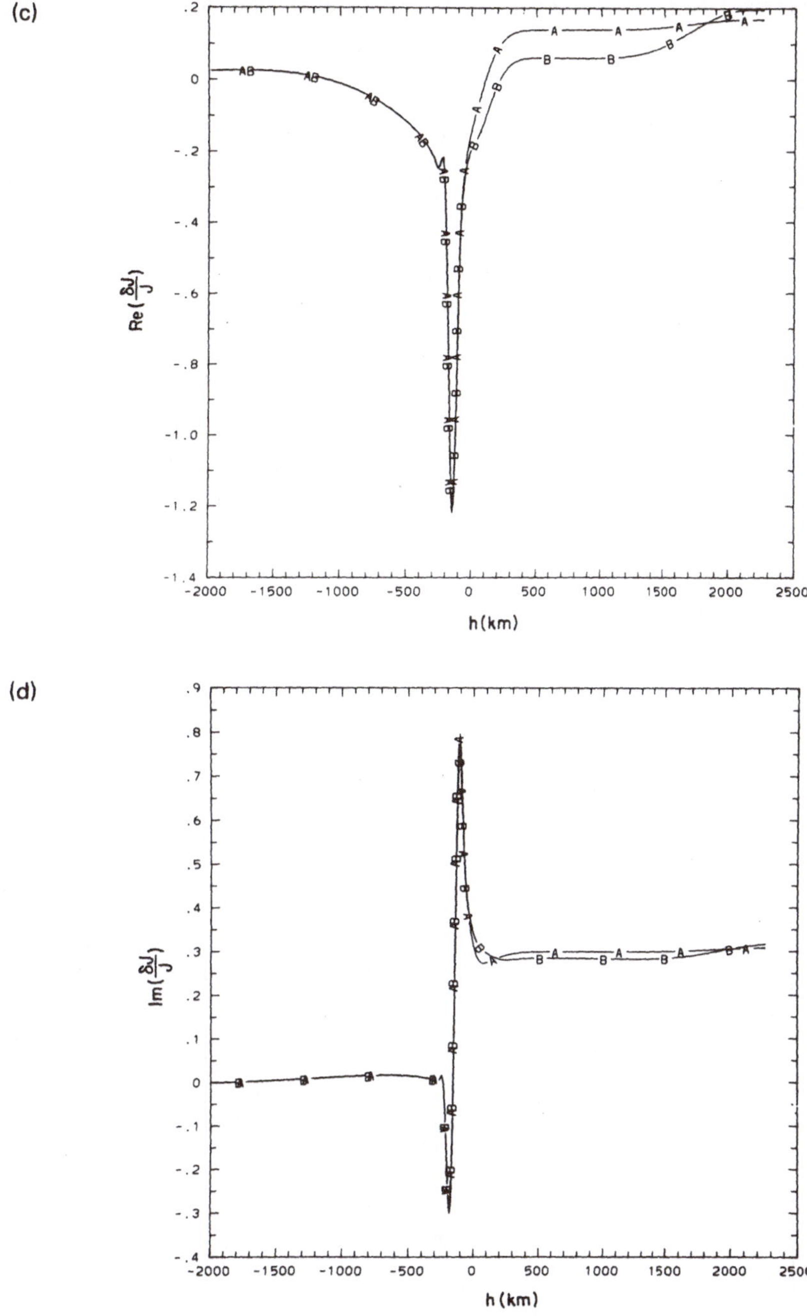

Fig. 3c–d.

Fig. 3a–d. Eigenfunctions in the Eddington approximation (A–A–A–) and with consistently iterated Eddington factors (B–B–B–) for the mode with frequency 4.18 mHz. Figures (a) and (b) show real and imaginary parts of the scaled displacement ζ, defined in Equation (4.4), and (c) and (d) real and imaginary parts of the relative perturbation $\delta J/J$ in the mean intensity.

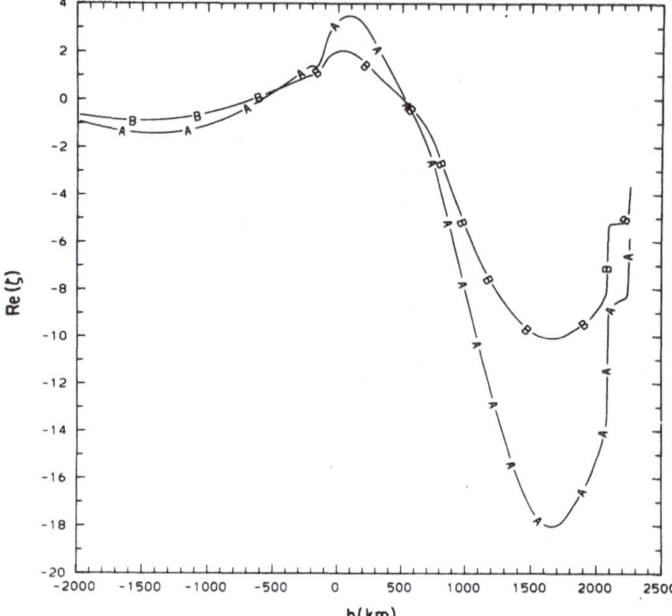

(b)

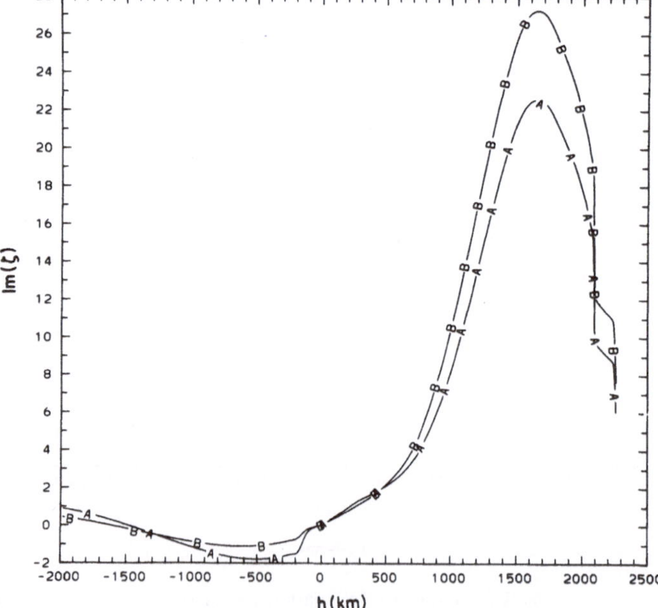

Fig. 4a–b.

(c)

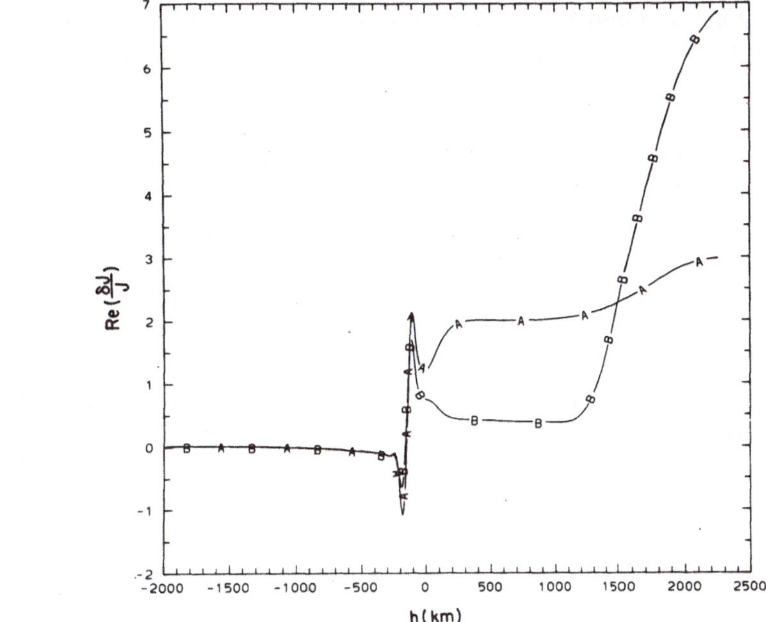

(d)

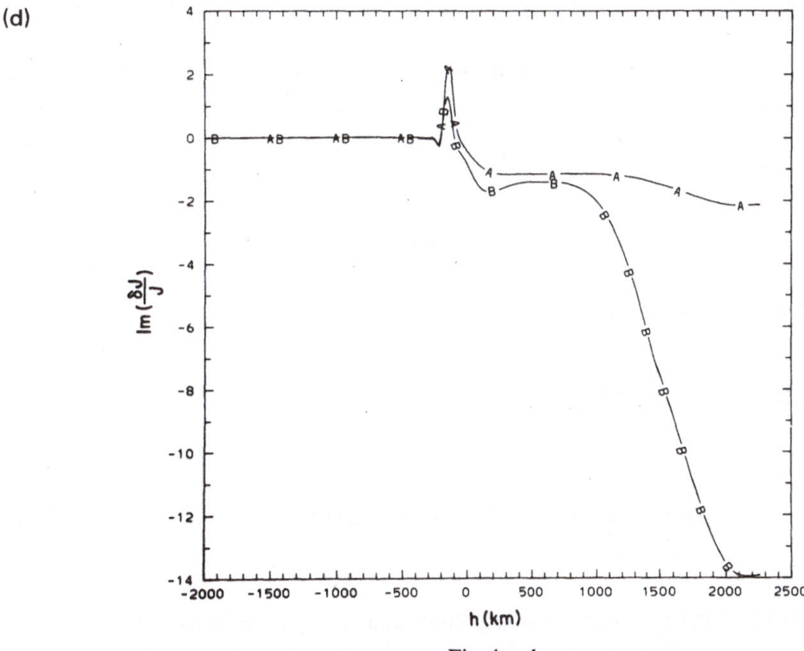

Fig. 4c–d.

Fig. 4a–d. Eigenfunctions for the chromospheric mode. See caption to Figure 3.

are presented on Figures 3 and 4, for the interior mode with frequency $\nu = 4.18$ mHz and the chromospheric mode, respectively. To indicate the distribution of the kinetic energy of pulsation we plot in each case

$$\zeta = \frac{\rho^{1/2}\delta r}{r_{\text{phot}}^{-3/2} E^{1/2}} , \tag{4.4}$$

where

$$E = \int_{r_b}^{r_s} r^2 \rho |\delta r|^2 \, dr \tag{4.5}$$

is related to the kinetic energy of pulsation. In addition we show $\delta J/J$. The solution was normalized such that $\delta r/r_e = 1$ at $r = r_{\text{phot}}$, where

$$r_e = \frac{\mathcal{R} T_{\text{eff}}}{g_s} \tag{4.6}$$

is related to the photospheric pressure scale height, $\mathcal{R}$ being the gas constant and g_s the surface gravity of the model.

For the interior mode shown on Figure 3 most of the pulsational energy is clearly associated with the inner cavity, the maximum in ζ being close to the turning point at about $h = 120$ km predicted on the basis of Figure 1. The effect of the treatment of radiation on ζ is so small as to be almost unnoticeable. A somewhat larger, but still fairly small, effect is seen for $\delta J/J$.

The chromospheric mode shown on Figure 4 has a pronounced maximum in ζ in the chromospheric cavity shown on Figure 1 and is evidently strongly reflected in the transition region. Not surprisingly, for this mode ζ is significantly affected by the treatment of the radiation field. Furthermore there is now a large difference in $\delta J/J$ between the two cases, which could conceivably have some observational effect. Significant differences are also found for the remaining perturbations.

Finally we may consider in more detail the results on the damping of the modes. It is easy to show from the oscillation equations that

$$2\omega_i \omega_r E = \operatorname{Im}(r^2 \delta r^* \delta p) \Big|_{r_b}^{r_s} - \\ - \operatorname{Re}\left\{ \frac{1}{\omega} \int_{r_b}^{r_s} (\Gamma_3 - 1) \frac{\delta \rho^*}{\rho} \, \delta(\operatorname{div} \mathcal{F}_R) r^2 \, dr \right\} , \tag{4.7}$$

where $\Gamma_3 - 1 = (\partial \ln T / \partial \ln \rho)_s$. Here the integrated term corresponds to the work done from the outside on the region considered or, equivalently, the acoustical energy flux across the boundaries, whereas the integral gives the contribution form the internal (positive or negative) dissipation. If the no-work condition (2.31) is assumed,

Equation (4.7) may be written as

$$\eta = \eta_s + W(r_s)\,, \tag{4.8}$$

where

$$\eta_s = \frac{\mathrm{Im}\,(r^2 \delta r^* \delta p)}{2\omega_r^2 E} \tag{4.9}$$

comes from the loss of acoustical energy at the outer boundary, and

$$W(r) = -\frac{\mathrm{Re}\left[\dfrac{1}{\omega}\displaystyle\int_{r_b}^{r} (\Gamma_3 - 1)\frac{\delta\rho^*}{\rho}\,\delta(\mathrm{div}\,\mathscr{F}_R) r'^2\, \mathrm{d}r'\right]}{2\omega_r^2 E} \tag{4.10}$$

is the normalized work integral.

If the bottom boundary condition (2.31) is used we may compute η as a continuous function of ν. The results are shown on Figure 5, both for the consistent treatment of the radiation field and using the Ando and Osaki approximation. For the exact case the figure also shows the ratio η_s/η, i.e. the relative contribution of the leakage of wave energy at the outer boundary to the total damping rate of the oscillation. As was also found by Ando and Osaki (1977) this contribution is never dominant for interior modes; for the chromospheric mode $\eta_s/\eta \approx 0.32$, and so here wave leakage accounts for about $\frac{1}{3}$ of the total damping.

The very large peak in η at around 5.2 mHz corresponds to a pronounced minimum in the normalized pulsational energy,

$$\mathscr{E}_{\mathrm{osc}} = \frac{E}{M \delta r(r_s)^2} \tag{4.11}$$

(M being the total mass of the Sun), so that here the energy of the mode is predominantly in the atmosphere; this also causes the peak in η_s/η. These frequencies are close to or above ν_c, and so there is no longer strict trapping in the chromosphere; but the shape of the eigenfunctions indicates that there may still be some resonance with the first overtone of the chromospheric cavity, and this is probably the reason for the enhanced damping. Similarly the small peak in η_s/η at 3.8 mHz might be caused by resonance with the fundamental chromospheric mode, and there is some indication that the peak in η at $\nu = 7.2$ mHz is associated with resonance with the second chromospheric overtone.

The apparent abrupt cut-off of power in the Birmingham whole-disk spectra at about 6 mHz presented by Isaak (1983) might be related to the resonance close to this frequency. There is clearly no sudden increase in damping at an 'acoustical cut-off frequency', and from the shape of the 'potential' in Figure 1 non would be expected; but

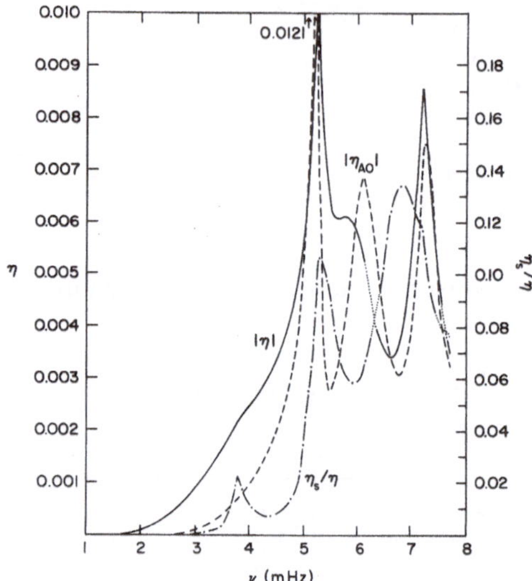

Fig. 5. Relative damping rates for oscillations extending over the entire envelope, as functions of the cyclic frequency ν. The curve labelled η (———) was obtained with consistent treatment of the radiation field, whereas for the curve labelled η_{AO} (– – –) we used the Eddington approximation and assumed $J = B$ in the equilibrium model (as did Ando and Osaki, 1975, 1977). These two curves use the left-hand scale. Finally the curve labelled η_s/η (–·–·–), using the right-hand scale, shows the relative contribution to the total damping rate from the leakage of wave energy into the corona. The dotted parts of the curves correspond to frequency regions where the near-vanishing of δK at points in the atmosphere caused convergence problems in the Eddington factor iteration.

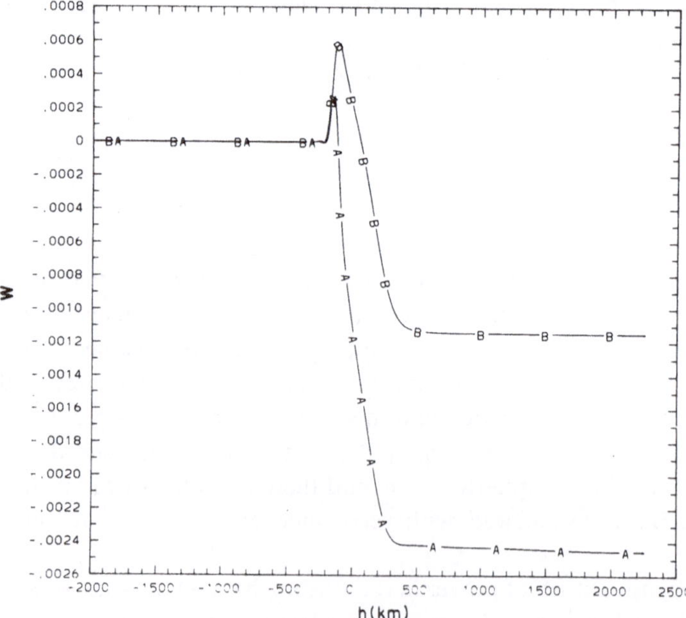

Fig. 6. The normalized work integral W (cf. Equation (4.10)) for the mode of frequency 4.18 mHz, with consistent treatment of the radiation field (A–A–A–) and in the Ando and Osaki approximation (B–B–B–).

it is still true, as pointed out by Isaak, that the position of the power cut-off might eventually be used as a diagnostic for the structure of the solar atmosphere, although not quite in the simple form envisaged by Isaak.

The damping rates found using the Ando and Osaki approximation are generally considerably smaller than those obtained when the approximation is not made. To illustrate this Figure 6 shows the behaviour of the work integral in the two cases. In both there is some excitation (which Ando and Osaki attributes to the κ mechanism) close to the outer edge of the convection zone. However this is weaker in the exact case, and here the damping in the lower atmosphere is also much stronger. The difference is solely an effect of assuming $\Delta_c = 0$; results obtained in the Eddington approximation but keeping the terms in Δ_c, are virtually indistinguishable form the exact case.

In the region of large damping $|\Delta_c|$ is quite small, less than about 0.05, and so it might at first seem surprising that it has such a significant effect on the stability of the oscillation. While there appears to be no obvious physical reason why this is so, it should be noticed that in Equation (2.35) Δ_c occurs in the combination $\omega_R \Delta_c$, which is to be compared with ω; in the region considered ω_R is much larger than ω (see also Figure 6 of Ando and Osaki, 1975), and so the terms in Δ_c may in fact dominate. The physical significance of this effect may, however, be questionable; it is certainly linked to our neglect of the perturbation in the convective and mechanical cooling rates (cf. Equation (2.8)), and so it should perhaps be regarded as a measure of the uncertainties introduced by this assumption.

5. Radiative Transfer in Non-Radial Oscillations

The general equations of non-radial, non-adiabatic oscillation are given by e.g. Ando and Osaki (1975) (see also Christensen-Dalsgaard, 1981), and will not be presented here; instead we concentrate on the part of the problem involving radiative transfer. To facilitate comparison with earlier work the spherical terms in the moment equations are kept; however the treatment of the radiation field is essentially plane-parallel.

The angular and temporal variation of the perturbations may be separated as, e.g.,

$$\delta p(r, \vartheta, \phi, t) = \mathrm{Re}[\delta p(r) Y_l^m(\vartheta, \phi) e^{-i\omega t}], \tag{5.1}$$

where r, ϑ, and ϕ are spherical polar coordinates; here Y_l^m is a spherical harmonic, and for simplicity we have used the same symbol for the perturbation and its amplitude. The displacement vector, no longer purely radial, may be written as

$$\delta \mathbf{r} = \mathrm{Re}\left\{\left[\xi_r(r) Y_l^m(\vartheta, \phi)\mathbf{a}_r + \right.\right.$$
$$\left.\left. + \xi_h(r)\left(\frac{\partial Y_l^m}{\partial\vartheta}\mathbf{a}_\vartheta + \frac{1}{\sin\vartheta}\frac{\partial Y_l^m}{\partial\phi}\mathbf{a}_\phi\right)\right]\exp(-i\omega t)\right\}, \tag{5.2}$$

where $\mathbf{a}_\vartheta$ and $\mathbf{a}_\phi$ are unit vectors in the ϑ and ϕ directions. In addition to the 'global' coordinates (r, ϑ, ϕ) it is useful to introduce a local Cartesian system of coordinates

(x, y, z) in a given point $(r_0, \vartheta_0, \phi_0)$; here $z = r - r_0$ and the x-axis may be chosen to be in the direction of the local horizontal component of the wave vector, so that, e.g.,

$$\delta p(r, \vartheta, \phi, t) = \mathrm{Re}\,\{\delta\tilde{p}(z) \exp[i(kx - \omega t)]\}\ , \tag{5.3}$$

where the local horizontal wave number k is related to l by

$$k = r_0^{-1}[l(l+1)]^{1/2}\ . \tag{5.4}$$

For radial oscillations the equations for the perturbation in the radiative intensity, Equations (2.19) or (A.4), were relatively simple, because of the rotational symmetry around the radial direction. For non-radial oscillations this symmetry is clearly destroyed, and we must consider the general equation of transfer

$$\mathbf{n}\cdot\nabla I(\mathbf{r}, \mathbf{n}) = \rho\kappa_s J + \rho\kappa_a B - \rho(\kappa_s + \kappa_a)I \equiv \rho A\ , \tag{5.5}$$

where $I(\mathbf{r}, \mathbf{n})$ is the specific intensity in the direction of the unit vector $\mathbf{n}$. For simplicity we have dropped the subscript ν. In the spherical case the components of $\mathbf{n}$, relative to a local coordinate system, changes with $\mathbf{r}$, and so $\mathbf{n}\cdot\nabla I$ involves derivatives of I with respect to $\mathbf{n}$ as well as $\mathbf{r}$. However the solar atmosphere is thin compared with the solar radius, and so the terms involving $\mathbf{n}$-derivatives may be expected to be small. If (x, y, z) is the local Cartesian system introduced above, and (n_x, n_y, n_z) are the components of $\mathbf{n}$ in this system, we shall neglect the derivatives of I with respect to n_x and n_y; but we keep the derivative with respect to n_z, as this gives rise to terms also found when the 3-dimensional Eddington approximation (Unno and Spiegel, 1966) is used. Then the left-hand side of Equation (5.5) may be written

$$\mathbf{n}\cdot\nabla I = n_z\frac{\partial I}{\partial z} + n_x\frac{\partial I}{\partial x} + n_y\frac{\partial I}{\partial y} + \frac{1}{r}(1 - n_z^2)\frac{\partial I}{\partial n_z}\ . \tag{5.6}$$

Assuming that all perturbations, including the Lagrangian perturbation δI of I, are of the form given in Equation (5.3), it is straightforward to find the perturbation of Equation (5.5):

$$n_z\frac{\partial\delta\tilde{I}}{\partial z} + ikn_x\delta\tilde{I} + \frac{1 - n_z^2}{r}\frac{\partial\delta\tilde{I}}{\partial n_z} =$$

$$= \delta(\widetilde{\rho A}) + \rho A\frac{\mathrm{d}\tilde{\xi}_r}{\mathrm{d}z} - (1 - n_z^2)\frac{\partial}{\partial z}\left(\frac{\tilde{\xi}_r}{r}\frac{\partial I}{\partial\mu}\right) + ikn_x\frac{\partial I}{\partial z}\tilde{\xi}_r\ . \tag{5.7}$$

We now replace $\partial/\partial z$ by $\partial/\partial r$, drop the tildes over the perturbed quantities and use the equation of continuity

$$\frac{1}{r^2}\frac{\mathrm{d}}{\mathrm{d}r}(r^2\xi_r) - \frac{l(l+1)}{r}\xi_h = -\frac{\delta\rho}{\rho} \tag{5.8}$$

to eliminate $d\xi_r/dr$. The perturbed equation of transfer is then finally

$$n_z \frac{\partial \delta I}{\partial r} + ikn_x \delta I + \frac{1-n_z^2}{r} \frac{\partial \delta I}{\partial n_z} = \rho \delta A + \rho A \left[\frac{l(l+1)}{r} \xi_h - \frac{2}{r} \xi_r \right] +$$

$$+ \frac{1}{r}(1-n_z^2) \frac{\partial I}{\partial n_z} \left[\frac{\delta \rho}{\rho} + \frac{3}{r} \xi_r - \frac{l(l+1)}{r} \xi_h \right] + ikn_x \frac{\partial I}{\partial r} \xi_r , \qquad (5.9)$$

where

$$\delta A = \delta \kappa_s J + \kappa_s \delta J + \delta \kappa_a B + \kappa_a \delta B - (\delta \kappa_s + \delta \kappa_a) I - (\kappa_s + \kappa_a) \delta I. \qquad (5.10)$$

It should be noticed that the derivation of the perturbed energy equation, Equation (2.15), did not require that the oscillations be radial. Thus this equation, as well as Equation (2.18), are valid also for non-radial oscillations, provided the assumptions (2.8) are still made.

It is convenient again to introduce moments of the radiation field. The mean intensity is, as in Equation (2.20),

$$\delta J(r) = \frac{1}{4\pi} \oint \delta I(r, \mathbf{n}) \, d\Omega . \qquad (5.11)$$

The first moment is now a vector with components (in the local coordinate system)

$$\delta H_i(r) = \frac{1}{4\pi} \oint n_i \delta I(r, \mathbf{n}) \, d\Omega , \qquad (5.12)$$

and the second moment is a tensor with components

$$\delta K_{ij}(r) = \frac{1}{4\pi} \oint n_i n_j \delta I(r, \mathbf{n}) \, d\Omega ; \qquad (5.13)$$

here the indices i and j take the values x, y, and z. From Equation (5.3) it is obvious that

$$\delta H_y = 0 , \qquad \delta K_{xy} = \delta K_{zy} = 0 ; \qquad (5.14)$$

in addition we clearly have

$$\delta K_{ij} = \delta K_{ji} \quad \text{for all} \quad i, j ,$$

$$\delta K_{xx} + \delta K_{yy} + \delta K_{zz} = \delta J . \qquad (5.15)$$

The three-dimensional Eddington approximation of Unno and Spiegel (1966) corresponds to assuming that δK_{ij} is isotropic, i.e.

$$\delta K_{ij} = 0 \quad \text{for} \quad i \neq j ,$$

$$\delta K_{xx} = \delta K_{yy} = \delta K_{zz} = \tfrac{1}{3} \delta J . \qquad (5.16)$$

We define the generalized Eddington factors by

$$f_{\mathrm{osc}} = \delta K_{zz}/\delta J\,, \tag{5.17}$$

$$\phi_{xx} = \delta K_{xx}/\delta K_{zz}\,, \tag{5.18}$$

and

$$\phi_{xz} = \delta K_{xz}/(ikr\delta K_{zz})\,. \tag{5.19}$$

The factor ikr was included in the definition of ϕ_{xz} because this quantity, as shown below, tends to an approximately real, non-zero constant in the limit of radial oscillations. It is evident that in the Eddington approximation $f_{\mathrm{osc}} = \frac{1}{3}$, $\phi_{xx} = 1$, and $\phi_{xz} = 0$.

By taking moments of Equation (5.9), using Equation (5.10), we now obtain

$$\begin{aligned}\frac{\mathrm{d}\delta K_{zz}}{\mathrm{d}r} = {} & -\rho\kappa\delta H_z + \rho\kappa\left[\frac{2}{r}\xi_r - \frac{l(l+1)}{r}\xi_h - \frac{\delta\kappa}{\kappa}\right]H + \\ & \underline{+\left[\frac{l(l+1)}{r}\phi_{xz} - \frac{1}{r}\left(3 - \frac{1}{f_{\mathrm{osc}}}\right)\right]\delta K_{zz} +} \\ & \underline{+\frac{J}{r}(3f_{\mathrm{eq}} - 1)\left[\frac{3}{r}\xi_r - \frac{l(l+1)}{r}\xi_h + \frac{\delta\rho}{\rho}\right]}\,,\end{aligned} \tag{5.20}$$

and

$$\begin{aligned}\frac{\mathrm{d}\delta H_z}{\mathrm{d}r} = {} & -\frac{2}{r}\delta H_z + \kappa_a\rho(J - B)\left[\frac{2}{r}\xi_r - \frac{l(l+1)}{r}\xi_h - \frac{\delta\kappa_a}{\kappa_a}\right] + \\ & + \frac{H}{r}\left[2\frac{\delta\rho}{\rho} + 6\frac{\xi_r}{r} - \frac{l(l+1)}{r}(2\xi_h + \xi_r)\right] + \rho\kappa_a(\delta B - \delta J) - \\ & - \frac{l(l+1)}{r^2\kappa\rho}\underline{\phi_{xx}}\delta K_{zz} - \\ & \underline{-\frac{l(l+1)}{r^2\kappa\rho}\left[r\phi_{xz}\frac{\mathrm{d}\delta K_{zz}}{\mathrm{d}r} + \left(r\frac{\mathrm{d}\phi_{xz}}{\mathrm{d}r} + 4\phi_{xz}\right)\delta K_{zz} +\right.} \\ & \underline{\left.+\left\{\frac{J}{r}(3f_{\mathrm{eq}} - 1) + \frac{1}{2}\frac{\mathrm{d}}{\mathrm{d}r}[J(3f_{\mathrm{eq}} - 1)]\right\}\xi_r\right]}\,,\end{aligned} \tag{5.21}$$

where Equation (5.20) may be substituted for $\mathrm{d}\delta K_{zz}/\mathrm{d}r$. When the underlined factors or terms are omitted, Equations (5.20) and (5.21) reduce to those obtained in the Eddington approximation (e.g. Christensen-Dalsgaard, 1981).

If the Eddington factors were known, Equations (5.20) and (5.21), together with the energy equation and the mechanical equations, would form a closed system. Thus, exactly as in the radial case, one may solve the problem by iterating for the Eddington factors. Given the solution for a set of trial Eddington factors Equation (5.9) may be integrated for δI on a suitable grid of directions $\mathbf{n}_j$, and new values of the Eddington factors obtained from Equations (5.11)–(5.13) and (5.17)–(5.19). This procedure is then iterated simultaneously with the iteration for the eigenfrequency.

The integration for the radiation field is now, however, considerably more complicated than in the radial case. To make it more efficient we generalize a technique first used by Logan and Hill (1980) to find the perturbation in the mean intensity. Neglecting the spherical term in $\partial\delta I/\partial n_z$, Equation (5.9) may be written as

$$n_z \frac{\partial \delta I}{\partial r} + ikn_x \delta I = -\rho\kappa(\delta I - \delta S), \tag{5.22}$$

which defines δS. This equation has the solution

$$\delta I(r, \mathbf{n}) = \int_{\tau}^{\mathscr{T}(n_s)} \exp\{[\tau - \tau' + in_x k(r - r')]/n_z\}\, \delta S(\tau', \mathbf{n})\, \mathrm{d}\tau'/n_z, \tag{5.23}$$

$$\mathscr{T}(n_z) = \begin{cases} \tau_s & \text{for} \quad n_z < 0, \\ \infty & \text{for} \quad n_z > 0, \end{cases} \tag{5.24}$$

where τ is the optical depth defined by

$$\frac{\mathrm{d}\tau}{\mathrm{d}r} = -\kappa\rho, \quad \tau \to 0 \quad \text{as} \quad r \to \infty, \tag{5.25}$$

$\tau' = \tau(r')$, τ_s is the value of τ at the surface of the model, and we have explicitly indicated that δS depends on $\mathbf{n}$; in fact

$$\delta S = \delta S_0 + \delta S_1 I + \frac{ikn_z}{\kappa\rho} \frac{\partial I}{\partial r} \xi_r, \tag{5.26}$$

where δS_0 and δS_1 are then independent of $\mathbf{n}$, whereas I is of course a function of n_z.

In spherical coordinates the vector n has components

$$n_z = \cos\hat{\vartheta}, \qquad n_x = \sin\hat{\vartheta}\cos\hat{\phi}, \qquad n_y = \sin\hat{\vartheta}\sin\hat{\phi}, \tag{5.27}$$

where $\hat{\vartheta}$ is the angle between the vertical and $\mathbf{n}$, and ϕ the angle between the projection of $\mathbf{n}$ on the horizontal plane and the horizontal wave vector. We now introduce moments of δI with respect to ϕ as

$$\overline{\delta I}_j(r, \mu) = \frac{1}{2\pi} \int_0^{2\pi} \cos^j \phi\, \delta I(r, \mathbf{n})\, \mathrm{d}\hat{\phi}, \tag{5.28}$$

where, in accordance with the notation in Section 2, we have written μ for n_z. From Equations (5.23) and (5.26) now follows that

$$\overline{\delta I}_j(r, \mu) = \int_{\tau}^{\mathcal{T}(\mu)} e^{(\tau - \tau')/\mu} \left[(\delta S_0 + \delta S_1 I) \mathcal{J}_j(x) + \right.$$
$$\left. + \frac{ik}{\kappa\rho} \frac{\partial I}{\partial r} \xi_r \sqrt{1 - \mu^2} \mathcal{J}_{j+1}(x) \right] d\tau/\mu , \tag{5.29}$$

where

$$x = k(r - r') \tan \hat{\vartheta} , \tag{5.30}$$

and

$$\mathcal{J}_j = \frac{1}{2\pi} \int_0^{2\pi} \cos^j \phi \, e^{ix \cos \phi} \, d\phi . \tag{5.31}$$

The $\mathcal{J}_j$ may clearly be expressed in terms of Bessel functions. For j less than 4 (which is all we shall need) the result is

$$\left.\begin{aligned} \mathcal{J}_0(x) &= J_0(x) , \\ \mathcal{J}_1(x) &= iJ_1(x) , \\ \mathcal{J}_2(x) &= \tfrac{1}{2}[J_0(x) - J_2(x)] , \\ \mathcal{J}_3(x) &= \frac{i}{4}[3J_1(x) - J_3(x)] . \end{aligned}\right\} \tag{5.32}$$

Finally the moments of δI with respect to $\mathbf{n}$ may be obtained as

$$\left.\begin{aligned} \delta J(r) &= \tfrac{1}{2} \int_{-1}^{1} \overline{\delta I}_0(r, \mu) \, d\mu , \\ \delta H_z(r) &= \tfrac{1}{2} \int_{-1}^{1} \mu \overline{\delta I}_0(r, \mu) \, d\mu , \\ \delta H_x(r) &= \tfrac{1}{2} \int_{-1}^{1} \sqrt{1 - \mu^2} \, \overline{\delta I}_1(r, \mu) \, d\mu , \\ \delta K_{zz}(r) &= \tfrac{1}{2} \int_{-1}^{1} \mu^2 \overline{\delta I}_0(r, \mu) \, d\mu , \\ \delta K_{xz}(r) &= \tfrac{1}{2} \int_{-1}^{1} \mu \sqrt{1 - \mu^2} \, \overline{\delta I}_1(r, \mu) \, d\mu , \\ \delta K_{xx}(r) &= \tfrac{1}{2} \int_{-1}^{1} (1 - \mu^2) \overline{\delta I}_2(r, \mu) \, d\mu . \end{aligned}\right\} \tag{5.33}$$

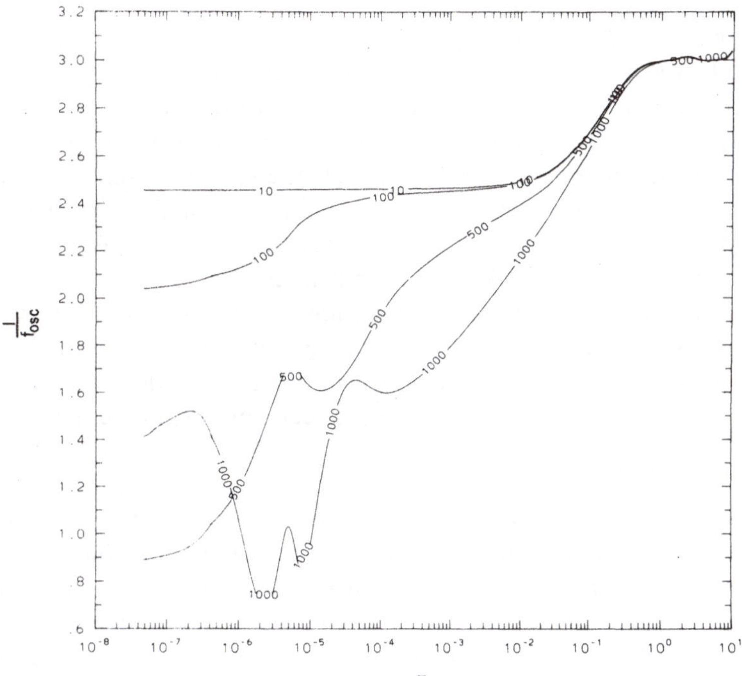

Fig. 7. The generalized Eddington factor $1/f_{osc} = \delta J/\delta K_{zz}$ calculated from the equilibrium source function, as a function of optical depth τ. The curves are labelled with the degree l.

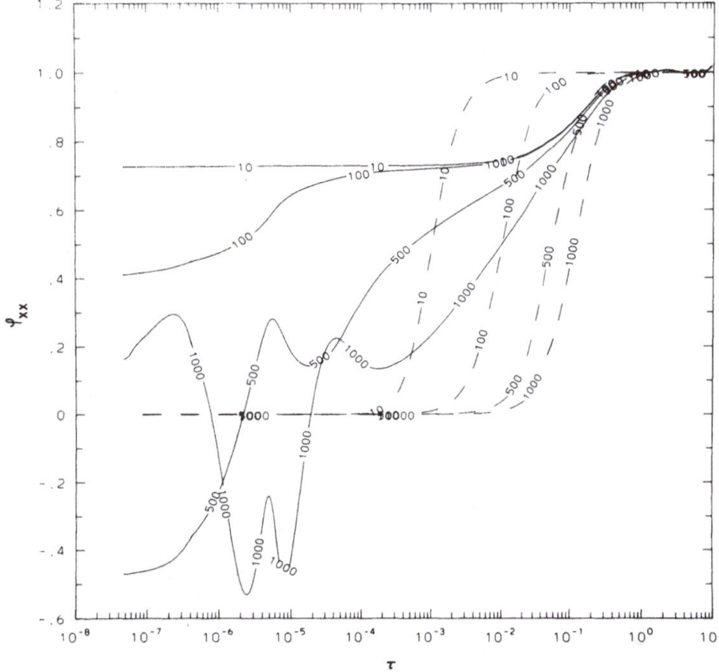

Fig. 8. The generalized Eddington factor $\phi_{xx} = \delta K_{xx}/\delta K_{zz}$. The dashed curves show the approximation suggested by Ando and Osaki (1977). See caption to Figure 7.

It should be noticed that $\mathcal{J}_1(x) \sim ix$ as $x \to 0$, and so δK_{xz} goes as ik as k tends to zero. This is why the factor ikr was taken out in the definition of ϕ_{xz} in Equation (5.19).

The calculation is now quite efficient. The integrals in Equation (5.29) are approximated by a product between vectors, containing components of the source function at discrete points in τ, and matrices derived from the kernels $\mathcal{J}_j$. To compute the matrices one must clearly evaluate a large number of Bessel functions and so this is fairly time consuming; however these matrices need only be evaluated once, for given k, and when they are known the calculation of the moments is quite fast.

We have yet to implement the iteration for the non-radial Eddington factors, and so we cannot comment on its convergence. However an initial estimate of the Eddington factors, which in any case should be useful for starting the iteration, can be obtained by replacing in Equation (5.29) $\delta S_0 + I\delta S_1$ by the equilibrium source function

$$S = \kappa^{-1}(\kappa_a B + \kappa_s J) \tag{5.34}$$

and neglecting the term in ξ_r. In the radial case, i.e. $k = 0$, the resulting f_{osc} is identical to f_{eq}, the equilibrium Eddington factor. Results for the model described in Section 3 are presented on Figure 7, 8, and 9 showing, respectively, $1/f_{\text{osc}}$, ϕ_{xx}, and ϕ_{xz} for four different values of l. The main effect of increasing l is seen to be a decrease of $1/f_{\text{osc}}$, ϕ_{xx}, and ϕ_{xz}. The reason is that for $l \neq 0$ oblique rays pass through regions of different phase, so that the contribution to the intensity is to some extent averaged out. Thus with

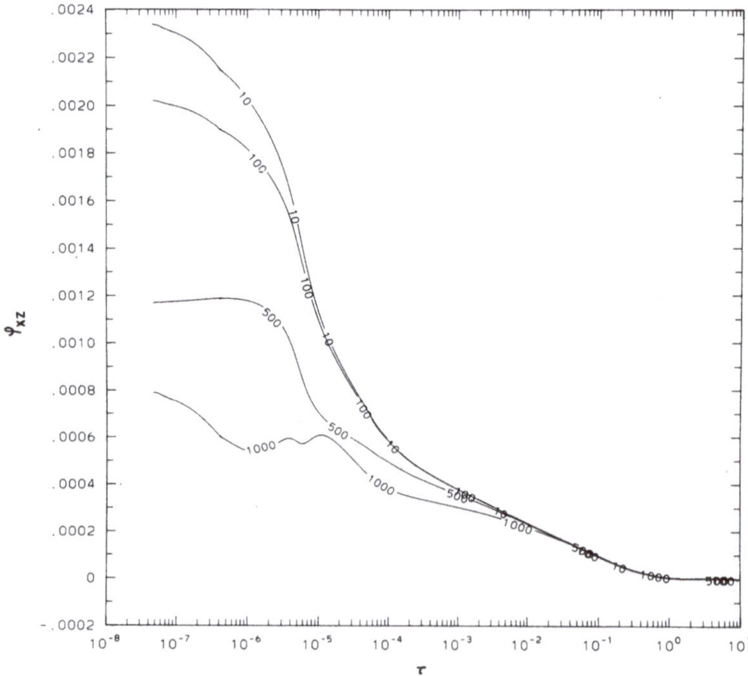

Fig. 9. The generalized Eddington factor $\phi_{xz} = \delta K_{xz}/(ikr\delta K_{xx})$. See caption to Figure 7.

increasing l δI becomes more strongly peaked in the radial direction, and this increases δK_{zz} relative to δJ, δK_{xx}, δK_{xz}. The reduction of the horizontal heat exchange relative to the Eddington approximation, corresponding to the reduction in ϕ_{xx}, was anticipated by Ando and Osaki (1977). They proposed a modification to the equations which corresponds, in our notation, essentially to approximating ϕ_{xx} by

$$\phi_{xx}^{(\mathrm{AO})} = \frac{1}{1 + l(l+1)/(\kappa\rho r)^2}\,; \tag{5.35}$$

these values are shown on Figure 7, 8, and 9 as dashed lines. Although the approximation leads to a decrease in ϕ_{xx} with decreasing τ or increasing l, the effect is clearly in general too large. In fact better approximations that are still computationally simple could presumably be derived from the ϕ_{xx} found here. However the full moment equations contain a number of additional terms, and detailed calculations, with a consistent iteration for the Eddington factors, will be needed to determine the effects of the departures from the Eddington approximation.

6. Discussion

The principal result of the present work is probably that for grey radiative transport in radial solar oscillations the Eddington approximation is generally adequate. In that sense the consistent formalism developed here is not required, at least in the grey case; however a detailed calculation was needed to demonstrate this.

The numerical results obtained in Section 4 only strictly apply to radial oscillations. For non-radial oscillations of low degree, however, the optical thickness across a horizontal wavelength is large, and the effect of the horizontal variations therefore small; thus here the results obtained in the radial case are probably still valid. The preliminary estimates for non-radial oscillations presented in Section 5 showed that when l is greater than about 100 the radiation field is increasingly affected by the horizontal variation, and so here the full non-radial case must be treated. We have so far not attempted to implement the iteration for the generalized, non-radial Eddington factors in the calculation; but it seems reasonable to hope that the convergence properties of this iteration will not be much worse than in the radial case.

Similar results were obtained by Kneer and Heasley (1979) for static perturbations in a simplified, grey atmosphere. As here the Eddington approximation was adequate for plane-parallel perturbations, but was found to be increasingly inaccurate with decreasing horizontal wavelength of the perturbation.

Perhaps the most serious among the approximations made here is that only radiation contributes to the Lagrangian perturbation in the local heating rate. We are now in the process of incorporating the perturbation in the convective flux, using the formalisms of Gough (1977) and Unno (1967). The mechanical heating in the atmosphere presents a greater challenge, as even the physical mechanism responsible is not definitively known. It might be possible to parameterize and then perturb the weak shock theory of, e.g., Ulmschneider (1971), and we intend to consider this; but the mechanical heating is likely to remain a serious difficulty in the theory of atmospheric solar oscillations. A related problem concerns the perturbation of the 'turbulent pressure'. As shown by Gough (1977) and Baker and Gough (1979), in the convection zone this can be treated within the framework of mixing length theory (although an equilibrium model which consistently incorporates turbulent pressure has yet to be constructed); but in the atmosphere we lack even a proper physical model for the turbulent pressure.

We have assumed that the model is spherically symmetric, and thus neglected the inhomogeneous nature of the solar atmosphere. This becomes an increasingly bad approximation with increasing height in the atmosphere. For the interior acoustic modes, which are trapped below the temperature minimum, the effects of inhomogeneities are probably not serious; in fact the observed sharp ridges in the $k - \omega$ diagram of, e.g., Deubner *et al.* (1979) and Rhodes *et al.* (1983) are evidence that for these modes the reflection in the atmosphere is not significantly affected by the inhomogeneities. These are almost certain to be important, however, in the chromosphere, and may invalidate the notion of a well defined chromospheric cavity as shown on Figure 1, and hence of a chromospheric mode. On the other hand it may be that modes whose horizontal wavelength is much longer than the scale of the inhomogeneities feel only the mean structure of the atmosphere, and that therefore our results may apply to such modes. More work on the behaviour of waves in an inhomogeneous medium is certainly needed to investigate this question.

In contrast to Ando and Osaki (1975, 1977) we found that all acoustic modes were stable. As shown in Section 4 this difference is largely caused by the fact that Ando and Osaki assumed $J = B$ in the equilibrium model. When making this approximation we also find instability of a number of modes, but in a smaller frequency range than Ando and Osaki. Although they only consider non-radial oscillations with $l \geq 10$ this is unlikely to be the cause of the difference; essentially all the damping and excitation of these modes take place very close to the surface where the vertical scale of the modes is small compared with the horizontal scale. Thus the difference is most likely caused by differences in the equilibrium model. Goldreich and Keeley (1977) also found stability of the acoustic modes of the Sun, although in their case an important contribution to the damping came from turbulent viscosity, treated in a highly simplified way (like Ando and Osaki they assumed $J = B$ and so they found instability when no turbulent viscosity was included). A similar conclusion was reached by Baker and Gough (cf. Gough, 1980), using Gough's (1977) treatment of the convective flux perturbation. Thus there now seems to be fairly strong evidence that acoustic modes are not self-excited in the Sun. However the calculated damping rates seem to be in conflict with the observed lifetimes of modes in the 5 min range of a few days (Grec *et al.*, 1980) or possibly up to a month (Claverie *et al.*, 1981; Woodard and Hudson, 1983). In fact the values of the relative damping rates η shown on Figure 4 correspond to a natural line width (full width at half maximum) of about 6 μHz at a frequency of 3 mHz, which is considerably larger than the observed line widths. Thus there appears to be problems in our understanding of the excitation and damping of these modes.

We intend to further extend the calculations. We have developed a programme to solve the non-grey equations in the atmosphere using the method devised by Rybicki (1971) which allows the inclusion of a fairly large number of frequency points; the atmospheric solution is then matched to a grey solution in the interior of the model. This procedure furthermore allows for the inclusion of at least a few spectral lines. Not only are these important from a diagnostic point of view but they may also have important dynamic effects; in fact radiative losses in the chromosphere occur predominantly in spectral lines (e.g. Giovanelli, 1978; Athay, 1981) and hence are probably severly underestimated in the present, grey calculation. We hope shortly to be able to report initial results of these calculations. By comparing the predictions of such increasingly detailed calculations with the large body of observational evidence that is becoming available it may become possible to determine how our theoretical treatment of the oscillations, and perhaps eventually the underlying atmospheric model, should be improved.

Acknowledgements

We are very grateful to A. Skumanich for supplying us with the atmospheric model and advising us on how to modify it. T. M. Brown, L. E. Cram, B. Durney, D. O. Gough, D. G. Hummer, J. W. Leibacher, B. W. Lites, B. W. Mihalas, D. Mihalas, B. Schmieder, R. Scuflaire, A. Skumanich, and J. Toomre are thanked for useful discussions, and we thank C. J. Durrant and J. D. Logan for reading and commenting on an earlier version of the manuscript. JC-D would like to thank Prof. P. Ledoux for hospitality at l'Institut d'Astrophysique, Liège.

Appendix: The Equations for Radial Oscillations in Spherical Geometry

The continuity equation is

$$\frac{1}{r^2}\frac{\mathrm{d}}{\mathrm{d}r}(r^2\delta r) = -\frac{\delta\rho}{\rho}\,. \tag{A.1}$$

Furthermore it is easy to show, by generalizing the analysis of e.g. Cox (1980) that the equation of motion is now

$$\frac{\mathrm{d}}{\mathrm{d}r}\left(\frac{\delta p}{p}\right) = \frac{\rho\tilde{g}}{p}\frac{\delta p}{p} + \frac{\rho}{p}\left[\omega^2 + \frac{2}{r}g(1+\lambda)\right]\delta r\,, \tag{A.2}$$

where $\lambda = \tilde{g}/g$ and we have assumed Equation (2.4).

The static equation of transfer in the spherical case is

$$\mu\frac{\partial I_\nu}{\partial r} + \frac{1}{r}(1-\mu^2)\frac{\partial I_\nu}{\partial\mu} = \rho[\kappa_{s,\nu}J_\nu + \kappa_{a,\nu}B_\nu - (\kappa_{s,\nu}+\kappa_{a,\nu})I_\nu] \tag{A.3}$$

(e.g. Hummer and Rybicki, 1971); under the same conditions as in Section 2 its Lagrangian perturbation is

$$\mu \frac{\partial \delta I_\nu}{\partial r} + \frac{1}{r}(1-\mu^2)\frac{\partial \delta I_\nu}{\partial \mu} + \rho\kappa_\nu \delta I_\nu =$$

$$= \rho[\kappa_{s,\nu}\delta J_\nu + \kappa_{a,\nu}\delta B_\nu + \delta\kappa_{s,\nu}B_\nu - (\delta\kappa_{s,\nu} + \delta\kappa_{a,\nu})I_\nu] -$$

$$- \frac{2}{r}\rho[\kappa_{s,\nu}J_\nu + \kappa_{a,\nu}B_\nu - (\kappa_{s,\nu} + \kappa_{a,\nu})I_\nu]\,\delta r +$$

$$+ \frac{1}{r}\left(\frac{\delta\rho}{\rho} + 3\frac{\delta r}{r}\right)(1-\mu^2)\frac{\partial I_\nu}{\partial \mu}\,, \tag{A.4}$$

where Equation (A.1) was used to eliminate the derivative of δr. By taking moments of Equation (A.4) we obtain

$$\frac{\mathrm{d}\delta H_\nu}{\mathrm{d}r} + \frac{2}{r}\delta H_\nu = \rho[\kappa_{a,\nu}(\delta B_\nu - \delta J_\nu) + \delta\kappa_{a,\nu}(B_\nu - J_\nu)] -$$

$$- \frac{2}{r}\rho\kappa_{a,\nu}(B_\nu - J_\nu)\,\delta r + \left(\frac{\delta\rho}{\rho} + 3\frac{\delta r}{r}\right)H_\nu\,, \tag{A.5}$$

and

$$\frac{\mathrm{d}\delta K_\nu}{\mathrm{d}r} + \frac{1}{r}(3\delta K_\nu - \delta J_\nu) = -\rho[(\kappa_\nu \delta H_\nu + \delta\kappa_\nu H_\nu] +$$

$$+ \frac{2}{r}\rho\kappa_\nu H_\nu \delta r + \frac{1}{r}\left(\frac{\delta\rho}{\rho} + 3\frac{\delta r}{r}\right)(3K_\nu - J_\nu)\,. \tag{A.6}$$

The energy equation, in the form of Equation (2.15) or (2.18), is unchanged.

As shown by Hummer and Rybicki (1971) in the static case, it is possible to generalize the variable Eddington factor method to spherical geometry; for the oscillations this could be done by integrating Equation (A.4), with the right-hand side found from a trial solution, and determining the Eddington factors from the moments of the resulting δI_ν. On the other hand the effort involved in solving the transfer equation is considerably greater in spherical geometry than in the plane-parallel case; furthermore the dominant spherical term in Equation (A.4) appears to be the last, especially at high oscillation frequencies where $\delta\rho/\rho$ and $\delta r/r$ increase rapidly with height in the atmosphere. We have approximated this term by finding $\partial I_\nu/\partial\mu$ from the plane-parallel equilibrium equation of transfer; from Equation (2.9) it follows that

$$\mu \frac{\partial}{\partial r}\left(\frac{\partial I_\nu}{\partial \mu}\right) = -\frac{\rho}{\mu}[\kappa_{s,\nu}J_\nu + \kappa_{a,\nu}B_\nu - (\kappa_{s,\nu} + \kappa_{a,\nu})I_\nu] - (\kappa_{s,\nu} + \kappa_{a,\nu})\frac{\partial I_\nu}{\partial \mu}\,, \tag{A.7}$$

and this equation can be solved by quadrature once I_ν has been determined. On the other hand we neglected the spherical term on the left-hand side of Equation (A.4). Although this procedure is not quite consistent, it seems to ensure, at least in the grey case, approximate agreement between the δH and δK calculated from δI, and the δH and δK resulting from solving the moment equations with the iterated Eddington factors; thus it may be adequate. We intend in future to make a consistent investigation of the effects of spherical geometry on the Eddington factors for the oscillations.

References

Ando, H. and Osaki, Y.: 1975, *Publ. Astron. Soc. Japan* **27**, 95.

Ando, H. and Osaki, Y.: 1977, *Publ. Astron. Soc. Japan* **29**, 581.

Athay, R. G.: 1981, in S. Jordan (ed.), *The Sun as a Star*, CNRS, NASA, p. 85.

Auer, L. H. and Mihalas, D.: 1970, *Monthly Notices Roy. Astron. Soc.* **149**, 65.

Auer, L. H., Heasley, G. N., and Milkey, R. W.: 1972, 'A Computational Program for the Solution of Non-LTE Transfer Problems by the Complete Linearization Method', KPNO Contr. No. 555.

Bahng, J. and Schwarzschild, M.: 1963, *Astrophys. J.* **137**, 901.

Baker, N. H. and Gough, D. O.: 1979, *Astrophys. J.* **234**, 232.

Baker, N. H. and Kippenhahn, R.: 1965, *Astrophys. J.* **142**, 868.

Baker, N. H. and Temesváry, S.: 1966, *Tables of Convective Stellar Envelopes*, Goddard Institute for Space Studies, New York.

Baker, N. H., Moore, D. W., and Spiegel, E. A.: 1971, *Ql. J. Mech. Appl. Math.* **24**, 391.

Brown, T. M. and Harrison, R. L.: 1980, in H. A. Hill and W. Dziembowski (eds.), *Lecture Notes in Physics* **125**, Springer, Heidelberg, p. 200.

Brown, T. M., Stebbins, R. T., and Hill, H. A.: 1978, *Astrophys. J.* **223**, 324.

Canfield, R. C.: 1976, *Solar Phys.* **50**, 239.

Christensen-Dalsgaard, J.: 1981, *Monthly Notices Roy. Astron. Soc.* **194**, 229.

Christensen-Dalsgaard, J.: 1982, *Monthly Notices Roy. Astron. Soc.* **199**, 735.

Christensen-Dalsgaard, J. and Gough, D. O.: 1980, *Nature* **288**, 544.

Christensen-Dalsgaard, J., Cooper, A. J., and Gough, D. O.: 1983, *Monthly Notices Roy. Astron. Soc.*, in press.

Claverie, A., Isaak, G. R., McLeod, C. P., van der Raay, H. B., and Roca Cortes, T.: 1981, *Nature* **293**, 443.

Cline, A. K.: 1974, *Comm. ACM* **17**, 218.

Cox, A. N. and Tabor, J. E.: 1976, *Astrophys. J. Suppl.* **31**, 271.

Cox, J. P.: 1980, *Theory of Stellar Pulsation*, Princeton University Press, Princeton, N. J.

Cram, L. E.: 1978, *Astron. Astrophys.* **70**, 345.

Cram, L. E., Keil, S. L., and Ulmschneider, P.: 1979, *Astrophys. J.* **234**, 768.

Davis Jr., C. G.: 1971, *J. Quant. Spectrosc. Rad. Transfer* **11**, 647.

Deubner, F.-L., Ulrich, R. K., and Rhodes Jr., E. J.: 1979, *Astron. Astrophys.* **72**, 177.

Dziembowski, W.: 1977, *Acta Astron.* **27**, 95.

Eggleton, P. P., Faulkner, J., and Flannery, B. P.: 1973, *Astron. Astrophys.* **23**, 325.

Fossat, E. and Ricort, G.: 1975, *Astron. Astrophys.* **43**, 253.

Giovanelli, R. G.: 1978, *Solar Phys.* **59**, 293.

Goldreich, P. and Keeley, D. A.: 1977, *Astrophys. J.* **211**, 934.

Gough, D. O.: 1977, *Astrophys. J.* **214**, 196.

Gough, D. O.: 1980, in H. A. Hill and W. Dziembowski (eds.), *Lecture Notes in Physics* **125**, Springer, Heidelberg, p. 273.

Gouttebroze, P. and Leibacher, J. W.: 1980, *Astrophys. J.* **238**, 1134.

Grec, G., Fossat, E., and Pomerantz, M.: 1980, *Nature* **288**, 541.

Hill, H. A., Rosenwald, R. D., and Caudell, T. P.: 1978, *Astrophys. J.* **225**, 304.

Hummer, D. G. and Rybicki, G. B.: 1971, *Monthly Notices Roy. Astron. Soc.* **152**, 1.

Isaak, G. R.: 1983, *Solar Phys.* **82**, 205 (this volume).

Kalkofen, W. and Ulmschneider, P.: 1977, *Astron. Astrophys.* **57**, 193.

Karp, A. H.: 1975, *Astrophys. J.* **199**, 461.
Keil, S. L.: 1980, *Astron. Astrophys.* **82**, 144.
Kneer, F. and Heasley, J. N.: 1979, *Astron. Astrophys.* **79**, 14.
Lamb, H.: 1909, *Proc. London Math. Soc.* **7**, 122.
Legait, A.: 1982a, *Astron. Astrophys.* **108**, 287.
Legait, A.: 1982b, submitted to *Astron. Astrophys.*
Levy, M.: 1974, *Astron. Astrophys.* **31**, 451.
Lites, B. W. and Chipman, E. G.: 1979, *Astrophys. J.* **231**, 570.
Logan, J. D. and Hill, H. A: 1980, *Space Sci. Rev.* **27**, 301.
Mein, P.: 1971, *Solar Phys.* **20**, 3.
Mihalas, B. W.: 1979, Thesis, University of Colorado.
Mihalas, B. W. and Toomre, J.: 1981, *Astrophys. J.* **249**, 349.
Mihalas, D.: 1978, *Stellar Atmospheres*, 2 Ed., Freeman, San Fransisco.
Mihalas, D.: 1980, *Astrophys. J.* **237**, 574.
Nordlund, A.: 1982, *Astron. Astrophys.* **107**, 1.
Noyes, R. W. and Leighton, R. B.: 1963, *Astrophys. J.* **138**, 631.
Rhodes Jr., E. J., Howard, R. F., Ulrich, R. K., and Smith, E. J.: 1983, *Solar Phys.* **82**, 245 (this volume).
Rybicki, G.: 1971, *J. Quant. Spectrosc. Rad. Transfer* **11**, 589.
Schmieder, B.: 1976, *Solar Phys.* **47**, 435.
Schmieder, B.: 1977, *Solar Phys.* **54**, 269.
Schmieder, B.: 1978, *Solar Phys.* **57**, 245.
Spiegel, E. A.: 1957, *Astrophys. J.* **126**, 202.
Stebbins, R. T., Hill, H. A., Zanoni, R., and Davis, R. E.: 1980, in H. A. Hill and W. Dziembowski (eds.), *Lecture Notes in Physics* **125**, Springer, Heidelberg, p. 381.
Ulmschneider, P.: 1971, *Astron. Astrophys.* **12**, 297.
Ulrich, R. K. and Rhodes, Jr., E. J.: 1977, *Astrophys. J.* **218**, 521.
Unno, W.: 1967, *Publ. Astron. Soc. Japan* **19**, 140.
Unno, W. and Spiegel, E. A.: 1966, *Publ. Astron. Soc. Japan* **18**, 85.
Vernazza, J. E., Avrett, E. H., and Loeser, R.: 1981, *Astrophys. J. Suppl.* **45**, 635.
Woodard, M. and Hudson, H.: 1983, *Solar Phys.* **82**, 67 (this volume).

A NEW METHOD FOR DETERMINING THE HELIUM ABUNDANCE IN THE SOLAR ATMOSPHERE*

(Invited Review)

G. R. ISAAK

Department of Physics, University of Birmingham, P.O. Box 363, Birmingham, B15 2TT, U.K.

Abstract. Recent observations of a cut-off frequency in the acoustic modes of the Sun (Claverie *et al.*, 1981b) should help determine the mean molecular weight and, thereby, the helium abundance in the visible layers of the solar atmosphere. A first preliminary result of $Y = 0.42 \pm 0.04$ is obtained for an assumed minimum photospheric temperature of 4400 K $\pm$ 200 K.

The determination of the helium abundance Y (by weight) in the Sun's interior would help elucidate the physics and astrophysics of the solar neutrino problem (Iben, 1969) and would presumably provide a firm upper bound to the cosmological helium abundance as no means of destroying helium in the Sun is known. The observation (Claverie *et al.*, 1979, 1980, 1981a; Grec *et al.*, 1980) of the structure in the global oscillations with periods near 5 min and their interpretation (Claverie *et al.*, 1979) in terms of two then current models (Iben and Mahaffy, 1976; Christensen-Dalsgaard *et al.*, 1979) suggested a low Z and low Y (Isaak, 1980). This has been questioned by Christensen-Dalsgaard and Gough (1980) but the facts on which they based their criticism are suspect. These authors adopted a mean spacing between neighbouring modes of the same kind of 136.0 μHz compared with the observed value of 135.2 $\pm$ 0.2 μHz obtained earlier and remeasured in 1980 (Claverie *et al.*, 1981a). The cut off frequency of 5.7 μHz observed by (Claverie *et al.*, 1981b) (see below) is substantially higher than the 5 mHz these authors assumed. Consequently the model frequencies near 3 mHz require corrections which are smaller than the authors claim.

It is, nevertheless, very likely that a clear picture will not emerge until lower order modes, particularly the fundamental radial mode and its overtones are observed (Brookes *et al.*, 1976) and unambiguously identified.

The determination of the helium abundance in the solar atmosphere would form a substitute for measurements of Y if one assumes that no segregation effects between the interior and the atmosphere occur. Even such measurements have been all but impossible (Unsöld, 1969).

The time-averaged composition of the solar wind (Hundhausen, 1972) is consistent with a $Y = 0.16$ but it is conceivable that selection effects exist in the solar wind and possibly even in the solar photosphere.

* Proceedings of the 66th IAU Colloquium: *Problems in Solar and Stellar Oscillations*, held at the Crimean Astrophysical Observatory, U.S.S.R., 1–5 September, 1981.

Solar Physics **82** (1983) 205–207. 0038–0938/83/0821–0205$00.45.

The aim of this note is to suggest a new method of determining the mean molecular weight of the upper photosphere and lower chromosphere, predominantly determined by hydrogen and helium abundances, and to report a first preliminary result using such a method.

It is well known (Bray and Loughhead, 1974) that an isothermal gravitationally stratified atmosphere has a Lamb acoustic cut off frequency $\nu_1 = \gamma g/4\pi v$ for vertically propagating waves, where γ is the ratio of specific heats, g is the acceleration due to gravity, and v is the local velocity of sound. The modifications due to radiative damping and the small magnetic field in the upper photosphere appear small (Bray and Loughhead, 1974) so long as one observes the cut-off frequency near the temperature minimum, thereby also approximating to the isothermal condition.

Recently it was discovered (Claverie *et al.*, 1981b) using the 7699 Å resonance line of neutral potassium that the acoustic modes centred on 3 mHz appear to have a cut-off frequency at $\nu_C = 5.7$ mHz.

According to de la Reza and Müller (1972) and Wiehr (1981), the part of the spectral line used in the velocity spectroscopy by the Birmingham group lies between about $\log \tau = -2$ and -1.6, τ being the optical depth. Below this frequency ν_C considerable power was seen in the mean velocity spectrum (averaged over 8 days) whereas above ν_C the power level is constant and corresponds to a velocity amplitude of 1.2 cm s^{-1} r.m.s. per channel.

This level appears to be due to finite photon statistics, instrumental noise and any residual noise due to the terrestrial atmosphere. This frequency ν_C was identified with the Lamb cut-off frequency ν_1.

The mean molecular weight m can be estimated from ν_C, g and the velocity of sound given by $v = (\gamma kT/m)^{1/2}$, k being Boltzmann's constant, and T the absolute temperature. γ and T can be estimated from a model of the visible layers of the solar atmosphere.

Although $\gamma = \frac{5}{3}$ to a good approximation (Ulmschneider, 1967), estimates of the temperature, near the temperature minimum, seem to range from 4600 K to 4200 K.

The resultant mean molecular weight ranges from 1.57 to 1.43, with corresponding Y values of 0.46 and 0.38 assuming $Z = 0.02$ and single ionization for C, N, O, etc. and no appreciable ionization for hydrogen and helium.

Thus for an adopted mean temperature of 4400 K $\pm$ 200 K the resultant helium abundance is found to be $Y = 0.42 \pm 0.04$.

It is interesting to point out that this is consistent with the result $Y = 0.38$ obtained by Unsöld (1969).

It is conceivable that two dimensional velocity spectroscopy of the Sun will achieve a sufficient resolution and stability in the near future to measure the velocity of sound propagation directly. An explosive event (flare, meteorite, comet) will produce a detectable shock wave the progress of which will be followed. Such a measurement might provide yet another method of determining the mean molecular weight in the visible layers of the Sun.

References

Bray, R. J. and Loughhead, R. E.: 1974, *The Solar Chromosphere*, Chapter 6, Chapman and Hall, London.
Brookes, J. R., Isaak, G. R., and van der Raay, H. B.: 1976, *Nature* **259**, 92.
Christensen-Dalsgaard, J., Gough, D. O., and Morgan, J. G.: 1979, *Astron. Astrophys.* **73**, 121.
Christensen-Dalsgaard, J. and Gough, D. O.: 1980, *Nature* **288**, 544.
Claverie, A., Isaak, G. R., McLeod, C. P., van der Raay, H. B., and Roca Cortes, T.: 1979, *Nature* **282**, 591.
Claverie, A., Isaak, G. R., McLeod, C. P., van der Raay, H. B., and Roca Cortes, T.: 1980, *Astron. Astrophys.* **91**, L9.
Claverie, A., Isaak, G. R., McLeod, C. P., van der Raay, H. B., and Roca Cortes, T.: 1981a, *Nature* **293**, 443.
Claverie, A., Isaak, G. R., McLeod, C. P., van der Raay, H. B., and Roca Cortes, T.: 1981b, *Solar Phys.* **74**, 51.
de la Reza, R. and Müller, E. A.: 1975, *Solar Phys.* **43**, 15.
Grec, G., Fossat, E., Pomerantz, M.: 1980, *Nature* **288**, 541.
Hundhausen, A. J.: 1972, *Coronal Expansion and the Solar Wind*, Springer, Berlin, p. 99.
Iben, I.: 1969, *Ann. Phys.* **54**, 164.
Iben, I. and Mahaffy, J.: 1976, *Astrophys. J. Letters* **209**, L39.
Isaak, G. R.: 1980, *Nature* **283**, 644.
Ulmschneider, P.: 1967, *Z. Astrophys.* **67**, 193.
Unsöld, A. O. J.: 1969, *Science* **163**, 1015.
Wiehr, E.: 1981, private communication, Göttingen.

160-min OSCILLATIONS OF THE SUN AS THE MEAN OF STUDY OF ITS INTERNAL STRUCTURE*

E. A. GAVRYUSEVA, YU. S. KOPYSOV, and G. T. ZATSEPIN

Institute for Nuclear Research of the Academy of Sciences of the USSR, 60th October Anniversary prospect, 7a, Moscow 117312, U.S.S.R.

Abstract. The investigation of the models of the contemporary Sun with a mixed core has shown that the amplitude of some gravity modes of oscillations of the star can be mainly concentrated in the central region. This phenomenon takes place if the node of the amplitude of radial displacement coincides with the boundary of the mixed core. In this case the core can be regarded as a driving generator of the oscillations, determining their period and phase. It is suggested as the explanation of the observational properties of the 160-min oscillation.

1. Introduction

In recent years there have become available new observational data in the solar physics. However such an important information about the Sun as the low high-energy neutrino flux (Davis, 1978) and the pulsation of the solar surface with 160-min period (Kotov *et al.*, 1976; Brookes *et al.*, 1976) did not find well grounded theoretical explanations. The radial oscillations can not have such long period. The nonradial oscillations with high spheroidal number l and low odd $l = 1, 3$ could not be registered in the experiment of Severny (Severny *et al.*, 1979). In the spectrum of quadrupole oscillations only moderately high order gravity g-modes correspond to 160-min pulsation. In this case according to usual theoretical considerations it is impossible to understand why the lower modes are not observed though they must be get excited first of all.

The wonderful property of 160-min pulsation consist in their possibility sometimes to disappear and then to appear again with the same phase as if they had not disappeared at all (Severny *et al.*, 1979). This property allowed to assume this oscillation to be concentrated in the inner regions of the Sun (Zatsepin *et al.*, 1980; Gavryuseva and Kopysov, 1981). In standard models of the Sun the amplitude of displacements is higher in the envelope than in the internal region. The standard models of the Sun have also difficulties in the interpretation of the other experimental data. For example the calculated flux of high energy solar neutrino is greater than observed by Davis *et al.* (1978). It is worth mentioning that the bulk experimental data about the Sun are connected to the contemporary state of the Sun. Therefore, there is sense to construct such models of the contemporary Sun which would be in accordance with the experimental results. In the models of the contemporary Sun with the mixed core the high energy neutrino flux is lower than in the standard models.

* Proceedings of the 66th IAU Colloquium: *Problems in Solar and Stellar Oscillations*, held at the Crimean Astrophysical Observatory, U.S.S.R., 1–5 September, 1981.

Solar Physics **82** (1983) 209–213. 0038–0938/83/0821–0209$00.75.

A great number of models of the contemporary Sun with various surface hydrogen abundances X_0 and with the different parts of mixed mass M_{mix} has been constructed (Gavryuseva, 1981). The core hydrogen abundance X_c for each model has been accepted constant throughout up to the core boundary and then increased up to X_0 as

$$X(M) = X_0 + (X_c - X_0) \exp \left\{ - \left(\frac{M - M_{\text{mix}}}{\Delta M} \right)^2 \right\} ,$$

where ΔM is the thickness of the transitional layer.

This article devotes to investigation of these models from helioseismological point of view.

2. Results

The small linear adiabatic oscillations have been studied originaly within the framework of multizone polytropic models (Zatsepin *et al.*, 1980; Gavryuseva and Kopysov, 1981) and later on models with different parts of the mixed mass M_{mix} for series of hydrogen concentrations in the envelope X_0 and thickness of the transitional layer ΔM.

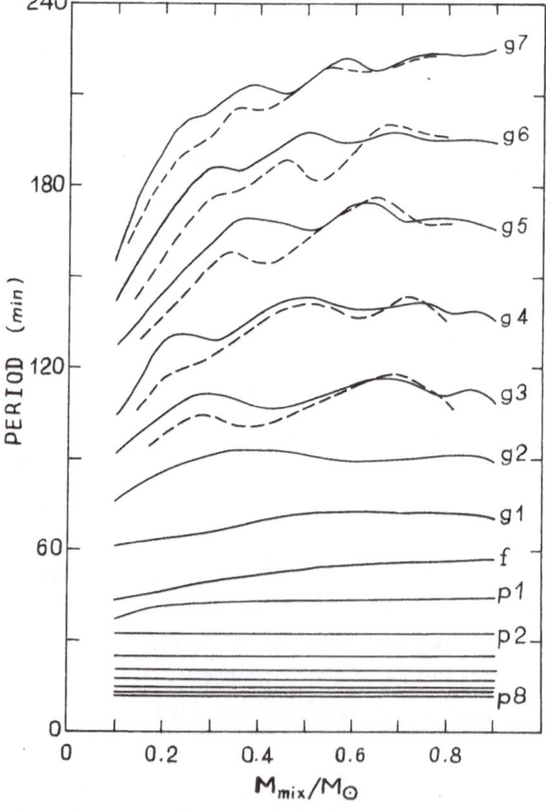

Fig. 1. The periods of quadrupole oscillations of the Sun as a function of the mixed core size for the envelope hydrogen abundance $X_0 = 0.7$ (continuous line) and $X_0 = 0.75$ (broken line).

The radial oscillations are practically not sensitive to variations of the parameters X_0 and M_{mix}. They actually occupy only the very upper layers of the Sun. In all models the amplitude of the radial displacement on the surface exceeds the maximum displacement in the central zones by 3 to 4 orders of magnitude. These oscillations have periods less or about one hour. Because of that they can not be used to explain the 160-min oscillation. The quadrupole oscillations are of two kinds. Acoustic modes have low periods. The behaviour of their eigenfunctions is similar to radial ones. The frequency and the form of eigenfunctions of long-periodical gravity modes are conditioned by structure of the Sun as a whole. Periods of *g*-modes of quadrupole oscillations, presented in Figure 1, are connected very closely with the values of M_{mix} parameter. The greater M_{mix} the longer the periods of gravity modes, especially of the high order ones. For M_{mix} smaller than 0.25 $M_\odot$ the 160-min period will belong to *g*6, *g*7 and so on by suitable selection of X_0. For values of M_{mix} in interval from 0.25 $M_\odot$ to 0.5 $M_\odot$ the hydrogen concentration in envelope can be selected so that the obtained model will have oscillations with 160-min periods which are identified as *g*5-mode. For $M_{mix} > 0.5\ M_\odot$ at any X_0 the *g*5-mode has a period some more than 160-min but the *g*4-mode has a

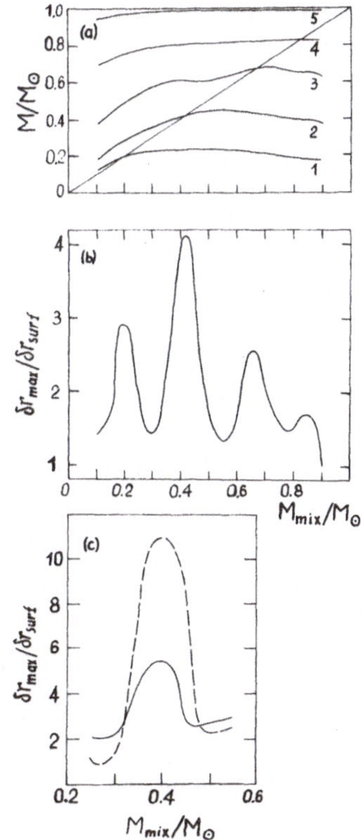

Fig. 2. (a) Position of the nodes of the radial displacement amplitude and (b) the ratio $\delta r_{max}/\delta r_{surf}$ for *g*5-mode of quadrupole oscillations of the Sun as a function of the mixed core size at $X_0 = 0.7$, $\Delta M = 0.05\ M_\odot$. (c) The same as (b) at $X_0 = 0.75$, $\Delta M = 0.05\ M_\odot$ (continuous line) and $\Delta M = 0.01\ M_\odot$ (broken line).

period less than that. The shape of eigenfunction of radial displacement of g-modes has proved to depend on the size of the mixed core. It is seen from Figure 2b, in which the ratio of the maximal amplitude of radial displacement inside the core δr_{max} for the $g5$-mode of quadrupole oscillations to that in the external layers δr_{surf} (δr_{surf} is usually maximal on the surface) is plotted as a function of M_{mix} for $X_0 = 0.7$. It is important to note that this ratio gets a maximum every time as some internal node finds itself at the boundary of the mixed core as seen from the comparison between Figures 2a and 2b. The behaviour of the positions of the nodes of the radial displacement amplitude by change M_{mix} is shown in Figure 2a. The more the difference in hydrogen concentration in core and in envelope and the thinner the transition layer ΔM the stronger this effect is expressed. Figure 2c, which shows the dependence of the ratio of δr_{max} to δr_{surf} of $g5$-mode on the size of the transition layer ΔM at $X_0 = 0.75$ and its comparison with Figure 2b confirm that.

This property of the amplitude of displacement allows to understand why only one of the high order g-modes of nonradial oscillations can be observed while the lower order modes can be absent. We suppose the losses of oscillation energy conditioned mainly by the viscous forces are most considerable in the convective envelope. These losses are proportional to the square of the amplitude displacement. If the core is a driving generator of the oscillations and their amplitude is great only there they will feebly damped in the external layer, because the dissipation of the oscillation energy will be weak thanks to small displacement. The more the amplitude in the internal regions of the Sun exceeds the amplitude on the surface the more probable is excitation of the oscillations in the core, which is due to thermonuclear driving proposed by Dilke and Gough (1972) (see also Christensen-Dalsgaard *et al.*, 1974).

On inspecting Figures 1 and 2b one can conclude that the mixed core could become a wave trap for the $g5$-mode of quadrupole oscillations only provided $M_{mix} \simeq 0.4\, M_\odot$. If one considers a model with $M_{mix} = 0.4\, M_\odot$ as a model of the contemporary Sun so, besides the $g5$-mode, the generation of the $g7$-mode with the period about 205 min, depending on X_0, is possible. $g6$, $g4$ and lower modes will be subject to strong dissipations in the external layer, because the amplitude of the radial displacement does not decrease but it increases to the surface.

In this model capture rate of the solar neutrino in a ^{37}Cl detector is changing from ≈ 3 SNU to 6.3 SNU for $S_{34}(0) = 0.29$–0.67 keV-barns.

The 160-min oscillation will be underlined also at $l = 2$ for $g7$-mode if mixed core occupies 0.1 solar mass, but the model of the Sun with $M_{mix} = 0.1\, M_\odot$ has very great high energy neutrino fluxes.

3. Conclusion

As has been shown above the solar interiors may become a selective trap for some gravity modes. This can take place if the Sun has a mixed core and the radial amplitude of the oscillation has the node at its boundary. In this case the oscillations of the Sun could be driven by nuclear reactions because of the radial amplitudes of the trapped mode are maximum inside the core. The explanation of properties of 160-min oscillation could come out in models of the contemporary Sun with the mixed mass equal to $M_{mix} \simeq 0.4\, M_{\odot}$. Their preferable exitation could be explained so that the oscillation mode with such a period being a high order g-mode of the whole Sun is one of the lowest modes of its core where driving of this oscillation occurs. Further observations would help to verify and specify our ideas about the internal structure of the Sun.

References

Brookes, J. R., Isaak, G. R., and van der Raay, H. B.: 1976, *Nature* **259**, 92.

Christensen-Dalsgaard, J., Dilke, F. W. W., and Gough, D. O.: 1974, *Monthly Notices Roy. Astron. Soc.* **169**, 429.

Davis, R., Jr.: 1978, *Proc. Inform. Conf. Brookhaven Nat. Lab.*, Upton, New York, Vol. 1, p. 1.

Dilke, F. W. W. and Gough, D. O.: 1972, *Nature* **240**, 262.

Gavryuseva, E. A.: 1981, *Soviet Physics – Lebedev Institute Report* **10**, 52.

Gavryuseva, E. A. and Kopysov, Yu. S.: 1981, *Astron. Zh.* **58**, 610.

Gavryuseva, E. A., Kopysov, Yu. S., and Zatsepin, G. T.: 1981, *Soviet Physics – Lebedev Institute Report* **10**, 46.

Kotov, V. A., Severny, A. B., and Tsap, T. T.: 1976, *Nature* **259**, 87.

Severny, A. B., Kotov, V. A., and Tsap, T. T.: 1979, *Astron. Zh.* **56**, 1137.

Zatsepin, G. T., Gavryuseva, E. A., and Kopysov, Yu. S.: 1980, *Rep. Acad. Sci. USSR* **251**, 1342.

ADIABATIC OSCILLATIONS OF SOLAR MODELS WITH A HIGH-Z CONVECTIVE CORE*

S. V. VORONTSOV and K. I. MARCHENKOV

Department of Theoretical Physics, Institute of Physics of the Earth, Academy of Sciences of the USSR, Moscow 123810, U.S.S.R.

Abstract. Normal mode spectra and neutrino counting rates are calculated for a set of chemically-inhomogeneous solar models. Each model has a core with a high concentration of heavy elements; high opacity makes the core convective. The structure of the envelope is that of the standard model. It is shown that (1) the spectrum of *g* modes becomes less densely separated than that of the standard model, which simplifies the problem of interpreting 160-min oscillations; (2) low neutrino counting rates may be achieved for a low initial helium concentration in the core; (3) the models do not contradict the frequency spacing of global 5-min oscillations.

1. Introduction

In connection with the solar neutrino problem, various non-standard models of the structure and evolution of the Sun have been proposed by different authors over the past decade. One of them is the solar model containing a core with an unusually high concentration of iron-group metals and a low initial concentration of helium, proposed by Hoyle (1975). The high opacity caused by these metals makes the core convective, and low neutrino emission arises, because of the high hydrogen concentration maintained at the center and because the convective mixing of ^{7}Be on a timescale less than ^{7}Be destruction time minimizes the importance of reaction $^7\mathrm{Be}\ (p, \gamma)^8\mathrm{B}$. The luminosity of such a model remains almost constant over the whole history of the Earth – the expectation is in better agreement with recent paleoclimatological data than is the standard model, which requires the solar constant to have been significantly lower during the Earth's early history.

In the present paper, a set of models of this type with different core masses is analysed. Theoretical normal mode spectra are calculated and compared with observations.

2. Models

The structure of the models beyond a convective core is taken to be that of the standard model. The standard model calculated by Abraham and Iben (1971) with $Z = 0.0149$, $Y = 0.253$, and the model of the outer convective zone constructed by Spruit (1974) with a depth of 198 000 km, are used. The part of the standard model below some given radius (R_{core}) is replaced by a adiabatic convective core of a different chemical composition.

* Proceedings of the 66th IAU Colloquium: *Problems in Solar and Stellar Oscillations*, held at the Crimean Astrophysical Observatory, U.S.S.R., 1–5 September, 1981.

Solar Physics **82** (1983) 215–222. 0038–0938/83/0821–0215$01.20.

The degree of ionization of heavy elements is assumed to be constant throughout the core. This assumption simplifies the calculations significantly because the density and pressure distributions in the core may be calculated using the Lane–Emden function $\theta(\xi)$ for polytrope index 1.5:

$$\rho = \rho_c \theta^{3/2} , \qquad p = p_c \theta^{5/2} , \tag{1}$$

$$\frac{10\pi r^4 p(r)}{GM^2(r)} = \frac{\theta^{5/2}}{(\mathrm{d}\theta/\mathrm{d}\xi)^2} , \qquad \frac{4\pi r^3 \rho(r)}{M(r)} = - \frac{\xi\theta^{3/2}}{\mathrm{d}\theta/\mathrm{d}\xi} . \tag{2}$$

The pressure and density distributions in the core are then determined by the conditions of pressure and mass continuity at the core boundary by using dimensionless expressions (2). The discontinuity in density at the core boundary was found to be dynamically stable for all the models; the value of $\rho(R_{\text{core}} - 0)/\rho(R_{\text{core}} + 0)$ is in the range from 1.06 for $M_{\text{core}} = 0.05\, M_\odot$ to 1.9 for $M_{\text{core}} = 0.9\, M_\odot$. At this stage the models are ready for the calculation of adiabatic normal mode spectra.

The condition of temperature continuity determines the mean molecular weight of the core material:

$$\bar{\mu}_{\text{core}} = \bar{\mu}_{\text{envelope}} \frac{\rho(R_{\text{core}} - 0)}{\rho(R_{\text{core}} + 0)} . \tag{3}$$

Three parameters of chemical composition X, Y, and Z of core material are then calculated from three conditions:

$$\begin{aligned} &X + Y + Z = 1 , \\ &2X + \tfrac{3}{4}Y + aZ = 1/\bar{\mu}_{\text{core}} , \\ &L = L_\odot . \end{aligned} \tag{4}$$

The degree of ionization of heavy elements is represented by parameter a which is in the range from $a = 0$ for zero ionization to $a = \frac{1}{2}$ for total ionization. The results, as will be shown, are only weakly dependent on the degree of ionization.

It was found that the condition of solar luminosity may be satisfied for a core mass up to $0.5\, M_\odot$ in the calculated series of models. After the values of X, Y, Z have been determined (there may be two solutions), the neutrino counting rates are computed.

The high mass-number elements in core material are assumed to be heavier than Mg, the usual C, N, O, and Ne being removed by α additions (Hoyle, 1975). Hence, there will be no ^{13}N or ^{15}O neutrinos from the core. Convective circulation time is short enough (Hoyle, 1975) to assume that all nuclear species (except deuterium) are distributed uniformly throughout the core. Nuclear reaction rates given by Fowler *et al.* (1974) were used in the computation of luminosity and neutrino fluxes.

3. Normal Mode Spectra

Adiabatic non-radial oscillations were calculated by numerical integrations of the usual fourth-order differential system, taking into account the perturbation of gravitational potential. The form of the equations and the method of calculation are the same as given in Vorontsov and Zharkov (1978).

Special attention was paid to the accurate computation of the radial eigenfunctions, because the modes associated with the discontinuity in density were expected to appear in the theoretical spectra. It was found, however, that these 'discontinuity' modes cannot be distinguished in the spectra for low l values because of their strong interaction with a number of g modes of close periods. The situation is illustrated by Figure 1, where radial displacement functions of quadrupole ($l = 2$) oscillations are shown for a model with $M_{\text{core}} = 0.4\, M_\odot$. The relative amplitudes at the core boundary versus mode number are shown in Figure 1b. This dependence has a maximum, but there is no unique 'discontinuity' mode. The situation for the models of a different core mass is the same. The 'discontinuity' modes become strongly pronounced in theoretical spectrum only for $l \approx 10$ and higher.

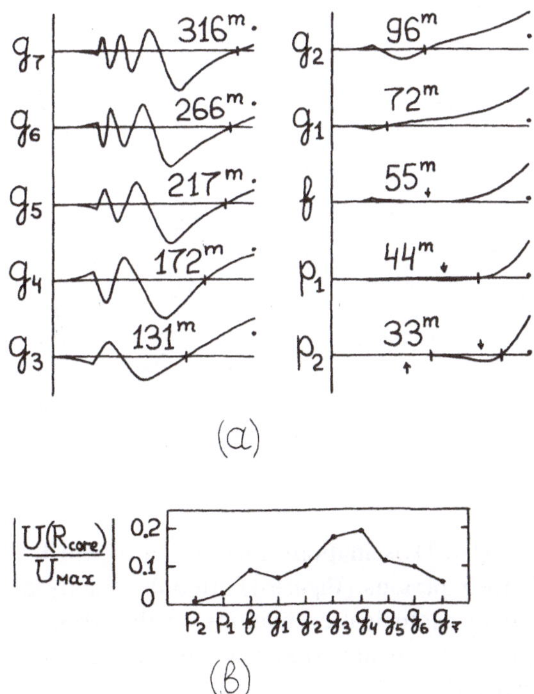

Fig. 1. The quadrupole ($l = 2$) oscillations for a model with $M_{\text{core}} = 0.4\, M_\odot$ ($R_{\text{core}} = 0.22\, R_\odot$). (a) The radial displacement functions $U(r)$ and periods in minutes. For each mode, the horizontal axis corresponds to the radius, from the center (left) to the surface. Relative amplitudes of tangential displacements at the surface are shown by dots. The extremes in eigenfunctions are shown by arrows when needed. (b) The relative amplitudes at the core boundary for different modes.

The periods of quadrupole oscillations for all the models are given in Figure 2. The limiting case $R_{\text{core}} = 0$ corresponds to the standard model. When the core radius increases, the spectrum of periods of g modes becomes more and more widely separated with respect to that of the standard model which, in general, simplifies the problem of the interpretation of 160-min solar oscillations. Possible identifications are shown in Figure 2 by arrows, but since solar luminosity may be achieved only for $M_{\text{core}} < 0.5\,M_\odot$, identification only with a g_3 or higher order g mode is allowed. The preferential excitation of a single g mode may be due to relatively higher amplitudes in the core (Figure 1b).

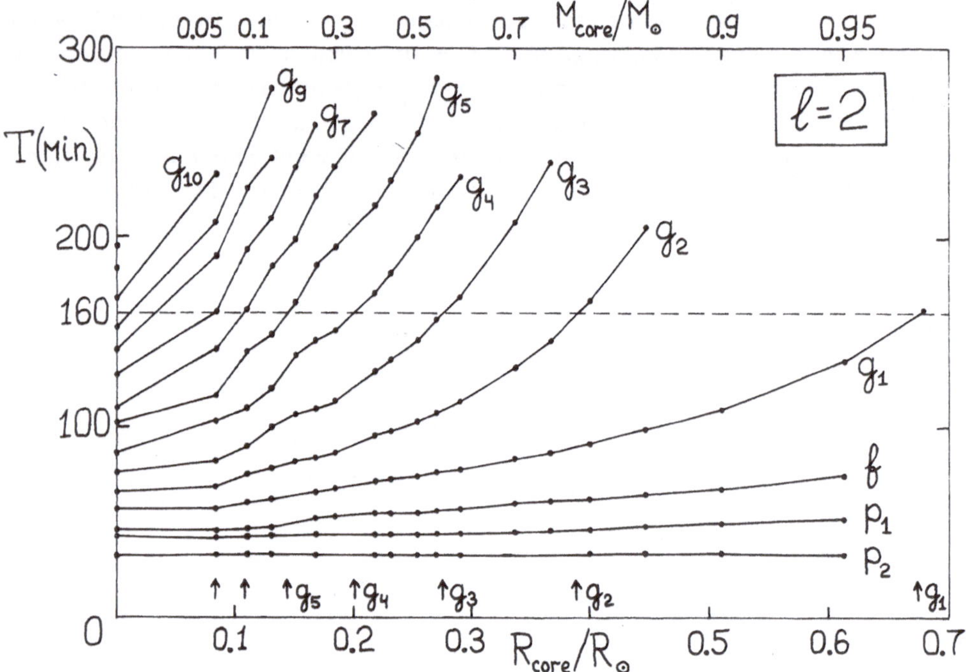

Fig. 2. Periods of quadrupole ($l = 2$) modes versus core radius. Possible identifications of the 160-min period are indicated.

The periods of dipole ($l = 1$) oscillations are given in Figure 3. The radial oscillations were also calculated; their periods (Figure 4) are only weakly-dependent on the core radius, like the periods of acoustical oscillations on the whole.

The high-order p mode frequency spacings were estimated using the asymptotic expression (Vandakurov, 1967)

$$\Delta\nu = \left[2\int_0^{R_\odot} \frac{\mathrm{d}r}{c}\right]^{-1}, \tag{5}$$

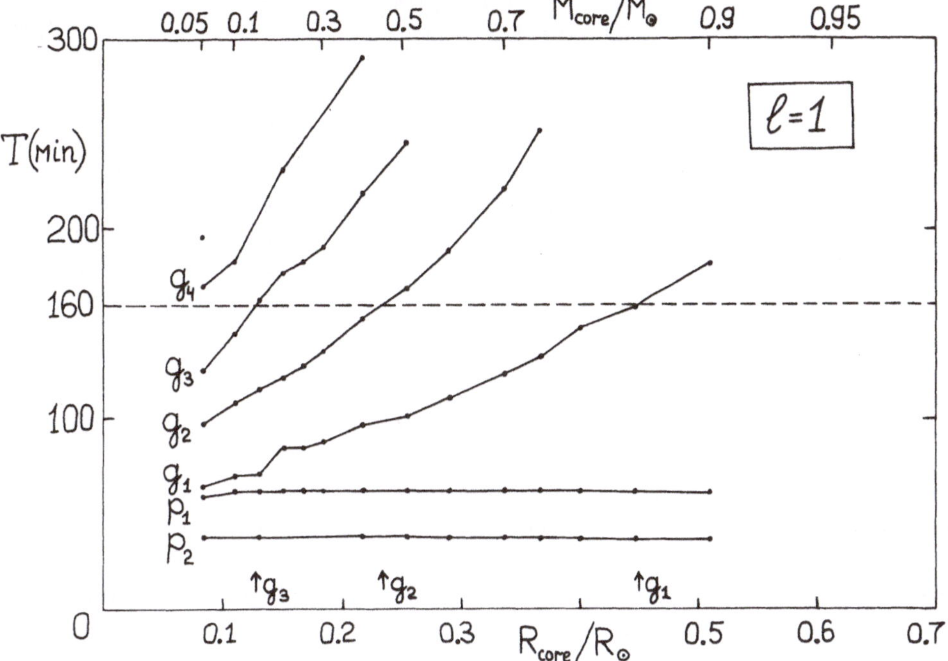

Fig. 3. Periods of dipole ($l = 1$) modes.

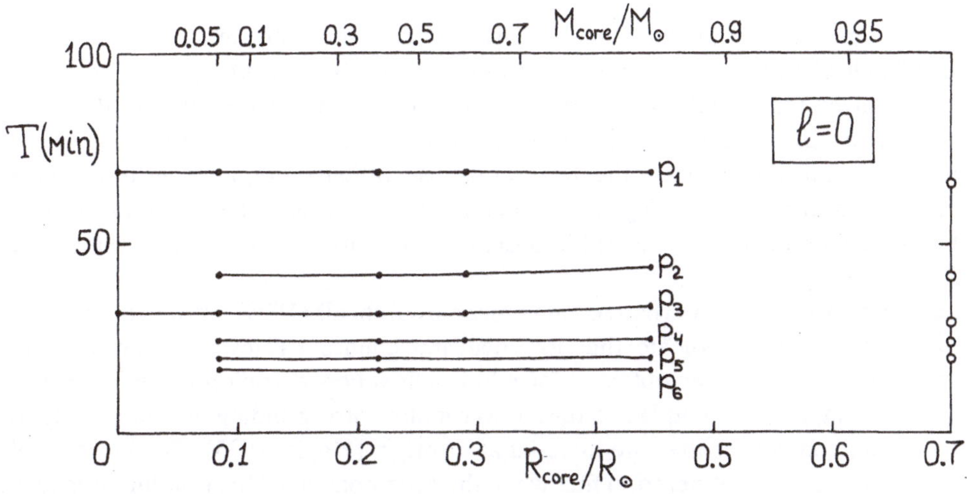

Fig. 4. Periods of radial ($l = 0$) modes.

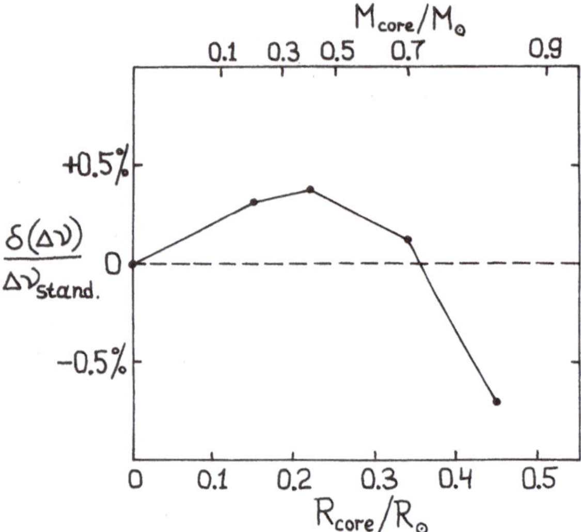

Fig. 5. The deviation of the high-order p-mode frequency spacing from that of the standard model.

where c is the adiabatic sound speed. The results are shown in Figure 5. Only the relative deviations of $\Delta\nu$ from the prediction of the standard model are given, because the influence of the solar atmosphere on the absolute value of $\Delta\nu$ was not included in the calculations. The small deviations obtained indicate that the models of the type under consideration show no evident contradiction with the frequency spacing of global 5-min solar oscillations, taking into account that the experimental spacing seems to be in agreement with the standard model predictions (Christensen-Dalsgaard and Gough, 1980).

4. Neutrino Counting Rates

The core compositions, luminosities and neutrino counting rates in ^{37}Cl detector are listed in Tables I and II for the models with $M_{core} = 0.4\,M_{\odot}$ and $M_{core} = 0.2\,M_{\odot}$. The results are given for the different degrees of ionization of heavy elements in the core. The presented values of $L/L_{\odot}$ deviate from unity in some cases because the same values of helium content were used for the different degrees of ionization in order to examine the influence of the assumed degree of ionization on the results. The limiting cases of total and zero ionization were tested in order to make sure that this influence is rather small.

The neutrino absorption cross-sections given by Bahcall (1979) were used in computations. The results indicate that low neutrino counting rates, comparable with observational limits, may be obtained for a low helium concentration in the core. The less Y values in Tables I and II are their lower limits corresponding approximately to a zero helium content for the core material at its origin (Hoyle, 1975). The increase of Y leads to the increase of neutrino flux while the accompanying effect on luminosity is rather small.

TABLE I

The model with $M_{core} = 0.4\, M_\odot$ ($R_{core} = 0.22\, R_\odot$, central temperature $T_c = 16.1 \times 10^6$ K, central density $\rho_c = 95$ g cm^{-3})

X	Y	Z	$L/L_\odot$	$\sum \phi_i \sigma_i$ (SNU)
		Zero ionization		
0.69	0.10	0.21	1.1	2.7
0.67	0.15	0.18	1.1	3.8
		60% ionization		
0.65	0.10	0.25	1.0	2.5
0.64	0.15	0.21	1.0	3.5
		Total ionization		
0.62	0.10	0.28	0.9	2.3
0.61	0.15	0.24	0.9	3.3

TABLE II

The model with $M_{core} = 0.2\, M_\odot$ ($R_{core} = 0.15\, R_\odot$, central temperature $T_c = 15.2 \times 10^6$ K, central density $\rho_c = 114$ g cm^{-3})

X	Y	Z	$L/L_\odot$	$\sum \phi_i \sigma_i$ (SNU)
		Zero ionization		
0.68	0.20	0.12	1.0	2.3
0.66	0.25	0.09	1.0	2.7
		60% ionization		
0.66	0.20	0.14	1.0	2.2
0.64	0.25	0.11	0.9	2.6
		Total ionization		
0.64	0.20	0.16	0.9	2.1
0.63	0.25	0.12	0.9	2.5

5. Conclusions

The set of models with a high-Z convective core which was examined is not unique: similar sets with a different helium concentration in the envelope and/or with a different structure of the outer convective zone may be constructed. Some simplifying assumptions, like that of homogeneous ionization of core material, were also used. Certain general conclusions for all the models of this type can, nevertheless, be formulated:

(a) the neutrino counting rates within the observational limits can be achieved for a low initial helium concentration;

(b) the gravity mode spectrum is more widely separated than that of the standard model, which can simplify the problem of interpretation of 160-min oscillations;

(c) the luminosity of such a model is approximately constant over the Earth's history, which is due to convective mixing of the products of nuclear reactions;

(d) the models do not contradict the usual interpretation of the global 5-min oscillations of the Sun.

References

Abraham, Z., and Iben, I.: 1971, *Astrophys. J.* **170**, 157.
Bahcall, J. N.: 1979, *Space Sci. Rev.* **24**, 227.
Christensen-Dalsgaard, J. and Gough, D. O.: 1980, *Nature* **288**, 544.
Fowler, W. A., Caughlan, G. R., and Zimmerman, B. A.: 1974, *OAP* **380**.
Hoyle, F.: 1975, *Astrophys. J.* **197**, L127.
Spruit, H. G.: 1974, *Solar Phys.* **34**, 277.
Vandakurov, Y. V.: 1967, *Soviet Astron.-AJ* **44**, 786.
Vorontsov, S. V. and Zharkov, V. N.: 1978, *Soviet Astron.-AJ* **55**, 84.

SOLAR MODELS WITH LOW OPACITY*

P. A. KUZURMAN and A. A. PAMYATNYKH

Astronomical Council of the Academy of Sciences of the USSR

Abstract. Evolutionary models of the present Sun with standard and artifically low opacity of stellar matter are obtained and adiabatic nonradial oscillations of the models are computed. The opacity $\varkappa$ in nonstandard model (which is the first in a series of future models with low $\varkappa$) was taken half as much as the Cox–Stewart opacity. The central temperature in such model is approximately 10% below the 'standard' values, which can explain the neutrino experiment results. Unlike solar models with very low heavy element abundance, the low $\varkappa$ model has approximately standard mass concentration and distribution of the matter in the outer layers – specifically, the standard characteristics of the convection zone. Hence, the spectrum of adiabatic oscillations is similar to that of the standard models and has the same capabilities for the explanation of the observed pulsations.

1. Introduction

The solar models with very low heavy element abundances in interiors (say, $Z = 0.002$) have low central temperature and can explain the experimental results on detecting solar neutrinos (see Bahcall *et al.*, 1973, and other references in Davis and Evans, 1978). However, these models have some defects, too. There is a very shallow convection zone in these models: it is only 30–100 thousand kilometers deep and has temperature at the bottom less than 10^6 K (Christensen-Dalsgaard *et al.*, 1979; see also Iben and Mahaffy, 1976). This does not agree with the interpretation of observed 5-min oscillations of small horizontal scale as acoustic modes of non-radial oscillations at high values of spherical harmonics (Rhodes *et al.*, 1977) and contradicts the hypothesis of thermonuclear destruction of lithium (Herbig, 1965; simple evaluations show that to explain the observed deficiency of lithium at normal abundance of berillium, the temperature at the bottom of the convection zone should approximately range from 2×10^6 to 3×10^6 K). The interpretation of 5-min oscillations of low degree (Grec *et al.*, 1980) also favours the standard solar models with a deep convection zone of about 200×10^3 km (Christensen-Dalsgaard and Gough, 1980; Shibahashi and Osaki, 1983). Moreover, the origin of models with low heavy element abundances in interiors and normal abundances on the surface requires a detailed investigation of interaction between accretion of interstellar matter and solar wind throughout the evolution of the Sun.

The main 'useful' effect of very low Z – low central temperature – could be reached at normal Z, if it turned out that solar interiors are more transparent than it follows from the radiative transfer law. Newman and Fowler (1976) considered solar models with postulated additional energy transport mechanisms. From the formal point of view this phenomenon corresponds to the reduction of effective opacity of the solar matter. Quite recently a very important physical argument for that reduction has appeared. Opher (1981) showed that the proper treatment of the plasmon – electron interactions leads

* Proceedings of the 66th IAU Colloquium: *Problems in Solar and Stellar Oscillations*, held at the Crimean Astrophysical Observatory, U.S.S.R., 1–5 September, 1981.

Solar Physics **82** (1983) 223–230. 0038–0938/83/0821–0223$01.20.

to an increase in the number of the high energy electrons compared with the Maxwell–Boltzmann distribution and, hence, to the opacity reduction approximately by half. The author emphasizes that his results contain no free parameters. Note that other mechanisms of the effective $\varkappa$ decrease are possible as well; for example, under

TABLE I

Solar model characteristics

Model Effective opacity	1 $\varkappa_{\text{table}}$	2 $\varkappa_{\text{table}}$	3 $\varkappa_{\text{table}}$	4 $\frac{1}{2}\varkappa_{\text{table}}$
Initial heavy element abundance, Z_0	0.01	0.02	0.03	0.02
Initial hydrogen abundance, X_0	0.77	0.73	0.693	0.86
Mixing length: pressure scale height, $\alpha = l/H_p$	1.24	1.41	1.52	1.285
Luminosity, $L/L_\odot$	1.0049	0.9997	1.0016	1.0090
Radius, $R/R_\odot$	1.0014	0.9985	0.9990	1.0021
Age (10^9 yr)	4.50	4.55	4.58	4.58
Central hydrogen abundance, X_c	0.39	0.34	0.30	0.46
Central temperature, T_c (10^6 K)	14.2	14.6	14.9	13.2
Central density (g cm^{-3})	127	135	142	133
Density at the bottom of convection zone (g cm^{-3})	0.115	0.185	0.241	0.178
Depth of convection zone ($R_\odot$)	0.244	0.279	0.297	0.274
Mass of convection zone ($M_\odot$)	0.0141	0.0245	0.0315	0.0230
Temperature at the bottom of convection zone, T_b (10^6 K)	1.72	2.11	2.36	1.89

the influence of propagation to the solar center of high frequency internal gravity waves generated at the bottom of the convection zone (Press, 1980)*. So, Opher's result is not confirmed, it is reasonable to investigate these models in more detail. Also, the decrease of effective opacity in stellar interiors can be useful in solution of some other astronomical problems, for example, of the problem of Cepheid masses, as it was investigated by Fricke *et al.* (1971).

* More detailed analysis by Press (1981, *Astrophys. J.* **245**, 286) and by Press and Rybicki (1981, *Astrophys. J.* **248**, 751) shows that internal waves may cause the increase of the effective opacity in radiatively stable regions.

At present we are investigating the main characteristics of models with artificially low opacity of solar matter. Some preliminary results are described below.

2. The Models

All models have solar luminosity, radius, and age of about 4.5–4.6 × 10^9 yr. Evolution from initial main sequence was calculated according to Schwarzschild's fitting method with the Cox and Stewart opacity tables and with treatment of convection according to the mixing – length theory. The rates of nuclear reactions of proton – proton and CN cycles were taken from Fowler *et al.* (1975). The main characteristics of models are listed in Table I and illustrated in Figure 1.

Models 1, 2, 3 are probably a reasonable set of standard models of the Sun. They have been calculated with opacity values from tables and the results may be compared with the results of similar calculations (Iben and Mahaffy, 1976; Dziembowski and Pamyatnykh, 1978; model A by Christensen-Dalsgaard *et al.*, 1979). Model 4 has been calculated with Z = 0.02 and with opacity coefficients which are half as much as the

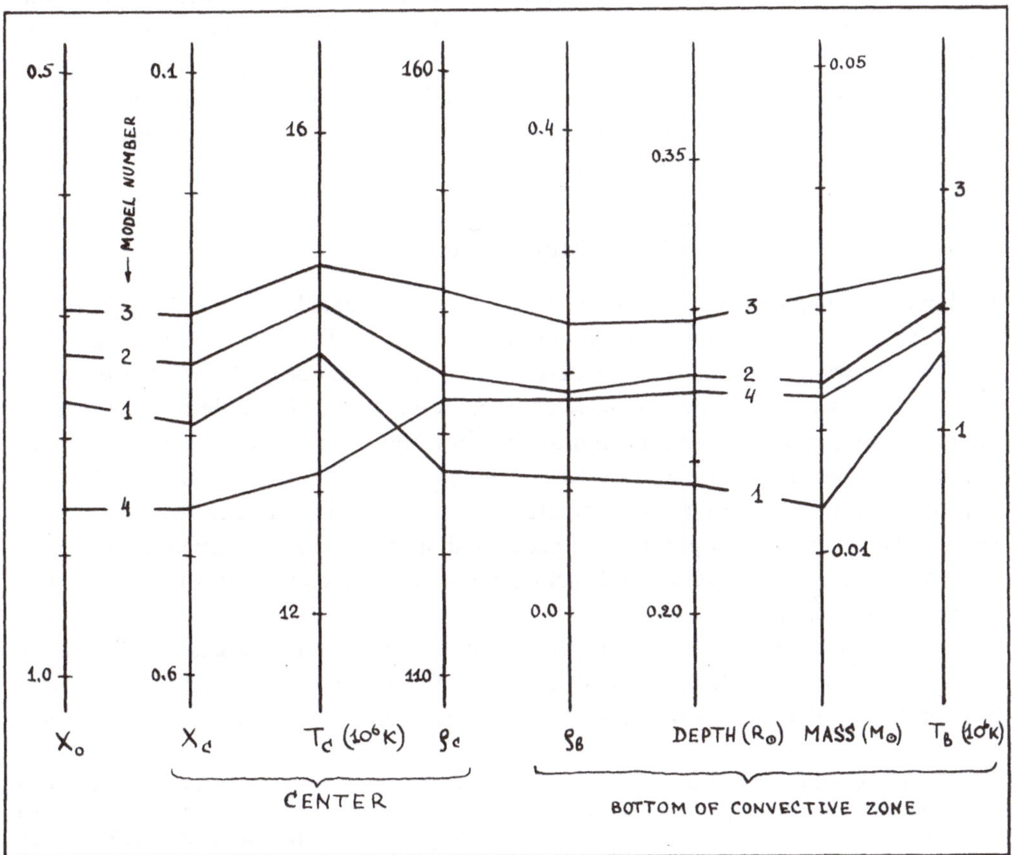

Fig. 1. Basic characteristics of different solar models. All scales are linear.

corresponding table values. This is the first in a series of future models with low $\varkappa$. Some parameters of this model may be compared with the results reported by Newman and Fowler (1976). In calculations of thermonuclear reactions we, for simplicity, did not take into account the branch *pp* III of $p-p$ cycle, whose role is negligible in the total energy generation, but which produces the main contribution (from decay of B^8) to the high energy neutrino flux. Therefore, in Table I there are no theoretical evaluations of the rate of neutrino capture by Cl^{37}. In models without mixing in interiors the high energy neutrino flux is approximately proportional to T_c^{13} (see, for example, Iben, 1969), and for agreement of the experimental results with theoretical predictions it is enough to decrease T_c by about $\sim 10\%$, which will reduce the flux ~ 3.5 times (for new data see, for example, short review by Peak, 1980). Model 4 has just this change of T_c in comparison with model 2, and besides both models have the same heavy element abundance: $Z = 0.02$.

The important difference between our model with low $\varkappa$ and models with low Z (say, $Z = 0.002$) is the standard distribution of matter and standard characteristics of the convective zone. These results are clearly displayed in Figure 1. For example, the central densities in models 2 and 4 do not differ by more than 2% and there is a similar difference in depths of the convection zones. So, the mechanical properties (specifically, the periods of the oscillations) of our model with low $\varkappa$ may be expected to be similar to those of the standard models. The lithium-berillium problem can also be solved in the models with low $\varkappa$ according to the hypothesis of the lithium thermonuclear depletion at the bottom of the convection zone*.

3. Oscillations of the Models

We have computed periods of several modes of adiabatic oscillations of low degree (both nonradial and radial) using the programs which were kindly placed at our disposal by W. Dziembowski (the programs are described by Dziembowski, 1977). The frequences of oscillations were computed as eigenvalues of the equations. The Eulerian perturbation of gravitational potential was taken into account. The outer boundary conditions were applied at optical depth $\tau = 0.01$ (the atmosphere was computed in the diffusion approximation). The results of computations are listed in Table II, where the conventional classification of gravity and acoustical modes is used. Frequences (in mHz), and not periods, for acoustical modes of high order are listed to simplify the comparison with other investigations.

Comparison of the periods of low order modes for different models shows that the periods of model 4 with low $\varkappa$ lie between those of standard models 1 and 2 as it should be expected from the above mentioned general considerations. Oscillations with periods

* In the course of Sun evolution from ZAMS to the present state the depth of the convection zone is continuously decreased. Lithium could be depleted at early phases of Sun evolution on the main sequence. Hence, the models of the Sun with $T_b \sim 2.0 \times 10^6$ K may be sufficient for solving the problem, since at early phases these models had $T_b \sim 2.5 \times 10^6$ K. The problem of lithium depletion and the effect of uncertainty of the mixing-length theory parameters was discussed, for example, by Ergma and Pamyatnykh (1971).

TABLE II

Solar models oscillations

(a) Oscillation periods of several gravity modes (min)

Model	$l = 1$			$l = 2$			
	g_1	g_2	nearest to 160^m	f	g_1	g_2	nearest to 160^m
1	69.44	96.36	g_5: 168.26	51.73	60.65	69.83	g_9: 160.20
2	66.53	91.86	g_5: 160.25	49.38	59.33	68.02	g_{10}: 165.28
3	64.29	88.34	g_5: 154.26	47.61	57.87	66.53	g_{10}: 159.06
4	67.59	93.44	g_5: 163.19	50.00	59.77	68.75	g_9: 155.49
							g_{10}: 168.40

(b) Oscillation periods of the low order acoustic modes (min)

Model	$l = 0$		$l = 1$		$l = 2$		$l = 3$	
	p_1	p_2	p_1	p_2	p_1	p_2	p_1	p_2
1	63.33	41.93	58.55	37.28	45.25	32.69	40.66	29.78
2	64.84	41.69	58.94	37.14	44.86	32.49	40.44	29.57
3	65.74	41.69	58.95	37.19	44.55	32.48	40.44	29.54
4	64.95	41.76	59.24	37.31	44.93	32.65	40.57	29.70

(c) Frequences of the high order acoustic modes of low degree (mHz)

Model	$l = 0$			$l = 1$		
	p_{20}	p_{21}	p_{22}	p_{20}	p_{21}	p_{22}
1	2.8999	3.0387	3.1785	2.9641	3.1032	3.2428
2	2.9253	3.0650	3.2056	2.9900	3.1298	3.2704
3	2.9297	3.0690	3.2096	2.9942	3.1340	3.2745
4	2.9093	3.0492	3.1892	2.9742	3.1139	3.2540

Model	$l = 2$			$l = 3$		
	p_{20}	p_{21}	p_{22}	p_{20}	p_{21}	p_{22}
1	3.0270	3.1669	3.3070	3.0859	3.2260	3.3671
2	3.0537	3.1944	3.3351	3.1133	3.2545	3.3959
3	3.0582	3.1989	3.3395	3.1182	3.2593	3.4007
4	3.0378	3.1779	3.3188	3.0970	3.2377	3.3795

of about 160 min are similar in all models: this period is corresponding to the dipole mode g_5 and the quadrupole mode g_9 or g_{10}. Hence, the problem of 160-min oscillations in models with low $\varkappa$ may be the same as in the 'standard' models.

The frequences of acoustic modes of high order are corresponding to the observed 5-min pulsations of large horizontal scale (Grec *et al.*, 1980). The results for models with low $\varkappa$ lie between those for standard models 1 and 2, too. However, as it can be seen from Table IIc, the frequency separation between the adjacent eigenfrequences for a given l is greater than the double difference (136.0 μHz) between peaks in power spectrum of the observed pulsations. For computed eigenmodes $p_{15} - p_{23}$ (in the Table IIc frequences only for $p_{20} - p_{22}$ are listed) with $l = 1$ we have the average frequency separation $\Delta_{nl}\nu = 138.2$ μHz for model 1, 139.0 μHz for model 2, 138.9 μHz for model 3, and 138.8 μHz for model 4. The differences between the calculated eigenfrequences and the observed values result partly from the neglecting of role of the atmosphere as it was shown by Christensen-Dalsgaard and Gough (1980). (We have computed the lower atmosphere, but very roughly.) Besides, the frequency separations $\Delta_{nl}\nu$ in our models are approximately 1 μHz greater than the corresponding values for model A by Christensen-Dalsgaard *et al.* (1979) also computed without the account of the atmosphere. Note that the depth of the convection zone in their model A with $Z = 0.02$ is lower than in our 'shallowest' model with $Z = 0.01$. It may be connected with differences in age: the age of the model A is 4.75×10^9 yr. In future we intend to investigate both the influence of the atmosphere and the effect of age (say, between 4.5 and 5.0×10^9 yr) in models with solar luminosity and radius. It is important to us here that the low $\varkappa$ models have, in fact, the 'standard' frequences and the frequency separations of acoustic modes of high order: e.g., the frequences listed in Table IIc for model 4 lie between those for models 1 and 2.*

4. Conclusions

Thus, our preliminary results show that the solar models with low $\varkappa$ and with normal heavy element abundance are potentially very useful for the explanation of such observed facts as the solar neutrino flux, the lithium and berillium abundances in solar atmosphere and 5-min oscillations of low and high degree. The defect of these models (as well as the models with low Z) is a fairly low initial helium abundance, $Y = 0.12$ in model 4. We hope, as it may be expected from the results by Newman and Fowler (1976), that the models with low $\varkappa$ in interiors and with standard $\varkappa$ in the envelope will have more reasonable values of the helium abundance, e.g., $Y = 0.2$.

In conclusion we have a comment about the frequency separation $\Delta_{nl}\nu$ between the frequences of the acoustic modes of high order for the standard models. In our models the frequency separation increases by about 0.8 μHz as Z increases from 0.01 to 0.02, i.e. as depth of the convection zone increases from 170×10^3 to 195×10^3 km (see Tables I and IIc). This result is similar qualitatively to the one reported by Christensen-

* See note added in proof.

Dalsgaard *et al.* (1979), Christensen-Dalsgaard and Gough (1980) and does not agree with the results by Scuflaire *et al.* (1981) (in any case, with the difference $\Delta_{nl}\nu$ between their models 1 and 2 whose convection zone parameters are similar roughly to those of our models 1 and 3). The decrease of $\Delta_{nl}\nu$ with the increase of the depth of the convection zone in the envelope models computed by Scuflaire *et al.* (1981) may be easily understood since these models do not fit the corresponding interior models. In such envelope models the average temperature decreases with the increase of the depth of the convection zone. It would lead to the decrease of the average sound speed and, thus, to the decrease of the value $\Delta_{nl}\nu$ (see the discussion of the $\Delta_{nl}\nu$ in Christensen-Dalsgaard *et al.*, 1979). However, for the fitting evolutionary models the situation will be more complicated: the larger is the depth of the convection zone the larger is, on the average, temperature in interiors (which leads to the increase of the sound speed and $\Delta_{nl}\nu$) and the higher is helium abundance (which leads to the decrease of the sound speed due to the effect of the molecular weight). It is possible, that for the very deep convection zones the frequency separation $\Delta_{nl}\nu$ varies in the direction reported by Scuflaire *et al.* (1981) due to the decrease of the average temperature in the very extensive convection zone. However, for moderately deep convection zones (as is the case for our models 1 and 2) we have the opposite effect. This remark explains the difference in $\Delta_{nl}\nu$ behaviour between our models and envelope models by Scuflaire *et al.* (1981).

Acknowledgements

We are very indebted to Dr W. Dziembowski for his programs for the computation of nonradial stellar oscillations. All computations were performed on the ES-1033 computer at the Computer Center of the Astronomical Council of the USSR Acad. Sci.

Note added in proof: Our new computations show that the difference between calculated and observed $\Delta_{nl}\nu$ (see Section 3) results mostly from too small number of discrete mass zones (~ 450) in our models: when we have $\sim 4 \times 450$ mass zones, $\Delta_{nl}\nu$ is in much better agreement with the observations and, for example, with the results by Shibahashi and Osaki (1983) where the same effect has been noted and discussed.

References

Bahcall, J. N., Huebner, W. F., Magee, N. H. Jr., Merts, A. L., and Ulrich, R. K.: 1973, *Astrophys. J.* **184**, 1.
Christensen-Dalsgaard, J. and Gough, D. O.: 1980, *Nature* **288**, 544.
Christensen-Dalsgaard, J., Gough, D. O., and Morgan, J. G.: 1979, *Astron. Astrophys.* **73**, 121.
Davis, R. J. and Evans, J. C. Jr.: 1978, in J. A. Eddy (ed.), *New Solar Physics*, Westview Press, Boulder, Colorado, p. 35.
Dziembowski, W.: 1977, *Acta Astron.* **27**, (2), p. 95.
Dziembowski, W. and Pamyatnykh, A. A.: 1978, in J. Rösch (ed.), *Pleins feux sur la physique solaire*, CNRS, Paris, p. 135.
Ergma, E. and Pamyatnykh, A. A.: 1971, *Nauchnye Informatsii* **20**, 71.
Fowler, W. A., Caughlan, G. R., and Zimmerman, B. A.: 1975, *Ann. Rev. Astron. Astrophys.* **13**, 69.
Fricke, K., Stobie, R. S., and Strittmatter, P. A.: 1971, *Monthly Notices Roy. Astron. Soc.* **154**, 23.

Grec, G., Fossat, E., and Pomerantz, M.: 1980, *Nature* **288**, 541.
Herbig, G. H.: 1965, *Astrophys. J.* **141**, 588.
Iben, I. Jr.: 1969, *Ann. Phys.* **54**, 164.
Iben, I. Jr. and Mahaffy, J.: 1976, *Astrophys. J.* **209**, L39.
Newman, M. J. and Fowler, W. A.: 1976, *Astrophys. J.* **207**, 601.
Opher, R.: 1981, *Astron. Astrophys.* **98**, 39.
Peak, L. S.: 1980, *Australian J. Phys.* **33**, 821.
Press, W. H.: 1980, Preprint No. 1313 of Center for Astrophys., Cambridge, Mass.
Rhodes, E. J. Jr., Ulrich, R. K., and Simon, G. W.: 1977, *Astrophys. J.* **218**, 901.
Scuflaire, R., Gabriel, M., and Noels, A.: 1981, *Astron. Astrophys.* **99**, 39.
Shibahashi, H. and Osaki, Y.: 1983, *Solar Phys.* **82**, 231 (this volume).

THEORETICAL EIGENFREQUENCIES OF SOLAR OSCILLATIONS OF LOW HARMONIC DEGREE *l* IN FIVE-MINUTE RANGE*

(Invited Review, Abstract)

H. SHIBAHASHI** and Y. OSAKI

Department of Astronomy, University of Tokyo, Bunkyo-ku, Tokyo 113, Japan

Abstract. We have calculated eigenfrequencies of radial and nonradial *p*-mode oscillations with low harmonic index l ($l = 0$, 1, 2, 3, and 4) for a standard solar model with normal composition and appoximately the correct age. It is found that theoretical eigenfrequencies calculated for our standard model agree approximately with observed peaks in the power spectra for the full-disk five minute oscillation of the Sun (Claverie *et al.*, 1980; Grec *et al.*, 1983; Scherrer *et al.*, 1983) in agreement with other recent works (Christensen-Dalsgaard and Gough, 1980; Scuflaire *et al.*, 1981). However, there still remains a slight discrepancy between theory and observations in such a sense that the theoretical eigenfrequencies are slightly lower than observations (see Figure 1).

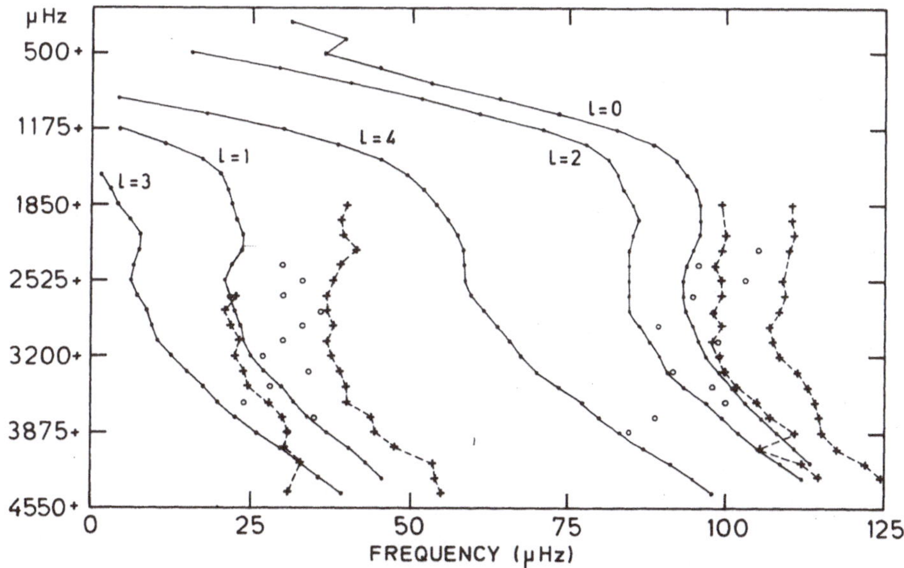

Fig. 1. Superposed frequency diagram for the 135 μHz steps. Dots, open circles, and crosses indicate the calculated theoretical eigenfrequencies, the observations by Claverie *et al.* (1980) in 1979, and those by Grec *et al.* (1983), respectively.

* Proceedings of the 66th IAU Colloquium: *Problems in Solar and Stellar Oscillations*, held at the Crimean Astrophysical Observatory, U.S.S.R., 1–5 September, 1981.

** Present address: Joint Institute for Laboratory Astrophysics, University of Colorado and National Bureau of Standard, Boulder, Colorado.

Solar Physics **82** (1983) 231–232. 0038–0938/83/0821–0231$00.30.

Eigenfrequencies of nonradial p-modes with high degree l for the same model are also calculated for comparison with observations of the conventional five-minute oscillation with shorter horizontal wavelength. It is found that theoretical eigenfrequencies lie slightly above the observed ridges in the diagnostic (k, ω)-diagram, which is in accord with Ulrich and Rhodes (1977) and Berthomieu *et al.* (1980) because our standard model has a slightly shallow convective zone. It remains to be seen whether improvement in the equilibrium model can remove this small discrepancy, concurrently with a better agreement between theory and observation for the whole-disk oscillations of low degree.

A full account of this work will be found in Shibahashi and Osaki (1981).

References

Berthomieu, G., Cooper, A. J., Gough, D. O., Osaki, Y., Provost, J., and Rocca, A.: 1980, in H. A. Hill and W. A. Dziembowski (eds.), *Nonradial and Nonlinear Stellar Pulsation*, Springer-Verlag, Berlin, p. 307.

Christensen-Dalsgaard, J. and Gough, D. O.: 1980, *Nature* **288**, 544.

Claverie, A., Isaak, G. R., McLeod, C. P., van der Raay, H. B., and Rocca-Cortes, T.: 1980, *Astron. Astrophys.* **91**, L9.

Grec, G., Fossat, E., and Pomerantz, M.: 1983, *Solar Phys.* **82**, 55 (this volume).

Scherrer, P. H., Wilcox, J. M., Christensen-Dansgaard, J., and Gough, D. O.: 1983, *Solar Phys.* **82**, 75 (this volume).

Scuflaire, R., Gabriel, M., and Noels, A.: 1981, *Astron. Astrophys.* **99**, 39.

Shibahashi, H., and Osaki, Y.: 1981, *Publ. Astron. Soc. Japan* **33**, 713.

Ulrich, R. K. and Rhodes, E. J. Jr.: 1977, *Astrophys. J.* **218**, 521.

ROTATIONAL SPLITTING OF SOLAR FIVE-MINUTE OSCILLATIONS OF LOW DEGREE*

(Invited Review, Abstract)

A. CLAVERIE, G. R. ISAAK, C. P. McLEOD, and H. B. VAN DER RAAY

Physics Department, University of Birmingham, U.K.

and

T. ROCA CORTES

Institute de Astrofisica de Canarias, La Laguna, Tenerife

Abstract. An analysis of 28 contiguous days of whole disk observations of the solar surface by means of optical resonant scattering in the K 769.9 nm line, taken at the Teide Observatory at Izana during July–August 1980, have thus far yielded two significant facts. Firstly when the results of an iterative sine-wave fitting procedure are considered in the period range 2–3 h, although the expected daily harmonics corresponding to 1/8, 1/10, 1/11, and 1/12 of a day are clearly seen the 1/9th contribution is significantly absent. It is suggested that this results from an interference between a signal of 160 min (1/9th of a day) with the daily harmonic. It is further pointed out that the observatories at which the 160 min oscillation has been seen, Crimea, Pic du Midi, and Stanford are all separated by integral numbers of 160 min, and thus the phase of the 160 min oscillation relative to the daily observation window is constant. However, the Teide Observatory is situated at a half integral number of 160 min periods relative to the others. Thus when constructive interference exist at the first three sites destructive interference will exist at the latter. It is thus concluded that the non-existence of a peak corresponding to the 1/9th harmonic of a day in the sine-wave fit data is strong indirect evidence for the existence of the 160 min signal.

An analysis of those same data in the 5 min region has revealed the now well established pattern of discrete frequencies and with the increased resolution obtainable from 28 contiguous days of data, clearly showed the existence of splitting by rotational effects. In all 33 discrete lines were considered in the frequency range 2.4–3.85 mHz, which could be divided up into 3 groups, each of 11 lines, corresponding to the $l = 0$, $l = 1$, and $l = 2$ modes. This definitive classification was possible as the lines are split into $(2l + 1)$ components yielding easily identifiable singlets, triplets and quintuplets. This first observation of the rotational splitting of solar oscillations gave the further information that the splitting of 0.75 ± 0.10 μHz indicated that the solar interior is rotating more rapidly than the observable surface (uniform rotation would yield a splitting of 0.4 μHz). A comparison of the widths of the individual peaks, with the

* Proceedings of the 66th IAU Colloquium: *Problems in Solar and Stellar Oscillations*, held at the Crimean Astrophysical Observatory, U.S.S.R., 1–5 September, 1981.

Solar Physics **82** (1983) 233–234. 0038–0938/83/0821–0233$00.30.

intrinsic resolution of the data string ($\sim 1/T$ where T = length of the data string), showed that these were consistent with a high Q value oscillation, a fact which is further confirmed by the very existence of singlets triplets and quintuplets in the data.

The exact value of the speed of internal rotation of the Sun can only be deduced from these data by a model dependent calculation. The simplest of these would suggest that if the core were almost equal to the solar diameter then it is rotating twice as fast as the surface, whereas if the core were only 15% of the solar diameter, it would be rotating at 9 times the surface rate.

Reference

Claverie, A., Isaak, G. R., McLeod, C. P., Van der Raay, H. B., and Roca Cortes, T.: 1981, *Nature* **293**, 443.

IS THERE AN OBLIQUE MAGNETIC ROTATOR INSIDE THE SUN?*

(Invited Review, Abstract)

G. R. ISAAK

Department of Physics, University of Birmingham, Birmingham, B15 2TT, UK

Abstract. The size of the rotational splitting recently observed (Claverie *et al.*, 1981) is correlated with the 12.2^d variation in the measurements of solar oblateness observed by Dicke (1976) and implies a convection zone of depth of $0.1\ R_{\odot}$. The near equality of amplitudes of global velocity oscillations (Claverie *et al.*, 1981) of the various m components of the $l = 1$ and $l = 2$ modes as seen from the Earth viewing the Sun nearly along the equator is unexpected for pure rotational splitting. It is suggested that a magnetic perturbation is present and an oblique asymmetric magnetic rotator with magnetic fields of a few million gauss is responsible. A more detailed account was submitted to *Nature*.

References

Claverie, A., Isaak, G. R., McLeod, C. P., van der Raay, H. B., and Roca Cortes, T.: 1981, *Nature* **292**, 443.
Dicke, R. H.: *Solar Phys.* **47**, 475.

* Proceedings of the 66th IAU Colloquium: *Problems in Solar and Stellar Oscillations*, held at the Crimean Astrophysical Observatory, U.S.S.R., 1–5 September, 1981.

Solar Physics **82** (1983) 235. 0038–0938/83/0821–0235$00.15.

THE STUDY OF VELOCITY OSCILLATIONS IN THE SOLAR PHOTOSPHERE USING THE VELOCITY SUBSTRACTION TECHNIQUE*

N. I. KOBANOV

Siberian Institute of Terrestrial Magnetism, Ionosphere and Radio Wave Propagation (SibIZMIR), USSR Academy of Sciences, Siberian Department, Irkutsk 33, P.O. Box 4, U.S.S.R.

Abstract. A method of measurement of local line-of-sight velocities in the solar atmosphere by means of polarization optics is described. No spurious signals due to instrumental displacements of the spectrum arise with this method. The sensitivity of the method obtained is 0.3 m s^{-1}, with a time constant $\tau = 5$ s and input aperture 1.4″ × 4.5″. Some preliminary results of the assessment of spatial characteristics of 5-min oscillations are included. Data are given to illustrate a center-to-limb variation of the spectrum of 5-min oscillations.

The past few years have witnessed an increasing interest in the study of line-of-sight velocities in the solar atmosphere. However, sensitivity of measurements is frequently not sufficient. For instance, when a conventional diffraction-grating spectrograph is used, due to accidental displacements of the spectrum caused by air turbulence within the spectrograph, and by thermal deformations, spurious signals reach 150–250 m s^{-1} (Dittmeyr, 1977; Brandt *et al.*, 1978). Such noise is absent when differential methods are employed (Kalinyak and Vasilyeva, 1971; Kotov *et al.*, 1978; Dittmeyr, 1977). This paper develops the differential method for the investigation of local quasi-periodic motions of the gas.

A scheme illustrating the passage of optical rays is presented in Figure 1. At the entrance slit of the spectrograph is a calcite plate and a phase plate $\lambda/4$. In calcite, each beam splits into two beams with orthogonal directions of the linear polarization. Points A and B of the original image correspond to points a, a' and b, b' at the output of the calcite. The $\lambda/4$ plate, oriented at 45° with respect to the linear polarization

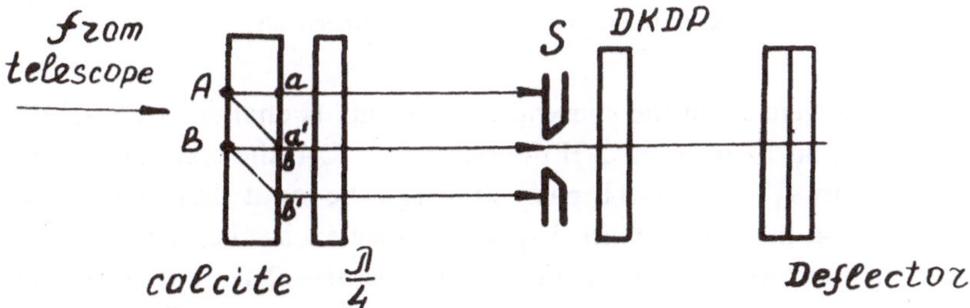

Fig. 1. Optical scheme of the device as used by the author.

* Proceedings of the 66th IAU Colloquium: *Problems in Solar and Stellar Oscillations*, held at the Crimean Astrophysical Observatory, U.S.S.R., 1–5 September, 1981.

Solar Physics **82** (1983) 237–243. 0038–0938/83/0821–0237$01.05.

directions of calcite, converts the linear polarization into a circular one. The entrance slit of the spectrograph from point A of the original image receives a beam with a right-hand circular polarization, and from point B, a beam with a left-hand circular polarization. If the line-of-sight velocities from the A and B elements are different, each spectral line will consist of two components which are circularly polarized in mutually opposite directions (Figure 2a). The distance between the components in Figure 2a is $\Delta\lambda_v \sim v_A - v_B$, where v_A, v_B are the line-of-sight velocities of the A and B elements, respectively. In its outward appearance, the picture resembles Zeeman spectral line splitting, with the exception that, in this case, we have 100% polarization of the components. Further measurements can be made by any method usually applied in recording the longitudinal component of the magnetic-field strength. In this treatment, we have used a scheme containing a deflector and a single photomultiplier (Lebedev and Grigoryev, 1976). The deflector consists of two calcite plates oriented so that a beam being ordinary for one plate becomes extraordinary for the other, and vice versa. The plates have the same thickness which is chosen so that the total deviation of beams along the dispersion of the spectrograph $\Delta\lambda_0$ is approximately equal to half the width of the spectral line in which the observations are being made. The polarization directions of the beams at the deflector output make an angle of 45° with the direction of grooves of the diffraction grating.

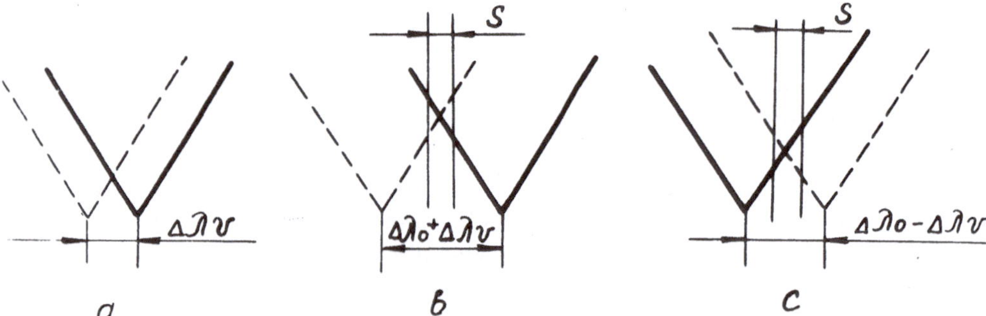

Fig. 2. Position of spectral components in the entrance slit plane of photometer. The spectral line contour is depicted simplified. (a) Original position (without deflector). (b, c) Positions of the components in the DKDP phase $+\lambda/4$ and $-\lambda/4$, respectively.

The spatial arrangement of the spectral components is changed by DKDP and the deflector, as shown in Figures 2b, c. If the phase is $+\lambda/4$, the components 'go away', and light of intensity $I_1 \sim \Delta\lambda_0 + \Delta\lambda_v$ passes through the input slit of the spectrograph. If the phase is $-\lambda/4$, the components displace to meet each other, and $I_2 \sim \Delta\lambda_0 - \Delta\lambda_v$. The difference in intensity for the modulation period is $\Delta I \sim 2\Delta\lambda_v$ and the output current of the multiplier is a variable signal at a frequency of modulation. Figures 2b, c clearly shows that simultaneous displacements of the two components due to instrumental factors would not have any appreciable influence upon the signal. One would think that there is no need for any Doppler compensator. However, this is not the case. In order to significantly decrease the possible influence due to instrumental polarization or to the difference in brightness of the A and B elements, it is desirable to provide a symmetrical arrangement of the spectral components with respect to the photometer slit. Otherwise, the simultaneous effect to these factors and instrumental displacements of the spectrum may give rise to a noise-like signal.

The instrumental polarization effect can be completely eliminated by means of a polaroid, placed in front of the calcite and oriented at an angle of 45° with respect to the directions of linear polarization of the beams in calcite. However, in this case the light intensity is decreased twice. Therefore, the polaroid càn not be used whenever the intensity of the original beam of the telescope is small. The use of a depolarizer composed of a stack of crystal plates or a $\lambda/2$ plate oriented at a definite angle, may turn out to be more advantageous in certain applications.

The calibration procedure is performed in the same way as for magnetographic measurements. A polaroid is placed before the calcite plate such that its main direction coincides with one of the directions of linear polarization in calcite. In this case, bifurcation of image is absent and only one of the circular components remains in the spectrum. By rotating the plane-parallel plate of the Doppler compensator through a fixed angle, the estimated displacement $\Delta\lambda_K$ of the spectral line is assigned and the amplitude of calibration is recorded. Then, the polaroid is removed, DKDP voltage de-energized and the noise, the mean square value of the amplitude of which is assumed to be the sensitivity, is registered. In order to rule out the effect of local velocities on the accuracy of calibration, this is carried out with a substantially defocussed image.

In the summer of 1980 and the spring of 1981, this method was used in observations with the telescope of the Sayan Observatory of SibIZMIR. Figure 3 is a photograph of

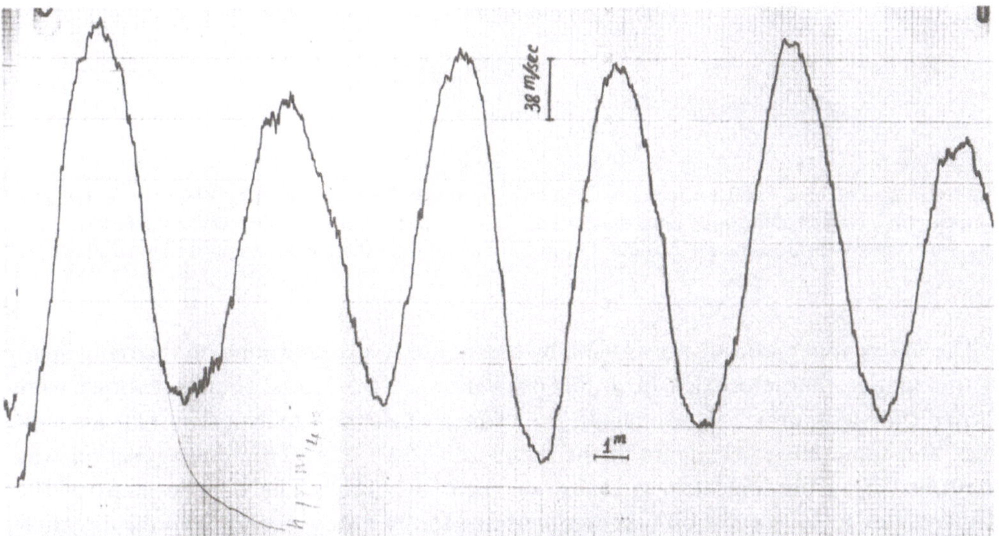

Fig. 3. Fragment of a record of 5-min oscillations of the line-of-sight velocity, as obtained with the differential method for $L = 12''$. Though the value of the velocity is not large on this record (peak to peak 250 m s^{-1}) the signal is very clear because the noise of the spectrograph has been compensated for.

part of a record of 5-min oscillations of the line-of-sight velocity observed in the quiet photosphere at the center of the disk. The observations were made in the Fe I 5250.2 Å line with a 1.4″ × 4.5″ aperture. The sensitivity of this record, determined as a mean square of the noise amplitude with the DKDP de-energized, is about 0.3 m s^{-1} for a time constant $\tau = 5$ s. This noise is quite inconspicuous on the record, while noise-like deviations reaching 10–15 m s^{-1} observed on some portions of the record, are the real signal due to image motion. Further sensitivity augmentation is possible through an increase in light intensity and signal integration time. For comparison, Figure 4 gives a fragment of a similar record taken using a conventional method with the aid of a Doppler compensator. On this record, the noise amplitude is as high as 250–300 m s^{-1} for $\tau = 1$ s.

The method described herein is primarily intended to measure line-of-sight velocities. However, it is easy to modify it for measurements of brightness relative variations of two elements of the solar surface in a chosen part of the continuum or in white light.

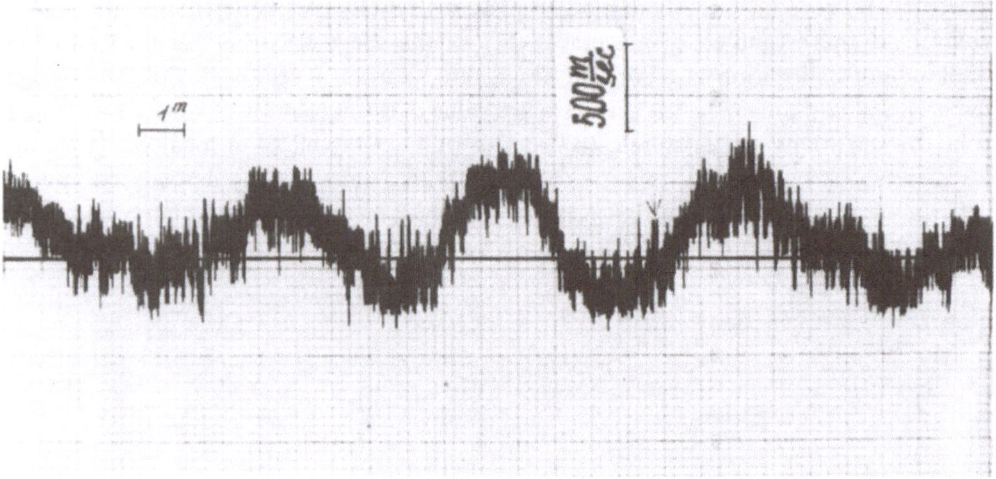

Fig. 4. Fragment of a similar record, as obtained by the author earlier with the aid of a Doppler compensator. A region of the quiet photosphere 1.4″ × 2.5″ at disk center was observed in the Fe I 5250.2 Å line. The velocity (peak to peak) is about 700 m s^{-1}. The noise of the spectrograph is about 250 m s^{-1}.

The differential method was applied by the author in observations on the refinement of the spatial characteristics of 5-min oscillations. Two kinds of observations were carried out: temporal and spatial. In the former case, the spectrograph entrance slit 'sees' the same image elements in the course of 1.5 to 3 hr. The image position was controlled by a photoelectric guide to an accuracy of 1″. The displacement of the elements due to solar rotation was compensated for by a slow scanning in the direction of rotation. Regions of a quiet photosphere at the disk center without pores, faculae, and strong magnetic fields were usually selected for observation. The temporal series performed differed from each other in distance L between the elements under investigation. For most observations, the direction was the same as that of the solar equator. A set of calcite plates available to us allowed us to discretely change L from 2″ to 40″. The length of the spectrograph entrance slit was varied from 2″ to 12″, while its width was kept constant, 1.4″. With this method, a minimum amplitude of 5-min oscillations must be observed for the elements, in which both the phase and amplitude of oscillations show the best matching. In this case, using the value of L it is possible to try to determine the spatial wavelength λ. An analysis of 24 observations (each of a 1.5 to 3 hr duration) yields a value of $\lambda \approx 30-32''$. Approximately the same period can be identified from the power spectra, calculated by the FFT method using results of spatial scan processing (Figure 5). Scans of 300–350″ length in the direction L were taken in quiet regions of the photosphere near the disk center, with apertures of 1.4 × 4.5″ and 1.4 × 6.5″.

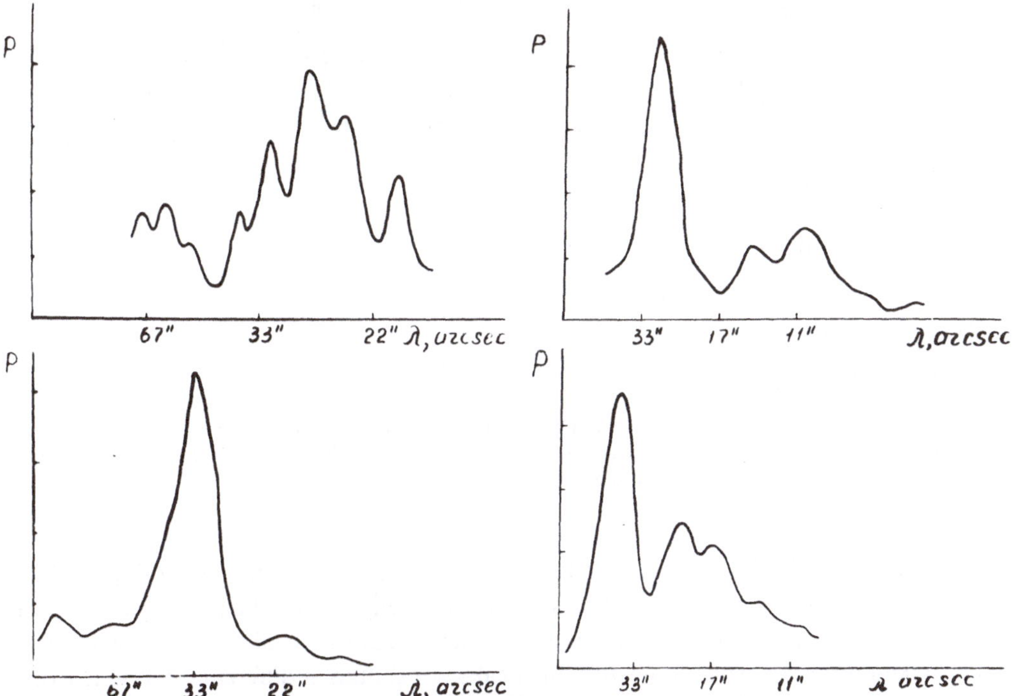

Fig. 5. Power spectra of individual scans. A mean length of one scan is 350″. The scanning rate is 1″ per sec. The scale P is arbitrary.

Of course, the statistic of these observations is insufficient, therefore the estimate we made should be regarded as a preliminary one. Results like ours have been obtained earlier by Fossat and Ricort (1973, 1975) and Deubner (1972).

The series in Figure 6 visualizes the manner of center-to-limb variation of the spectral composition of 5-min oscillations. The spectra were obtained from numerical treatment of three time series (each of 2.5 to 3 hr) taken in the 5250.2 Å line at L = 12″ and with a 1.4 × 6.5″ spectrograph entrance slit; the first series corresponds to the disk center,

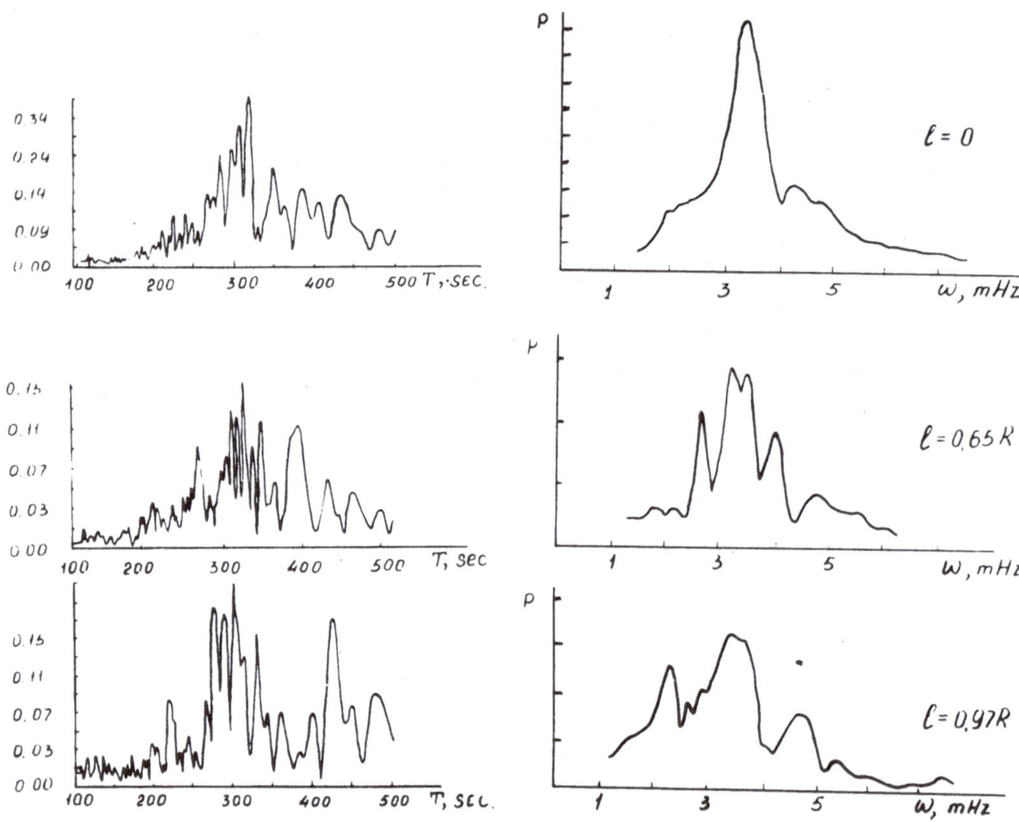

Fig. 6. Center-to-limb variations of the spectrum of 5-min oscillations. On the left are correloperiodograms as calculated by the method of Kopecký and Kuklin (1971); on the right, power spectra. Arbitrary units along the ordinate axis are used for all spectra.

the second to 0.65 R distance between the center and the N-pole, and the third to 0.97 R. Regions for observations were chosen near the central meridian. Typically, with increasing distance from the center there is an increase in side peaks with respect to the main maximum; at the same time, the side peaks are shifted in different directions from their central position, thereby expanding slightly the frequency band. More significant variations refer to the low-frequency region of the spectrum. It is interesting that on some histograms showing the distribution of the number of oscillations with a period, like the one obtained earlier by Howard (1967), a similar behavior of center-to-limb variations can be observed. If these are indeed the real variations of the structure of the spectrum of 5-min oscillations rather than projection effects, for example, then they possibly should be taken into account also in studies on low-frequency global oscillations, being made in integral light with the differential method.

Acknowledgements

I would like to thank Prof. V. E. Stepanov and Dr V. M. Grigoryev for helpful discussions and V. G. Mikhalkovsky for his assistance with the translation and typing of the manuscript.

References

Brandt, P. N., Deubner, F. L., and Schröter, E. H.: 1978, *Proceedings of the Workshop on Solar Rotation*, p. 126.

Deubner, F. L.: 1972, *Solar Phys.* **22**, 263.

Dittmeyr, P. H.: 1977, PHD thesis, Stanford University.

Fossat, E. and Ricort, G.: 1973, *Solar Phys.* **28**, 311.

Fossat, E. and Ricort, G.: 1975, *Astron. Astrophys.* **43**, 243.

Howard, R.: 1967, *Solar Phys.* **2**, 3.

Kalinjak, A. A. and Vasilyeva, G. I.: 1971, *Solar Phys.* **16**, 37.

Kopecký, M. and Kuklin, G. V.: 1971, in *Issledovaniya po geomagnetizmu, aeronomii i fizike Solntsa* **2**, 167.

Kotov, V. A., Severny, A. B., and Tsap, T. T.: 1978, *Monthly Notices Roy. Astron. Soc.* **183**, 61.

Lebedev, N. N. and Grigoryev, V. M.: 1976, *Solar Phys.* **48**, 417.

A NEW SYSTEM FOR OBSERVING SOLAR OSCILLATIONS AT THE MOUNT WILSON OBSERVATORY*

I. *System Design and Installation*

EDWARD J. RHODES, JR.

Department of Astronomy and Earth and Space Sciences Institute, University of Southern California, Los Angeles, CA 90007, U.S.A.

and

Space Physics Section, Jet Propulsion Laboratory, California Institute of Technology, Pasadena, CA 91109, U.S.A.

ROBERT F. HOWARD

Mount Wilson and Las Campanas Observatories, Carnegie Institution of Washington, Pasadena, CA 91101, U.S.A.

ROGER K. ULRICH

Department of Astronomy, University of California at Los Angeles, Los Angeles, CA 90024, U.S.A.

and

EDWARD J. SMITH

Space Physics Section, Jet Propulsion Laboratory, California Institute of Technology, Pasadena, CA 91109, U.S.A.

Abstract. In this paper we describe a new observing system which is currently nearing completation at the Mount Wilson Observatory. This system has been designed to obtain daily measurements of solar photospheric and subphotospheric rotational velocities from the frequency splitting of non-radial solar *p*-mode oscillations of moderate to high degree (i.e. $l > 150$). The completed system will combine a 244 × 248 pixel CID camera with a high-speed floating point array processor, a 32-bit minicomputer, and a large-capacity disc storage system. We are integrating these components into the spectrograph of the 60-foot solar tower telescope at Mount Wilson in order to provide a facility which will be dedicated to the acquisition of oscillation data.

1. Introduction

The observations described in the previous paper (Rhodes *et al.*, 1983) were obtained as a portion of the on-going program of visitor research conducted at the Kitt Peak National Observatory. As such these observations could only be obtained when the

* Proceedings of the 66th IAU Colloquium: *Problems in Solar and Stellar Oscillations*, held at the Crimean Astrophysical Observatory, U.S.S.R., 1–5 September, 1981.

Solar Physics **82** (1983) 245–258. 0038–0938/83/0822–0245$02.10.

McMath telescope was available for this particular project. In order to provide similar observations with enough frequency to study temporal fluctuations in the solar rotation rate and in the excitation and decay of the p-mode oscillations, an observing system must be provided which is nearly dedicated to the acquisition of solar oscillation observations. Such a system is now nearing completion at the Mount Wilson Observatory. In this paper we will give a brief description of this system and outline its current state of development.

2. Scientific Goals

As described in the preceding paper (Rhodes *et al.*, 1983), the solar non-radial p-mode oscillations can be utilized to measure the rate of rotation of the Sun both within the photosphere and over a range of distances below it. A technique for doing so with moderate and high-degree modes was described by Rhodes (1977), by Rhodes *et al.* (1979), and by Deubner *et al.* (1979). By repeatedly observing a rectangular area at the center of the solar disk for intervals of up to 12 hr in duration, we can separate out those oscillations traveling eastward around the Sun from those traveling westward.

When the Sun is accurately kept centered in the telescope's field of view for such long durations, the p-mode ridges are shifted by an amount, $\Delta\omega$, which depends directly upon both the rotational velocity, V_{drift}, and the horizontal wavenumber, k_h, as follows:

$$\Delta\omega = V_{\text{drift}} k_h .$$

Thus, by measuring $\Delta\omega$ as a function of k_h, one can measure the rotational velocity of the wave pattern across the telescope's field of view.

The first such observations were obtained by Franz Deubner during 1977 at the Sacramento Peak Observatory with the diode array on the tower telescope there. These results were published by Deubner *et al.* (1979). In that paper rotational velocities, V_H, normalized to the surface ($Z_{\text{eff}} = 0$) velocity were shown as functions of depth, Z_{eff}, for three different days. Near the surface the three days gave essentially the surface rotation rate, while below a depth of 8000 km there was some suggestion of an increase in rotational velocity with depth. However, the observational scatter became larger below 8000 km and the three days' measurements were insufficient to conclusively demonstrate the existence of more rapid rotation below the photosphere. Furthermore, earlier at this colloquium both Deubner (1983) and Rhodes *et al.* (1983) presented more recent observations of p-mode frequency splitting which did not show any evidence for such a rotational velocity increase. On the other hand, Claverie *et al.* (1981) recently presented evidence of rapid interior rotation based on the frequency splitting of dipolar and quadrupolar p-modes in the five-minute band. If the interpretation suggested by Claverie *et al.* (1981) is correct, then the solar angular velocity must increase at some depth below the photosphere.

Apart from such questions of the reality of a radial gradient in the angular velocity, the technique described above is also useful for obtaining absolute rotational velocities in a manner that is completely independent of existing Dopplershift techniques.

Preliminary absolute rotation rates were discussed by Rhodes *et al.* (1983) and additional analysis of the data they presented is currently in progress.

In view of the somewhat contradictory nature of observations of both the absolute photospheric rotation rate and of the sub-photospheric rotation rate, we believe that it is important to obtain additional observations of the type presented by Deubner *et al.* (1979) and by Rhodes *et al.* (1983). Specifically, we hope to study the following problems: (1) What is the absolute, non-normalized rotational velocity determined for each day in the photosphere? (2) Are there daily variations in this rate? (3) Is there a true radial gradient in rotational velocity? (4) If so, does it extend as deep as 20 000 km? (5) Are there temporal fluctuations in the deepest observable rotational rates? (6) Can we see any evidence for the rapid interior rotation observed by Claverie *et al.* (1981)?

With the system we are describing here, we plan to obtain rotational velocity vs. depth maps similar to those shown by Deubner *et al.* (1979), on a daily basis for large fractions of each year. We are aware that Gough (1978) has criticized the technique employed by Ulrich *et al.* (1979) in the computation of the effective depths at which various p-modes are sensitive to the rotational velocity. Specifically, Gough (1978) criticized the assumption of a monotonic increase in angular velocity with sub-photospheric depth which Ulrich *et al.* employed. Gough further argued that when he employed a theoretical angular velocity profile which was not monotonic with depth, he found the effective depths calculated with the Ulrich *et al.* technique to be multi-valued.

In answer to this criticism we note the following points: (1) The method used by Ulrich *et al.* employed a Taylor series expansion of the solar angular velocity profile. While they truncated this expansion to include only the constant and linear terms in the numerical computations they presented, they pointed out that effective depths corresponding to the higher-order terms could also be calculated after which a least-squares analysis could be carried out to determine the magnitudes of these terms should future observational data suggest a need to do so. (2) The more recent observational data of Deubner (1983) and Rhodes *et al.* (1983) do not support the existence of any radial gradient in the angular velocity and instead suggest that the alternative inversion techniques described by Gough (1978) are not necessary. (3) The inversion technique described by Gough (1978) employs a linear combination of eigenfunctions to obtain a contribution function which is sharply localized. As Gough (1983) has himself pointed out at this colloquium, this linear combination includes both positive and negative coefficients and hence can be quite sensitive to observational errors in the frequency shifts. Gough (1983) has gone on to outline a modification of his 1978 technique which is designed to allow for such observational uncertainties; however, this revised method has not yet been employed with actual data.

By obtaining a $k_h - \omega$ diagram each day, we will be able to search for evidence of the temperature and horizontal flow perturbations described at this colloquium by Hill *et al.* (1983). For those intervals of time when day-to-day fluctuations are small we should be able to improve upon the error brackets presented by Deubner *et al.* (1979) by combining the results of successive days.

Furthermore, with this system we will be able to study the statistical properties of the

p-mode oscillations themselves. In particular, we plan to compare the oscillation amplitudes and the observed frequencies on different days in order to search for possible changes with time. Beating between unresolved eigenmodes can certainly produce amplitude variations; however, it should be possible to test this hypothesis by studying a long sequence of observations and then combining data from several sequential days of observation.

3. The 60-Foot Telescope

The new system is currently being installed at the 60-foot solar tower telescope at Mount Wilson. This telescope is the instrument with which magnetic fields were discovered on the Sun (Hale, 1908) and with which the solar '5-minute' oscillations were discovered (Leighton *et al.*, 1962). The Sun is acquired with two movable flat mirrors which are located in the dome at the top of the tower. With this coelostat arrangement the solar image does not rotate during the cource of a day. The light from these two mirrors is aimed down the tower to either of two alternative sets of optics. In the first optical arrangement a 12-inch doublet lens is located just below the two flat mirrors. This lens has a 60-foot focal length and forms a 6.8-inch diameter solar image at the entrance slit of the spectrograph which is located at the bottom of the tower.

In the second optical arrangement the light follows an off-axis folded path. A 10-inch diameter off-axis paraboloid mirror is located just to the side of the optical axis of the tower at the level of the spectrograph entrance slit. The light from the flat mirrors at the top of the tower is reflected upward by this off-axis mirror to a third flat mirror which is located 15 feet above it. The converging beam is reflected back downward by the flat mirror and finally forms a 3.4-inch diameter image at the spectrograpth entrance slit. The three aperture masks which are used with the paraboloid mirror have diameters of 5, 7, and 10 inches, resulting if $f/72$, $f/48$, and $f/36$ beams, respectively.

Currently, the 60-foot telescope and its spectrograph are being used to continue two daily sequence of calcium k-line and Hα spectroheliograms which were begun in 1915. The acquisition of these spectoheliograms requires only an average of 30 min per day. Consequently, the telescope and spectrograph can be dedicated to the acquisition of oscillation observations for the remainder of every clear day.

4. Guider Assembly

One of the critical requirements of the rotation-measurement techniques we are using is that of very accurate guiding of the solar image. Specifically, the limbs of the solar image must be kept fixed with respect to the spectrograph entrance slit to within a tolerance of less than one or two arcseconds for durations of up to 12 hr at a time. To obtain such stability we have installed a new guider assembly at the 60-foot telescope. The new guider is a smaller copy of the four-limb guider currently in operation at the 150-foot telescope, which is a well-tested design. Each of the two axes of the guider functions in the following way: The light from opposite ends of a diameter of the solar image strikes the cathode of a single photomultiplier tube alternately as a chopper admits each beam. A servo amplifier nulls the a.c. signal which results from an offset of the image by tilting one of the flat mirrors at the top of the tower. The d.c. component of the photomultiplier signal is kept constant by varying the high voltage to the tube so that the characteristics of the servo do not vary with sky transparency. This type of guider has advantages over the multiple-detector variety where the image may drift as the characteristics of the individual detectors vary with time.

In operation a beam splitter sends a part of the main telescope beam to the guider assembly. The light from four limb positions is measured and kept balanced by the guider electronics. Additionally, the guider head now has stepper motors and encoders on each of the two orthogonal axes. These items were added to the guider assembly to provide the necessary scanning capability for this project. Under the control of an LSI−11 the guider head will move across the large plate on which it is mounted. The signal generated by the guider will then move one of the mirrors at the top of the telescope by just the correct amount to move the solar image across the spectrograph entrance slit. A 256-pixel Reticon linear array will be interfaced to the LSI−11 to serve as a limb-position sensor on the spectrograph to keep a record of any drifts of the solar image during each day. These limb positions will then be used to register successive scans during the post-observing data reduction process.

5. The Spectrograph

The spectrograph to be used is a vertical pit spectrograph having a 13-foot focal length. It is located within a 25-foot deep concrete pit which is located within the earth at the bottom of the tower. The bottom of this pit is colder than the ambient room temperature at the top. The collimating and camera lenses and the grating are suspended from the top of the spectrograph itself in a girder-enclosed cage. This carge has been baffled and enclosed in order to interfere with the formation of air currents within the pit.

The spectrograph assembly is motorized and is used daily as a spectroheliograph, however, in our studies we will not move the spectrograph during the acquisition of the observations. Instead, we will use the instrument solely as a spectrograph. In addition, the cage assembly containing the spectrograph optics can be firmly attached to the side of the spectrograph pit during each day's oscillation observations in order to eliminate any vibrations which might be present in the spectrograph itself.

The spectrograph grating is a 5 × 6 inch Babcock grating having a groove spacing of 600 per millimeter. The grating is blazed for the second order yellow ($\lambda 5800$). The dispersion is 1.54 Å mm^{-1} in the second order at $\lambda 5576$.

6. The Oscillation Observing System

A. Scanning geometry and frame rates

One of the design goals of this system is the ability of obtaining three-dimensional (x, y, time) radial velocity maps during each day for later spatial compression and analysis. For each position of the entrance slit on the Sun, the spectrograph will disperse a small portion of the solar spectrum onto a CID array camera which will be part of this system. In this way we will be able to record spectral intensities at 244 different wave-lengths for each of 248 one-arc sec-wide positions along the spectrograph entrance slit. By recording this spectrum and then using the guider to rapidly step the solar range across the entrance slit, we will be able to obtain an $I(\lambda, x,$ and $y)$ map once every scan, where I is the intensity at wavelength λ at location x, y on the solar disk.

The geometry of the area which will be scanned in this fashion will be that of a large rectangle. The 248 one-arc sec-wide pixels which will be aligned parallel to the spectrograph's entrance slit on the CID array will be oriented in a north–south direction. The guider will be used to scan the image in the east–west direction. Hence, for a slit width of four arc sec, a 1000 arc sec distance along the solar equator will be obtained in 256 readouts of the CID array camera.

Preliminary light-level tests we have carried out at Mount Wilson with the CID camera suggest that we should be able to reach 50% of the chip's saturation level in 50 to 100 ms, depending upon time of day, time of year, magnification of the spectrum on the CID chip, and cleanliness of the optics in the telescope and spectrograph. Adding in the 50 ms readout time to read 7440 pixels (i.e. 30 rows of 248 pixels each), we should get a combined exposure and readout time of 100 to 150 ms. Allowing a few seconds for backward scanning between successive frames, we should be able to scan across the rectangle on the disk in 30 to 45 s. These frame rates are faster than the 60 s rate used at Kitt Peak by Rhodes *et al.* (1981). The frame time may be cut further by 15 s by modifying the camera circuitry so that a portion of the array can be readout simultaneously with the exposure of the remainder of the array. This modification would also eliminate the need for the cylindrical shutter.

B. Absolute velocity reference provision

By observing two telluric lines located on either side of the reference solar line, it is in principle possible to remove spectrograph drifts down to 10 to 20 m s^{-1}.

Two such solar lines are the Fe I line λ6302 and the Ni I line λ6213. Both of these lines have been employed at Kitt Peak by Rhodes, Harvey, and Duvall. The first, λ6302, is partially blended with one of the references lines and is not too useful. The second, λ6213, a clean line which is free of blends. This line was used in May, 1980, to obtain the p-mode power spectra presented by Rhodes *et al.* (1981). In these observations the telluric lines were also recorded but for the spectra which were published in Rhodes *et al.* the telluric lines were not actually used to calculate absolute wavelengths. Rather, the mean of each velocity scan was subtracted before the power spectrum was computed. Other, unpublished power spectra have been generated by Rhodes *et al.* in lines such as the Na D lines, λ8542, Mg B λ5173, λ6314, and Hα without absolute velocity reference information. Thus, it is our belief that, for studies of the frequencies of p-modes in the 5-min band, an absolute velocity reference is not essential. On the other hand, for studies of the excitation and decay of these modes and for the low-frequency portion of the $k_h - \omega$ plane such a reference system may be essential. If we find that the telluric lines are not adequate for this purpose, then we will work to incorporate a laboratory reference line into our system.

C. Optical path

In order to allow for both the continuation of the existing series of spectroheliograms and the oscillation observations, the schematic design shown in Figure 1 was selected. In this design the CID camera assembly is mounted in an enclosure which is attached to the side of the main spectrograph. Through the use of a narrow-periscope assembly, light can be deflected to the CID detector array which will in turn be located within a liquid-nitrogen-cooded dewar. By allowing the whole dewar-periscope assembly to move on rails in the direction of dispersion of the spectrograph (into and out of the plane of Figure 1), we can position it so that it is completely out of the path of the light which is used to expose the photographic plates placed above the spectrograph's exit slit each morning.

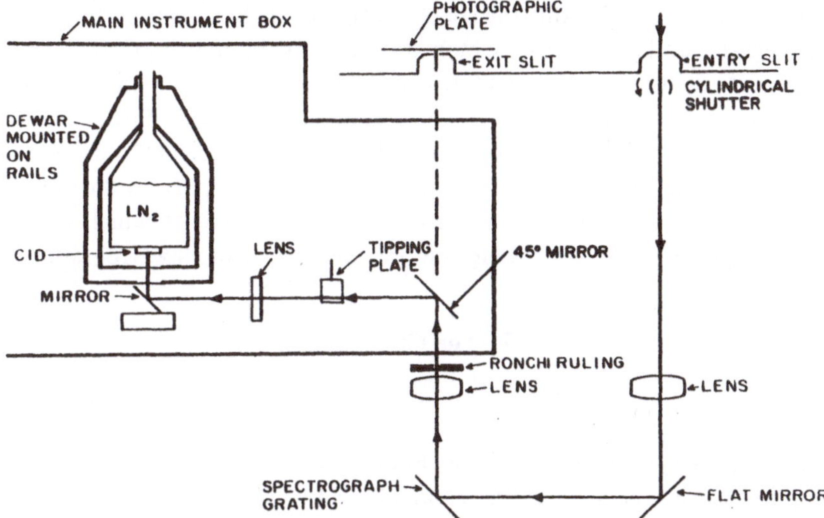

Fig. 1. Schematic design of the new system. The CID camera is mounted inside of an enclosure which is mounted in the side of the existing spectrograph. The light from the spectrograph can be directed toward either the CID camera for oscillation studies or the exit slit for photographic spectroheliograms. The various optical components are described in the text.

Other components of the system included in Figure 1 are: (1) a high-speed rotating cylindrical shutter which provides the short (~ 50–100 ms) exposure times needed for this project; (2) an anamorphic lens which magnifies the spectrum along the dispersion direction while leaving the spatial dimension unchanged; (3) a tipping plate which can be used to keep the spectrum centered on the CID detector array during each run; and (4) a small glass Ronchi ruling which can be swung into the light beam just above the exit lens of the Littrow spectrograph in order to provide flat-field illumination of the CID array when needed. The Ronchi rulings are parallel to those of the greating so that at the focus of the spectrograph many overlapping orders effectively smear out the spectrum lines. Also, since the CID chip is only 1 cm on each side and is located at a distance of roughly 13 feet from the location of the Ronchi ruling, it subtends an angle of only 0.2 degrees which is small enough that the light is uniform over the area of the chip.

D. Current installation status

The instrument enclosure and moveable stage described above was machined at U.C.L.A. and at the Jet Propulsion Laboratory, was assembled at the Jet Propulsion Laboratory and is now permanently mounted on the side of the spectrograph. It covers a hole which was cut into the side of the spectrograph to allow for the movement of the mechanical stage and the periscope assembly inside of it. The stage assembly was designed to provide rotational symmetry about the centerline of the downward-looking CID chip itself for ease in lining up the solar spectrum on the chip, while also providing for focus travel and motion along the dispersion of the spectrograph. When the stage and periscope are located within the spectrograph, the light arrives at the periscope from the spectrograph collimating lens which is located at the bottom of the cage assembly which supports the grating. The light is then reflected to a second 45° flat mirror which is located within the stage assembly. This second mirror reflects the light upward where it is imaged upon the downward-looking CID chip in the dewar. In normal operation, curtains are attached to the sides of a large plate which separates the enclosure from the spectrograph. These curtains serve to prevent air currents which might build up within the enclosure from disturbing the optical path of the spectrograph.

7. The CID Camera

A. Design

The CID camera system was built by Photometrics, Ltd. of Tucson, Arizona. The CID chip itself is a General Electric Model D 244 × 248 pixel charge-injection device. As previously described this chip is located in a dewar which is cooled to the temperature of liquid nitrogen to minimize thermal noise. Also part of the camera head is a small electronics box which is mounted on the dewar. This box includes the 12-bit high speed analog-to-digital converter, the analog circuits, the clock drivers, and the line drivers and receivers for the cables connecting the camera to the controller and to the array processor.

This camera is similar in overall design to the 244 × 248 pixel camera currently in operation at Kitt Peak, since it was designed by the same man, Richard Aikens of Photometrics, who designed and built the Kitt Peak system (Aikens, 1980). However, this camera has been improved over the Kitt Peak camera in several respects: (1) it is faster than the Kitt Peak system, with a per-pixel read time of only seven microseconds; (2) it provides in-camera row-crosstalk correction; (3) it contains enough memory to store a single bias frame (i.e. a low-intensity, artifically-exposed frame) within the camera; and (4) it subtracts the bias frame pixel-by-pixel from each data frame before outputting the digitized pixels. Thus, this system provides high-speed cleaned-up 12-bit digital data at a video rate. Only a flat-field frame need be stored outside the camera for use in subsequent photometric corrections.

The linearity of both General Electric model B and D CID chips has been measured accurately. Aikens (1980) has presented a plot showing that a model B CID in use at Kitt Peak was linear to better than 0.05% for light levels ranging between 10 and 95% of that chip's saturation level. It is the non-linear behavior below the 10% threshold which makes it necessary to acquire and subtract a bias frame as discussed above. More recently, Aikens (1982) has measured the linearity of model D chips identical to the one installed in the Mount Wilson camera and has likewise found no deviations larger than 0.05% for light levels ranging between 10 and 95% of the saturation of this chip.

The G.E. model D chip in use at Mount Wilson is a slightly re-designed version of the model B chip in use at Kitt Peak. The design of the earlier chip was altered to eliminate a small odd-even pixel difference which was present in the model B chips when they were illuminated with a sharp intensity gradient. The model D chip employs a strictly periodic electrode spacing between all pixels in a given row to eliminate this problem.

On both types of CID chips a small amount of signal leakage exists among adjacent pixels in a given row. This so-called row-crosstalk is removed from the Mount Wilson camera's video data by custom-designed signal-processing circuitry which is built into the electronics of this camera as mentioned above.

The random noise generated within the camera itself has been measured both in Tucson and at Mount Wilson by recording a low-level bias frome on the chip and then subtracting this bias frame from itself with the camera's non-destructive readout mode. By doing this over many successive frames, an rms noise of only one least-significant-bit of the A/D converter was measured in Tucson. Since the saturation signal level is close to 4096, an rms signal-to-noise ratio of about 4000 to 1, is possible with this type of detector. (The peak-to-peak noise was measured in the same tests to be about four or five times larger.) Since most measurements will be made at less than the saturation signal level, the signal-to-noise ratio will generally be less than the above upper limit. The model B CID chip is use at Kitt Peak has obtained a signal-to-noise ratio of 1500 to one (Aikens, 1982) in actual operation.

Because of the relatively noisy electromagnetic environment at Mount Wilson (due to the close procimity of numerous radio and television transmitters to the Observatory), we have retested the CID chip at the telescope. Our preliminary test results show an rms noise of two to three A/D converter units with a corresponding decrease in signal-to-noise ratio. However, these preliminary tests were carried out prior to any extensive efforts at shielding the camera. Since merely placing the camera inside the enclosure on the side of the spectrograph reduced the noise level by a factor of three, we are confident that additional shielding efforts will further reduce the measured noise so that it will approach the levels measured in Tucson. In any event, when we average the 248 pixels along the spectrograph slit in order to convert the three-dimensional

(x, y, time) data arrays into two-dimensional (x, time) array, we will realize an approximate 16-fold increase in the signal-to-noise ratio.

All of the optical components have been installed in the system and the Na D1 line has been focused on the CID chip. An analog display of this portion of the solar spectrum is displayed in Figure 2. This figure shows the spectrum as it is displayed on

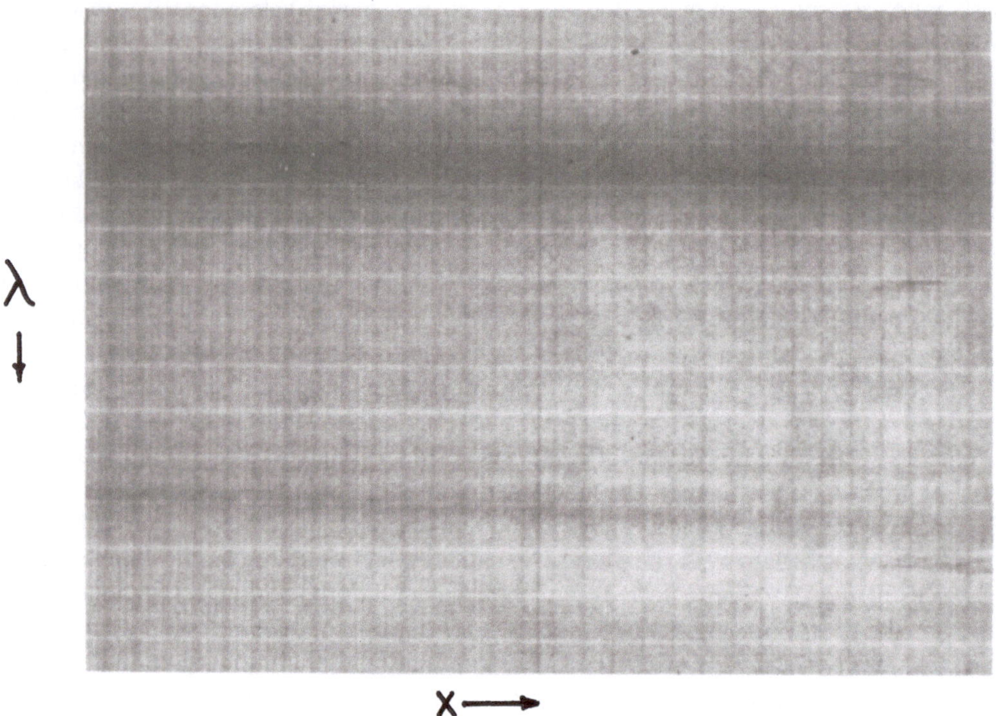

Fig. 2. A monitor display of a portion of the solar spectrum near the Na I D1 line is shown. The spectral intensities are shown as various grey levels. The directions of increasing wavelength and position on the Sun are shown a the edges of the display. The bright lines are the reseau pattern on the monitor and are not defects in the camera.

the Tektonix 604 persistence monitor located beside the spectrograph. The spectral lines are shown as a grey-level display. The original digital data generated by the CID camera has been re-coverted into an analog signal display on the monitor. The evenly-spaced bright and dark lines in the picture are not defects generated by the CID but are simply a grid located on the monitor faceplate. Furthermore, this display was not corrected for diode sensivity changes since the flat field corrections are made latter in the array processor. The uncorrected display will be used mainly for focusing and alignment purposes during actual observations.

8. The Data Processing System

A. Data path

The data path of this system is shown schematically in Figure 3. The CID camera system is controlled by an LSI–11 microcomputer which functions solely as a control microprocessor. The LSI–11 obtains interrupt pulses from the rotating cylindrical shutter and commands the camera to alternately build up exposures or read them out. It also commands the bias-frame exposures and flatfield exposures. The LSI–11 also controls the guider as it scans the solar image and controls the tipping plate and Ronchi grating motors.

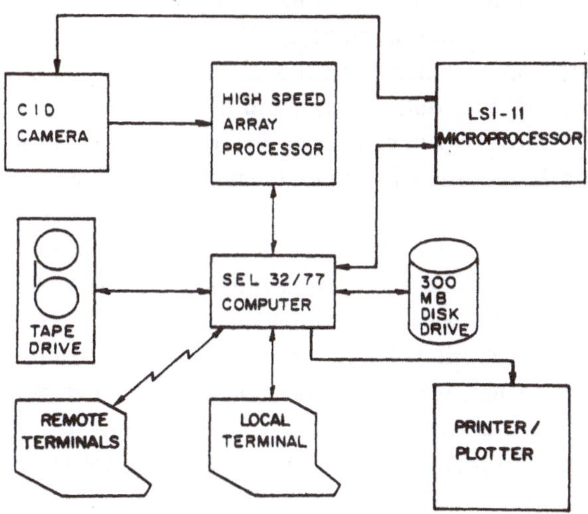

Fig. 3. The data path for the system is shown schematically. The camera spectral data is passed into the array processor for real-time conversion into line-of-sight Doppler velocities. The velocities are then passed to the SEL 32/77 computer for collection into two-dimensional images before being stored on the disc storage system. The LSI–11 serves as the control processor for synchronizing the camera, shutter, guider and array processor. The entire system also can be monitored and operated remotely via dial-up telephone line connections.

From the CID camera the data passes through an auxiliary I/O port into a high-speed CSPI MAP 300 array processor. The camera is designed to signal the I/O device upon the completion of each exposure and to signal it as each pixel within a single frame is ready to be output to the array processor. The CSPI MAP 300 processor is designed for high-speed asynchronous applications and is configured with memory on two different internal buses. Thus, the CID camera system can send one picture of data of the MAP 300, while the processor is computing with the previously-transmitted picture's data. This double-buffered approach allows for a very high speed, yet it is completely synchronized to the operation of the camera on a frame-by-frame basis.

The array processor will take as input from the CID camera a set of thirty or forty different spectral intensities at each of the 248 spatial locations along the entrance slit. The processor will next correct the raw data for the previously-stored flat-field irregularities and then convert the line profiles into Doppler velocities. Several different methods of line center determination can be used with the floating point array functions provided in the MAP 300. We will empirically determine which profile-fitting technique is most useful for our purposes.

Following the conversion of the spectral intensities into 248 Doppler velocities for each CID frame, the velocities will then be sent to the Systems Engineering Laboratories (SEL) 32/77 computer before ultimately being stored on an Ampex 300 megabyte disk storage system. The SEL 32/77 is a 32-bit minicomputer capable of providing up to 16 separate users one-megabyte partitions. The SEL 32/77 we have currently has 1.25 megabytes of 600-nanosecond MOS memory. The SEL 32/77 will store each set of velocities on the disk system and then at the conclusion of each day's observing, it will read the data back from the disk, convert three-dimensional (x, y, time) arrays into two-dimensional (x, time; or y, time) arrays and then compute the two- or three-dimensional power spectra which will result from each day's worth of data.

The 300 megabyte capacity of the disk system was seleced to provide sufficient storage for a complete 12-hour observing sequence. In one 248×256 element frame there will be nearly 64 000 pixels. By storing the velocities as 16-bit halfword integers we will require roughly 128 kilobytes to store a frame. At a frame rate of 2 frames per minute, as discussed earlier, we will require 256 kilobytes per minute. Since there are 720 min in 12 hr, we should be able to store one full run in roughly 180 megabytes. Furthermore, the velocity measurements themselves will not require 16 bits of storage and so it should be possible to store an 8-bit intensity pixel and an 8-bit velocity pixel in each 16-bit halfword. In this way we could obtain power spectra of both velocity and intensity images each day. Alternatively, we could combine two 8-bit velocity pixels in each halfword and obtain power spectra in two separate spectral lines each day. Upon computation of the power spectra, the SEL 32/77 will then calculate the solar rotational velocities for different locations on the $k_h - \omega$ plane. Finally, the power spectrum and rotational velocities from each day's run will be stored on magnetic tape with two Kennedy model 9100, 75 inch-per-second, 9-track tape drives. The complete system will be under the local terminal control of the observer at the telescope and under remote control from one of several remote terminal stations. A Printronix P300 printer/plotter is also available for generating simple contour and gray-level displays of the power spectra as well as computational printouts. A Tektronix 4006 graphics terminal is also available for higher-resolution plotting in real time. A Tektronix 4014 graphics terminal, hard copy unit, flat-bed plotter, and line printer are also available at a remote workstation on the U.S.C. campus. All plots generated at Mount Wilson can be sent to U.S.C. for inspection there. Other terminals which provide text-editing capability alone are located at the Mount Wilson offices in Pasadena, at the Jet Propulsion Laboratory, and at U.C.L.A.

B. CURRENT INSTALLATION STATUS

The computing hardware has all been installed in a new data room which was added to the 60-foot telescope for this project.

The I/O port connecting the CID camera to the array proccessor has been successfully operated under program control. Digitized video data has been transferred from the camera into the array processor before being transferred into the SEL 32/77 computer. A tape containing the digital data was subsequently generated and taken to the Mount Wilson offices in Pasadena, California and displayed there with other equipment.

The LSI–11 has been successfully linked to the SEL computer and programs and data have been transferred between the two. The new guider and its scanning hardware and control circuitry have all been installed. The cylindrical shutter has been built and mounted on the spectrograph.

9. Tasks Remaining

The remaining tasks are the completion of control circuitry to interface the guider, shutter, and camera to the LSI–11 which will provide the real time control signals; the completion of the real-time operating software, the array processor software, and the final data reduction software which will convert the time series of velocity measurements into $k_h - \omega$ power spectra.

Acknowledgements

The enclosure, shutter, and mechanical stage were all designed by Thomas Andrews of the Jet Propulsion Laboratory.

James Wilkie, Leland Barnes, and Harland Epps of the UCLA Astronomy Department were responsible for the fabrication of the mechanical stage assembly. The fabrication of all other parts was carried out at JPL. Maynard Clark of the Mount Wilson Observatory was responsible for the construction of the new guider and for its interface to the LSI–11. He was also responsible for the design and construction of the control circuitry for the shutter, tipping plate, and Ronchi grating mechanism. Richard Aikens of Photometrics, Ltd. designed the CID camera system and its interface to the array processor. Jeffrey Mannan of the Orbis Corp. was responsible for the installation, integration, and operation of the SEL 32/77 computer. John Boyden of the Mount Wilson Observatory was responsible for the interface of the LSI–11 to the SEL 32/77 and for all of the real-time system software. Bradley Wood and Dan McKenna of the UCLA Astronomy Department assisted in the evaluation of the suitability of the CID camera for this work. In addition Dan McKenna suggested the periscope optical path design which allowed the new equipment to be mounted within the existing spectrograph. Extensive discussions with Jack Harvey and Tom Duvall at Kitt Peak regarding the operation and software control of their CID camera system have been invaluable in the development of the detailed design of this system.

We wish to acknowledge that financial support for this project has been provided by the following: NSF grants ATM 7820420, AST 7820404, and ATM 8009469: NASA grants NSG 7407 and NAGW–13: the JPL Director's Discretionary Fund (Project No. 730–00169–0–3280); and Carnegie Institution of Washington. The Jet Propulsion Laboratory is operated by the California Institute of Technology, for the National Aeronautics and Space Administration, under contract NAS7–100.

References

Aikens, R. S.: 1980, 'A Large CID for Use in Astronomy', AURA Engineering Technical Report No. 66, Kitt Peak National Observatory, Tucson, Arizona.

Aikens, R. S.: 1982, private communication.

Claverie, A., Isaak, G. R., McLeod, C. P., van der Raay, H. B., and Roca Cortes, T.: 1981, *Nature* **293**, 443.

Deubner, F.-L.: 1983, *Solar Phys.* **82**, 103 (this volume).

Deubner, F.-L., Ulrich, R. K., and Rhodes, E. J., Jr.: 1979, *Astron. Astrophys.* **72**, 177.

Gough, D. O.: 1978, in G. Belvedere and L. Paterno (eds.), *Proc. Workshop on Solar Rotation*, University of Catania, Catania, Italy, p. 255.

Gough, D. O.: 1983, *Solar Phys.* **82**, 7 (this volume).

Hale, G. E.: 1908, *Astrophys. J.* **28**, 315.

Hill, F., Toomre, J., and November, L. J.: 1983, *Solar Phys.* **82**, 411 (this volume).

Leighton, R. G., Noyes, R. W., and Simon, G. W.: 1962: *Astrophys. J.* **135**, 474.

Rhodes, E. J., Jr.: 1977, Ph.D. Thesis, U.C.L.A.

Rhodes, E. J., Jr., Deubner, F.-L., and Ulrich, R. K.: 1979, *Astrophys. J.* **227**, 629.

Rhodes, E. J., Jr., Ulrich, R. K., Harvey, R. W., and Duvall, T. L., Jr.: 1981, in R. B. Dunn (ed.), *Solar Instrumentation: What's Next?*, Proceedings of Sacramento Peak Observatory Workshop, Sunspot, N.M., p. 37.

Rhodes, E. J., Jr., Harvey, J. W., and Duvall, T. L., Jr.: 1983, *Solar Phys.* **82**, 111 (this volume).

RESONANT COUPLING BETWEEN SOLAR GRAVITY MODES*

(Invited Review)

W. DZIEMBOWSKI

N. Copernicus Astronomical Center, 00-716 Warsaw, Poland

Abstract. It is shown that in consequence of the parametric resonance, g modes of low spherical harmonic degree l are strongly coupled to the modes of high degree. The coupling limits the growth of low l modes to very small amplitudes. For g_1, $l = 1$ mode, the final amplitude of the radial velocity is of the order of 10 cm s^{-1}. A mixing of solar core as a result of a finite-amplitude development of linear instability of this mode is thus highly unlikely.

1. Introduction

The observational evidence for gravity-mode excitation in the Sun is still considered as controversial. Nevertheless, there has been a significant amount of work devoted to the study of the theoretical properties of such modes. This interest was stimulated by Dilke and Gough's (1972) suggestion that the overstability of g modes may lead to mixing in the core, as well as by the potential importance of the detection of these modes for probing the solar interior (e.g. Hill and Caudell, 1979).

Following Dilke and Gough's work, numerical calculations of solar g modes have been made by several groups (Christensen-Dalsgaard *et al.*, 1974; Boury *et al.*, 1975; Shibahashi *et al.*, 1975; Saio, 1980). The common conclusion of those papers is that there is a significant driving effect in the core via ε-mechanism for the g_1 and g_2 modes corresponding to $l = 1$ harmonics. However, the question of whether there is an actual instability will be possible to answer only when a credible theory of the interaction between convection and oscillation is available.

The present paper deals with nonlinear effects in solar gravity modes. Such effects must be investigated if we want to determine whether the development of instability may lead to mixing in the Sun's interior or to predict the amplitudes of radial velocity variation caused by oscillation. The study of nonlinear effects may also help us to understand why, if the 160 min oscillation represents a gravity mode, there has only been one such mode detected so far.

It may be expected that the effect limiting the amplitude growth of an unstable g mode is the parametric resonance. This is the instability of an oscillating system with the angular frequency ω_1 relative to the growth of two modes with angular frequencies ω_2 and ω_3 such that $\omega_2 + \omega_3 \approx \omega_1$. These two newly-excited modes derive their energy from the originally-excited mode and the mutual interaction between all three leads, under certain conditions, to a limitation of the amplitudes.

* Proceedings of the 66th IAU Colloquium: *Problems in Solar and Stellar Oscillations*, held at the Crimean Astrophysical Observatory, U.S.S.R., 1–5 September, 1981.

Solar Physics **82** (1983) 259–266. 0038–0938/83/0822–0259$01.20.

The importance of parametric (resonance) instability in the stellar pulsation theory was first pointed out by Vandakurov (1965). Now a more detailed theory of this phenomenon also covering the equilibrium solution for the interacting modes, is available (Dziembowski, 1982).

2. Onset of the Parametric Instability

The criterion of instability to the growth of modes 2 and 3 in the presence of an excited mode 1 may be written in the following form:

$$Q_1 > \frac{1}{\nu} \left\{ \Delta\sigma^2 \left[1 - \left(\frac{\gamma_2 - \gamma_3}{\gamma_2 + \gamma_3} \right)^2 \right] + 4\gamma_2 \gamma_3 \right\}^{1/2} . \tag{1}$$

In this formula Q_1 is the r.m.s. amplitude of the relative distortion of stellar surface $\Delta R/R$, due to the excitation of mode 1. It is assumed that $Q_1 \ll 1$. $\Delta\sigma = \sigma_1 - \sigma_2 - \sigma_3$ is the frequency mismatch, where σ is the angular frequency in units $\sqrt{4\pi G \langle \rho \rangle}$ and $\langle \rho \rangle$ is the mean stellar density. $\gamma_{2,3}$ are the linear nonadiabatic damping rates in the same units. We assume that $\gamma_{2,3} > 0$. By ν we denoted a normalized coupling coefficient. Its explicit form (Dziembowski, 1982) containing integrals of products of the eigenfunctions for the involved modes is very complicated and will not be reproduced here.

At this point it is not important what was the cause of mode 1 excitation, but it is essential for the validity of the criterion that the present growth (or damping) rate is small in the sense that $|\gamma_1| \ll Q_1 \nu$. The criterion also holds in the cases when some other modes are excited, provided that they are not coupled to any of the three modes considered.

Let us assume now that mode 1 is a gravity mode of low radial order k_1 and low degree l_1, and ask what are the characteristics of the pairs of modes that may be excited as a result of the parametric resonance at the lowest amplitudes of mode 1. As seen from Equation (1), the coupling coefficient must be nonzero and possibly large, while the frequency mismatch and the damping rates of modes 2 and 3 must possibly be small.

The nonzero values of ν occur only for certain combinations of the spherical harmonics, $Y_{l,j}$, of the involved modes. Choosing $l_3 \geq l_2$, the condition that $\nu \neq 0$ may be written as

$$j_1 = j_2 + j_3 ,$$

$$l_3 = \begin{cases} l_2 + 1 & \text{if } l_1 = 1 , \\ l_2 \ \text{ or } \ l_2 + 2 & \text{if } l_1 = 2 , \\ l_2 + 1 \ \text{ or } \ l_2 + 3 & \text{if } l_1 = 3 , \end{cases} \tag{2}$$

and so on.

Moreover, it turns out that the coupling coefficient is large only, at least in the present case, if the radial order of modes 2 and 3 do not differ very much. As we shall see, the expected values of $l_{2,3}$ and $k_{2,3}$ are large. For such modes we have $k \sim l/\sigma$ which implies in conjunction with Equation (2) and the resonance condition ($\Delta\sigma \approx 0$) that

$$k_2 \approx k_3 , \qquad \sigma_2 \approx \sigma_3 \approx \sigma_1/2$$

for the relevant modes. Thus, for a given l_2 we have to consider a few values of l_3 and a few pairs with frequencies close to $\sigma_1/2$.

Consider now the r.h.s. of Equation (1) as a function of l_2. For large l_2, the value of ν is virtually independent of l_2 (Dziembowski, 1982; see also Table I in Section 4).

The distance between the consecutive radial modes, $\delta\sigma$, is given by

$$\delta\sigma \approx \frac{\beta}{l}\left(\frac{\sigma}{\sigma_1/2}\right)^2 , \quad \text{where} \quad \beta = \text{const.} \tag{3}$$

Thus, the expected frequency mismatch decreases with l_2 like l_2^{-1}. On the other hand, for damping rates we have (Dziembowski, 1982)

$$\gamma \approx \alpha \left(\frac{\sigma_1/2}{\sigma}\right)^2 l^2 , \tag{4}$$

where α^{-1} is the thermal time-scale in the units $(4\pi G \langle\rho\rangle)^{-1}$ for the region where the g modes are trapped. For the Sun α is in the range 10^{-12}–10^{-11}. Clearly, there is a certain optimum range of the l_2-values for the instability to occur and as α is so small, in our case this range falls at rather high values (10^2–10^3). In such a situation, a very precise fitting of the resonance is possible.

It would be unreasonable to rely on the numerical determination of the eigenfrequencies with an accuracy better than, say, 10^{-4}. Therefore, adopting a probabilistic approach we look for the probability P that the instability occurs at a given amplitude Q_1. For this purpose we assume $\sigma_2 = \sigma_3 = \sigma_1/2$ and ignore the difference between γ_2 and γ_3. Using now Equation (4) in Equation (1) we get for the amplitude of mode 1 at the onset of the instability

$$Q_1 = \frac{1}{\nu}(\Delta\sigma^2 + 4\alpha^2 l_2^4)^{1/2} . \tag{5}$$

The probability, $P_{l,i}$, that the instability occurs for given $l_2 = l$ and for given pair of modes 2 and 3, i, is

$$P_{l,i} = \begin{cases} \dfrac{\sqrt{Q_1^2 \nu_i^2 - 4\alpha^2 l^4}}{\beta l^{-1}} & \text{for} \quad l < l_c \\ 0 & \text{for} \quad l > l_c , \end{cases} \tag{6}$$

where we used Equation (3) for $\delta\sigma$ and denoted

$$l_c = \sqrt{\frac{Q_1 \nu_i}{2\alpha}} . \tag{7}$$

Assuming the independence of $P_{l,i}$, which is actually not true but usually gives a good approximation, we get

$$P = 1 - \prod_l \left[\prod_i (1 - P_{l,i}) \right] \approx 1 - \exp\left(- \sum_l \sum_i P_{l,i} \right) .$$

Replacing the summation over l by the integral, we obtain

$$\sum_l P_{l,i} \approx \frac{Q_1 \nu_i}{\beta} \int_0^{l_c} l \sqrt{1 - \left(\frac{l}{l_c}\right)^4} \, dl =$$

$$= \frac{Q_1^2 \nu_i^2}{2\alpha\beta} \int_0^1 x \sqrt{1 - x^4} \, dx = \frac{\pi}{16} \frac{Q_1^2 \nu_i^2}{\alpha\beta} \tag{8}$$

and finally

$$P = 1 - \exp\left(- \frac{\pi}{16} \frac{Q_1^2}{\alpha\beta} \sum_i \nu_i^2 \right) . \tag{9}$$

3. Equilibrium Amplitudes of the Interacting Modes

The initial growth of the amplitudes of modes 2 and 3 is exponential. This, however, is quickly terminated, once the amplitudes are high enough.

In the case of $\gamma_1 < 0$ (linear instability) and $\gamma_{2,3} > 0$, the equilibrium solution exists. However, it is stable only in a certain range of parameters $\Delta\sigma/\gamma_1$, $\gamma_{2,3}/\gamma_1$ and only then does it describe the final stage of the parametric resonance phenomenon.

In the case of our interest, namely $|\gamma_1| \ll \gamma_2 \approx \gamma_3$, the stability criteria (Wersinger *et al.*, 1980; Dziembowski, 1982) are reduced to $\Delta\sigma > 2\gamma_2$, which, using Equations (4), (5) and (7), may be written as

$$l \leq l_c (\tfrac{1}{2})^{1/4} .$$

Performing similar integration as in Equation (8) but in the limits $[0, l_c(\frac{1}{2})^{1/4}]$ it is easy to calculate that the conditional probability that the instability leads to the stable equilibrium is 0.82.

The equilibrium amplitude of mode 1 is

$$Q_1 = \frac{2}{\nu} \sqrt{\gamma_2 \gamma_3 \left[1 + \left(\frac{\Delta\sigma}{\gamma_2 + \gamma_3 + \gamma_1} \right)^2 \right]} \approx \frac{1}{\nu} \sqrt{\Delta\sigma^2 + 4\gamma_2^2} ; \tag{10}$$

thus it is approximately the same as at the onset of the instability. For the remaining two amplitudes we have

$$Q_k = Q_1 \sqrt{-\frac{I_1 \gamma_1 \sigma_1}{I_k \gamma_k \sigma_k}}, \qquad k = 1, 2 \tag{11}$$

with $I_k = \int \rho |\mathbf{h}_k|^2 \, d^3x$, where $\mathbf{h}_k$ is the displacement eigenvector.

Since the expected values of l_2 and l_3 are large, these modes are effectively trapped in the radiative cores and, consequently, their surface amplitudes are exceedingly small. More relevant measure of the amplitude for such modes is the maximum ratio of the amplitude of the horizontal displacement $|\mathbf{h}_H|$ to the radial half-wavelength λ denoted by ζ. Using Equation (11) we have

$$\zeta = \max(|\mathbf{h}_H|/\lambda) = \mu Q_1 \sqrt{-\gamma_1}, \tag{12}$$

where μ may be considered as constant, because for $\sigma \approx \sigma_1/2$ we have $\lambda \sim l_2^{-1}$ and $\gamma_2 \sim l_2^2$. This constant must be calculated numerically. As long as $\zeta \ll 1$, the motion may be regarded as purely oscillatory and the mixing cannot be expected.

4. A Numerical Example

We shall provide here an estimate of the maximum amplitude for the g_1, $l = 1$ mode for the solar model calculated by Dziembowski and Pamjatnykh (1978). The model is characterized by the following parameters:

envelope composition $X = 0.775$, $Z = 0.01$,

$X_c = 0.41$, $T_c = 145 \times 10^6$, $\rho_c = 133$,

bottom of the convective envelope at $r = 0.77\, R_\odot$.

The g_1, $l = 1$ mode has nondimensional frequency $\sigma = 1.345$, that corresponds to a period of 71.6 min. The mode exhibits the net driving in the radiative core and if the convection effects are ignored, its excitation rate is $-\gamma_1 = 3.4 \times 10^{-12}$.

The damping rates for high-order modes with frequencies close to $\sigma_1/2$ are accurately described by Equation (4) with $\alpha = 6.9 \times 10^{-12}$, while the distance between the consecutive radial modes by Equation (3) with $\beta = 0.18$.

The normalized coupling coefficients, v_i, were calculated according to the formulae given by Dziembowski (1982). The values of v_i for modes 2 and 3 having frequencies close to $\sigma_1/2$ at some chosen l_2 are given in Table I. The were evaluated for $j_1 = 0$ and $j_3 = 0$. The latter choice corresponds to the maximum of v. In the case $j_1 = \pm 1$ the maximum occurs for $j_2 = \mp l_2$ and is by a factor $\sqrt{2}$ higher. In the same table we given values of the frequency mismatch, $\Delta\sigma$, as obtained from the numerical calculations. It is important to observe the constancy of v with varying l_2 and a rapid decline, on the average, of v with increasing $|k_2 - k_3|$.

To evaluate the probability of the onset of parametric instability according to

TABLE I

The coupling coefficient (v) and frequency mismatch ($\Delta\sigma$) for some pairs of modes coupled to the g_1, $l = 1$ mode

k_2	k_3	v	$\Delta\sigma$
$l_2 = 100$	$l_3 = 101$		
300	300	12.58	1.73 E−3
300	299	14.07	−8.21 E−3
299	300	5.63	−7.14 E−3
301	299	8.30	1.71 E−3
299	301	0.98	1.73 E−3
301	298	2.66	−1.12 E−4
298	301	0.25	−8.05 E−4
302	298	0.53	1.67 E−3
298	302	0.21	1.71 E−3
302	297	0.26	−1.60 E−4
297	302	0.35	−1.09 E−4
303	297	0.18	1.61 E−3
297	303	0.21	1.68 E−3
303	296	0.21	−2.27 E−4
296	303	0.29	−1.56 E−4
304	296	0.10	1.53 E−3
296	304	0.14	1.62 E−3
304	295	0.15	−3.12 E−4
295	304	0.21	−2.22 E−4
305	295	0.06	1.45 E−3
395	305	0.05	1.54 E−3
305	294	0.11	−4.16 E−4
294	305	0.12	−3.07 E−4
$l_2 = 200$	$l_3 = 201$		
596	596	12.41	7.82 E−4
596	595	14.00	−1.28 E−4
595	596	5.53	−1.25 E−4
597	595	8.22	8.03 E−4
595	597	0.84	7.82 E−4
597	594	2.63	−1.35 E−4
$l_2 = 500$	$l_3 = 501$		
1484	1484	12.48	1.97 E−4
1484	1483	14.04	−1.67 E−4
1483	1484	5.64	−1.67 E−4
1485	1483	8.26	1.96 E−4
1483	1485	0.90	1.97 E−4
1485	1482	2.66	−1.69 E−4

Equation (9), the values of v for the first six pairs were used along with the values of α and β quoted above. The results are shown in Table II. It is seen that the instability is likely to occur already at the amplitude Q_1 of the order of 10^{-7}. The values of $l_{\max}$ given in this table are the values of l at the maximum of the integrand in Equation (8)

TABLE II

Probability of the parametric instability (P), l_2 at the maximum of probability (l_{max}), and the corresponding frequency mismatch $\Delta\sigma$ as a function of amplitude of the g_1, $l = 1$ mode (Q_1)

Q_1	P	l_{max}	$\Delta\sigma$
2.5 E−8	0.044	121	2.9 E−7
5.0 E−8	0.165	172	5.7 E−7
1.0 E−7	0.514	243	1.1 E−6
2.0 E−7	0.944	343	2.3 E−6
3.0 E−7	0.998	420	3.4 E−6

for the largest v_i. It is easy to see that $l_{max} = (\frac{1}{3})^{1/4} l_c$. The corresponding values of $\Delta\sigma$ were calculated with the use of Equation (5).

As we already noted in the previous section, the equilibrium amplitudes Q_1 are practically the same as at the onset of instability. We therefore conclude that the maximum amplitude for the g_1, $l = 1$ mode is not likely to exceed 2×10^{-7}. At this level, the nonresonant nonlinear effects are completely negligible. For instance, as the maximum value of the relative pressure perturbation, $\Delta p/p$, is of the same order as Q_1, the nonlinear effects in the excitation rate γ_1 may be safely ignored.

The motion is purely oscillatory, as the parameter ζ for all three modes involved is much smaller than unity. For modes 2 and 3 it was found that

$$\zeta = 9 \times 10^7 Q_1 \sqrt{-\gamma_1} = 1.7 \times 10^2 Q_1 ,$$

thus it is of the order of 10^{-5}.

It is easy to convert the amplitude Q_1 to the amplitude of the radial velocity for the whole-disc measurements ΔV_{rad}. In such measurements, only the modes with the equatorial symmetry, i.e., with $j_1 = \pm 1$, are visible. We have

$$\Delta V_{rad} = \sqrt{2} \times \sqrt{3/2} \times 0.55 \times 7 \times 10^{10} \times 1.5 \times 10^{-3} Q_1 \text{ cm s}^{-1} =$$

$$= 1.0 \times 10^6 Q_1 \text{ cm s}^{-1} .$$

In this formula, factor $\sqrt{2}$ follows from the difference in v between $j = 0$ and $j = 1$ cases, factor $\sqrt{3/2}$ converts the r.m.s. amplitude to the actual amplitude, 0.55 is an averaging factor for radial velocity, 7×10^{10} cm $\approx R_\odot$, 1.5×10^{-3} s^{-1} = ω_1. Thus, the expected radial velocity amplitude is of the order of 10 cm s^{-1}.

5. Conclusions

We have seen that the resonant coupling of g_1, $l = 1$ to higher-order g modes limits its amplitude to an unexpectedly low level. It is believed that with the accuracy to ± one order of magnitude our estimated value of Q_1 is valid for all low-order and low-degree g modes in all conventional solar modes.

The reason why the nonlinear effects are important at so low amplitudes is that there is a large number of highly adiabatic modes that may strongly interact with the unstable modes. This large number of possibilities permits us to tune the resonance very precisely.

In view of the presented results, the hypothesis that the finite amplitude development of the linear instability of the $l = 1$ g modes leads to mixing in the solar core seems highly implausible.

At its present stage, the theory of mode coupling does not cause any difficulty to the interpretation of the 160 min oscillation in terms of the excitation of a low-degree g mode. The observed amplitudes of 0.56 m s^{-1} (Severny *et al.*, 1979) and 0.22 m s^{-1} (Scherrer *et al.*, 1979) are certainly within the limits of uncertainty of our estimate. The apparent absence of other modes corresponding to different j at the same l and k may be consequence of the fact that their frequencies fall closer to the perfect resonance condition $\Delta\sigma = 0$, with some pairs of high-order g modes. The probability of such an event remains to be calculated.

Acknowledgements

I would like to thank Drs Yu. V. Vandakurov and D. O. Gough for valuable discussions. The numerical calculations reported in this paper were carried out with the PDP11/45 computer donated to the N. Copernicus Center by the U.S. Academy of Sciences.

References

Boury, A., Gabriel, M., Noels, A., Scuflaire, R., and Ledoux, P.: 1975, *Astron. Astrophys.* **41**, 279.

Christensen-Dalsgaard, J., Dilke, F. W. W., and Gough, D. O.: 1974, *Monthly Notices Roy. Astron. Soc.* **169**, 429.

Dilke, F. W. W. and Gough, D. O.: 1972, *Nature* **240**, 262.

Dziembowski, W.: 1982, *Acta Astron.* **32**, (in press).

Dziembowski, W. and Pamjatnykh, A. A.: 1978, in M. J. Rösch (ed.), *Pleins feux sur la physique solaire*, CNRS, p. 135.

Hill, H. A. and Caudell, T. P.: 1979, *Monthly Notices Roy. Astron. Soc.* **168**, 327.

Saio, H.: 1980, *Astrophys. J.* **240**, 685.

Scherrer, P. H., Wilcox, J. M., Severny, A. B., Kotov, V. A., and Tsap, T. T.: 1979, Stanford Univ. IPR Report No. 798.

Severny, A. B., Kotov, V. A., and Tsap, T. T.: 1979, *Astron. Zh.* **56**, 1137.

Shibahashi, H., Osaki, Y., and Unno, W.: 1975, *Publ. Astron. Soc. Japan* **27**, 401.

Vandakurov, Y. V.: 1965, *Proc. Acad. Sci. U.S.S.R.* **164**, 525.

Wersinger, J. M., Finn, J. M., and Edward Ott: 1980, *Phys. Fluids* **23**, 1142.

DERIVATION OF THE AMPLITUDE EQUATIONS OF ACOUSTIC MODES OF AN UNSTABLE SEMI-INFINITE POLYTROPE*

(Invited Review)

JEAN-PIERRE POYET

Observatoire du Pic-du-Midi et de Toulouse, 14 Avenue Edouard Belin, 31400 Toulouse, France

Abstract. The coupling of non-radial pulsations and convection is studied on a simple example: an unstable semi-infinite polytrope. An expansion is proposed and the equations that should describe correctly the pulsations are isolated. The additional complications of the real solar case are discussed at the end.

1. Introduction

The purpose of this study is to try to establish what kinds of equations are obeyed by the solar pulsations and to illustrate the method by a simple example. One of the main causes of the awkwardness of the solar problem is the presence of a convective zone, given the lack of any theory of convection so far. Moreover, the fuzziness and the scarcity of the observations make it difficult to test even a simple empirical theory such as the mixing length theory, as is well illustrated by Gough and Weiss (1976). Fortunately, the increasing accuracy of present measurements is beginning to give information about the evolution of the amplitude of individual pulsation modes (see Grec *et al.*, 1983) and their correlation time (Grec *et al.*, 1980). As a consequence, it is time to try to write precise equations and to settle certain ambiguities about the present state of the treatment of the interaction of convection and pulsation in a star.

In the case of a star in purely radial pulsation, it is clear how to separate the pulsating motion from the convective one: a suitable time average generally suffices (Gough, 1976; Unno, 1967, 1977; Poyet and Spiegel, 1979). In the case of non-radial pulsations the situation is less clear and the problem is really well-defined only when a separation of scale is assumed: the convective motions being supposed to arise on scales much smaller than the pulsating motions. One can then perform a two-scale development of the problem and average over the small convective scales (Unno *et al.*, 1979). In this limit only, the concept of turbulent viscosity, introduced to model the effects of the convection zone, can be justified physically. However this is not the case in the Sun since it is generally believed by people studying its convective zone that some cells of about the size of the convection zone ($\approx 200\,000$ km) must be present and there are theoretical reasons which might lead one to believe in the presence of even larger scales (in linear theory, it is found, in Boussinesq convection between plane parallel layers, that the

* Proceedings of the 66th IAU Colloquium: *Problems in Solar and Stellar Oscillations*, held at the Crimean Astrophysical Observatory, U.S.S.R., 1–5 September, 1981.

Solar Physics **82** (1983) 267–296. 0038–0938/83/0822–0267$04.50.

convective instability arises first at zero wavenumber for boundary conditions with fixed flux; see Hurle *et al.*, 1967). So theoretically, it should be assumed that convection covers scales from the granules ($l \approx 1000$, where l is the usual index of the spherical harmonics), or even from the subgranular scales that are likely to be turbulent, to the solar circumference, and consequently that convective and pulsating scales in the Sun are comparable. Thus there is no easy physical way to separate them and an arbitrary definition must be adopted. Given any type of well-defined linear problem, either of the type of the simple polytropic example that will be treated later or coming from a numerical program, the modes will be classified according to their time-dependence $e^{i\sigma t}$, where, in general, $\sigma = \sigma_R + i\sigma_I$ with σ_R and σ_I real. If $|\sigma_I| \ll |\sigma_R|$, the mode will be considered as a pulsating one. If this is not true, the modes will be treated as convective ones (overstable if $|\sigma_R|$ is different from zero). As an illustration, for a superadiabatic polytrope (unstable to convection) the f and p modes will be considered as pulsating modes, while the g^- and toroidal modes (with zero frequencies) will be considered as belonging to the convective modes.

Another outstanding problem of the subject is the perturbation of the convective flux. It is generally ignored because it involves a time-dependent theory of convection. In the solar case, one can imagine trying to tackle the problem because, despite the non-linearity due to the convection, the pulsations are of such small amplitudes that they must behave almost linearly in some sense. The question is then to attempt to isolate this almost linear problem theoretically and use all of its nice linear properties to understand the pulsation-convection coupling. When observers interpret the peaks in their spectra as normal modes of the Sun, they do nothing else but assume the existence of these linear properties, despite the messy couplings with the convection. Given these facts, there are two ways one can think of, to start the problem. On one hand, some convective terms in the mean equations that one wishes to start with could be kept. Then one is faced with the rather difficult problem of an unknown mean state to calculate further. This would generally force one to close the convection equations (via the mixing-length theory for instance) already in the mean state. This procedure has two disadvantages: the first one is that one has to perturb a bad theory (the mixing length theory, for example) to calculate even the linear problem and that there is no unique way of doing it; the second one is that the closure is on the mean state of the convection and it is well known, in usual turbulence theory, that such theories give results which are rather worse than those obtained by trying to close on higher order correlations. On the other hand, it seems more natural to perturb exact equations and then to try to close the problem. This is the method adopted in what follows, where the exact equations for the convection are kept and then perturbed. A very simple example will be treated, embodying, in our opinion, the conceptual difficulties of the problem. The additional difficulties of the real solar problem will be commented on at the end.

Before starting the calculation for an unstable polytrope, it is useful to review the motivations for doing such a non-linear calculation. One of the issues about the Sun, that should be settled by a good non-linear theory, is the nature of the excitation of the pulsating modes. It has sometimes been believed, after the unsuccessful attempt by

Goldreich and Keeley (1977a, b) to excite the modes stochastically by turbulent convection, that the κ-mechanism was at work, as confirmed by the calculation of Ando and Osaki (1975a, b). However the uncertainties in the method used by Goldreich and Keeley; the fact that a more accurate treatment of radiative transfer stabilizes the modes found unstable by Ando and Osaki (see the article by Christensen-Dalsgaard and Frandsen, 1983); and the behaviour of the amplitude of the modes identified in the South Pole experiment (see the article by Grec *et al.*, 1983) induce one to think that indeed the convection zone is exciting the modes. Two things must be explained by the proponents of the κ-mechanism as an excitation process: why doesn't the Sun pulsate with the amplitude of a Cepheid and, more importantly, why is there a lifetime of the modes of two days in Fossat's results (one would rather expect a stationnary solution with the κ-mechanism)? However it is not excluded that some contribution to the excitation of the pulsation comes from the κ-mechanism and, later on, the inclusion of such an effect in a realistic calculation will be discussed. In any case, a more realistic treatment of the convection is needed and we shall concentrate on this problem here.

2. The Unstable Polytrope

We consider an adiabatic problem and take as origin the surface of the star ($z = z_s$) with a z-axis oriented positively inward. We suppose that there is no viscosity or thermal conductivity and that γ, the usual ratio of specific heats, is constant. We neglect any radiative effect and suppose that the gravity is constant. The fluid hydrodynamical equations for such a problem are

$$\frac{\partial \rho}{\partial t} + \nabla \cdot (\rho \mathbf{v}) = 0 , \tag{2.1}$$

$$\rho \frac{\mathrm{d} v_i}{\mathrm{d} t} = - \frac{\partial p}{\partial x_i} + \rho g \delta_{i3} , \qquad i = 1, 2, 3, \tag{2.2}$$

$$\frac{dp}{\mathrm{d}t} = c^2 \frac{\mathrm{d}\rho}{\mathrm{d}t} = - c^2 \rho \nabla \cdot \mathbf{v} = - \gamma p \nabla \cdot \mathbf{v} , \tag{2.3}$$

$$c^2 = \gamma \frac{p}{\rho} , \tag{2.4}$$

where ρ, $\mathbf{v}$, and p are the density, velocity, and pressure in the medium. Indices 1 and 2 for i are for the horizontal velocity and 3 is for the vertical velocity. An infinite plane paralel layer is considered. One starts by defining a polytropic state when $\mathbf{v} = 0$. A reference layer, $z = z_0$, is chosen for this polytrope such that at $z = z_0$, $T_0 = \Theta$, and $p_0 = P$, where the index zero designates polytropic variables. It follows that

$$T_0 = \beta z , \tag{2.5}$$

$$p_0 = P\left(\frac{z}{z_0}\right)^{m+1} = Kz^{m+1}\,, \tag{2.6}$$

$$\rho_0 = \frac{K}{R\beta} z^m\,, \tag{2.7}$$

$$K = \frac{P}{z_0^{m+1}}\,, \tag{2.8}$$

$$m = \frac{g}{R\beta} - 1\,, \tag{2.9}$$

$$c_0^2 = \gamma \frac{p_0}{\rho_0} = \gamma R\beta z = \frac{\gamma g}{m+1} z\,, \tag{2.10}$$

where β is the constant static temperature gradient, R the constant of the perfect gases, m the polytropic index, and c_0 the sound speed for the polytrope. A polytrope of thickness d is considered, so that

$$z_s < z < z_s + d\,, \tag{2.11}$$

and the thickness d, when it is finite, can be taken for z_0 for example. The case of the infinite polytrope that is treated later will be given simply by taking $z_s = 0$ and $d \to \infty$. The equations are nondimensionalized by choosing the length z_0 and an arbitrary time τ as units.

It is now convenient to introduce potential variables. Potential pressure π and temperature θ are defined by

$$\pi = \left(\frac{p}{P}\right)^{R/c_p}\,, \tag{2.12}$$

$$T = \Theta\pi\theta\,. \tag{2.13}$$

Remembering that

$$\frac{R}{c_p} = \frac{\gamma - 1}{\gamma}\,, \tag{2.14}$$

using θ and π instead of T and p and eliminating ρ from the equations via the equation of state of perfect gases

$$p = \rho RT\,, \tag{2.15}$$

one obtains the following equations:

$$\lambda_a M^2 \frac{\mathrm{d}\mathbf{v}}{\mathrm{d}t} = -\theta\nabla\pi + \lambda_a \delta_{i3}\,, \tag{2.16}$$

$$\frac{\mathrm{d}\pi}{\mathrm{d}t} = -\pi(\gamma - 1)\nabla \cdot \mathbf{v}\,, \tag{2.17}$$

$$\frac{\mathrm{d}\theta}{\mathrm{d}t} = 0\,, \tag{2.18}$$

with

$$\lambda_a = \frac{g}{c_p\beta}\,, \tag{2.19}$$

$$M^2 = \frac{(z_0/\tau)^2}{gz_0}\,, \tag{2.20}$$

where λ_a is the ratio of the adiabatic and static temperature gradients and M^2 can be interpreted as the ratio of a typical velocity to a free-fall velocity and is analogous to a Mach number squared. Such equations have been used by Ogura and Phillips (1962) to derive the anelastic approximation from the convection equations in the atmosphere. That approximation is obtained by an asymptotic expansion to $O(M^2)$ of the preceding equations. The main property of the anelastic equations is that the acoustic waves have been filtered out and that they should be an accurate representation of purely convective phenomena. Consequently, it will be interesting to trace their role, in the final equations derived later, by considering the M^2 parameter. (For the derivation of this approximation in a realistic stellar case, see Gough (1969).)

One might wonder, at this stage, about the usefulness of potential variables. On top of the fact that they simplify somewhat the initial equations by giving only quadratic non-linearities, it will be seen that the linear problem for the complete velocity and thermodynamic fields makes them the natural variables to utilize.

The polytropic state can be rewritten with the new potential variables, taking into account the units of time and length τ and z_0, as

$$\pi_0 = z^{m+1-[(m+1)/\gamma]}\,, \tag{2.21}$$

$$\theta_0 = z^{[(m+1)/\gamma]-m}\,, \tag{2.22}$$

$$c_0^2 = \frac{\gamma - 1}{\lambda_a M^2}\,z\,. \tag{2.23}$$

In terms of these variables, the hydrostatic mean state equation becomes

$$\theta_0\,\frac{\mathrm{d}\pi_0}{\mathrm{d}z} = \lambda_a\,. \tag{2.24}$$

A quantity $f(x, y, z, t)$ will be expressed as

$$f(x, y, z, t) = f_0(z) + f'(x, y, z, t)\,, \tag{2.25}$$

where f_0 describes the polytropic state. From Equations (2.16), (2.17), and (2.18), one can get equations for the primed quantities,

$$\frac{\partial \mathbf{v}'}{\partial t} + \frac{\theta_0}{\lambda_a M^2} \nabla \pi' + \frac{1}{\lambda_a M^2} \frac{\mathrm{d}\pi_0}{\mathrm{d}_z} \delta_{i3} \theta' = -\mathbf{v}' \cdot \nabla \mathbf{v}' - \frac{\theta' \nabla \pi'}{\lambda_a M^2}, \tag{2.26}$$

$$\frac{\partial \pi'}{\partial t} + \pi_0(\gamma - 1) \nabla \cdot \mathbf{v}' + \frac{\mathrm{d}\pi_0}{\mathrm{d}z} w' = -\mathbf{v}' \cdot \nabla \pi' - (\gamma - 1) \pi' \nabla \cdot \mathbf{v}', \tag{2.27}$$

$$\frac{\partial \theta'}{\partial t} + \frac{\mathrm{d}\theta_0}{\mathrm{d}z} w' = -\mathbf{v}' \cdot \nabla \theta', \tag{2.28}$$

where w' is the vertical component of the velocity. The left-hand side of (2.26), (2.27), and (2.28) constitutes the linear problem. By Fourier-transforming it horizontally, one obtains

$$\frac{\partial \mathbf{v}_\mathbf{k}}{\partial t} + \frac{\theta_0}{\lambda_a M^2} \left(- i\mathbf{k} \pi_\mathbf{k} + \frac{\partial \pi_\mathbf{k}}{\partial z} \delta_{i3} \right) + \frac{1}{\lambda_a M^2} \frac{\mathrm{d}\pi_0}{\mathrm{d}_z} \delta_{i3} \theta_\mathbf{k} =$$

$$= - \frac{1}{2\pi} \iint_{-\infty}^{+\infty} \mathrm{d}\mathbf{k}' \left[i(\mathbf{k} - \mathbf{k}') \cdot \mathbf{u}_{\mathbf{k}'} + w_{\mathbf{k}'} \frac{\partial}{\partial z} \right] \mathbf{v}_{\mathbf{k}-\mathbf{k}'} -$$

$$- \frac{1}{2\pi} \frac{1}{\lambda_a M^2} \iint_{-\infty}^{+\infty} \mathrm{d}\mathbf{k}' \, \theta_{\mathbf{k}'} \left[i(\mathbf{k} - \mathbf{k}') \pi_{\mathbf{k}-\mathbf{k}'} + \frac{\partial}{\partial z} \pi_{\mathbf{k}-\mathbf{k}'} \delta_{i3} \right], \tag{2.29}$$

$$\frac{\partial \pi_\mathbf{k}}{\partial t} - \pi_0(\gamma - 1) i\mathbf{k} \cdot \mathbf{u}_\mathbf{k} + \left[\frac{\mathrm{d}\pi_0}{\mathrm{d}z} + \pi_0(\gamma - 1) \frac{\partial}{\partial z} \right] w_\mathbf{k} =$$

$$= - \frac{1}{2\pi} \iint_{-\infty}^{+\infty} \mathrm{d}\mathbf{k}' \left[i(\mathbf{k} - \mathbf{k}') \cdot \mathbf{u}_{\mathbf{k}'} + w_{\mathbf{k}'} \frac{\partial}{\partial z} \right] \pi_{\mathbf{k}-\mathbf{k}'} -$$

$$- \frac{(\gamma - 1)}{2\pi} \iint_{-\infty}^{+\infty} \mathrm{d}\mathbf{k}' \, \pi_{\mathbf{k}'} \left[i(\mathbf{k} - \mathbf{k}') \cdot \mathbf{u}_{\mathbf{k}-\mathbf{k}'} + \frac{\partial}{\partial z} w_{\mathbf{k}-\mathbf{k}'} \right], \tag{2.30}$$

$$\frac{\partial \theta_\mathbf{k}}{\partial t} + \frac{\mathrm{d}\theta_0}{\mathrm{d}z} w_\mathbf{k} = - \frac{1}{2\pi} \iint_{-\infty}^{+\infty} \mathrm{d}\mathbf{k}' \left[i(\mathbf{k} - \mathbf{k}') \cdot \mathbf{u}_{\mathbf{k}'} + w_{\mathbf{k}'} \frac{\partial}{\partial z} \right] \theta_{\mathbf{k}-\mathbf{k}'} . \tag{2.31}$$

Here $\mathbf{k}$ is the horizontal wavenumber vector with components $(k_x, k_y, 0) \cdot \mathbf{v_k}$, $\pi_\mathbf{k}$ and $\theta_\mathbf{k}$ are the Fourier-transforms of $\mathbf{v}'$, π', and θ', $\mathbf{u_k}$ is the Fourier-transform of the horizontal velocity with components $(u_{1_\mathbf{k}}, u_{2_\mathbf{k}}, 0)$, and $w_\mathbf{k}$ is the Fourier-transform of the vertical velocity.

3. The Linear Problem

3.1. The general polytropic case

The general problem of waves in a polytrope has been studied by Lamb (1932) in the stable case, by Spiegel and Unno (1962) in the unstable case and has been applied to the solar case by Gough (1978). When one considers the fit showed by Gough (1978) between the polytropic case and the solar case, one is struck by the fact that, at least for the category of modes he considers (high l acoustic modes), the fit is not that bad and that one of the main reasons for such a fit must be the stratified nature of the medium which imposes roughly, its characteristics on the modes. This is another reason to use potential variables which are variables that include physically part of the stratification effects. Some more complete expressions, in the case of a finite polytrope, will be given here along with the values for the π' and θ' fields, but the usual linear problem will not be detailed. For any variable f', a solution of the form

$$f'(x, y, z, t) = F(z) \exp[i(\sigma t - \mathbf{k} \cdot \mathbf{x})] \tag{3.1}$$

is considered, where σ and $\mathbf{k}$ are the dimensional frequency and horizontal wavenumber vector (of components k_x and k_y and modulus k).

A general equation can be written for the divergence of the velocity field,

$$\chi = \frac{\partial u'}{\partial x} + \frac{\partial v'}{\partial y} + \frac{\partial w'}{\partial z}, \tag{3.2}$$

or its horizontal Fourier-transform $\chi_\mathbf{k}$,

$$\chi_\mathbf{k} = - -i\mathbf{k} \cdot \mathbf{u_k} + \frac{\partial w_\mathbf{k}}{\partial z}, \tag{3.3}$$

$$z \frac{\partial^2 \chi_\mathbf{k}}{\partial z^2} + (m+2) \frac{\partial \chi_\mathbf{k}}{\partial z} + \left[\frac{m+1}{\gamma} M^2 \sigma^2 - k^2 z - \frac{k^2}{M^2 \sigma^2}(1 - \lambda_a)\right] \chi_\mathbf{k} = 0. \tag{3.4}$$

Defining $\psi_\mathbf{k}$ such that

$$\psi_\mathbf{k} = e^{kz} \chi_\mathbf{k}, \tag{3.5}$$

one obtains the usual equation (Lamb, 1932)

$$\zeta \frac{\partial^2 \psi_\mathbf{k}}{\partial \zeta^2} + (m + 2 - \zeta) \frac{\partial \psi_\mathbf{k}}{\partial \zeta} - \alpha \psi_\mathbf{k} = 0, \tag{3.6}$$

where

$$\zeta = 2kz\,, \tag{3.7}$$

and

$$2\alpha = -\frac{m+1}{\gamma}\,\omega^2 + (1-\lambda_a)\frac{1}{\omega^2} + m + 2\,, \tag{3.8}$$

$$\omega^2 = \frac{M^2\sigma^2}{k}\,. \tag{3.9}$$

Equation (3.8) is the differential equation of confluent hypergeometric functions. Two solutions are $M(\alpha, m+2, \zeta)$ and $(1/\zeta^{m+1})M(\alpha - m - 1, -m, \zeta)$, where $M(a, b, \zeta)$ is the confluent hypergeometric function of the first kind, sometimes called Kummer's function, and defined by

$$M(a, b, \zeta) = \sum_{j=0}^{\infty} \frac{(a)_j}{(b)_j}\frac{\zeta^j}{j!}\,, \tag{3.10}$$

where

$$(a)_0 = 1 \quad \text{and} \quad (a)_j = a(a+1)\ldots(a+j-1)\,. \tag{3.11}$$

The two solutions are independent, their Wronskian being

$$\mathscr{W}[1, 2] = -\frac{(m+1)e^{\zeta}}{\zeta^{m+2}}\,. \tag{3.12}$$

ζ varies from $2kz_s$ to $2k(z_s + d)$ and, in the case of an infinite polytrope, ζ will vary from 0 to ∞. Once $\chi_{\mathbf{k}}$ is known, it is fairly easy to deduce that the vertical velocity $W_{\mathbf{k}}$ is given by (Spiegel and Unno, 1962)

$$(\omega^4 - 1)W_{\mathbf{k}} = \frac{\gamma e^{-\zeta/2}}{(m+1)k}\left\{\omega^2\left[-\zeta\frac{\partial\psi_{\mathbf{k}}}{\partial\zeta} - (m+1)\psi_{\mathbf{k}}\right] + \tfrac{1}{2}\zeta\psi_{\mathbf{k}}(\omega^2+1)\right\}\,. \tag{3.13}$$

Some manipulations with the confluent hypergeometric functions can be used (see Appendix A) to cast the velocity into the more convenient form

$$\begin{aligned} W_{\mathbf{k}} = &-A\,\frac{\gamma}{2k}\,e^{-\zeta/2}\left[\frac{1}{\omega^2+1}M(\alpha, m+1, \zeta) + \frac{1}{\omega^2-1}M(\alpha-1, m+1, \zeta)\right] + \\ &+ B\,\frac{\gamma}{2k}\,\frac{e^{-\zeta/2}}{m(m+1)\zeta^m}\left[\frac{1}{\omega^2+1}(\alpha-m-1)M(\alpha-m, -m+1, \zeta) + \right. \\ &\left. + \frac{1}{\omega^2-1}(\alpha-1)M(\alpha-m-1, -m+1, \zeta)\right]\,, \end{aligned} \tag{3.14}$$

where A and B are arbitrary constants. Now, if boundary conditions are applied to $W_{\mathbf{k}}$ at the top and the bottom of the layer, a complicated equation determining the eigenvalues ω^2 follows.

3.2. The infinite polytrope

This equation simplifies considerably in the case of an infinite polytrope for which ζ varies from 0 to ∞. We choose the simplest boundary conditions

$$W_{\mathbf{k}}(0) = W_{\mathbf{k}}(\infty) = 0\,, \tag{3.15}$$

and consider from now on only the case of an infinite polytrope. The boundary condition at $\zeta = 0$ forces $B = 0$ in (3.14) and, using asymptotic expansions for the confluent hypergeometric functions when $\zeta \to \infty$, Spiegel and Unno (1962) established that the equation for the eigenvalues becomes simply

$$\alpha = 1 - n\,, \qquad n = 1, 2, 3, \text{etc.}\,, \tag{3.16}$$

where α is given by (3.8). One obtains a quadratic equation for ω^2 whose solution is (Gough, 1978)

$$\omega_{n_p}^2 = \frac{\gamma}{m+1}\left\{n_p + \frac{m}{2} + \left[\left(n_p + \frac{m}{2}\right)^2 - \frac{m+1}{\gamma}(\lambda_a - 1)\right]^{1/2}\right\}, \tag{3.17}$$

$$n_p = 1, 2, 3, \text{etc.}\,, \tag{3.18}$$

$$\omega_{n_g}^2 = \frac{\gamma}{m+1}\left\{n_g + \frac{m}{2} - \left[\left(n_g + \frac{m}{2}\right)^2 - \frac{m+1}{\gamma}(\lambda_a - 1)\right]^{1/2}\right\}, \tag{3.19}$$

$$n_g = 1, 2, 3, \text{etc.} \tag{3.20}$$

From now on, the polytrope is supposed to be unstable, i.e. $\lambda_a < 1$, which is seen from (2.19) to be the familiar condition of superadiabaticity. The squared frequencies ω_n^2 are positive and associated with the stable acoustic modes p_n. The squared frequencies ω_n^2 are real and negative and associated with the unstable g_n^- modes whose strong interaction will give rise to turbulent convection. As far as the frequencies themselves are concerned, a special notation is required to differentiate the positive and negative parts of each frequency. One defines from (2.9)

$$\sigma_{\mathbf{k},\,+n_p} = \frac{\sqrt{k}}{M}\,|\omega_{n_p}|\,, \tag{3.21}$$

$$\sigma_{\mathbf{k},\,-n_p} = -\frac{\sqrt{k}}{M}\,|\omega_{n_p}|\,, \tag{3.22}$$

$$n_p = 1, 2, 3, \text{etc.} \tag{3.23}$$

Since the ω_{ng}^2 are strictly negative, it is convenient to define

$$\overline{\omega}_{ng}^2 = -\omega_{ng}^2 \tag{3.24}$$

and

$$\sigma_{\mathbf{k},\,+n_g} = \frac{i\sqrt{k}}{M}\,|\overline{\omega}_{n_g}|\,, \tag{3.25}$$

$$\sigma_{\mathbf{k},\,-n_g} = -\frac{i\sqrt{k}}{M}\,|\overline{\omega}_{n_g}|\,, \tag{3.26}$$

$$n_g = 1, 2, 3, \text{etc.} \tag{3.27}$$

The modes corresponding to (3.21), (3.22), (3.25), and (3.26) will be denoted respectively by p_{n_p}, p_{-n_p}, $g_{n_g}^-$, $g_{-n_g}^-$. Before going on with the calculation of the eigenfunctions, it is time to mention here that the above procedure has left out two types of modes. These are modes which have zero divergence. The first type has

$$\omega_{n_f}^2 = 1\,, \tag{3.28}$$

and is the familiar class of f-mode (or Kelvin mode) which will be associated with the index n_f where n_f is zero symbolically,

$$\sigma_{\mathbf{k},\,n_f} = \frac{\sqrt{k}}{M}\,, \tag{3.29}$$

$$\sigma_{\mathbf{k},\,-n_f} = -\frac{\sqrt{k}}{M}\,. \tag{3.30}$$

The other type has zero frequency and corresponds to the toroïdal modes of the spherical case; and index n_t will later be associated with them. They are such that $\omega_{n_t}^2 = 0$.

One can now calculate the eigenvectors associated with each type of mode and start with the p and g modes. The calculation is much simplified by the fact that α is a negative integer. In this limit the confluent hypergeometric functions are expressible in terms of generalized Laguerre polynomials,

$$M(1-n, m+1, \zeta) = \frac{(n-1)!}{(m+1)_{n-1}}\,L_{n-1}^m(\zeta)\,, \tag{3.31}$$

where the notation $(m+1)_{n-1}$ is the same as the one used in (3.11). The Laguerre polynomials are given by

$$L_n^m(\zeta) = e^{\zeta}\,\frac{\zeta^{-m}}{n!}\,\frac{\mathrm{d}^n}{\mathrm{d}\zeta^n}(e^{-\zeta}\zeta^{n+m})\,. \tag{3.32}$$

Using (3.14) with $B = 0$ and $A = \omega_n^2 - 1$, one gets

$$W_{\mathbf{k},j} = -\frac{\gamma}{2k} e^{-\zeta/2} Q_{|j|}(\zeta), \tag{3.33}$$

where

$$Q_{|j|}(\zeta) = \frac{|j|!}{(m+1)_{|j|}} \left[L^m_{|j|}(\zeta) + \frac{\omega^2_{|j|} - 1}{\omega^2_{|j|} + 1} \frac{m + |j|}{|j|} L^m_{|j|-1}(\zeta) \right], \tag{3.34}$$

where j stands for $n_p, n_g, -n_p$, or $-n_g$. From now on $|j|$ will mean that the quantity that depends on j is independent of the sign of the frequency. From (3.5) and the normalization used in (3.33), one gets

$$\chi_{\mathbf{k},j} = e^{-\zeta/2} \frac{(|j| - 1)!}{(m+2)_{|j|-1}} L^{m+1}_{|j|-1}(\zeta). \tag{3.35}$$

Using (3.33), (3.35), the linear part of Equations (2.26), (2.27), (2.28), and the following identity between generalized Laguerre polynomials:

$$\zeta L^{m+1}_{n-1}(\zeta) = -\zeta \frac{\mathrm{d}}{\mathrm{d}\zeta} L^m_n(\zeta) = (n+m) L^m_{n-1}(\zeta) - n L^m_n(\zeta), \tag{3.36}$$

it is easy to express the total eigenvector as

$$\varphi_{\mathbf{k},j} = \begin{bmatrix} U_{\mathbf{k},|j|} \\ V_{\mathbf{k},|j|} \\ W_{\mathbf{k},|j|} \\ \Pi_{\mathbf{k},j} \\ \Theta_{\mathbf{k},j} \end{bmatrix} = e^{-\zeta/2} \begin{bmatrix} -i \dfrac{k_x}{k} \dfrac{\gamma}{2k} P_{|j|}(\zeta) \\ -i \dfrac{k_y}{k} \dfrac{\gamma}{2k} P_{|j|}(\zeta) \\ -\dfrac{\gamma}{2k} Q_{|j|}(\zeta) \\ -i \dfrac{M^2}{k} \sigma_{\mathbf{k},j} \dfrac{(\gamma - 1)(m+1)}{(2k)^{m+1-[(m+1)/\gamma]}} \zeta^{m-[(m+1)/\gamma]} P_{|j|}(\zeta) \\ -\dfrac{i}{\sigma_{\mathbf{k},j}} \dfrac{(m+1-m\gamma)}{(2k)^{[(m+1)/\gamma]-m}} \zeta^{[(m+1)/\gamma]-m-1} Q_{|j|}(\zeta) \end{bmatrix}, \tag{3.37}$$

where

$$P_{|j|}(\zeta) = \frac{|j|!}{(m+1)_{|j|}} \left[L^m_{|j|}(\zeta) - \frac{\omega^2_{|j|} - 1}{\omega^2_{|j|} + 1} \frac{m + |j|}{|j|} L^m_{|j|-1}(\zeta) \right], \tag{3.38}$$

and j has the same significance as before. The eigenvector can be expressed with the polynomials of degree $|j|$, $P_{|j|}$, and $Q_{|j|}$. The use of potential variables gives simple expressions, simply proportional to $P_{|j|}$ and $Q_{|j|}$, for the thermodynamic part of the eigenvector which then looks like the velocity part. This will, later on, ease the task of finding orthogonality relations. The use of any of the original thermodynamic variables would not have had such an effect and would have given a combination of $P_{|j|}$ and $Q_{|j|}$ in the thermodynamic part of the eigenvector. As far as the f-modes are concerned, it can be easily verified that they have an analogous expression to (3.37) with $j = 0$ if one adopts the conventions

$$\omega_0^2 = 1 , \tag{3.39}$$

$$(m + 1)_0 = 1 , \tag{3.40}$$

$$\left[\frac{\omega_j^2 - 1}{|j|}\right]_{j=0} L^m_{-1}(\zeta) = 0 . \tag{3.41}$$

One gets the eigenvectors $\boldsymbol{\varphi}_{\mathbf{k},\, \pm n_f}$ by replacing j by $\pm n_f$ in (3.37) and defining

$$P_{|n_f|}(\zeta) = P_0(\zeta) = L_0^m(\zeta) = 1 , \tag{3.42}$$

$$Q_{|n_f|}(\zeta) = Q_0(\zeta) = L_0^m(\zeta) = 1 . \tag{3.43}$$

These modes have zero divergence

$$\chi_{\mathbf{k},\, \pm n_f} = 0 . \tag{3.44}$$

The toroidal modes can be found directly from the linear part of (2.29), (2.30), and (2.31) and have the following expression:

$$\boldsymbol{\varphi}_{\mathbf{k},\, t} = f(z) \begin{bmatrix} 1 \\ -\dfrac{k_x}{k_y} \\ 0 \\ 0 \\ 0 \end{bmatrix} , \tag{3.45}$$

where $f(z)$ is an arbitrary function of z. Their divergence is also zero:

$$\chi_{\mathbf{k},\, n_t} = 0 . \tag{3.46}$$

One must now try to build some orthogonality relation between all the modes. One very useful result is the completeness of the system of polynomials $\{e^{-\zeta/2}\zeta^{m/2}L_n^m(\zeta)\}$, $n = 0, 1, 2,$ etc., with respect to square integrable functions on $[0, +\infty]$ (see for example

Sansone, 1959). This allows us to build at once the following toroidal eigenvectors;

$$\varphi_{\mathbf{k}, n_t} = e^{-\zeta/2} L_{n_t}^m(\zeta) \begin{bmatrix} k_y \\ -k_x \\ 0 \\ 0 \\ 0 \end{bmatrix}, \tag{3.47}$$

where $n_t = 0, 1, 2$, etc.

3.3. Properties of the linear modes

The first problem that arises in building the non-linear theory with an expansion over the eigenmodes is the completeness problem. When one speaks of completeness properties, one should mention the vector space and the norm with respect to which completeness is defined. In stellar oscillation theory, completeness means, generally, completeness of the operator acting on the Lagrangian displacements $\boldsymbol{\xi}$ (which forms a Hilbert space) with respect to the norm

$$\langle \xi_1, \xi_2 \rangle = \int_0^M \xi_1 \cdot \xi_2^* \, \mathrm{d}M_r, \tag{3.48}$$

where ξ_1 and ξ_2 are arbitrary displacement eigenvectors, M is the mass of the star, M_r the mass inside radius r, and * is complex conjugation. The results, so far, have been established in the frame of adiabatic theory with 'zero boundary conditions' and use fairly sophisticated techniques of functional analysis. Kaniel and Kovetz (1967) showed the existence of an expansion theorem; Eisenfeld (1969) established completeness in the convectively stable case. More recently, Dyson and Schutz (1979) generalized these results to the case of a differentially rotating body. Their proof appears to us as the most careful and most general one since it includes the convective case. In the non-rotating case the only condition on their completeness result is that the square of the local Brunt-Väisäla frequency has a lower bound (possibly negative!) over the whole star. Some more general results connected with these problems can be found in Eisenfeld (1968) and Weinberger (1968). In the present case, (3.48) is connected with the usual orthogonality relation between the velocity eigenmodes,

$$\int_0^\infty \rho_0 \mathbf{v}_i \cdot \mathbf{v}_k^* \, \mathrm{d}z = \delta_{ik}, \tag{3.49}$$

for a suitable normalization of the eigenvectors, and is due to the orthogonality relation of the generalized Laguerre polynomials

$$\int_0^\infty e^{-\zeta} \zeta^m L_n^m(\zeta) L_l^m(\zeta) \, \mathrm{d}\zeta = \frac{\Gamma(n+m+1)}{n!} \delta_{nl}; \qquad n, l \geq 0 \tag{3.50}$$

(ρ_0 is proportional to ζ^m). Equation (3.50) has the immediate consequence

$$\int_0^\infty \zeta^m e^{-\zeta}[P_n(\zeta)P_l(\zeta) + Q_n(\zeta)Q_l(\zeta)]\,\mathrm{d}\zeta =$$

$$= \frac{2n!\,\Gamma(n+m+1)}{[(m+1)_n]^2}\,\delta_{nl}\left[1 + \left(\frac{\omega_n^2 - 1}{\omega_n^2 + 1}\right)^2 \frac{m+n}{n}\right], \tag{3.51}$$

where n and l are any of the indices n_p, n_g, and n_f. (The formula is valid for n_f if one follows the convention that n_f is associated with the value 0 and $[(\omega_{|n_f|}^2 - 1)^2]/n_f = 0$.) Let $\bar{\boldsymbol{\varphi}}$ be the velocity part of the eigenvector $\boldsymbol{\varphi}$, then the following relation ensues from (3.51), (3.17), and (3.19):

$$\int_0^\infty \zeta^m \bar{\boldsymbol{\varphi}}_{\mathbf{k},n} \cdot \bar{\boldsymbol{\varphi}}_{\mathbf{k},l}^*\,\mathrm{d}\zeta = 0 \quad \text{if} \quad n \neq l, \tag{3.52}$$

where n and l are any of the indices n_p, n_g, n_f, or n_t, and

$$\int_0^\infty \zeta^m |\bar{\boldsymbol{\varphi}}_{\mathbf{k},n}|^2\,\mathrm{d}\zeta = \frac{\gamma^2}{2k^2}\,\frac{|n|!\,\Gamma(|n|+m+1)}{[(m+1)_{|n|}]^2}\left[1 + \left(\frac{\omega_{|n|}^2 - 1}{\omega_{|n|}^2 + 1}\right)^2 \frac{m+|n|}{|n|}\right], \tag{3.53}$$

for n being an index n_p, n_g, n_f ((3.53) is the same if one considers the negative frequency modes), and

$$\int_0^\infty \zeta^m |\bar{\boldsymbol{\varphi}}_{\mathbf{k},n_t}|^2\,\mathrm{d}\zeta = k^2\,\frac{\Gamma(n_t+m+1)}{n_t!} \tag{3.54}$$

for all indices n_t.

What is needed is a similar orthogonality relation for the eigenvector $\boldsymbol{\varphi}$. The problem is that the completeness result is valid only for the velocity part of the eigenvector $\boldsymbol{\varphi}$ which is insensitive to the sign of the frequency. (Only the thermodynamic part is sensitive to this sign.) Completeness is generally linked with the self-adjointness of the equations, and the equations for the velocity (after elimination of the thermodynamic variables) are self-adjoint with respect to the scalar product (3.49). However, it is fairly easy to prove that for no scalar product of the form

$$(\boldsymbol{\varphi}, \boldsymbol{\psi}) = \int_0^\infty \left[z^m(\varphi_1\psi_1^* + \varphi_2\psi_2^* + \varphi_3\psi_3^*) + \alpha_{44}\varphi_4\psi_4^* + \alpha_{55}\varphi_5\psi_5^*\right]\mathrm{d}z\,, \tag{3.55}$$

where α_{44} and α_{55} are arbitrary functions of z, is the differential operator $\mathscr{L}(\mathbf{k}, z, \partial/\partial z)$ involved in the linear part of (2.29), (2.30), and (2.31), and which satisfies the eigenvalue equation

$$\frac{\partial \varphi_{\mathbf{k}}}{\partial t} + \mathscr{L}\left(\mathbf{k}, z, \frac{\partial}{\partial z}\right) \varphi_{\mathbf{k}} = 0 , \tag{3.56}$$

self-adjoint (one can find a bilinear form satisfying (3.55) and making $\mathscr{L}(\mathbf{k}, z, \partial/\partial z)$ self-adjoint in appearance but this bilinear form is not positive definite). One can think of two ways to tackle the problem.

3.3.1. *The Use of New Unknown Variables*

The inspection of the eigenvector (3.37) reveals that in fact each component, when n varies, forms a complete basis in z-space. One can then use the scalar product (3.52), (3.53), and (3.54) for the velocity part and associate a variable $A_{\mathbf{k},n}(t)$ with it and similarly associate variables $B_{\mathbf{k},n}(t)$, and $C_{\mathbf{k},n}(t)$ with the pressure and the temperature. One gets the expansions

$$\bar{\boldsymbol{\varphi}}_{\mathbf{k}} = \sum_n A_{\mathbf{k},n}(t) \bar{\boldsymbol{\varphi}}_{\mathbf{k},n}(z) , \tag{3.57}$$

$$\bar{\varphi}_{\mathbf{k},4} = \sum_n B_{\mathbf{k},n}(t) \varphi_{\mathbf{k},n,4} , \tag{3.58}$$

$$\bar{\varphi}_{\mathbf{k},5} = \sum_n C_{\mathbf{k},n}(t) \varphi_{\mathbf{k},n,5} . \tag{3.59}$$

This method will certainly provide completeness but it has the drawback of having several variables.

3.3.2. *The Use of the Adjoint Problem*

We *choose* the following scalar product in the five-dimensional space of the $\boldsymbol{\varphi}$,

$$(\varphi_{\mathbf{k}}, \psi_{\mathbf{k}}) = \int_0^\infty [\alpha(\varphi_{\mathbf{k},1} \psi^*_{\mathbf{k},1} + \varphi_{\mathbf{k},2} \psi^*_{\mathbf{k},2} + \varphi_{\mathbf{k},3} \psi^*_{\mathbf{k},3}) + \\ + \alpha_4 \varphi_{\mathbf{k},4} \psi^*_{\mathbf{k},4} + \alpha_5 \varphi_{\mathbf{k},5} \psi^*_{\mathbf{k},5}] \, \mathrm{d}z , \tag{3.60}$$

with

$$\alpha = z^m , \tag{3.61}$$

$$\alpha_4 = z^{-m + [(2(m+1))/\gamma]} , \tag{3.62}$$

$$\alpha_5 = z^{-[(2(m+1))/\gamma] + 3m + 2} . \tag{3.63}$$

As already mentioned in (3.56), the eigenvectors are solution of

$$\frac{\partial \varphi_{\mathbf{k}}}{\partial t} + \mathscr{L}_{\mathbf{k}} \varphi_{\mathbf{k}} = 0 , \tag{3.64}$$

where $\mathscr{L}_{\mathbf{k}}$ is the five-dimensional operator

$$\mathscr{L}_{\mathbf{k}} = \begin{bmatrix} 0 & 0 & 0 & -\dfrac{\theta_0}{\lambda_a M^2} ik_x & 0 \\ 0 & 0 & 0 & -\dfrac{\theta_0}{\lambda_a M^2} ik_y & 0 \\ 0 & 0 & 0 & \dfrac{\theta_0}{\lambda_a M^2}\dfrac{\partial}{\partial z} & \dfrac{1}{\lambda_a M^2}\dfrac{\mathrm{d}\pi_0}{\mathrm{d}z} \\ -ik_x\pi_0(\gamma-1) & -ik_y\pi_0(\gamma-1) & \dfrac{\mathrm{d}\pi_0}{\mathrm{d}z} + \pi_0(\gamma-1)\dfrac{\partial}{\partial z} & 0 & 0 \\ 0 & 0 & \dfrac{\mathrm{d}\theta_0}{\mathrm{d}z} & 0 & 0 \end{bmatrix}. \tag{3.65}$$

The adjoint operator $\mathscr{L}_{\mathbf{k}}^{\dagger}$ is defined by

$$(\mathscr{L}_{\mathbf{k}}\boldsymbol{\varphi}_{\mathbf{k}}, \boldsymbol{\psi}_{\mathbf{k}}) = (\boldsymbol{\varphi}_{\mathbf{k}}, \mathscr{L}_{\mathbf{k}}^{\dagger}\boldsymbol{\psi}_{\mathbf{k}}) + [M(\boldsymbol{\varphi}_{\mathbf{k}}, \boldsymbol{\psi}_{\mathbf{k}})]_0^{\infty}, \tag{3.66}$$

where $M(\boldsymbol{\varphi}_{\mathbf{k}}, \boldsymbol{\psi}_{\mathbf{k}})$ is the bilinear concomitant. The adjoint operator $\mathscr{L}_{\mathbf{k}}^{\dagger}$ is given by

$$\mathscr{L}_{\mathbf{k}}^{\dagger} = \begin{bmatrix} 0 & 0 & 0 & \dfrac{\alpha_4}{\alpha}\pi_0(\gamma-1)ik_x & 0 \\ 0 & 0 & 0 & \dfrac{\alpha_4}{\alpha}\pi_0(\gamma-1)ik_y & 0 \\ 0 & 0 & 0 & \dfrac{\alpha_4}{\alpha}\dfrac{\mathrm{d}\pi_0}{\mathrm{d}z} - \dfrac{(\gamma-1)}{\alpha}\left[\dfrac{\mathrm{d}}{\mathrm{d}z}(\alpha_4\pi_0) + \alpha_4\pi_0\dfrac{\partial}{\partial z}\right] & \dfrac{\alpha_5}{\alpha}\dfrac{\mathrm{d}\theta_0}{\mathrm{d}z} \\ \dfrac{\alpha}{\alpha_4}\dfrac{\theta_0}{\lambda_a M^2} ik_x & \dfrac{\alpha}{\alpha_4}\dfrac{\theta_0}{\lambda_a M^2} ik_y & -\dfrac{1}{\alpha_4\lambda_a M^2}\left[\dfrac{\mathrm{d}}{\mathrm{d}z}(\alpha\theta_0) + \alpha\theta_0\dfrac{\partial}{\partial z}\right] & 0 & 0 \\ 0 & 0 & \dfrac{\alpha}{\alpha_5}\dfrac{1}{\lambda_a M^2}\dfrac{\mathrm{d}\pi_0}{\mathrm{d}z} & 0 & 0 \end{bmatrix}, \tag{3.67}$$

and the bilinear concomitant by

$$M(\boldsymbol{\varphi}_{\mathbf{k}}, \boldsymbol{\psi}_{\mathbf{k}}) = \alpha_4\pi_0(\gamma-1)\varphi_{\mathbf{k},3}\psi^*_{\mathbf{k},4} + \alpha\frac{\theta_0}{\lambda_a M^2}\varphi_{\mathbf{k},4}\psi^*_{\mathbf{k},3}. \tag{3.68}$$

The boundary condition (3.15) makes $\varphi_{\mathbf{k},3}$ vanish at $z = 0$ and $z = \infty$; the same condition will be imposed on the adjoint eigenvectors, solutions of

$$\mathscr{L}_{\mathbf{k}}^{\dagger} \varphi_{\mathbf{k}}^{\dagger} = i\sigma_{\mathbf{k}}^{*} \varphi_{\mathbf{k}}^{\dagger} , \tag{3.69}$$

and, as a result, the bilinear concomitant (3.68) becomes equal to zero.

It is easy now to calculate the eigenvectors of the adjoint operator by introducing solutions of the form (3.1). The calculation is simplified by the fact that the differential system obtained by eliminating the pressure and the temperature is self-adjoint with respect to the eigenvalues $\sigma_{\mathbf{k},n}^2$. As a consequence, one obtains the same equation for the eigenvalues and the same 'velocity part' for the eigenvectors. The only part left to calculate is the adjoint pressure and temperature which are given by (3.67) and (3.69). One gets for the adjoint eigenvectors

$$\varphi_{\mathbf{k},j}^{\dagger} = \begin{bmatrix} U_{\mathbf{k},|j|}^{\dagger} \\ V_{\mathbf{k},|j|}^{\dagger} \\ W_{\mathbf{k},|j|}^{\dagger} \\ \Pi_{\mathbf{k},j}^{\dagger} \\ \Theta_{\mathbf{k},j}^{\dagger} \end{bmatrix} = e^{-\zeta/2} \begin{bmatrix} -i \dfrac{k_x}{k} \dfrac{\gamma}{2k} P_{|j}(\zeta) \\ -i \dfrac{k_y}{k} \dfrac{\gamma}{2k} P_{|j}(\zeta) \\ -\dfrac{\gamma}{2k} Q_{|j|}(\zeta) \\ -i\sigma_{\mathbf{k},j}^{*} \dfrac{\gamma}{(\gamma-1)2k^2} \dfrac{\zeta^{m-1-[(m+1)/\gamma]}}{(2k)^{m-1-[(m+1)/\gamma]}} P_{|j|}(\zeta) \\ \dfrac{i}{\sigma_{\mathbf{k},j}^{*}} \dfrac{\gamma}{2kM^2} \dfrac{\zeta^{-m-2+[(m+1)/\gamma]}}{(2k)^{-m-2+[(m+1)/\gamma]}} Q_{|j|}(\zeta) \end{bmatrix} . \tag{3.70}$$

The toroidal modes and f modes are the same as in (3.39)–(3.44), and (3.45) (with the same conventions). From (3.64), (3.66), and (3.69), one gets the following properties:

$$\mathscr{L}_{\mathbf{k}} \varphi_{\mathbf{k},j} = -i\sigma_{\mathbf{k},j} \varphi_{\mathbf{k},j} , \tag{3.71}$$

$$\mathscr{L}_{\mathbf{k}}^{\dagger} \psi_{\mathbf{k},l}^{\dagger} = i\sigma_{\mathbf{k},l}^{*} \psi_{\mathbf{k},l}^{\dagger} , \tag{3.72}$$

so that, using (3.66), one obtains

$$(\sigma_{\mathbf{k},j} - \sigma_{\mathbf{k},l})(\varphi_{\mathbf{k},j}, \psi_{\mathbf{k},l}^{\dagger}) = 0 , \tag{3.73}$$

which implies, when $\sigma_{\mathbf{k},j}$ and $\sigma_{\mathbf{k},l}$ are different, the orthogonality property

$$(\varphi_{\mathbf{k},j}, \psi_{\mathbf{k},l}^{\dagger}) = 0 \quad \text{if} \quad \sigma_{\mathbf{k},j} \neq \sigma_{\mathbf{k},l} . \tag{3.74}$$

Let now the eigenvectors be normalized such that

$$(\boldsymbol{\Phi}_{\mathbf{k},j}, \boldsymbol{\Psi}_{\mathbf{k},l}^{\dagger}) = \delta_{jl} . \tag{3.75}$$

For this, one needs to calculate

$$(\varphi_{\mathbf{k},j}, \psi^{\dagger}_{\mathbf{k},j}) = \int_0^{\infty} e^{-\zeta} \left\{ \frac{\zeta^m}{(2k)^m} \frac{\gamma^2}{4k^2} [(P_{|j|}(\zeta))^2 + (Q_{|j|}(\zeta))^2] + \right.$$

$$\left. + \frac{\zeta^{m-1}}{(2k)^m} \frac{\gamma}{2k^2} \left[(m+1)\omega^2_{|j|} P^2_{|j|}(\zeta) + \frac{(m\gamma - m - 1)}{\omega^2_{|j|}} Q^2_{|j|}(\zeta) \right] \right\} \frac{d\zeta}{2k}, \tag{3.76}$$

if j is any of $n_p, n_g, n_f, -n_p, -n_g, -n_f$. For the toroidal modes, one gets

$$(\varphi_{\mathbf{k},n_t}, \Psi^{\dagger}_{\mathbf{k},n_t}) = \int_0^{\infty} e^{-\zeta} \frac{\zeta^m}{(2k)^m} k^2 [L^m_{n_t}(\zeta)]^2 \frac{d\zeta}{2k}. \tag{3.77}$$

The calculation of (3.76) is tedious but straightforward; one has to use the dispersion relation (3.16) and the following identities for Laguerre polynomials

$$L^m_j(\zeta) = \sum_{h=0}^{j} L^{m-1}_h(\zeta), \tag{3.78}$$

and

$$\int_0^{\infty} e^{-\zeta} \zeta^{m-1} L^m_j(\zeta) L^m_l(\zeta)\, d\zeta = \frac{\Gamma[\mathrm{Inf}(j,l) + m + 1]}{[\mathrm{Inf}(j,l)]!\, m}. \tag{3.79}$$

One gets

$$(\varphi_{\mathbf{k},j}, \psi^{\dagger}_{\mathbf{k},j}) = N(\mathbf{k}, j) = \frac{1}{(2k)^{m+1}} \frac{\gamma^2}{k^2} \frac{|j|!\,\Gamma(|j| + m + 1)}{[(m+1)_{|j|}]^2} \times$$

$$\times \left\{ \left(1 + \frac{|j|}{m}\right) \left[1 + \left(\frac{\omega^2_{|j|} - 1}{\omega^2_{|j|} + 1} \right)^2 \frac{m + |j|}{|j|} \right] - \right.$$

$$\left. - \left(\frac{\omega^2_{|j|} - 1}{\omega^2_{|j|} + 1} \right) \left[\frac{m + 1 - m\gamma + (m+1)\omega^4_{|j|}}{m\gamma\omega^2_{|j|}} \right] \right\}, \tag{3.80}$$

for j belonging to $\pm n_p, \pm n_g, \pm n_f$, and

$$(\varphi_{\mathbf{k},n_t}, \psi^{\dagger}_{\mathbf{k},n_t}) = N(\mathbf{k}, n_t) = \frac{1}{(2k)^{m+1}} \frac{k^2 \Gamma(n_t + m + 1)}{n_t!}. \tag{3.81}$$

This suggests that we define

$$\Phi_{\mathbf{k},j} = \varphi_{\mathbf{k},j} [N(\mathbf{k}, j)]^{-1/2} \tag{3.82}$$

and

$$\Psi^{\dagger}_{\mathbf{k},j} = \psi_{\mathbf{k},j}[N(\mathbf{k},j)]^{-1/2} \tag{3.83}$$

for any index j, so that (3.75) is satisfied. (If $N(\mathbf{k},j)$ is negative, one takes one of the imaginary roots.)

It is now possible to develop on the basis of the eigenvectors $\boldsymbol{\Phi}_{\mathbf{k},n}$. The general vector $\boldsymbol{\Phi}_{\mathbf{k}}$, solution of (2.29), (2.30), and (2.31), is written

$$\boldsymbol{\Phi}_{\mathbf{k}} = \sum_j A_{\mathbf{k},j}(t)\,\boldsymbol{\Phi}_{\mathbf{k},j}(z)\,, \tag{3.84}$$

where, by convection, a summation over index j covers all the indices $\pm n_p$, $\pm n_g$, $\pm n_f$, and n_t. The linear part of the equation, given by (3.64), becomes

$$\frac{\partial \boldsymbol{\Phi}_{\mathbf{k}}}{\partial t} + \mathscr{L}_{\mathbf{k}}\boldsymbol{\Phi}_{\mathbf{k}} = \sum_j (\dot{A}_{\mathbf{k},j} - i\sigma_{\mathbf{k},j}A_{\mathbf{k},j})\,\boldsymbol{\Phi}_{\mathbf{k},j}(z)\,, \tag{3.85}$$

and the equations (2.29), (2.30), and (2.31) are now

$$\sum_j (\dot{A}_{\mathbf{k},j} - i\sigma_{\mathbf{k},j}A_{\mathbf{k},j})\,\overline{\boldsymbol{\Phi}}_{\mathbf{k},j}(z) = -\frac{1}{2\pi}\sum_{j,l}\iint_{-\infty}^{+\infty} \mathrm{d}\mathbf{k}' \left\{\left[i(\mathbf{k}-\mathbf{k}')\cdot\hat{\boldsymbol{\Phi}}_{\mathbf{k}',j} + \right.\right.$$

$$\left.\left. + \,\Phi_{\mathbf{k}',j,3}\frac{\partial}{\partial z}\right]\overline{\boldsymbol{\Phi}}_{\mathbf{k}-\mathbf{k}',l}\right\} A_{\mathbf{k}',j}A_{\mathbf{k}-\mathbf{k}',l} - \frac{1}{2\pi}\frac{1}{\lambda_a M^2}\sum_{j,l}\iint_{-\infty}^{+\infty}\mathrm{d}\mathbf{k}'\,\times$$

$$\times\left\{\Phi_{\mathbf{k}',j,5}\left[i(\mathbf{k}-\mathbf{k}')\,\Phi_{\mathbf{k}-\mathbf{k}',l,4} + \frac{\partial}{\partial z}\Phi_{\mathbf{k}-\mathbf{k}',l,4}\,\delta_{i3}\right]\right\} A_{\mathbf{k}',j}A_{\mathbf{k}-\mathbf{k}',l}\,, \tag{3.86}$$

$$\sum_j (\dot{A}_{\mathbf{k},j} - i\sigma_{\mathbf{k},j}A_{\mathbf{k},j})\,\Phi_{\mathbf{k},j,4} = -\frac{1}{2\pi}\sum_{j,l}\iint_{-\infty}^{+\infty}\mathrm{d}\mathbf{k}'\left\{\left[i(\mathbf{k}-\mathbf{k}')\cdot\hat{\boldsymbol{\Phi}}_{\mathbf{k}',j} + \right.\right.$$

$$\left.\left. + \,\Phi_{\mathbf{k}',j,3}\frac{\partial}{\partial z}\right]\Phi_{\mathbf{k}-\mathbf{k}',l,4}\right\} A_{\mathbf{k}',j}A_{\mathbf{k}-\mathbf{k}',l} - \frac{(\gamma-1)}{2\pi}\sum_{j,l}\iint_{-\infty}^{+\infty}\mathrm{d}\mathbf{k}'\left\{\Phi_{\mathbf{k}',j,4}\,\times\right.$$

$$\left.\times\left[i(\mathbf{k}-\mathbf{k}')\cdot\hat{\boldsymbol{\Phi}}_{\mathbf{k}-\mathbf{k}',l} + \frac{\mathrm{d}}{\mathrm{d}z}\Phi_{\mathbf{k}-\mathbf{k}',l,3}\right]\right\} A_{\mathbf{k}',j}A_{\mathbf{k}-\mathbf{k}',l}\,, \tag{3.87}$$

$$\sum_j (\dot{A}_{\mathbf{k},j} - i\sigma_{\mathbf{k},j} A_{\mathbf{k},j}) \Phi_{\mathbf{k},j,5} = -\frac{1}{2\pi} \sum_{j,l} \iint_{-\infty}^{+\infty} \mathrm{d}\mathbf{k}' \times$$

$$\times \left\{ \left[i(\mathbf{k} - \mathbf{k}') \cdot \boldsymbol{\Phi}_{\mathbf{k}',j} + \Phi_{\mathbf{k}',j,3} \frac{\mathrm{d}}{\mathrm{d}z} \right] \Phi_{\mathbf{k}-\mathbf{k}',l,5} \right\} A_{\mathbf{k}',j} A_{\mathbf{k}-\mathbf{k}',l}, \tag{3.88}$$

where $\overline{\boldsymbol{\Phi}}_{\mathbf{k},j}$ is the velocity part of the eigenvector and $\hat{\boldsymbol{\Phi}}_{\mathbf{k},j}$ the horizontal velocity vector. The evolution equation of the amplitude of the mode n follows by taking the outer product with $\boldsymbol{\Psi}^{\dagger}_{\mathbf{k},n}$,

$$\dot{A}_{\mathbf{k},n} - i\sigma_{\mathbf{k},n} A_{\mathbf{k},n} = \sum_{j,l} \iint_{-\infty}^{+\infty} \mathrm{d}\mathbf{k}'\, C^{jln}_{\mathbf{k},\mathbf{k}'} A_{\mathbf{k}',j} A_{\mathbf{k}-\mathbf{k}',l}, \tag{3.89}$$

where $C^{jln}_{\mathbf{k},\mathbf{k}'}$ are the coupling coefficients given in Appendix B. In general, in the development (3.84), the $A_{\mathbf{k},j}(t)$ are complex coefficients. If one is assured of completeness, the system of Equations (3.89) for all $\mathbf{k}$ and n is equivalent to the initial non-linear equations. Given the completeness result about the velocity vector, completeness in the present case seems to depend only on the fact that, in the linear problem, the relations which link $\Pi_{\mathbf{k},n}$ and $\Theta_{\mathbf{k},n}$ to the velocity have coefficients which do not cancel and it seems to be the case here.

4. The Non-Linear Problem

At this point, the modes must be classified according to the criterion of Section 1. So the f and p modes (respectively g and toroïdal) will be considered as pulsating (resp. convective) modes and are associated symbolically with the index p (resp. c). The equations for both kinds of modes become

$$\frac{\mathrm{d}}{\mathrm{d}t} A_{\mathbf{k},p} - i\sigma_{\mathbf{k},p} A_{\mathbf{k},p} = \sum_{c_1,c_2} \iint_{-\infty}^{+\infty} \mathrm{d}\mathbf{k}'\, C^{c_1 c_2 p}_{\mathbf{k},\mathbf{k}'} A_{\mathbf{k}',c_1} A_{\mathbf{k}-\mathbf{k}',c_2} +$$

$$+ \sum_{c_1,p_2} \iint_{-\infty}^{+\infty} \mathrm{d}\mathbf{k}'\, C^{c_1 p_2 p}_{\mathbf{k},\mathbf{k}'} A_{\mathbf{k}',c_1} A_{\mathbf{k}-\mathbf{k}',p_2} +$$

$$+ \sum_{p_1,c_2} \iint_{-\infty}^{+\infty} \mathrm{d}\mathbf{k}'\, C^{p_1 c_2 p}_{\mathbf{k},\mathbf{k}'} A_{\mathbf{k}',p_1} A_{\mathbf{k}-\mathbf{k}',c_2} +$$

$$+ \sum_{p_1,p_2} \iint_{-\infty}^{+\infty} \mathrm{d}\mathbf{k}'\, C^{p_1 p_2 p}_{\mathbf{k},\mathbf{k}'} A_{\mathbf{k}',p_1} A_{\mathbf{k}-\mathbf{k}',p_2}, \tag{4.1}$$

$$\frac{d}{dt}A_{\mathbf{k},c} + \tilde{\sigma}_{\mathbf{k},c}A_{\mathbf{k},c} = \sum_{c_1,c_2} \iint_{-\infty}^{+\infty} d\mathbf{k}' \; C_{\mathbf{k},\mathbf{k}'}^{c_1 c_2 c} A_{\mathbf{k}',c_1} A_{\mathbf{k}-\mathbf{k}',c_2} +$$

$$+ \sum_{c_1,p_2} \iint_{-\infty}^{+\infty} d\mathbf{k}' \; C_{\mathbf{k},\mathbf{k}'}^{c_1 p_2 c} A_{\mathbf{k}',c_1} A_{\mathbf{k}-\mathbf{k}',p_2} +$$

$$+ \sum_{p_1,c_2} \iint_{-\infty}^{+\infty} d\mathbf{k}' \; C_{\mathbf{k},\mathbf{k}'}^{p_1 c_2 c} A_{\mathbf{k}',p_1} A_{\mathbf{k}-\mathbf{k}',c_2} +$$

$$+ \sum_{p_1,p_2} \iint_{-\infty}^{+\infty} d\mathbf{k}' \; C_{\mathbf{k},\mathbf{k}'}^{p_1 p_2 c} A_{\mathbf{k}',p_1} A_{\mathbf{k}-\mathbf{k}',p_2} . \tag{4.2}$$

In these equations, indices c, c_1, and c_2 (resp. p, p_1, and p_2) are associated with convective (resp. pulsating) modes. $\sigma_{\mathbf{k},p}$ is real and $\tilde{\sigma}_{\mathbf{k},c}$ is the imaginary part of $\sigma_{\mathbf{k},n_g}$ or $\sigma_{\mathbf{k},-n_g}$ (it is zero for toroïdal modes). All the different-types of coupling have been shown and, as far as interpretation goes, (4.1) is the system of non-linear equations that governs the pulsation. The terms on the right-hand side of (4.1) have a clear physical meaning. The first term is the forcing by the convection and it is very much like a 'Lighthill type' forcing term, though Lighthill (1952) did his calculation for a homogeneous case (see also Lighthill, 1959, 1962; Proudman, 1952; and Pierce and Coroniti, 1966; for a possible example of this type of coupling in geophysics). The second and third terms will have a net damping effect on the pulsation (positive or negative!) and the last term is the wave-wave interaction which gives rise to wave-turbulence.

To simplify these equations, one must exploit the fact that the pulsation is 'almost linear'. This means that we shall try to deal with the pulsation problem as with a wave turbulence problem (i.e. a weakly non linear problem) whereas the convection will stay a fully non-linear problem. Wave-turbulence problems are a familiar class of problems in water-wave theory and in plasma physics. The general problem of weak interaction of surface waves has been studied by Benney and Saffman (1966), Hasselman (1966, 1968), and reviewed by Phillips (1981).

In plasma physics, the evolution of wave correlations has been studied by Davidson (1967). One of the fundamental differences between these cases and the solar case is the inhomogeneity: in general, the wave problems are homogeneous and the linear properties of the problem are simpler than in Section 3. The other difference is that turbulence, when introduced, is generally considered as an external source (in the problem of wind-driven surface waves for instance) and is never perturbed by the waves. In the solar case, it is the goal of the theory to perturb the convection with the oscillations; as a result one has to expect a much more complicated problem.

Some conditions on $A_{\mathbf{k},j}$ come from the fact that the vector

$$\boldsymbol{\Phi} = \begin{bmatrix} u' \\ v' \\ w' \\ \pi' \\ \theta' \end{bmatrix} = \frac{1}{2\pi} \iint_{-\infty}^{+\infty} \boldsymbol{\Phi}_{\mathbf{k}}(z, t) e^{i\mathbf{k}\cdot\mathbf{x}} \, \mathrm{d}\mathbf{k} = \frac{1}{2\pi} \sum_j \iint_{-\infty}^{+\infty} A_{\mathbf{k},j}(t) \boldsymbol{\Phi}_{\mathbf{k},j}(z) e^{i\mathbf{k}\cdot\mathbf{x}} \, \mathrm{d}\mathbf{k} \tag{4.3}$$

is real. These are

$$\text{for } p \text{ and } f \text{ modes: } A_{-\mathbf{k},-j} = A^*_{\mathbf{k},j}, \tag{4.4}$$

$$\text{for } g \text{ and toroidal modes: } A_{-\mathbf{k},j} = A^*_{\mathbf{k},j}. \tag{4.5}$$

There are two ways to treat equations (4.1) and (4.2). Either one chooses to expand first with respect to a small parameter and then to carry out statistical averages to obtain equations for the evolution of the correlations or one does these in reverse order. In both cases, one comes up with the same equations though there is a certain amount of debate in the literature about the domain of validity of the expansions which seems to depend on the path followed. In any case, to obtain a uniform development, one has to use multi-time expansions, i.e. perturb the frequencies of the pulsation modes. Here, the method is basically the same as in wave turbulence theory, except that strongly unstable modes (the amplitudes $A_{\mathbf{k},c}$) are now present and they require special treatment. It will be seen that indeed, contrary to the case of wave turbulence, these convective terms dominate the problem. The way to really deal with the problem here is to compute first the different coupling coefficients: there is an infinity of them and it might look a hopeless task, but it is simplified by the symmetries with respect to the indices that these coefficients have, and the fact that one can apply well established asymptotic formulae for the eigenfrequencies and eigenvectors for certain ranges of indices. The key question of the coupling is the ordering of Equations (4.1) and (4.2) when one develops them with respect to the pulsation amplitude (which is small). One has also to understand why it is small. Only when one has defined the expansion procedure and the scaling will it be possible to interpret the expression 'perturbation of the convection by the pulsation' from the equations and there are *a priori* several ways to do it. We shall propose one that looks reasonable to us and corresponds to what we think about the physics of the coupling, but others may be imagined. The settlement of this question demands the computation of the respective strengths of the coupling coefficients. We shall not go into the detailed calculation of these coupling coefficients here (it involves rather heavy numerical calculations which are not the object of this paper). The following method of scaling the coupling coefficients is introduced: one introduces ε (the small bookkeeping parameter of the problem) per index p of the coupling coefficients.

So one gets

$$C^{c_1c_2c}_{\mathbf{k},\mathbf{k}'} = D^{c_1c_2c}_{\mathbf{k},\mathbf{k}'} , \tag{4.6}$$

$$C^{c_1p_2c}_{\mathbf{k},\mathbf{k}'} = \varepsilon D^{c_1p_2c}_{\mathbf{k},\mathbf{k}'} , \tag{4.7}$$

$$C^{p_1p_2c}_{\mathbf{k},\mathbf{k}'} = \varepsilon^2 D^{p_1p_2c}_{\mathbf{k},\mathbf{k}'} , \tag{4.8}$$

$$C^{p_1p_2p}_{\mathbf{k},\mathbf{k}'} = \varepsilon^2 D^{p_1p_2p}_{\mathbf{k},\mathbf{k}'} , \tag{4.9}$$

and so on ... One looks for a solution of the form

$$A_{\mathbf{k},c} = A^{(0)}_{\mathbf{k},c} + \varepsilon A^{(1)}_{\mathbf{k},c} + \varepsilon^2 A^{(2)}_{\mathbf{k},c} + \dots , \tag{4.10}$$

$$A_{\mathbf{k},p} = A^{(0)}_{\mathbf{k},p} + \varepsilon A^{(1)}_{\mathbf{k},p} + \varepsilon^2 A^{(2)}_{\mathbf{k},p} + \dots . \tag{4.11}$$

It is also necessary to introduce a procedure to get rid of resonant terms in the 'oscillatory type equations'. There are several ways to do it. Davidson (1967) uses, for example, a two-timing method. Here the frequencies are perturbed according to a procedure that is akin to the Poincaré–Lindstedt method. The new time $s_{\mathbf{k},p}$ such that

$$t = s_{\mathbf{k},p}(1 + \varepsilon\sigma^{(1)}_{\mathbf{k},p} + \varepsilon^2\sigma^{(2)}_{\mathbf{k},p} + \dots) \tag{4.12}$$

is introduced. $\sigma^{(1)}_{\mathbf{k},p}$, $\sigma^{(2)}_{\mathbf{k},p}$, etc. are chosen so as to kill the resonances appearing on the right-hand side of Equation (4.1). There are other methods to deal with this problem but they all give essentially the same results: they kill the secular terms so that uniformly valid expansions can be obtained. (4.12) establishes a relation between the new shifted frequencies of the system and the finite amplitudes of the perturbations.

The first equation that appears to order ε^0 is

$$\frac{\mathrm{d}}{\mathrm{d}t} A^{(0)}_{\mathbf{k},c} + \tilde{\sigma}_{\mathbf{k},c} A^{(0)}_{\mathbf{k},c} = \sum_{c_1,c_2} \iint_{-\infty}^{+\infty} \mathrm{d}\mathbf{k}'\, D^{c_1c_2c}_{\mathbf{k},\mathbf{k}'} A^{(0)}_{\mathbf{k}',c_1} A^{(0)}_{\mathbf{k}-\mathbf{k}',c_2} . \tag{4.13}$$

This is a strongly non-linear problem and an answer to it necessarily asks for a theory of convection (for instance, some closure scheme on the BBGKY hierarchy of equations for the correlations derived from (IV.13) such as the eddy damped quasi-normal markovian scheme (Orszag, 1970) or the test field model (Kraichnan, 1971, 1972); we shall try closures of this type in a later paper). This problem will be supposed to have been solved; in fact, it is the problem on which one wants to put constraints from pulsation theory. (4.13) describes convection in the unperturbed star. The other equations to order ε^0 are

$$\frac{\mathrm{d}}{\mathrm{d}s_{\mathbf{k},p}} A^{(0)}_{\mathbf{k},p} - i\sigma_{\mathbf{k},p} A^{(0)}_{\mathbf{k},p} = 0 , \tag{4.14}$$

with solutions

$$A^{(0)}_{\mathbf{k},p} = \alpha_{\mathbf{k},p} e^{i\sigma_{\mathbf{k},p} s_{\mathbf{k},p}} . \tag{4.15}$$

To order ε, the following equations are obtained:

$$\frac{d}{dt} A^{(1)}_{\mathbf{k},c} + \tilde{\sigma}_{\mathbf{k},c} A^{(1)}_{\mathbf{k},c} = \sum_{c_1, c_2} \iint_{-\infty}^{+\infty} d\mathbf{k}' \, D^{c_1 c_2 c}_{\mathbf{k},\mathbf{k}'} (A^{(0)}_{\mathbf{k}',c_1} A^{(1)}_{\mathbf{k}-\mathbf{k}',c_2} + A^{(1)}_{\mathbf{k}',c_1} A^{(0)}_{\mathbf{k}-\mathbf{k}',c_2}) +$$

$$+ \sum_{c_1, p_2} \iint_{-\infty}^{+\infty} d\mathbf{k}' \, D^{c_1 p_2 c}_{\mathbf{k},\mathbf{k}'} (A^{(0)}_{\mathbf{k}',c_1} A^{(0)}_{\mathbf{k}-\mathbf{k}',p_2}) +$$

$$+ \sum_{p_1, c_2} \iint_{-\infty}^{+\infty} d\mathbf{k}' \, D^{p_1 c_2 c}_{\mathbf{k},\mathbf{k}'} (A^{(0)}_{\mathbf{k}',p_1} A^{(0)}_{\mathbf{k}-\mathbf{k}',c_2}), \tag{4.16}$$

$$\frac{d}{ds_{\mathbf{k},p}} A^{(1)}_{\mathbf{k},p} - i\sigma_{\mathbf{k},p} A^{(1)}_{\mathbf{k},p} = i\sigma_{\mathbf{k},p} \sigma^{(1)}_{\mathbf{k},p} A^{(0)}_{\mathbf{k},p} +$$

$$+ \sum_{c_1, c_2} \iint_{-\infty}^{+\infty} d\mathbf{k}' \, D^{c_1 c_2 p}_{\mathbf{k},\mathbf{k}'} A^{(0)}_{\mathbf{k}',c_1} A^{(0)}_{\mathbf{k}-\mathbf{k}',c_2} . \tag{4.17}$$

Equation (4.17) describes the forcing of the pulsation by the convective terms. $\sigma^{(1)}_{\mathbf{k},p}$ is chosen in such a way that it kills the resonant part of these convective terms. Equation (4.17) is generally the type of equation treated under various forms in the literature to describe the coupling between convection and pulsation (see Goldreich and Keeley, 1977b). However one can barely speak of 'coupling', since convection acts as a simple forcing term. Equation (4.16) is the equation for the perturbation of the convection. It would be more correct, at this stage, to reason directly on correlations, since perturbing the unstable equation (4.2) is a bit ambiguous. One could suppose that $\langle A_{\mathbf{k}',p'} A_{\mathbf{k},p'} \rangle$ reaches a steady state and then perturb the correlation. But the results, as far as equations and expansions are concerned, remain the same. Equation (4.16) is very different in nature from Equation (4.13) despite the resemblance of the left-hand sides: there is no non-linear term on the right-hand side of (4.16). The first term is linear in the amplitude of the perturbed convection and will act as a damping (positive or negative) while the two remaining terms induce pulsation-convection coupling. Consequently, Equation (4.16) will be easier to deal with than (4.13). To order ε^2, one gets the equation

$$\frac{d}{ds_{\mathbf{k},p}} A^{(2)}_{\mathbf{k},p} - i\sigma_{\mathbf{k},p} A^{(2)}_{\mathbf{k},p} = i\sigma_{\mathbf{k},p} \sigma^{(1)}_{\mathbf{k},p} A^{(1)}_{\mathbf{k},p} + i\sigma_{\mathbf{k},p} \sigma^{(2)}_{\mathbf{k},p} A^{(0)}_{\mathbf{k},p} +$$

$$+ \sum_{c_1, c_2} \iint_{-\infty}^{+\infty} d\mathbf{k}' \, D^{c_1 c_2 p}_{\mathbf{k},\mathbf{k}'} (A^{(0)}_{\mathbf{k}',c_1} A^{(1)}_{\mathbf{k}-\mathbf{k}',c_2} + A^{(1)}_{\mathbf{k}',c_1} A^{(0)}_{\mathbf{k}-\mathbf{k}',c_2}) +$$

$$+ \sum_{c_1, p_2} \iint_{-\infty}^{+\infty} \mathrm{d}\mathbf{k}' \; D^{c_1 p_2 p}_{\mathbf{k},\mathbf{k}'} A^{(0)}_{\mathbf{k}', c_1} A^{(0)}_{\mathbf{k}-\mathbf{k}', p_2} +$$

$$+ \sum_{p_1, c_2} \iint_{-\infty}^{+\infty} \mathrm{d}\mathbf{k}' \; D^{p_1 c_2 p}_{\mathbf{k},\mathbf{k}'} A^{(0)}_{\mathbf{k}', p_1} A^{(0)}_{\mathbf{k}-\mathbf{k}', c_2} . \qquad (4.18)$$

Equation (4.18) is the equation we really want. To this order, $\sigma^{(2)}_{\mathbf{k},p}$ is chosen to kill the resonances of the right-hand side of the equation. One can now say that the pulsation and the convection have really perturbed each other and that the resulting effect for the pulsation is Equation (4.18). So, in our opinion, Equations (4.13), (4.14), (4.16), (4.17), and (4.18) are the equations applicable to the solar case. One can see that, to order ε^2, with the scaling chosen, there is still no interaction of the pulsation with itself. Equation (4.12), to order ε^2, gives the frequency shifts as a function of ε. As already emphasized, one can write equations for the correlations between the different amplitudes and it is these equations that have to be dealt with numerically. But, as far as establishing the equations is concerned, the physical interpretation of the different terms that appear in the expansion is more straightforward if one deals directly with the amplitudes.

5. Conclusion and Prospects

The polytropic problem is a much simplified version of the full solar problem. In the solar case, five facts complicate the problem. First one has to use a solar model with a convective zone calculated by mixing length theory as the static state. Second, the linear problem of the calculation of eigenmodes and eigenfrequencies has more complicated boundary conditions: one has to allow for running waves above the cutoff frequency which act as a drain of energy on the system and one generally has to put a model atmosphere on top of the solar model to have realistic boundary conditions. Third, the eigenmodes are more varied; there are also g^+ modes that will be put into the wave category (according to the criterion of Section 1). Fourth, some overstable pulsations can be found, for p-modes for instance, due to κ-mechanism (Ando and Osaki, 1975). Finally one has to operate in spherical geometry. None of these facts makes it more difficult, in theory, to deal with the problem along the lines of Sections 3 and 4. In the problem of non-radial pulsation, perturbation of the convective flux has generally been ignored. So one would use as an eigenbasis the eigenvectors and eigenvalues calculated without coupling and combine them in the non-linear calculation of Section 4, where the perturbation of the convective flux effectively appears. From a numerical point of view, the eigenvectors and eigenfunctions can be calculated one by one for low l and one can use the various asymptotic formulae known for the others. Two cases must be considered to treat the problem of overstability. If it is a pulsation mode which is overstable, then one can scale the unstable part by an ε and have it interact in the next order, as in the usual wave-turbulence problem (the system is weakly non-linear). If it is a convective mode that is overstable, one can keep convective frequencies with imaginary and real parts in the zeroth order problem of the convection and expand on the convective modes with these frequencies.

The procedure of Section 4 links the convection spectrum to the pulsation spectrum. There are three ways to deal with it. First one can try to keep only a few significant modes of each type, truncate, and study the dynamical system that follows. Second, one can try arbitrary theories (mixing length or more recent closures) on Equation (4.13) to see the consequences on the pulsation modes. Third, one can try to use all the dynamical information measured so far on the modes (see Grec *et al.*, 1983; Bos and Hill, 1983) and the more complete information that will be provided by DISCO (see Bonnet, 1983) to derive constraints on the convection spectrum in the solar case. It is a complicated inverse scattering problem, but, as mentioned earlier, this would be the way to understand solar convection a bit better. Moreover more constraints could be derived by applying the same method to the erratically varying white dwarfs that have recently been observed. One can note that the convective equations can be simplified further by differentiating between the equations where $\tilde{\sigma}_{\mathbf{k},c} > 0$ and those where $\tilde{\sigma}_{\mathbf{k},c} < 0$. In the first case, the mode is strongly damped and one can put it equal to zero. These modes can be eliminated and this brings in new coupling coefficients for the modes with $\tilde{\sigma}_{\mathbf{k},c} < 0$. What we have in mind is that the interaction of these remaining modes produces a spectrum that is perhaps analogous to what is found in the ocean: broad peaks (corresponding to the different cells) superposed on a continuum.

In conclusion, the nature of the non-linear problem in Section 4 should be emphasized. It is homogeneous in $\mathbf{k}$ (i.e. horizontally) and the inhomogeneity appears through the coupling coefficients. Because of this very inhomogeneity, it might turn out that the turbulence problem has simpler features than the usual homogeneous turbulence problems because the coupling coefficients will be imposed roughly by the stratification and one has to expect strong selection rules coming from this. The energy transfers might then be markedly different due to the compressible character of the problem.

Acknowledgements

I wish to thank D. O. Gough especially, for his encouragements and advice and an anonymous referee for his helpful comments. Moreover I am indebted to the French 'Ministère des Relations Extérieures' for financing my Crimean trip, the Observatoire de Toulouse et Pic-du-Midi for financial support and the director of the Institut d'Astrophysique de Paris, Jean Audouze, for receiving me in his Institute while doing this research.

Appendix A

The two following relations concerning confluent hypergeometric functions will be useful in what follows:

$$\frac{\mathrm{d}}{\mathrm{d}\zeta} M(a, b, \zeta) = \frac{a}{b} M(a + 1, b + 1, \zeta), \tag{A.1}$$

$$M(a, b, \zeta) = M(a + 1, b, \zeta) - \frac{\zeta}{b} M(a + 1, b + 1, \zeta) . \tag{A.2}$$

From this, it is easy to deduce the identities

$$\zeta \frac{\mathrm{d}}{\mathrm{d}\zeta} M(a, b, \zeta) = aM(a + 1, b, \zeta) - aM(a, b, \zeta) , \tag{A.3}$$

$$\zeta M(a, b, \zeta) = (a - b)M(a - 1, b, \zeta) + (b - 2a)M(a, b, \zeta) + aM(a + 1, b, \zeta) . \tag{A.4}$$

The term between brackets of Equation (3.13) is now calculated for each of the independent solutions found for $\psi_{\mathbf{k}}$. One starts with $M(\alpha, m + 2, \zeta)$:

$$H_1 = \omega^2 \left[-\zeta \frac{\partial M(\alpha, m + 2, \zeta)}{\partial \zeta} - (m + 1)M(\alpha, m + 2, \zeta) \right] + \\ + \tfrac{1}{2}\zeta(1 + \omega^2)M(\alpha, m + 2, \zeta) . \tag{A.5}$$

Using (A.3) and (A.4) to express the first and third terms, one gets

$$H_1 = \frac{\alpha}{2}(1 - \omega^2)M(\alpha + 1, m + 2, \zeta) + \left[\frac{m}{2}(1 - \omega^2) + 1 - \alpha \right] \times \\ \times M(\alpha, m + 2, \zeta) + \tfrac{1}{2}(1 + \omega^2)(\alpha - m - 2)M(\alpha - 1, m + 2, \zeta) . \tag{A.6}$$

(A.6) can now be cast into the form

$$H_1 = \frac{1 - \omega^2}{2} \{\alpha M(\alpha + 1, m + 2, \zeta) + (m + 1 - \alpha)M(\alpha, m + 2, \zeta)\} + \\ + \frac{1 + \omega^2}{2} \{(1 - \alpha)M(\alpha, m + 2, \zeta) + (\alpha - m - 2)M(\alpha - 1, m + 2, \zeta)\} . \tag{A.7}$$

Using the identity

$$(1 + a - b)M(a, b, \zeta) - aM(a + 1, b, \zeta) + (b - 1)M(a, b - 1, \zeta) = 0 , \tag{A.8}$$

to express the first and fourth terms in (A.7), (A.7) finally becomes

$$H_1 = (m + 1) \left\{ \frac{1 - \omega^2}{2} M(\alpha, m + 1, \zeta) - \frac{1 + \omega^2}{2} M(\alpha - 1, m + 1, \zeta) \right\} . \tag{A.9}$$

Applying the same kind of transformation to

$$H_2 = \omega^2 \left\{ -\zeta \frac{\partial}{\partial \zeta} \left[\frac{M(\alpha - m - 1, -m, \zeta)}{\zeta^{m+1}} \right] - \right.$$

$$\left. - \frac{(m+1)}{\zeta^{m+1}} M(\alpha - m - 1, -m, \zeta) \right\} +$$

$$+ \tfrac{1}{2}(1 + \omega^2) \frac{M(\alpha - m - 1, -m, \zeta)}{\zeta^m}, \tag{A.10}$$

one obtains

$$H_2 = \frac{\omega^2}{\zeta^m} \frac{\partial}{\partial \zeta} M(\alpha - m - 1, -m, \zeta) + \tfrac{1}{2}(1 + \omega^2) \frac{M(\alpha - m - 1, -m, \zeta)}{\zeta^m}. \tag{A.11}$$

Using (A.1) and (A.8), one obtains the identity

$$b \frac{\mathrm{d}}{\mathrm{d}\zeta} M(a, b, \zeta) = bM(a, b, \zeta) + (a - b) M(a, b + 1, \zeta), \tag{A.12}$$

which can be used to transform the first term in (A.11),

$$H_2 = \frac{1}{\zeta^m} \left\{ \frac{(1 - \omega^2)}{2} M(\alpha - m - 1, -m, \zeta) + \frac{(\alpha - 1)\omega^2}{m} \times \right.$$

$$\left. \times M(\alpha - m - 1, -m + 1, \zeta) \right\}. \tag{A.13}$$

Using (A.8) to transform the first term, one gets

$$H_2 = \frac{1}{\zeta^m} \left\{ \frac{1 + \omega^2}{2} \frac{(\alpha - 1)}{m} M(\alpha - m - 1, -m + 1, \zeta) - \right.$$

$$\left. - \left(\frac{1 - \omega^2}{2} \right) \frac{(\alpha - m - 1)}{m} M(\alpha - m, -m + 1, \zeta) \right\}. \tag{A.14}$$

When one regroups H_1 and H_2 given by (A.9) and (A.14) with (3.13), one gets (3.14) for the velocity.

Appendix B

The expression for the coupling coefficient $C^{jln}_{\mathbf{k},\mathbf{k}'}$ is the following:

$$C^{jln}_{\mathbf{k},\mathbf{k}'} = -\frac{1}{2\pi}\int_0^\infty \mathrm{d}z\, z^m \Big\{ i(\mathbf{k}-\mathbf{k}')\cdot \hat{\boldsymbol{\Phi}}_{\mathbf{k}',j}(\overline{\boldsymbol{\Phi}}_{\mathbf{k}-\mathbf{k}',l}\cdot \overline{\boldsymbol{\Psi}}^{\dagger}_{\mathbf{k},n}) +$$

$$+ \Phi_{\mathbf{k}',j,3}\left[\left(\frac{\mathrm{d}}{\mathrm{d}z}\overline{\boldsymbol{\Phi}}_{\mathbf{k}-\mathbf{k}',l}\right)\cdot \overline{\boldsymbol{\Psi}}^{\dagger}_{\mathbf{k},n}\right]\Big\} -$$

$$-\frac{1}{\lambda_a M^2}\frac{1}{2\pi}\int_0^\infty \mathrm{d}z\, z^m\, \Phi_{\mathbf{k}',j,5}\Big[i(\mathbf{k}-\mathbf{k}')\cdot \hat{\boldsymbol{\Psi}}^{\dagger}_{\mathbf{k},n}\, \Phi_{\mathbf{k}-\mathbf{k}',l,4} +$$

$$+\left(\frac{\mathrm{d}}{\mathrm{d}z}\Phi_{\mathbf{k}-\mathbf{k}',l,4}\right)\Psi^{\dagger}_{\mathbf{k},n,3}\Big] -$$

$$-\frac{1}{2\pi}\int_0^\infty \mathrm{d}z\, z^{-m+[(2(m+1))/\gamma]}\Big[i(\mathbf{k}-\mathbf{k}')\cdot \hat{\boldsymbol{\Phi}}_{\mathbf{k}',j}\, \Phi_{\mathbf{k}-\mathbf{k}',l,4}\, \Psi^{\dagger}_{\mathbf{k},n,4} +$$

$$+ \Phi_{\mathbf{k}',j,3}\left(\frac{\mathrm{d}}{\mathrm{d}z}\Phi_{\mathbf{k}-\mathbf{k}',l,4}\right)\Psi^{\dagger}_{\mathbf{k},n,4}\Big] -$$

$$-\frac{(\gamma-1)}{2\pi}\int_0^\infty \mathrm{d}z\, z^{-m+[(2(m+1))/\gamma]}\, \Phi_{\mathbf{k}',j,4}\, \Psi^{\dagger}_{\mathbf{k},n,4}\Big[i(\mathbf{k}-\mathbf{k}')\cdot \hat{\boldsymbol{\Phi}}_{\mathbf{k}-\mathbf{k}',l} +$$

$$+\left(\frac{\mathrm{d}}{\mathrm{d}z}\Phi_{\mathbf{k}-\mathbf{k}',l,3}\right)\Big] -$$

$$-\frac{1}{2\pi}\int_0^\infty \mathrm{d}z\, z^{-[(2(m+1))/\gamma]+3m+2}\Big[i(\mathbf{k}-\mathbf{k}')\cdot \hat{\boldsymbol{\Phi}}_{\mathbf{k}',j}\, \Phi_{\mathbf{k}-\mathbf{k}',l,5}\, \Psi^{\dagger}_{\mathbf{k},n,5} +$$

$$+ \Phi_{\mathbf{k}',j,3}\left(\frac{\mathrm{d}}{\mathrm{d}z}\Phi_{\mathbf{k}-\mathbf{k}',l,5}\right)\Psi^{\dagger}_{\mathbf{k},n,5}\Big], \tag{B.1}$$

where the symbols $\hat{\boldsymbol{\Psi}}^{\dagger}$ and $\overline{\boldsymbol{\Psi}}^{\dagger}$ have the same meanings as those introduced for $\boldsymbol{\Phi}$. Ingeneral, performing the integrals in z will give rise to some selection rules between j, l, and n.

References

Ando, H. and Osaki, Y.: 1975, *Publ. Astron. Soc. Japan* **27**, 581.

Ando, H. and Osaki, Y.: 1977, *Publ. Astron. Soc. Japan* **29**, 221.

Benney, D. J. and Saffman, P. G.: 1966, *Proc. Roy. Soc. London* **A289**, 301.

Bonnet, R. M.: 1983, *Solar Phys.* **82**, 487 (this volume).

Bos, R. J. and Hill, H. A.: 1983, *Solar Phys.* **82**, 89 (this volume).

Christensen-Dalsgaard, J. and Frandsen, S.: 1983, *Solar Phys.* **82**, 165 (this volume).

Davidson, R. C.: 1967, *J. Plasma Phys.* **1**, 341.

Dyson, J. and Schutz, B.: 1979, *Proc. Roy. Soc. London* **A368**, 389.

Eisenfeld, J.: 1968, *J. Math. Anal. Appl.* **23**, 58.

Eisenfeld, J.: 1969, *J. Math. Anal. Appl.* **26**, 357.

Goldreich, P. and Keeley, D. A.: 1977a, *Astrophys. J.* **211**, 934.

Goldreich, P. and Keeley, D. A.: 1977b, *Astrophys. J.* **212**, 243.

Gough, D. O.: 1969, *J. Atmospheric Sci.* **26**, 448.

Gough, D. O.: 1976, *Astrophys. J.* **214**, 196.

Gough, D. O.: 1978, in G. Belvedere and L. Paterno (eds.), *Proc. Workshop on Solar Rotation*, Univ. Catania Press, p. 255.

Gough, D. O. and Weiss, N. O.: 1976, *Monthly Notices Roy. Astron. Soc.* **176**, 589.

Grec, G., Fossat, E., and Pomerantz, M.: 1980, *Nature* **288**, 541.

Grec, G., Fossat, E., and Pomerantz, M. A.: 1983, *Solar Phys.* **82**, 55 (this volume).

Hasselman, K.: 1966, *Proc. Roy. Soc. London* **A289**, 77.

Hasselman, K.: 1968, in M. Holt (ed.), *Basic Developments in Fluid Dynamics*, Vol. 2, Academic Press.

Hurle, D. T. J., Jakeman, E., and Pike, E. R.: 1967, *Proc. Roy. Soc. London* **A296**, 469.

Kaniel, S. and Kovetz, A.: 1967, *Phys. Fluids* **10**, 1186.

Kraichnan, R. H.: 1971, *J. Fluid Mech.* **47**, 513.

Kraichnan, R. H.: 1972, *J. Fluid Mech.* **56**, 287.

Lamb, H.: 1932, *Hydrodynamics*, 6th edition, Dover.

Lighthill, M. J.: 1952, *Proc. Roy. Soc. London* **A211**, 564.

Lighthill, M. J.: 1954, *Proc. Roy. Soc. London* **A222**, 1.

Lighthill, M. J.: 1962, *Proc. Roy. Soc. London* **A267**, 147.

Ogura, Y. and Phillips, N. A.: 1962, *J. Atmospheric Sci.* **19**, 173.

Orszag, S. A.: 1970, *J. Fluid. Mech.* **41**, 363.

Phillips, O. M.: 1981, *J. Fluid. Mech.* **106**, 215.

Pierce, A. D. and Coroniti, S. C.: 1966, *Nature* **210**, 1209.

Poyet, J.-P. and Spiegel, E. A.: 1979, *Astron. J.* **84**, 1918.

Proudman, I.: 1952, *Proc. Roy. Soc. London* **A214**, 119.

Sansone, G.: 1959, *Orthogonal Functions*, Interscience Publishers, New-York.

Spiegel, E. A. and Unno, W.: 1962, *Publ. Astron. Soc. Japan* **14**, 28.

Unno, W.: 1967, *Publ. Astron. Soc. Japan* **19**, 140.

Unno, W.: 1977, in E. A. Spiegel and J.-P. Zahn (eds.), *Problems of Stellar Convection*, Springer-Verlag, p. 315.

Unno, W., Osaki, Y., Ando, H., and Shibahashi, H.: 1979, *Non-Radial Oscillations of Stars*, University of Tokyo Press.

Weinberger, H. F.: 1968, *J. Math. Anal. Appl.* **21**, 506.

KOLMOGOROV UNSTABLE STELLAR OSCILLATIONS*

J. PERDANG**

Institute of Astronomy, Cambridge, CB3 OHA, England

and

Institut d'Astrophysique, Cointe-Ougrée, B-4200, Belgium†

Abstract. We survey the mathematics of non-linear Hamiltonian oscillations with emphasis being laid on the more recently discovered Kolmogorov instability. In the context of radial adiabatic oscillations of stars this formalism predicts a Kolmogorov instability even at low oscillation energies, provided that sufficiently high linear asymptotic modes have been excited.

Numerical analysis confirms the occurrence of this instability. It is found to show up already among the lowest order modes, although high surface amplitudes are then required ($|\delta r|/R \sim 0.5$ for an unstable fundamental mode – first harmonic coupling). On the basis of numerical evidence we conjecture that in the Kolmogorov unstable regime the enhanced coupling due to internal resonance effects leads to an equipartition of energy over all interacting degrees of freedom. We also indicate that the power spectrum of such oscillations is expected to display two components: A very broad band of overlapping pseudo-linear frequency peaks spread out over the asymptotic range, and a strictly non-linear $1/f$-noise type component close to the frequency origin.

It is finally argued that the Kolmogorov instability is likely to occur among non-linearly coupled non-radial stellar modes at a surface amplitude much lower than in the radial case. This lends support to the view that this instability might be operative among the solar oscillations.

1. Motivation

It has recently been observed that the SCLERA power spectra of solar oscillations (Brown *et al.*, 1978) are not incompatible with the presence of highly non-linear turbulent-like motions at the Sun's surface (Perdang, 1981; Blacher and Perdang, 1981b). Since the relative radial amplitudes of the reported motions are extremely small ($|\delta r|/R < 10^{-5}$) most solar theorists, invoking the principle that small causes have small effects, are tempted to discard the suggestion that the non-linear coupling among the solar linear modes might have any serious influence on the actual oscillations. However a trivial illustration pinpoints a way to invalidate this rule in the context of interacting oscillators.

Take the coupled oscillator equations

$$\begin{aligned} \ddot{x} + \omega_x^2 x &= \varepsilon X(x, y)\,, \\ \ddot{y} + \omega_y^2 y &= \varepsilon Y(x, y)\,, \end{aligned} \tag{1}$$

where X and Y are non-linear functions and ε is a small parameter, with initial conditions $x = y = 1, \dot{x} = \dot{y} = 0$; suppose for instance $X(x, y) = y^3 + \cdots$. In a standard

* Proceedings of the 66th IAU Colloquium: *Problems in Solar and Stellar Oscillations*, held at the Crimean Astrophysical Observatory, U.S.S.R., 1–5 September, 1981.

** Chercheur Qualifié FNRS, Belgium.

† Permanent address.

Solar Physics **82** (1983) 297–321. 0038–0938/83/0822–0297$03.75.

perturbation scheme the first correction to the harmonic oscillator solution in x due to the contribution y^3 is of order $\varepsilon/(\omega_x^2 - \omega_y^2)$; therefore if a sufficiently sharp resonance between both linear frequencies takes place, then the perturbation procedure suggests that a finite correction can be produced even though the coupling parameter ε is infinitesimal. By the same token, since butterfly effects typically characterise unstable situations, the above example makes it plausible that internal resonances in weakly coupled oscillators may trigger some instability.

The main part of this paper is dedicated to a review of the rigorous mathematical information now available on the behaviour of non-linear Hamiltonian oscillators. Under special circumstances, among which the approximate internal resonances spotted above rank as necessary conditions, sinusoidal oscillations bifurcate towards a more irregular type of motion. In the available phase-space the latter occupy a zone of higher dimensionality*. This transition is referred to as the Kolmogorov instability.

For exploratory reasons we analyse here the purely radial adiabatic stellar oscillations in the framework of this theory, the latter being trivially recast into a Hamiltonian form. Although the phenomenon of Kolmogorov unstable oscillations is far more likely to occur among non-radial stellar motions, our numerical experiments disclose that this instability can occur already in the radial case, and, perhaps rather unexpectedly, under favourable circumstances even among the lowest order radial modes.

We illustrate that this instability reveals itself most easily in the power spectrum of the surface displacement of the star: While prior to the transition, in the 'regular' regime, the spectrum shows just a few peaks in all of our experiments, each peak transforms into an irregular band displaying a highly complex substructure when the instability sets in; studies of the same motion at several spectral resolutions show that the bands disclose a statistically hierarchical fine-structure; this confers the spectrum of Kolmogorov unstable motions a noisy aspect. The time-behaviour of such motions lies in between regular, deterministic, and irregular, random or noisy variability, the degree of randomness being related to the width of the bands.

In this paper no attempt has been made to apply our numerical experiments to specific stellar situations. We merely point out that since non-radial oscillations are most vulnerable by this instability, Kolmogorov unstable motions on the Sun's surface deserve serious investigation. A conclusive numerical approach of this question, allowing a meaningful comparison of theoretical power spectra with their observational counterparts requires a non-linear coupling formulation involving a few hundred linear modes.

We wish to stress finally that while Hamiltonian oscillations are now reasonably well understood, there exists so far, to be best of our knowledge, no general theory of dissipative motions. The question then of how dissipation characteristically affects Kolmogorov unstable oscillations remains open.

* Stellar vibrational instability gives rise to a similar change of dimensionality in phase-space: prior to the instability, the star's being in equilibrium shrinks the orbit to a point (0-dimensional) in phase-space; at the transition – a Hopf bifurcation in mathematical parlance – the point explodes into a closed curve (1-dimensional): the star is now periodically oscillating.

2. The Coupled Harmonic Oscillator-Approximation to Non-Linear Stellar Oscillations

To generate a set of adiabatic oscillation equations of a star which lends itself to a discussion in the framework of point-mechanics we rely on the energy principle. The potential energy V of the star, made up of the sum of the gravitational and the internal energy, and the kinetic energy of the internal motions K, are given by the following expressions:

$$V = -G \int_M \mathrm{d}m\, m/r(m, t) + \int_M \mathrm{d}m\, u(r(m, t), \mathrm{d}r/\mathrm{d}m)\,, \tag{2}$$

$$K = \tfrac{1}{2} \int_M \mathrm{d}m\, (\mathrm{d}r(m, t)/\mathrm{d}t)^2\,. \tag{3}$$

In these relations u denotes the specific internal energy which under local conservation of entropy (assumption of adiabatic motions) and mass becomes a function of the local radius $r(m, t)$ and its derivative with respect to the mass variable. All other notations are standard. In this form the total energy allows us to fully describe the radial motions of a star in the neighbourhood of a state of minimum potential energy. To this end it suffices to apply the expansion procedure adopted in Demaret *et al.* (1978) in a slightly modified version. Since the eigenfunctions of the linear radial adiabatic oscillation problem of a star

$$\xi_n(m) = \delta r_n(m)/r(m)\,, \qquad n = 1, 2, \ldots, \tag{4}$$

form a complete set, any radial displacement (satisfying physically reasonable smoothness conditions) can be expanded in the form

$$r(m, t) = r_E(m)\left[1 + \sum_n \xi_n(m) q_n(t)\right], \tag{5}$$

where $r_E(m)$ denotes the local equilibrium radius and $q_n(t)$ represents a set of weights attached to the linear amplitude distributions. We require the eigenfunctions to obey the usual normalisation

$$\int_M \mathrm{d}m\, r(m)^2\, \xi_k(m^*)\, \xi_l(m) = \delta_{kl}\,. \tag{6}$$

On substitution into Equations (2) and (3) we obtain for the total energy, if we set

$$p_n = \mathrm{d}q_n/\mathrm{d}t\,, \qquad n = 1, 2, \ldots \tag{7}$$

$$\begin{aligned} H(q_n, p_n) &= V_E + \tfrac{1}{2}\sum_k |p_k|^2 + \tfrac{1}{2}\sum_k \omega_k^2 |q_k|^2 + \frac{1}{3!}\sum_{klm} V^{(3)}_{klm} q_k q_l q_m + \cdots \\ &= V_E + H^{(2)}(q_n, p_n) + V^{(3)}(q_n) + \cdots\,. \end{aligned} \tag{8}$$

The factor V_E denotes the potential energy of the equilibrium configuration. The coefficients ω_k are the frequencies of the linear oscillations. The expansion coefficients $V^{(3)}_{klm}$ can be derived directly from the formulae given in Demaret *et al.* (1978). The formal expansion (8) can be interpreted as a Hamiltonian describing an infinite number of non-linearly coupled harmonic oscillators, of generalised positions q_n and momenta p_n, $n = 1, 2, \ldots$. Since in any numerical application of this formalism we are bound to cut-off the expansion of the radius (5) at some finite number F of linear modes, we shall restrict our theoretical discussion to the latter situation. The Hamiltonian (8) then refers to the motion of F non-linearly coupled harmonic oscillators. Under the change of variables

$$q_n \to \varepsilon q_n\,, \qquad p_n \to \varepsilon p_n\,, \qquad H \to \varepsilon^{-2} H\,, \tag{9}$$

where ε is a small book-keeping parameter measuring the order of magnitude of the amplitudes of the motion, or equivalenty the order of magnitude of the oscillation energy, we have

$$H(q_n, p_n) = H^{(2)}(q_n, p_n) + \varepsilon H^{(3)}(q_n) + \cdots\,, \tag{8$'$}$$

where $H^{(2)}$ is the Hamiltonian of the uncoupled linear oscillations; the terms involving the small parameter ε describe the non-linear coupling, each $H^{(k)}$, $k = 3, 4, \ldots$ being a homogeneous polynomial of degree k in the coordinates q_n. The equilibrium potential V_E is independent of the q_n, p_n; it leads to no contribution to the motion we investigate and has therefore been discarded.

The Hamiltonian formalism of stellar oscillations dates back to Woltjer (1935, 1937, 1943; cf. also Rosseland, 1949). This author was particularly concerned with the determination of analytically expressible corrections to the harmonic oscillator solutions. In the next section we shall see that such procedures cease to be justified mathematically when certain resonance conditions in the harmonic oscillators are fulfilled.

3. Application of PBSKAM-Theory to Stellar Oscillations

We survey in this section a few mathematical results of point-mechanics which prove to be directly relevant to the stellar oscillation problem. These developments originated with Poincaré (P) in the last century, were continued by Birkhoff (B) and Siegel (S), and culminated in a celebrated theorem first formulated by Kolmogorov (K) and later proved by Arnold (A) and Moser (M) (the KAM theorem).

Consider first the Hamiltonian obtained if $\varepsilon = 0$ in Equation (8$'$). Introduce a new set of canonical variables φ_n, J_n, $n = 1, 2, \ldots, F$, referred to as angle-action variables, defined as follows:

$$\begin{aligned} \omega_n^{1/2} q_n &= -(2J_n)^{1/2} \sin\varphi_n\,, \\ \omega_n^{-1/2} p_n &= +(2J_n)^{1/2} \cos\varphi_n\,, \qquad n = 1, 2, \ldots, F\,. \end{aligned} \tag{10}$$

In terms of the new variables the Hamiltonian depends on the actions J_n alone:

$$H = F(J_n)\,. \tag{11}$$

The corresponding Hamiltonian equations can be integrated explicitly:

$$\begin{aligned} J_n(t) &= J_n^0\,, \qquad \varphi_n(t) = \varphi_n^0 + \Omega_n t \qquad (\mathrm{mod}\, 2\pi)\,, \\ \Omega_n &\equiv \partial H/\partial J_n\,, \qquad n = 1, 2, \ldots, F\,, \end{aligned} \tag{11'}$$

where J_n^0 and φ_n^0 are the initial conditions. Since the φ_n^0 are angles, our convention is to define the latter in $0 \leq \varphi_n < 2\pi$; the notation $\varphi\,(\mathrm{mod}\,2\pi)$ indicates that any value of φ originally not in the range 0 to 2π is recast into that interval by adding or subtracting a multiple of 2π. In the special case just envisaged $H = H^{(2)}$ and $\Omega_n = \omega_n$ (independent of J_n). More generally, given an arbitrary Hamiltonian $H(q_n, p_n)$, it is said to be *integrable* if and only if a canonical transformation to angle-action variables exists such that the transformed Hamiltonian depends on the actions alone. For an integrable system the general solution can explicitly be written down (Equation (11′)); expressed in the original variables q_n, p_n the motion is then given as an F-uple Fourier series

$$q_n(t) = \sum_{k_1 k_2 \ldots k_F} A^{(n)}_{k_1 k_2 \ldots k_F} \exp i(k_1\Omega_1 + k_2\Omega_2 + \cdots + k_F\Omega_F)t \tag{11''}$$

and a similar expression for $p_n(t)$, $n = 1, 2, \ldots, F$, with $k_1, k_2, \ldots = 0, \pm 1, \pm 2, \ldots$. Under conditions of analyticity of the Hamiltonian the expansion coefficients obey

$$|A^{(n)}_{k_1 k_2 \ldots k_F}| \leq A \exp - B\,|k|\,, \qquad |k| = \sum_{i=1}^{F} |k_i|\,, \tag{11'''}$$

where A and B are positive constants independent of the k_i. Functions of type (11″), (11‴) are known as *quasi-periodic* functions, and the corresponding motion is said to be a quasi-periodic motion.

Since total energy E is conserved during the motion, we represent the latter in terms of its angle-action variables (11′) on the $2F-1$-dimensional energy 'surface', which we parametrise by the coordinates $\varphi_1, \varphi_2, \ldots, \varphi_F, J_1, J_2, \ldots, J_{F-1}$. For a 2-oscillator system ($F = 2$) we have illustrated the motion of an integrable Hamiltonian system in Figure 1. The energy 'surface' is represented by the box $0 \leq \varphi_1 < 2\pi$, $0 \leq \varphi_2 < 2\pi$, $0 \leq J_1 \leq J_1^M$, J_1^M being the maximum action J_1 compatible with the value of the energy. For given initial conditions φ_1^0, φ_2^0, J_1^0, J_2^0 $(= J_2(J_1^0, E))$, the orbit is confined to the square $ABCD$ at the position $J_1 = J_1^0$ parallel to the angle plane φ_1, φ_2; the trajectory is a straight line with the property that each time it touches an edge and disappears, it reappears on the opposite edge with same slope, at the projection of the point of disappearance. Such a square with opposite edges being identified ($AB = DC$ and $AD = BC$) has the geometrical structure of a doughnut (cf. Figure 1); therefore it is referred to as a 2-dimensional torus. In the general case of an integrable Hamiltonian system of F degrees of freedom, the motion (11′) is likewise said to evolve on an F-dimensional torus of the $2F-1$-dimensional energy manifold.

If the frequencies Ω_n, $n = 1, 2, \ldots, F$, are rationally dependent, or resonant, of order N, i.e. if a set of integers $k_1, k_2, \ldots, k_F$ exists such that

$$\sum_{n=1}^{F} k_n\Omega_n = 0 \quad \text{with} \quad |k| = N\,, \tag{12}$$

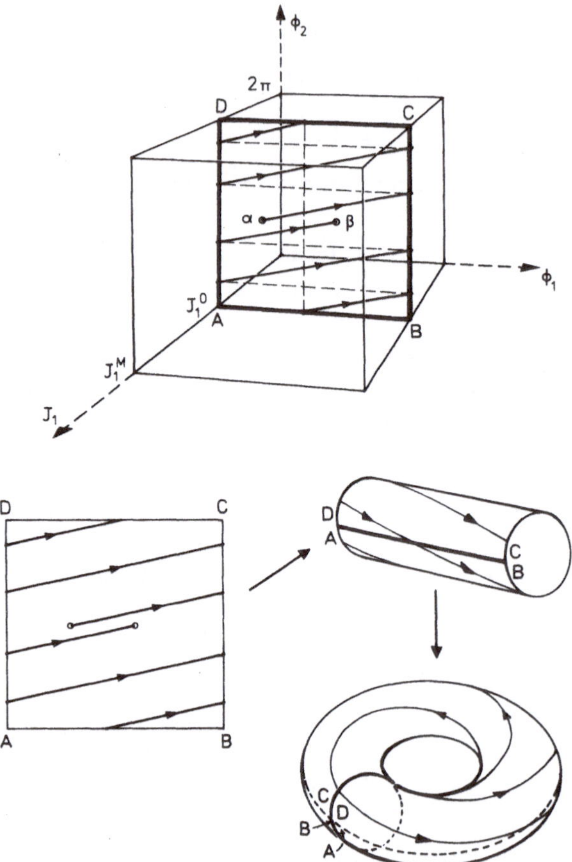

Fig. 1. The energy box $H(\varphi_1, \varphi_2, J_1, J_2) = E = c^{st}$, $0 \leq \varphi_1 < 2\pi$, $0 \leq \varphi_2 < 2\pi$, $0 \leq J_1 \leq J_1^M$ for an integrable system of 2 degrees of freedom; the orbit α, β is carried by the square *ABCD* (equivalent to a torus, cf. bottom of figure).

then the orbit in the energy manifold is closed, so that the motion is periodic; the motion-carrying torus is then said to be a resonant torus. If for no set of integers $k_1, k_2, \ldots, k_F$, whatever N, relation (12) can be satisfied, then the orbit will eventually go through any region, chosen as small as we like, on the torus; such an orbit covers the whole torus.

For integrable Hamiltonian systems the energy manifold is stratified into invariant tori: Any point of this manifold belongs to one and only one torus.

Integrability is not automatically shared by all Hamiltonian systems. This point was recognised by Poincaré (1890) who proved that the 3-body problem of celestial mechanics is precisely not integrable. Poincaré also seems to have been aware that integrability is in fact an exceptional property of Hamiltonian systems. In geometric terms the very existence of non-integrable Hamiltonian systems means that there are orbits which do not lie on tori; equivalently there are motions which do not admit of multiple Fourier expansions of type (11″), (11‴).

A refinement of Poincaré's result is due to Birkhoff (1927) who proved the following theorem:

If the Hamiltonian is given by a formal power series in ε (Equation (8′)), with the frequencies ω_n entering the harmonic part $H^{(2)}$ being rationally independent, then a formal canonical transformation exists, $q_n, p_n \to Q_n, P_n$, $n = 1, 2, \ldots, F$, such that

$$H = F(J_n) = \sum_k \omega_k J_k + \tfrac{1}{2} \sum_{kl} \omega_{kl} J_k J_l + \cdots \tag{13}$$

is a formal power series in the variables J_n defined by

$$\omega_n J_n = \tfrac{1}{2}(Q_n^2 \omega_n^2 + P_n^2), \qquad n = 1, 2, \ldots, F. \tag{13′}$$

The proof as given in Arnold (1963a, b) consists in the explicit construction of a sequence of canonical transformations

$$\begin{array}{lll} q_n, p_n \to q_n', p_n' & \text{generator:} & S(q_n, p_n'), \\ q_n', p_n' \to q_n'', p_n'' & & S'(q_n', p_n''), \\ \ldots & & \ldots . \end{array} \tag{14}$$

The generators of these transformations are determined by requiring that $S(q_n, p_n')$ eliminates the non-integrable contribution of order ε in the formal series of the Hamiltonian, $S'(q_n', p_n'')$ produces a vanishing order ε^2 contribution, etc. These generators are sought in the form of multiple Fourier expansions; the Fourier coefficients then involve denominators $\sum_{n=1}^F k_n \omega_n$, $k_n = 0, \pm 1, \pm 2, \ldots$ ($|k| \neq 0$). One finds that the generator S is formally defined if no resonances of order ≤ 4 occur among the linear frequencies; this generator then reduces the full Hamiltonian H (Equation (8′)) to the form:

$$H(q_n, p_n) = \left[\sum_{n=1}^{F} \omega_n J_n + \tfrac{1}{2} \sum_{n,k}^{F} \omega_{nk} J_n J_k \right] + O(\varepsilon^2) = H_0(J_n) + O(\varepsilon^2), \tag{15}$$

where the $O(\varepsilon^2)$ contribution does not depend on the action variables alone. To eliminate higher order non-integrable components in the formal series (15) higher order resonances are to be excluded as well.

The construction of this sequence of generators breaks down once the frequencies of the harmonic part $H^{(2)}$ are resonant to some order N. This proves that resonances are responsible for destroying the integrability of the full Hamiltonian (Equation (8′)). But since for any set of frequencies $\omega_1, \omega_2, \ldots, \omega_F$ one can always find a set of integers $k_1^*, k_2^*, \ldots, k_F^*$, $k_n^* = 0, \pm 1, \pm 2, \ldots$ $|k^*| \neq 0$, such that $|\sum_{n=1}^F k_n^* \omega_n| \leq \eta$, η being any preassigned precision, it becomes doubtful whether the series of generators and therefore also the formal series (13) are ever convergent.

The question of convergence or divergence was settled by Siegel (1954) who proved that the formal series (13) is generically divergent. If we choose at random a Hamiltonian among the class of Hamiltonians given by the series expansion (8′) with coefficients of the polynomials $H^{(k)}$ in some finite interval, say $(-1, +1)$, then the probability of hitting an integrable Hamiltonian is zero. The typical property of a Hamiltonian is to be non-integrable.

This conclusion directly pertains to the weakly non–linear stellar oscillations: For general stellar models we have no reason to expect the oscillation Hamiltonian (Equation (8.8′)) to possess the atypical property of integrability. Therefore solutions of the stellar oscillations in the form of Fourier type expansions (11″) as assumed in Woltjer's procedure, and more recently in the iterative technique adopted by Simon (1972), are not justified *a priori* in the presence of multi-mode coupling.

Non-integrability means that not all motions are carried by tori. What is physically relevant, however, is to know how frequent the orbits on tori are as compared to the totality of trajectories of a non-integrable Hamiltonian system, or more precisely to have information on the volume occupied by the regular, quasi-periodic motions in comparison with the whole volume of the energy manifold. A partial answer to this question is provided by the notorious **KAM** theorem (Kolmogorov, 1957):

If a Hamiltonian of a system of F degrees of freedom is given in the form

$$H(\varphi_n, J_n, \varepsilon) = H_0(J_n) + \varepsilon H_1(\varphi_n, J_n, \varepsilon)\,, \tag{16}$$

where

(1) H is real analytic in all of its arguments, for $0 < \varepsilon < \varepsilon_0$, for J_n defined in some open region J of the F-dimensional action space, and for the angles $0 \leq \varphi_n < 2\pi$, $n = 1, 2, \ldots, F$, as well as periodic in the latter (of period 2π);

(2) let $J_n = J_n^0$, $n = 1, 2, \ldots, F$, in J, characterise an invariant torus of the non-perturbed Hamiltonian H_0 such that the frequencies

$$\Omega_n(J_m) = \partial H_0/\partial J_n \quad \text{at} \quad J_n = J_n^0\,, \quad n = 1, 2, \ldots, F \tag{17}$$

obey the non-quasi-resonance condition

$$\left|\sum_{n=1}^{F} k_n \Omega_n\right| \geq c\,|k|^{-\alpha} \quad \text{for any set of integers} \quad k_n = 0, \pm 1, \pm 2, \ldots \quad (|k| \neq 0) \tag{18}$$

for some positive constants c and α, as well as the non-degeneracy condition

$$\mathrm{dtm}(\partial^2 H_0/\partial J_m \partial J_n) \neq 0 \quad \text{at} \quad J_n = J_n^0\,. \tag{18′}$$

Then, provided that ε_0 is sufficiently small:

(1) there exists a deformed invariant torus

$$\begin{aligned} \varphi_n(t) &= \varphi_n^0(t) + \varepsilon\phi(\varphi_n^0(t), \varepsilon)\,, \\ J_n(t) &= J_n^0 + \varepsilon\Lambda(\varphi_n^0(t), \varepsilon)\,, \end{aligned} \tag{19}$$

with ϕ and Λ real analytic functions of their arguments and periodic in the $\varphi_n^0(t)$, i.e. the angles of the non-perturbed orbit;

(2) if Γ is the open region in phase-space over which hypothesis (1) holds, and K the region filled out by the regular solutions (19), then K is closed and nowhere dense; moreover it covers most of Γ.

The latter stipulation means that the volume of the zone $\Gamma - K$ occupied by non-quasiperiodic solutions can be made as small as we like if the coupling ε is small enough.

Region K is referred to as the Kolmogorov set. The first full proof of this theorem was given by Arnold (1963a). A perhaps more intuitive proof based on a procedure going back to Poincaré (1912), namely the method of the surface of section, or of area preserving Poincaré mappings, is due to Moser (1962, cf. also 1973). In essence, and for $F = 2$, this procedure amounts to studying the sequence of successive intersections of the orbit in the 3-dimensional energy manifold in the coordinate basis q_1, q_2, p_2, by the plane (surface of section) $q_1 = 0$ with $p_1 > 0$. Denote by $q_2^{(i)}, p_2^{(i)}$ the ith intersection point of the orbit with the surface of section; the transformation that carries $q_2^{(j)}, p_2^{(j)}$ into $q_2^{(j+1)}, p_2^{(j+1)}$, $j = 1, 2, \ldots$ is the Poincaré map. A motion on a torus (cf. Figure 1) shows up in the surface of section as a sequence of points all distributed along a closed curve (which can degenerate into a point). A closed curve being topologically equivalent to a circle, the simplest Poincaré map that simulates all topological features of any quasi-periodic Hamiltonian solution is the 'twist map'

$$\begin{aligned} r^{(i+1)} &= r^{(i)}, \\ \varphi^{(i+1)} &= \varphi^{(i)} + \Omega(r^{(i)}), \end{aligned} \tag{20}$$

where φ, r are polar coordinates of the intersection points q_2, p_2 in the surface of section; the twist map transforms the circle of radius $r = r^{(i)}, i = 1, 2$, into itself. Any conceivable slight perturbation of the Hamiltonian deforms the corresponding Poincaré map (20) as follows:

$$\begin{aligned} r^{(i+1)} &= r^{(i)} + \varepsilon R(r^{(i)}, \varphi^{(i)}), \\ \varphi^{(i+1)} &= \varphi^{(i)} + \Omega(r^{(i)}) + \varepsilon\phi(r^{(i)}, \varphi^{(i)}). \end{aligned} \tag{20'}$$

(The Hamiltonian character of the motion requires area conservation of the map (20′), $r^2 \,\mathrm{d}r\, \mathrm{d}\varphi = c^{\mathrm{st}}$, so that the functions R and ϕ are not independent.) Moser proved that if

$$\mathrm{d}\Omega/\mathrm{d}r \neq 0 \tag{21}$$

and

$$|n\Omega - m2\pi| \geq c\,|n|^{-\alpha} \quad \text{for any set of integers } n, m \tag{22}$$

for some positive numbers c and α, then the disturbance to the twist map generates again a closed curve in the surface of section that remains close to the original circle for ε sufficiently small. Note that conditions (21) and (22) duplicate the non-degeneracy and non-quasi-resonance requirements.

It was already known to Poincaré and Birkhoff that the twist map is unstable under slight perturbations (20′) once $\Omega = (m/n)2\pi$; under those conditions the original invariant circle is blown up, an even number of points on it remaining however fixed. The latter are alternatively stable and unstable: A stable fixed point has the property that the map (20′) carries all points close to it into points that remain close to it; unstable fixed points of the Poincaré map have neighbouring points in the surface of section that do not stay close to those points under the transformation (20′). An unstable fixed point either (a) has the sequence of successive image points of any point in its neighbourhood lying on a closed *curve*, or (b) there are points around it which under the iterated map (20′) fill out an *area* in the surface of section. Alternative (a) occurs if the perturbed twist map simulates an integrable Hamiltonian system; it produces an explosion of the original torus into a series of second generation tori. Alternative (b) is the typical case; it tells us that the resonant tori, which densely cover the energy manifold, acquire a certain thickness along the action axes: these tori, of dimension F, explode into configurations of higher dimensionality $\leq 2F - 1$.

The motions carried by the Kolmogorov set K are quasi-periodic (Equations (11′), (11″)). The orbits lying outside K are 'Kolmogorov unstable' in Chirikov's terminology (Izrailev and Chirikov, 1966), or 'stochastic' (Zaslavskii and Chirikov, 1972), or 'chaotic'. So far the precise mathematical characteristics of these motions are not yet known.

We observe that for a system of $F = 2$ coupled oscillators at low ε viewed in the energy box of Figure 1, any exploded torus is necessarily sandwiched between two invariant tori of the Kolmogorov set. Therefore, the actions of Kolmogorov unstable motions are confined to narrow intervals (for $F = 2$). This suggests that such motions still bear some resemblance with quasi-periodic motions.

If $F > 2$, an F-torus of the $2F - 1$-energy manifold no longer cuts the latter into two disconnected bits; therefore, the complementary set of the Kolmogorov tori i.e. the zone carrying the Kolmogorov unstable motions, can now become connected. The actions are then allowed to drift through the whole energy manifold. This phenomenon is known as the Arnold diffusion. Nekhoroshev (1977) proved that under special 'steepness' requirements of the non-perturbed Hamiltonian (Equation (16)), the actions $J_n(t)$ obey

$$|J_n(t) - J_n^0| < \varepsilon^b \quad \text{if} \quad 0 \leq t \leq T = \varepsilon^{-1} \exp(\varepsilon^{-a}), \qquad (23)$$
$$n = 1, 2, \ldots, F,$$

where $a, b > 0$ depend on the non-perturbed Hamiltonian, and J_n^0 represents the action of the solution of the integrable, non-perturbed Hamiltonian H_0. (The steepness hypothesis generalizes the stability condition $|\partial H_0/\partial J| > 0$ in a system with one degree of freedom).

We now adapt these results to the problem of stellar oscillations (Equation (8′)). First observe that if we introduce action-angle variables defined by Equation (10) into the oscillation Hamiltonian, then the unperturbed part becomes

$$H^{(2)}(q_n, p_n) = F(J_n) = \sum_{k=1}^{F} \omega_k J_k, \qquad (24)$$

with $\omega_n = \Omega_n$ (Equation (11′)). This integrable Hamiltonian violates the non-degeneracy requirement (18′), so that we cannot just use the harmonic oscillator approximation as the unperturbed system. However, we have seen in the analysis of Birkhoff's theorem that a canonical transformation exists reducing our stellar oscillation

Hamiltonian to the form (15); the terms between square brackets representing again an integrable Hamiltonian are now regarded as the non-perturbed Hamiltonian. The condition securing the existence of this transformation is given by

$$\sum_{n=1}^{F} k_n\omega_n \neq 0\,, \qquad 0 < |k| = \sum_{n=1}^{F} |k| \leq 4\,. \tag{25}$$

For arbitrary generic stellar models we can assume that $\mathrm{dtm}\,\omega_{ij} \neq 0$, so that the non-degeneracy requirement (18′) is now satisfied. Moreover, provided that the non-quasi-resonance condition (18) is obeyed, we fulfil the hypotheses of the KAM theorem.

The latter requirement demands that even approximate resonances

$$\sum_{n=1}^{F} k_n\omega_n \simeq 0 \tag{20}$$

have to be excluded among the linear oscillations. The precision to which this equality has to be fulfilled depends on the expansion parameter ε: small ε values demand a high accuracy in order to violate the KAM conditions (cf. next section).

We discuss now the possibility of resonances among radial stellar modes. In the first place, if we concentrate on the *sufficiently low* frequency part of the *linear spectrum*, the non-quasi-resonance condition is not violated in generic models and at a sufficiently low level of non-linearity ε. Exceptions occur in atypical models, constructed through an *ad hoc* selection of the model parameters to generate resonances: for instance among polytropes, by adjusting the index n it is possible to produce low order resonances in the low-frequency spectrum (cf. Simon, 1972). Such models have an almost zero probability to occur in reality. Therefore we have the following property:

(A) In a generic stellar model, if the radial modes of sufficiently low order are non-linearly coupled, and the oscillation energy is low enough, then most of the motions of this non-linear stellar oscillator remain close to the oscillations of the linear modes (KAM secures the closeness of the solutions to the motions of a non-linear integrable oscillator described by the non-perturbed Hamiltonian H_0 (Equation (15)); but for sufficiently small non-linearity ε, or equivalently sufficiently small oscillation energy, the motion of the latter oscillator remains as close as we like to the linear oscillations).

Consider next the sufficiently *high asymptotic* part of the *linear spectrum*. The linear frequencies in the asymptotic regime obey a representation formula

$$(N \to \infty): \quad \omega_N = N\Omega_a + \Omega_0 + \Omega_1/N + O(1/N^2)\,, \tag{27}$$

where Ω_a, Ω_0, Ω_1, ... are model constants. If ω_N, ω_{N+1} are two successive asymptotic frequencies, and if we express frequencies in units $\omega_N = 1$, we have

$$\omega_{N+1} - \omega_N = (1/N) + O(1/N^2)\,. \tag{28}$$

By choosing N large enough we have a resonance of order 2 to any preassigned degree of precision. This shows that we violate the non-quasi-resonance conditions of KAM in any stellar model, on condition that we couple non-linearly adjacent modes of the asymptotic spectrum. Therefore:

(B) In any stellar model, if we fix an oscillation energy (chosen sufficiently small), and if we couple non-linearly radial asymptotic modes of sufficiently high order N, we have no guarantee that the motion of this non-linear oscillator remains close to the linear harmonic oscillations of the uncoupled modes.

4. Empirical Data on the Kolmogorov Instability

The strict mathematical theory reviewed in the previous section does not answer the following questions:

(1) What does a Kolmogorov unstable motion look like?

(2) Under what conditions does the energy box carry a non-negligible fraction of Kolmogorov unstable tori?

(3) How does the Kolmogorov instability influence the energy exchange among modes?

These questions have been investigated by semi-analytical techniques and by direct numerical experiments.

(1) It has been argued that since by definition such motions cannot be represented by multiple Fourier series (Equations (11″), (11‴)), any phase-space coordinate $q_n(t)$, $p_n(t)$, and therefore also any linear combination of the latter, must give rise to a highly structured power spectrum (Blacher and Perdang, 1981a). In fact, it has been found that power spectra of Kolmogorov unstable motions invariably have a complex structure (Noid *et al.*, 1977; Powell and Percival, 1979); a quasi-periodic motion in contrast has a spectrum typically displaying just a few fine lines. A detailed analysis of the unstable motions of the Hénon–Heiles coupled harmonic oscillators ($F = 2$) (Hénon and Heiles, 1964) with a resonance $\omega_1 = \omega_2$ in the harmonic approximation shows that the power spectra typically display two conspicuous features (Blacher and Perdang, 1981a): a broad resonance band at frequency $\omega \sim \omega_1 = \omega_2$, and a second lower broad band at the combination frequency $|\omega_1 - \omega_2|$ (origin). The very existence of the first band tells us that a pseudo-periodicity survives in the Kolmogorov unstable regime. The spread in this band shows that this periodicity is not well-defined: If one views the profile of the band as a probability distribution of frequencies, then the motion can randomly switch from one frequency in the band to another; this is precisely observed in the analysis of the time-behaviour of the phase coordinates (Blacher and Perdang, 1981b). The second band near the origin can be given a similar probabilistic interpretation: it confers the motion an irregular long time-scale variability which manifests itself as an irregular amplitude modulation of the short time-scale pseudo-periodicity.

These heuristic results picture a Kolmogorov unstable motion as a blend of a deterministic, regular, component (reminiscent of a linear mode), and a purely random, irregular component; the degree of randomness is measured by the band-widths of the power spectra.

The appearance of a finite natural width of the frequency peaks under Kolmogorov instability is not surprising. It merely reflects the finite thickness of the exploded tori.

In fact, from Equation (17) we can say that each frequency $\Omega_k(J_n)$ of the integrable Hamiltonian H_0 explores an interval $\Delta\Omega_k$ roughly given by $\Delta\Omega_k \sim \sum_{l=1}^{F} |\partial^2 H_0/\partial J_k \partial J_l| \, |\Delta J_l|$ on the exploded torus $J_n \sim J_n^0$ of thickness ΔJ_n, $n = 1, 2, \ldots, F$.

We should mention also that the fractal dimension d of the (renormalised) bands of a power spectrum of Kolmogorov unstable motions obeys $d \gtrsim 1$, while the power spectrum of regular motions has a dimension $d = 1$; for the Hénon–Heiles unstable oscillations an approximate numerical technique devised to estimate this parameter (Perdang, 1981; Blacher and Perdang, 1981a) yields values in the range $1.25 \lesssim d \lesssim 1.5$.

(2) As regards the onset of an 'observable' Kolmogorov instability, i.e. the occurrence of Kolmogorov unstable motions over a fraction of phase-space of finite volume, it is found empirically that a mere violation of the KAM conditions is not sufficient to guarantee this phenomenon. The following empirical results are relevant in this connection.

The notorious experiments by Hénon and Heiles (1964) dealing with two harmonic oscillators of same frequency $\omega_1 = \omega_2$, coupled nonlinearly through the potential $V_3(q_1, q_2) = q_1^2 q_2 - \frac{1}{3} q_2^3$ have established that the transition towards (observable) Kolmogorov instability sets in abruptly, at some threshold energy E_T; the latter in turn is a fraction ($> \frac{1}{2}$) of the escape energy v_c, i.e. the energy above which the equipotential curves cease to be closed. Below this threshold the probability of hitting an unstable solution is zero. For oscillation energies $> E_T$ a sizeable fraction of phase-space becomes populated by stochastic solutions. These experiments show that the single resonance $\omega_1 = \omega_2$ in an $F = 2$ oscillator is not sufficient to generate Kolmogorov instability at low oscillation energy. The concept of a threshold energy has been clarified by Walker and Ford (1969) and developed by Zaslavskii and Chirikov (1972) (see also Chirikov, 1979), who pointed out that an overlap of exploded resonant tori is required to generate stochastic oscillations. We sketch this idea for an $F = 2$ oscillator. Expand the perturbing Hamiltonian (cf. Equation (16)) in a Fourier series of the angle variables:

$$H(q_1, q_2, p_1, p_2) = H_0(J_1, J_2) + \varepsilon \sum_{n_1, n_2} H^c_{n_1, n_2}(J_1, J_2) \times \times \cos(n_1 \varphi_1 + n_2 \varphi_2) + \ldots \tag{29}$$

From the equations of motions observe that a given Fourier component $H^c_{n_1 n_2}(J_1, J_2)$ leads to a non-negligible contribution to J_1 or J_2 provided that

$$\varepsilon H^c_{n_1 n_2}(J_1, J_2)/[n_1 \Omega_1(J_1, J_2) + n_2 \Omega_2(J_1, J_2)] = O(1), \tag{30}$$

i.e., when a small divisor compensates for the small value of ε.

If just a single Fourier component $H^c_{n_1 n_2}$ is non-zero in the expansion (29), the full Hamiltonian is seen to remain integrable. At fixed energy E the denominator in (30) becomes small on some torus J_1^0 of the unperturbed Hamiltonian; this torus then explodes under the influence of the perturbation, (cf. the discussion of the twist map).

However, since the new Hamiltonian remains integrable, the explosion merely manifests itself by the appearance of second generation tori (Figure 2). The latter are confined to an interval ΔJ_1 near J_1^0 of the action axis of the non-perturbed Hamiltonian, given by

$$\Delta J_1 \sim \varepsilon H^c_{n_1 n_2}(J_1^0, J_2^0)/[n_1 \Omega_1(J_1^0, J_2^0) + n_2 \Omega_2(J_1^0, J_2^0)] \,, \tag{31}$$

as flows from the equation of motion.

Suppose next that the Fourier series involves two factors $H^c_{n_1 n_2}$ and $H^c_{n'_1 n'_2}$, all other expansion factors being zero. The second factor $H^c_{n'_1 n'_2}$ will then play a non-negligible part on a torus $J_1'^0$ over which $n'_1 \Omega_1(J_1'^0, J_2'^0) + n'_2 \Omega_2(J_1'^0, J_2'^0)$ becomes small; this torus suffers the same fate as torus J_1^0; it breaks up into subtori covering again an interval $\Delta J'_1$, (given by a relation of type 31) provided that both unperturbed tori J_1^0, $J_1'^0$ were sufficiently far away from each other; under those conditions each resonance acts as if it existed alone.

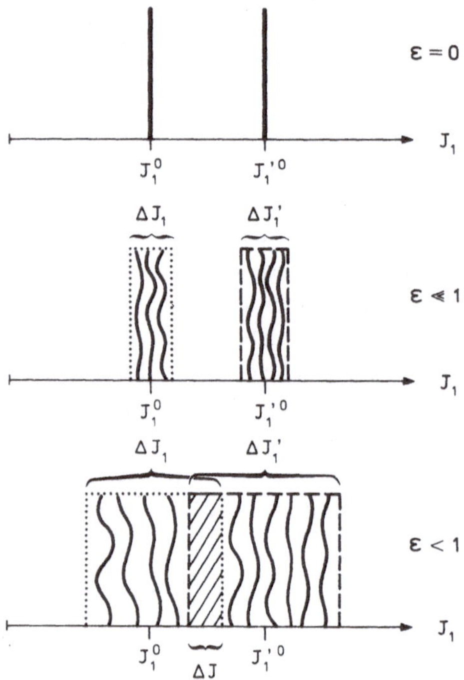

Fig. 2. Action of a non-integrable perturbation on two nearby resonant tori at positions J_1^0 and $J_1'^0$ in the energy box; the wavy area is populated by Kolmogorov tori; the shaded overlapping area lodges the Kolmogorov unstable orbits.

If however the exploded tori overlap (shaded area in Figure 2) the previous argument breaks down; within the region of overlapping resonances the integrability property is essentially lost. Empirically one observes that the condition

$$|J_1^0 - J_1'^0| \sim \tfrac{1}{2}|\Delta J_1 + \Delta J'_1| \tag{32}$$

approximately determines the onset of stochasticity (cf. Walker and Ford, 1969). The width ΔJ_1 of an exploded torus increases with ε, or equivalently with the energy fed into the oscillation; the existence of a threshold energy E_T in Hénon and Heiles's experiments then follows directly from the condition of overlap (32).

Numerical experiments on the effect of a nonlinear coupling of 3 harmonic oscillators of linear frequencies $\omega_1 : \omega_2 : \omega_3 = 1 : 2 : 3$ have been performed by Ford and Lunsford (1971). Their experiments suggest that under this multiple resonance the threshold energy for stochasticity is arbitrarily small. This observation is compatible with the concept of overlapping resonances: the contributions

$$H_{3,0,-1} \cos(3\varphi_1 - \varphi_3) + H_{2,1,0} \cos(2\varphi_1 - \varphi_2)$$

in the Fourier expansion of the perturbed Hamiltonian simultaneously lead to small divisors for $\varepsilon \to 0$ (or at zero energy).

(3) If a large number F of harmonic oscillations are interacting, the eventual distribution of energy over these oscillators becomes an important issue. Fermi, Pasta and Ulam (1955) in an experiment in which harmonic oscillators are coupled to simulate a non-linear string, find that the motions remain quasi-periodic, and that no significant energy exchange takes place. Ford (1961) emphasises that their negative result is a consequence of the lack of approximate resonances among the lower order linear frequencies of the string*. Repeating a modified version of this experiment in which the frequencies of the harmonic oscillators are chosen to satisfy resonance conditions, Ford and Waters (1963) observe a relaxation towards thermalization of their oscillators: eventually the time interval over which any oscillator of the system has an energy between E and $E + \mathrm{d}E$ obeys a Boltzmann law.

Another variant of the Fermi–Pasta–Ulam experiment is due to Hirooka and Saitô (1969). These authors analyse a 2-dimensional lattice of oscillators, simulating non-linear oscillations of membranes. Under those conditions approximate resonances arise automatically. If a resonant linear oscillator is excited, the energy first remains trapped by this mode during an 'induction period'; then this mode decays and an eventual tendency towards equipartition is observed.

In the light of the heuristic information of the present section, the rigorous results (A) and (B) (Section 3) on the behaviour of adiabatic non-linear radial stellar oscillations can be specified further:

(A′) If a sufficiently high amount of oscillation energy is fed into the lower oscillation modes (violation of KAM through a high factor ε), the width of the ever present exploded tori (Equation (31)) can become appreciable, so that an overlapping of neighbouring exploded tori can take place. Therefore, Kolmogorov unstable oscillations are expected to occur among the lowest modes, provided that the energy input is large enough. The experiments discussed under (3) then suggest that an efficient diffusion of the oscillation energy towards higher modes should take place.

The phenomenon of enhanced energy diffusion among the stellar modes in the presence of non-linear resonance effects has actually been observed by Papaloizou (1973a, b)

* In the Fermi–Pasta–Ulam experiments lowest order modes alone had initially been excited; Izrailev and Chirikov (1966) point out that an initial excitation of sufficiently high modes would have favoured the occurrence of stochastic motions.

(B′) If one or several neighbouring asymptotic modes are initially excited, the quasi-resonances (28′) imply that an overlap of exploded tori can occur at a very low threshold. Kolmogorov unstable oscillations should then be the rule rather than the exception. From the experiments listed under (1) one expects a power spectrum with a sequence of overlapping bands at the asymptotic frequencies $1 \sim \omega_{N+1}, \omega_{N+1} \sim \omega_{N+2}, \ldots, \omega_{N+n} \sim \omega_{N+n+1}, \ldots$ (in relative units $\omega_N = 1$) possibly merging into one very broad band; moreover, near the origin a single broad peak should appear, due to the second order resonances

$$\omega_{\text{res}} \sim \omega_{N+n+1} - \omega_{N+n} = \frac{1}{N} + O\left(\frac{1}{N^2}\right), \qquad n = 0, 1, 2 \ldots; \tag{33}$$

in absolute units $\omega_{\text{res}} = \Omega_a + O(1/N)$; cf. Equation (28).

The surprising observation is that each asymptotic neighbouring pair of energised modes provides a power contribution at essentially the same frequency (33). The power spectrum of such a motion is then expected to show a conspicuous band peaked near the origin and joining a low level very broad band.

5. Numerical Experiments on Nonlinear Oscillations in Stars

This section is intended to demonstrate that the mathematical conclusions (A) and (B) as well as the informed guesses (A′) and (B′) do in fact hold in the stellar context. Moreover, we wish to get a quantitative idea of the orders of magnitude of the amplitude of the surface displacement under which the Kolmogorov instability sets in. We are also interested in the specific form of the time-behaviour of the surface displacement as well as of its power spectrum in this instability regime.

The numerical analysis is performed in the framework of the standard polytrope of index $n = 3$. Since this model is fairly representative for a whole class of stars, our conclusions are hoped to be 'typical' for stellar oscillations.

We shall briefly report here on just a few experiments. Technical details and a variety of numerical illustrations will be published elsewhere (Perdang and Blacher, 1982a, b).

In all our calculations the power series of the oscillation Hamiltonian (Equations (8), (8′)) was terminated after the cubic interaction $V^{(3)}$. The numerical analysis was then carried through as if this truncated expansion represented the exact Hamiltonian.

2-mode interaction

For any pair (i, j) among the lowest radial modes $(i, j = 0, 1, \ldots, 9)$ (0: fundamental; 1: first harmonic; ...) the equipotential curves (cf. Equation (8))

$$\begin{aligned} V(q_i, q_j) &= \frac{1}{2!} \sum_{k=i,j} (\omega_k^2 q_k^2 + p_k^2) + \frac{1}{3!} \sum_{k,l,m=i,j} V_{klm} q_k q_l q_m \\ &= v = c^{\text{st}} \end{aligned} \tag{34}$$

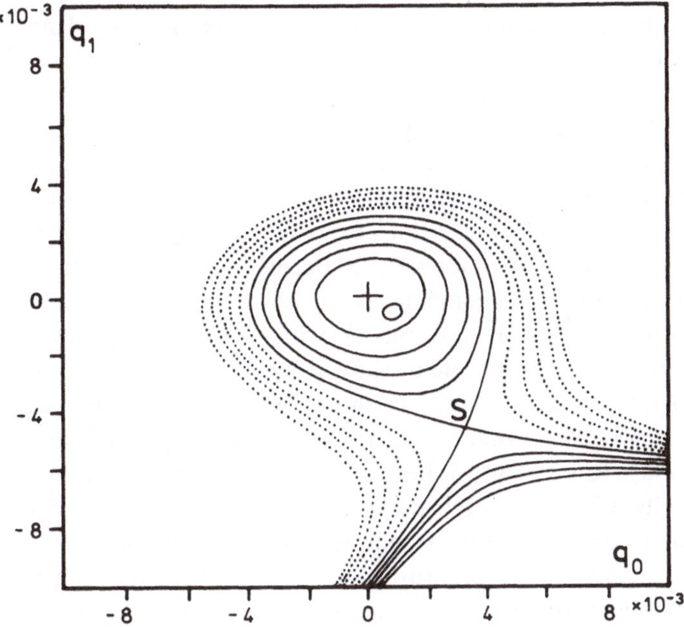

Fig. 3. Shape of the equipotential surface $V = c^{st}$ for the fundamental mode – first harmonic coupling.

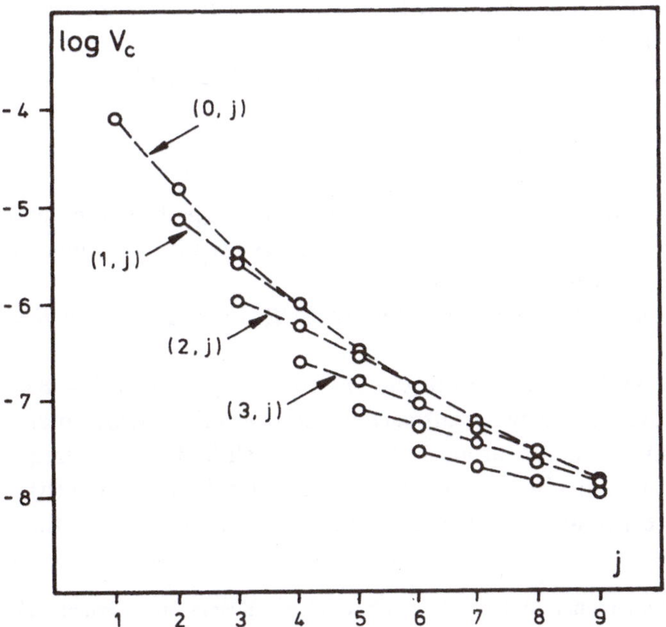

Fig. 4. Escape energy v_c for (i, j) couplings (v_c in units GM^2/R).

in the neighbourhood of the stable state 0 ($q_i = q_j = 0$) of the q_i, q_j plane have the typical form shown in Figure 3. For low values of v the curves are elliptical; they remain closed for $0 < v < v_c$, where v_c is a critical value such that the equipotential $V(q_i, q_j) = v_c$ goes through a saddle point S of the energy surface; for $v > v_c$, the equipotentials become open curves in the neighbourhood of the stable equilibrium. The value v_c thus defines an escape energy, just as in the Hénon–Heiles oscillator. The shape of this potential well of non-linear two-mode interactions in stellar oscillations shows that the latter are mechanically modelled by the motion of a marble in a sauce-boat.

Figure 3 refers to the (0, 1) coupling; in the (0, j) coupling $j = 2, 3, \ldots$ the potential surface becomes steeper in the q_j-direction with increasing j values, while the saddle point S shifts towards the origin and towards the q_j-axis. The critical energy v_c decreases rapidly with j (cf. Figure 4). Since in the Hénon–Heiles experiment the threshold $E_T = O(v_c) < v_c$, we expect that in the stellar case Kolmogorov instability takes place at a threshold close to v_c. Figure 4 then leads us to conjecture that the threshold energy decreases rapidly with the orders (i, j) of the coupled modes; if it is allowed to extrapolate this diagram into the asymptotic range, the coupling (N, $N + 1$) generates Kolmogorov instability at an arbitrarily small threshold, provided that N is large enough.

In the (0, 1) *coupling* the frequencies are non-resonant ($\omega_1/\omega_0 = 1.355$). From result (A) we know that the oscillations remain regular at low oscillation energy $E \ll v_c$. At sufficiently high energy a Kolmogorov instability is suspected to set in (A′). In fact at the oscillation energy $E = v_c$ a large fraction of the energy manifold is numerically found to carry unstable motions; Kolmogorov instability of these solutions has been established via the standard method of the surface of section: the sequence of successive intersection points of the same orbit in the energy manifold covers a finite *area* on the surface of section (instead of a *curve*).

Figure 5 exhibits an example of such a solution. We display:

(a) the relative stellar surface displacement

$$\delta r/R = \xi_0 q_0 + \xi_1 q_1 \qquad (|\delta r|/R < 0.5) \tag{35}$$

($\xi_0 = 23.2$; $\xi_1 = -87.7$: surface values of the linear eigenfunctions under the normalisation (6); the maximum allowed q_0 and q_1-values are given by the coordinates of the saddle point; hence the inequality);

(b) a lower resolution power spectrum of $\delta r/R$ (integration time $T = 819.2$ in units $\omega_0 = 1$); and

(c) a high resolution spectrum of the low-frequency range ($T' = 8T$).

The time behaviour clearly reveals several pseudo-periods which manifest themselves in the spectrum as bands around the (slightly shifted) linear frequencies ω_0, ω_1, $\omega_0 + \omega_1$, $\omega_1 - \omega_0$. An observer would presumably take spectrum (b) as evidence for exact periods, especially if his resolution is still poorer (note that here $T \sim 130$

Fig. 5. Two-mode interaction (0, 1): (a) time-behaviour of a Kolmogorov unstable surface displacement $\delta r/R$; (b) corresponding power spectrum at lower resolution; (c) low-frequency end of the power spectrum at high resolution.

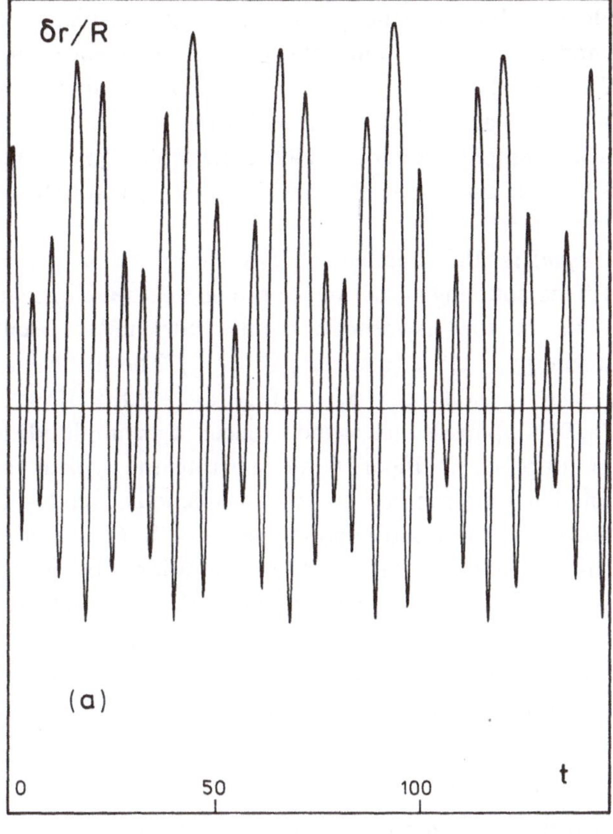
δr/R
(a)
0
50
100
t

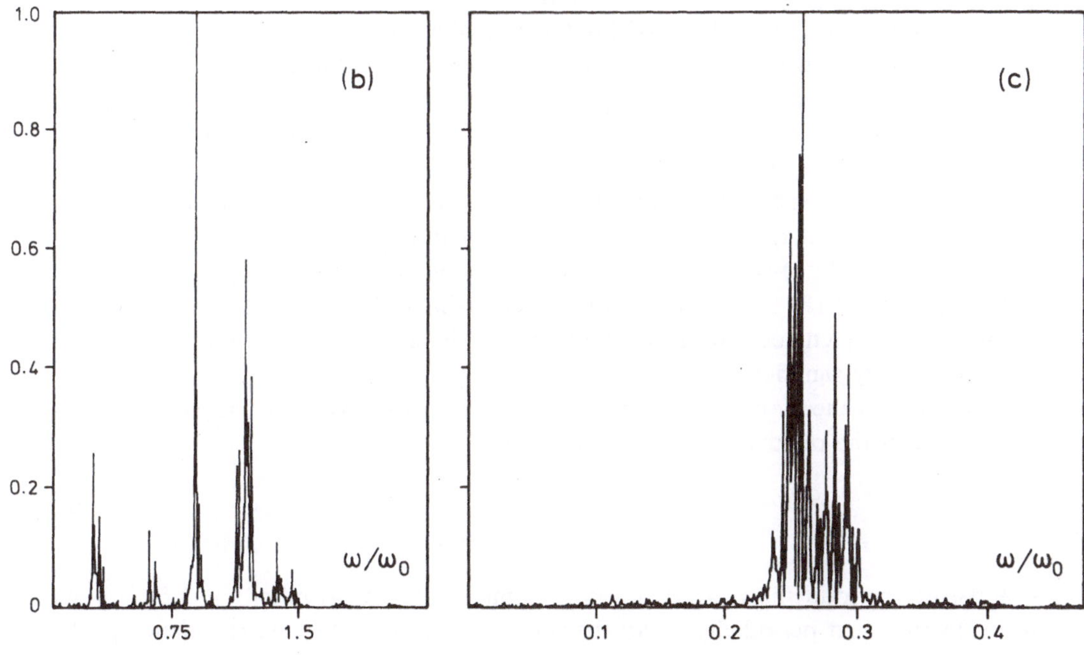
1.0
0.8
0.6
0.4
0.2
0
(b)
ω/ω0
0.75
1.5
(c)
ω/ω0
0.1
0.2
0.3
0.4

fundamental mode periods). Near the origin we have no conspicuous band (cf. (c)) due to the lack of an approximate resonance of order 2 in the linear frequencies; the broad band around 0.3 ω_0 is however reminiscent of the power spectra of Hénon–Heiles stochastic oscillations (Blacher and Perdang, 1981a).

The relative surface amplitude under which these Kolmogorov unstable oscillations occur is close to 0.5 (cf. Equation (35)), so that a (0, 1) instability is not a common phenomenon in real stars.

In the (8, 9) *coupling* the frequencies are closer to a resonance of order 2 ($\omega_9/\omega_8 = 1.098$). This coupling is selected to provide us some insight into the effect of 2-mode interactions in the asymptotic regime. Stochastic oscillations are now encountered at an energy $\sim 0.9\, v_c$ (hence $E_T \lesssim 0.9\, v_c$). The surface amplitude for these motions obeys $|\delta r/R| < 0.08$ ($\xi_8 = 5.7 \times 10^3$, $\xi_9 = -7.8 \times 10^3$). The Kolmogorov instability in the (8, 9) coupling is thus found to occur at an amplitude level 6 times lower than in the (0, 1) coupling. In Figure 6 we illustrate the typical features of such a stochastic solution. The time-run (a) shows a conspicuous pseudo-period $P \sim 9$ (units $\omega_8 = 1$) which shows up as a broad band in spectrum (b) ($T = 819.2$ in units $\omega_8 = 1$). In the high resolution spectrum (c) ($T' = 8T$) an additional band near the origin does materialise, as expected (B′). The general shape of this latter spectrum already bears some analogy with the structure of the SCLERA solar spectra.

A tentative extrapolation of the previous figures suggests that for modes of asymptotic order around 100, the relative surface amplitude level for stochasticity is $\sim 10^{-3}$.

Multiple-mode interaction

The equipotential surfaces for a coupling $(i_1, i_2, \ldots, i_F)$ between $F(>2)$ linear modes $i_1, i_2, \ldots, i_F$ in the neighbourhood of the stable equilibrium state $0(q_{i_1} = q_{i_2} = \ldots q_{i_F} = 0)$ are again given by an expression of form (34) where k, l, and m are now ranging over $i_1, i_2, \ldots, i_F$. For $V(q_{i_1}, q_{i_2}, \ldots, q_{i_F}) = v < v_c$ these surfaces are closed; for $V = v > v_c$ they become open; the critical surface $V = v_c$ possesses a multiple point S. A motion of energy less than v_c remains bounded; if its energy exceeds v_c it can escape; it is however worth noting that the time interval over which an orbit remains trapped in a potential pocket at energy $v > v_c$ increases sharply with F (Perdang and Blacher, 1982b).

In the (0, 1, 2, 3)-*coupling* the linear frequencies obey $\omega_0 : \omega_1 : \omega_2 : \omega_3 = 1 : 1.35 : 1.75 : 2.17$, so that we have no approximate low order (≤ 4) resonances among these frequencies. Therefore stochastic motions are not expected to occur at a threshold energy significantly lower than v_c.

At the critical energy v_c Kolmogorov instability is again observed. Figure 7 provides an example of the relative surface displacement

$$\begin{aligned} \delta r/R &= \xi_0 q_0 + \xi_1 q_1 + \xi_2 q_2 + \xi_3 q_3 \\ (\xi_2 &= 2.36 \times 10^2\,, \qquad \xi_3 = -5.21 \times 10^2)\,. \end{aligned} \tag{36}$$

Two pseudo-periods show clearly up in the time run (a), a periodicity ~ 3.5 (in units $\omega_0 = 1$) and a beat periodicity ~ 20; they correspond to the highest broadened peaks

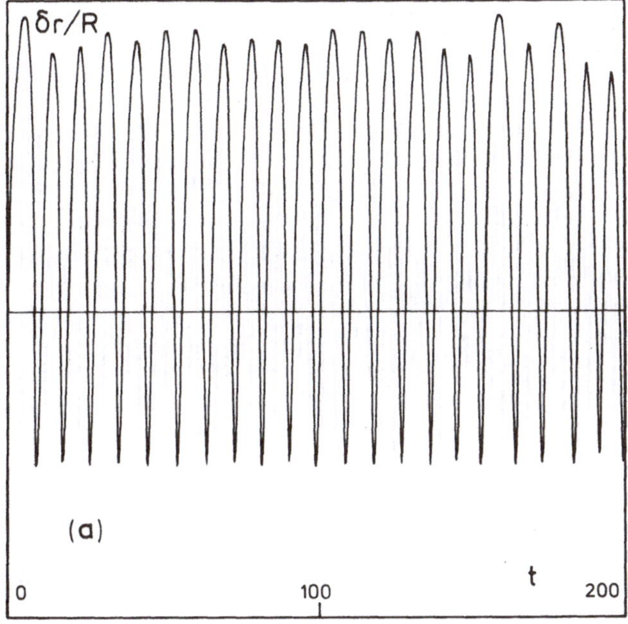

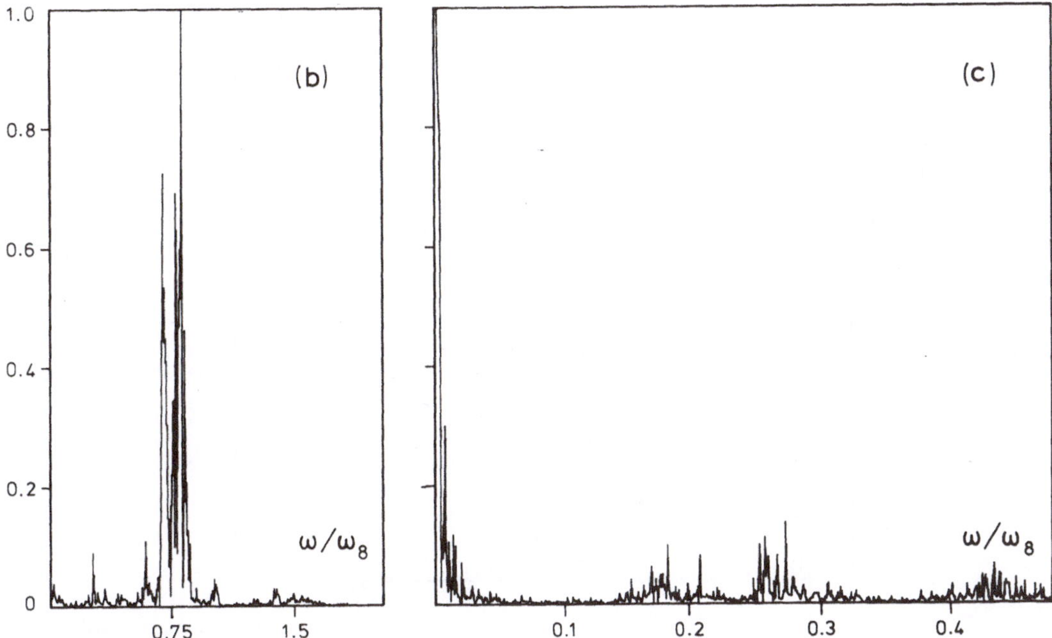

Fig. 6. Two-mode interaction (8, 9): (a) time behaviour of a Kolmogorov unstable surface displacement $\delta r/R$; (b) corresponding lower resolution power spectrum; (c) low-frequency end at higher resolution.

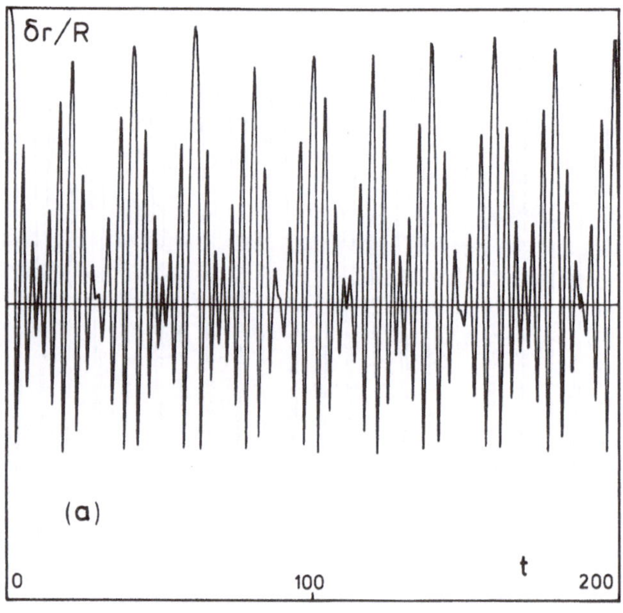

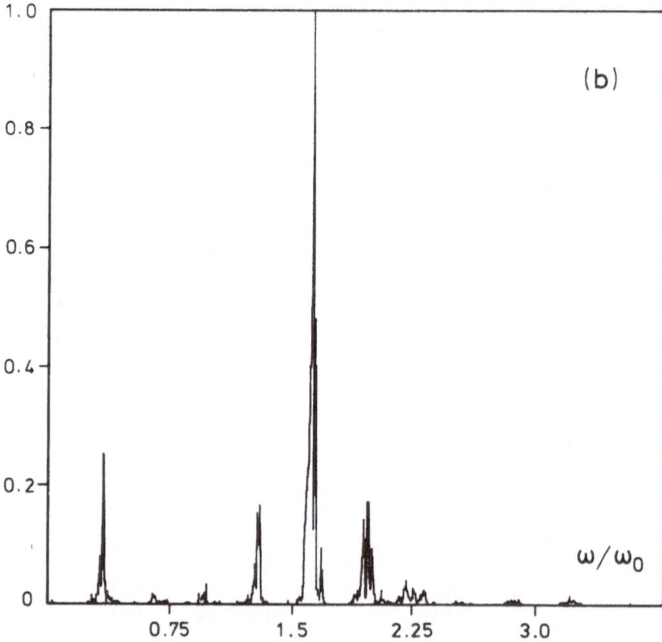

Fig. 7. Four-mode interaction (0, 1, 2, 3): (a) Kolmogorov unstable surface displacement $\delta r/R$; and (b) corresponding lower resolution power spectrum.

of the power spectrum (b) ($T = 819.2$). At low resolution, say $T'' \sim \frac{1}{4}T$, the broadened peaks in (b) appear as fine lines; such a spectrum would mistakenly be interpreted as being due to harmonic, or quasi-periodic motions (cf. Equations (11″), (11‴)). At high resolution ($T' = 8T$) each peak appears as a broad structured band analogous to the feature shown in Figure 5c. The fractal dimensions of all bands in the power spectra analysed are found to lie in the range $1.3 \lesssim d \lesssim 1.5$.

6. Conclusion and Outlook

The main result of the present analysis is the numerical proof that stellar adiabatic radial motions about a dynamically stable equilibrium state can exhibit the Kolmogorov instability. While in lower order mode couplings large surface displacements are required to produce this instability ($|\delta r|/R \sim 0.5$ for the (0, 1) coupling), the amplitude of the surface displacements decreases steadily with the orders of the coupled modes, so that in the asymptotic frequency range this instability can set in fairly easily.

It is thought – although this point remains to be proved in the stellar context – that the physically most relevant role of this instability is to secure an efficient energy transfer between all modes, which eventually leads to an equipartition of oscillation energy between all coupled degrees of freedom. If this conjecture holds a statistical approach to the distribution of the surface amplitudes as a function of frequency becomes meaningful under such stochasticity conditions. We might add that following remark (B′) one has to expect then a very broad-band asymptotic spectrum, due to the blown up and overlapping linear asymptotic frequency peaks at $\omega_N, \omega_{N+1}, \ldots,$ together with an intrinsically non-linear component, namely a band centred at the zero-frequency (or perhaps closer to the frequency ω_{res}; cf. Equation (33)), resulting from a piling up of power at the quasi-resonance of order 2; presumably, similar bands related to higher order resonances will appear, centered at about $2\,\omega_{\text{res}}, 3\,\omega_{\text{res}}, \ldots$ whose heights are rapidly decreasing with order; the overlapping of these bands gives the power spectrum the characteristic structure of '$1/f$-noise' in the low frequency region.

The readymade exploratory analysis of this paper pertains to all approximately adiabatic stellar oscillation phenomena. Obviously each specific type of variability requires a tailormade application of the theory. This is in particular so in the context of solar oscillations: since observation reveals non-radial motions, an extension of the present formalism to the non-radial case is needed. The very fact that a Hamiltonian formalism continues to hold means that the main theoretical conclusions (A and B) survive with however several modifications.

Resonances and approximate resonances are much more likely to occur among the non-radial modes than among the radial ones just as they are more probable in a membrane than in a string. In fact we now have 3 distinct types of resonance:

(1) A $(2l + 1)$-fold degeneracy, due to the radial symmetry of the equilibrium state, for any frequency (acoustic or gravity mode) of degree $l \neq 0$; hence we have an infinity of exact resonances of order 2; provided that the potential is such that eigenfunctions $\xi_{k,l,m}$ of same radial order k and same degree l but of different azimuthal number m

become coupled in the non-linear regime then these resonances favour the Kolmogorov instability; (surface amplitudes lower than in the radial oscillations will be required for this instability to set in).

(2) Sharp resonances of order 2 are known to occur among gravity modes associated with two nearby radiative zones separated by a convective shell (cf. Ledoux and Perdang, 1980).

(3) From the representation formula for acoustic frequencies of asymptotically large radial order $k = N$ and low degree l one has

$$(N \to \infty): \qquad \omega_{N,l} = \omega_{N-1,l+2} + O(1/N^2) \tag{37}$$

(cf. Christensen-Dalsgaard and Gough, 1980, for an estimate of the accuracy of this relation in the context of the Sun) so that a quasi-resonance much tighter than for adjacent radial asymptotic modes obtains.

It seems therefore reasonable that in the same asymptotic range the onset of Kolmogorov unstable non-radial oscillations will occur at a very much lower amplitude level than for radial oscillations. Unfortunately, since the interaction potential for non-radial modes, and in particular the location of its critical point closest to the origin is unknown, we are not in a position to make any quantitative estimate of the surface amplitude needed to generate such stochastic oscillations. In any event, if the acoustic oscillations are Kolmogorov unstable, one expects a power spectrum of the surface displacement with a $1/f$-noise shape near the frequency origin, which should extend over several times the resonance frequency ω_{res}; the tail of this distribution joins a wide band of broadened overlapping pseudo-linear asymptotic frequency peaks (pseudo-linear in the sense that their centres are slightly shifted towards the left by the coupling effect).

For currently favoured solar models the second order resonance frequency $\omega_{res} \simeq 0.136$ mHz (Christensen-Dalsgaard and Gough, 1980); this amounts to a periodicity slightly in excess of 2hr.

The SCLERA solar power spectra have an overall structure in agreement with the theoretically expected shape of a Kolmogorov unstable power spectrum. This is indicative that a detailed numerical investigation of non-linear non-radial acoustic mode couplings is needed before a convincing identification of the peaks can be performed.

Acknowledgements

This paper was completed during a stay at the Institute of Astronomy, Cambridge. The author wishes to thank the members of this institution for the hospitality extended to him. He is particularly indebted to Douglas Gough who presented this paper at the 66th IAU Colloquium. He gratefully acknowledges a NATO travel grant through the Luxembourg Ministère de l'Education Nationale.

References

Arnold, V. I.: 1963a, *Russian Math. Surveys* **18** (6), 85.
Arnold, V. I.: 1963b, *Russian Math. Surveys* **18** (5), 9.
Arnold, V. I.: 1976, *Méthodes Mathématiques de la Mécanique Classique*, éd. Mir Moscow, App. 7.

Birkhoff, G. D.: 1927, *Dynamical Systems*, American Mathematical Society, Providence, Rhode Island (Revised edition by J. Moser, 1966).
Blacher, S. and Perdang, J.: 1981a, *Physica* **3D**, 512.
Blacher, S. and Perdang, J.: 1981b, *Monthly Notices Roy. Astron. Soc.* **19**, 109 P.
Brown, T. M., Stebbins, R. T., and Hill, H. A.: 1978, *Astrophys. J.* **223**, 324.
Chirikov, B.: 1979, *Physics Reports* **52**, 265.
Christensen-Dalsgaard, J. and Gough, D. O.: 1980, *Nature* **288**, 544.
Demaret, J., Dzuba, V., and Perdang, J.: 1978, *Astron. Astrophys.* **70**, 287.
Fermi, E., Pasta, J., and Ulam, S.: 1955, Los Alamos Scientific Laboratory Report LA-1940.
Ford, J.: 1961, *J. Math. Phys.* **2**, 387.
Ford, J. and Lunsford, G. H.: 1971, *Phys. Rev.* **A1**, 59.
Ford, J., and Waters, J.: 1963, *J. Math. Phys.* **4**, 1293.
Hénon, M. and Heiles, C.: 1964, *Astron. J.* **69**, 73.
Hirooka, H. and Saitô, N.: 1969, *J. Phys. Soc. Japan* **26**, 624.
Izrailev, F. M. and Chirikov, B. V.: 1966, *Soviet Phys. Dokl.* **11**, 30.
Kolmogorov, A. N.: 1957, 'Théorie Genérale des Systèmes Dynamiques et Mécanique Classique', Proc. Int. Congress of Math., Amsterdam (Appendix D in R. Abraham, 1967, *Foundation of Mechanics*, Benjamin, New York)
Ledoux, P. and Perdang, J.: 1980, *Bull. Soc. Math. Belgique* **32**, 135.
Moser, J.: 1962, Nachr. der Akad. der Wissensch. in Göttingen Math.-Phys. Kl., 1.
Moser, J.: 1973, *Stable and Random Motions in Dynamical Systems*, Princeton Univ. Press.
Nekhoroshev, N. N.: 1977, *Russian Math. Surveys* **32** (6), 1.
Noid, D. W., Koszykowski, M. L., and Marcus, R. A.: 1977, *J. Chem. Phys.* **67**, 404.
Papaloizou, J. C. B.: 1973a, *Monthly Notices Roy. Astron. Soc.* **162**, 143.
Papaloizou, J. C. B.: 1973b, *Monthly Notices Roy. Astron. Soc.* **162**, 169.
Perdang, J.: 1981, *Astrophys. Space Sci.* **74**, 149.
Perdang, J. and Blacher, S.: 1982a, *Astron. Astrophys.* **112**, 35.
Perdang, J. and Blacher, S.: 1982b, in preparation.
Poincaré, H.: 1890, *Acta Math.* **13**, 1.
Poincaré, H.: 1912, *Rendic. Circ. Mat. Palermo* **33**, 375.
Powell, G. E. and Percival, I. C.: 1979, *J. Phys. A.: Math. Gen.* **12**, 2053.
Rosseland, S.: 1949, *The Pulsation Theory of Variable Stars*, Clarendon Press, Oxford, Sections 4.3, 4.4, 4.5; Chapter 7.
Siegel, C. L.: 1954, *Math. Ann.* **128**, 144.
Simon, N.: 1972, *Astron. Astrophys.* **21**, 45.
Walker, G. H. and Ford, J.: 1969, *Phys. Rev.* **188**, 416.
Woltjer, J.: 1935, *Monthly Notices Roy. Astron. Soc.* **95**, 260.
Woltjer, J.: 1937, *Bull. Astron. Inst. Netherlands* **8**, 193.
Woltjer, J.: 1943, *Bull. Astron. Inst. Netherlands* **9**, 435.
Zaslavskii, G. M. and Chirikov, B. V.: 1971, *Soviet Physics Uspekhi* **14**, 549.

ON THE EXCITATION OF OSCILLATIONS OF THE SUN (NUMERICAL MODELS)*

A. G. KOSOVICHEV and A. B. SEVERNY

Crimean Astrophysical Observatory, p/o Nauchny, 334413, Crimea, U.S.S.R.

Abstract. Numerical solutions of the general time-dependent gas-dynamical equations in linear adiabatic approximation are given for initial conditions imitating: (a) a central perturbation, (b) a boundary perturbation (in the convective envelope), and (c) a 'shrinking' of the Sun as a whole. For a variety of models of the Sun it is found that at the surface the radial component v_r of velocity is much greater than the tangential component v_t, and that the period T of stationary oscillations does not exceed 131^m. The appearance at the surface of a g mode with period 160^m is found to be improbable.

With the initial conditions adopted, a propagating wave is produced which is reflected successively from the centre to the periphery and back, producing 5-min oscillations at the surface of the Sun. Expansion of this wave into separate modes leads to a power spectrum qualitatively similar to that observed.

1. Introduction

160^m oscillations can be obtained as an eigenmode of a standard model of the Sun only among higher order g modes ($n = 9$–12 for $l = 2$; e.g., Christensen-Dalsgaard and Gough, 1976; Iben and Mahaffy, 1976; Vorontzov and Zharkov, 1978; and others). This is difficult to reconcile with the fact that usually lower modes are first and most easily excited. Therefore an explanation in terms of g modes seems to be artificial. Lower-order modes ($g_2 - g_4$) might explain the 160^m periodicity if the Sun has suffered permanent or transient mixing (e.g. model H of Christensen-Dalsgaard and Gough, 1980; Schatzman *et al.*, 1981, model with turbulent diffusion; and Christensen-Dalsgaard *et al.*, 1974, transiently mixed models). It is noteworthy that these mixed models also offer a possible explanation of the low output of neutrinos from the Sun.

Since, in our opinion, the problem of solar oscillations should be coupled with the problem of their excitation, we have taken a different approach from previous authors, and have considered numerically which oscillations can result from different types of excitation. The latter we represent by different initial perturbations.

2. Statement of the Problem

The general equations of motion, in linear adiabatic approximation, are:

$$\frac{\partial \mathbf{v}}{\partial t} = -\frac{1}{\rho_0}\nabla p' + \frac{\rho'}{\rho_0^2}\nabla p_0 + \nabla \phi' ,$$

* Proceedings of the 66th IAU Colloquium: *Problems in Solar and Stellar Oscillations*, held at the Crimean Astrophysical Observatory, U.S.S.R., 1–5 September, 1981.

Solar Physics **82** (1983) 323–329. 0038–0938/83/0822–0323$01.05.

$$\frac{\partial \rho'}{\partial t} + \mathrm{div}(\rho_0 \mathbf{v}) = 0 ,$$

$$\frac{\partial p'}{\partial t} - \gamma \frac{p_0}{\rho_0} \frac{\partial \rho'}{\partial t} + \left(\frac{\mathrm{d}p_0}{\mathrm{d}r} - \gamma \frac{p_0}{\rho_0} \frac{\mathrm{d}\rho_0}{\mathrm{d}r} \right) v_r = 0 , \tag{1}$$

$$\nabla^2 \phi' = -4\pi G \rho' .$$

These have been solved numerically for polytropes of indices $n = 0$, 1.5, and 3, and for several more realistic models of the Sun: the standard model of Dziembowski and Pamyatnykh (1978), model H of Christensen-Dalsgaard and Gough (1980), and the model of Schatzman *et al.* (1981) with turbulent diffusion. The boundary conditions are as usual: the regularity condition at the centre $r = 0$ (when relevant) and the absence of external forces and the continuity of gravitational potential and its first derivative at the photosphere $r = R_\odot$.

Time-dependent computations from an initial state of rest that is perturbed slightly from the equilibrium configuration were performed. Three kinds of initial perturbation were considered:

(a) one localized near the centre (imitating a fluctuation in energy generation):

$$\frac{\rho'}{\rho_0} = \varepsilon \left(\frac{r}{a} \right)^l \exp\left[-0.5 \left(\frac{r - r_0}{a} \right)^2 \right] P_l(\cos\theta) , \qquad \frac{p'}{p_0} = \gamma \frac{\rho'}{\rho_0} ; \tag{2}$$

$$\varepsilon = 10^{-10} , \qquad a = 0.1\, R_\odot , \qquad r_0 = 0 ; \tag{3}$$

(b) one localized near the surface (imitating a perturbation in the convection zone), which is of the same form as (2) but with

$$\varepsilon = 10^{-9} , \qquad a = 0.05\, R_\odot , \qquad r_0 = 0.85\, R_\odot ; \tag{4}$$

(c) a perturbation distributed throughout the body of the Sun (imitating an overall 'shrinking'):

$$r' = r - \xi(r) , \qquad \xi(r) = \xi_0 (r/R_\odot)^2 P_2(\cos\theta) ,$$

$$\frac{\rho'}{\rho_0} = -\frac{1}{\rho_0} \xi(r) \nabla \rho_0 , \tag{5}$$

with

$$\xi_0 = 1\ \mathrm{m} . \tag{6}$$

The excitation of oscillations by a pulsating core of radius $0.1\, R_\odot$ with periods 7^{m}–160^{m} has also been considered. In this case the perturbation is expressed as a boundary condition at $r = 0.1\, R_\odot$ (see Kosovichev and Severny, 1981).

3. Results

Results of the calculations for quadrupole oscillations of the $n = 3$ polytrope for cases (a), (b), (c) are presented on Figures 1 to 3, respectively. Plotted is the response to the initial conditions near the centre, at $r = 0.1\ R_\odot$, and near the surface, at $r = 0.995\ R_\odot$.

In case (a) the central perturbation produces a wave with amplitude $v_r = 1.4 \times 10^{-5}$ m s^{-1} propagating outwards to the surface, at which the amplitude has increased to 0.24 m s^{-1}. Then the wave is reflected back to the centre. The successive reflections from the centre and the surface lead to a stationary regime of oscillations with a period T of successive appearances of the perturbation at the surface given by

$$T = 2 \int_0^{R_\odot} \frac{\mathrm{d}r}{c_s} = 131^{\mathrm{m}} . \tag{5}$$

An important result is that near the surface $v_r \gg v_t$: radial motion, apparently associated with p modes, predominates. In the central regions, however, $v_r \sim v_t$, which suggests the presence of g modes.

In case (b) two waves appear, one propagating to the centre, and the other to the surface. In this case the resulting pattern of oscillations is qualitatively similar to the first.

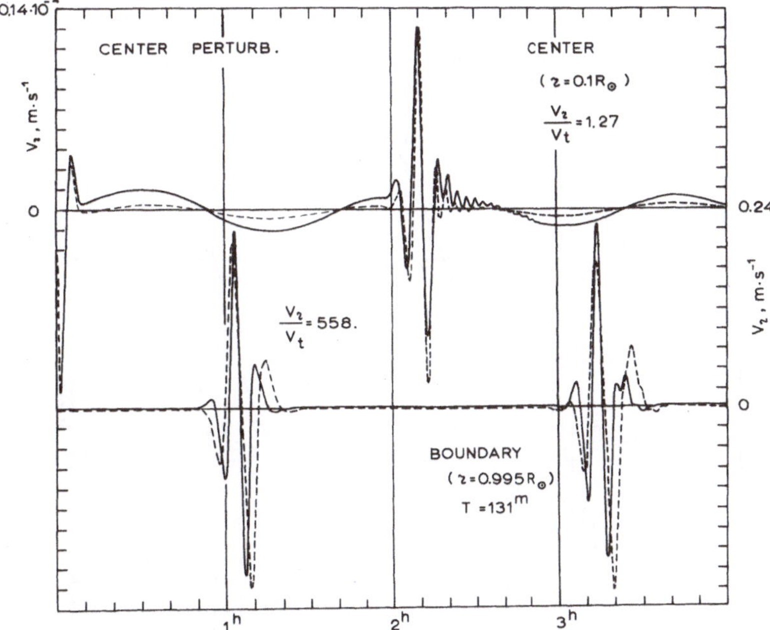

Fig. 1. Evolution of the radial velocity v_r (solid line) and the tangential velocity v_t (dashed line) near the centre of the Sun ($r = 0.1\ R_\odot$) and near the surface ($r = 0.995\ R_\odot$), after the initial perturbation (a) with $l = 2$.

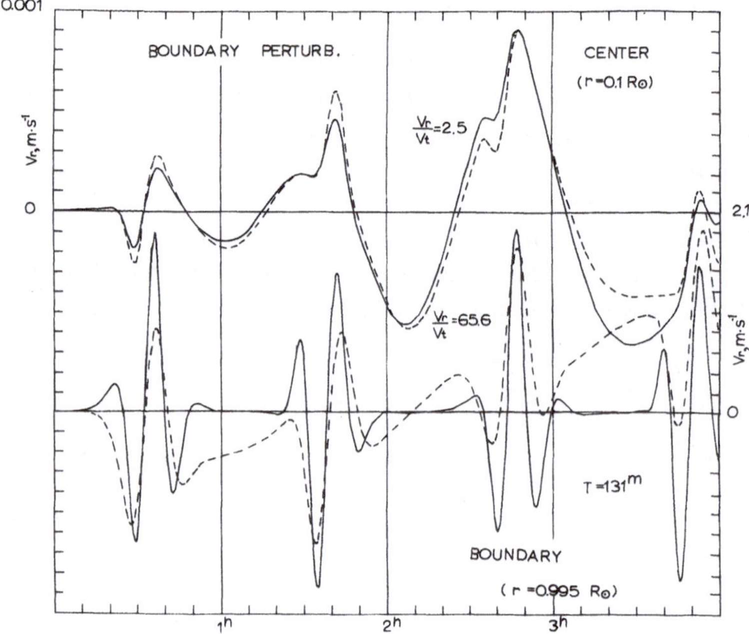

Fig. 2. Similar to Figure 1, but for initial perturbation (b).

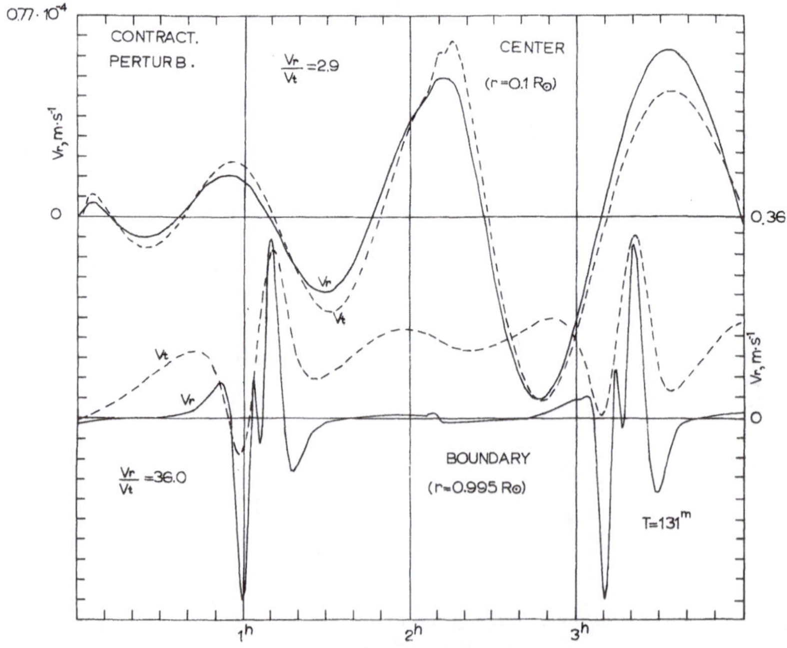

Fig. 3. Similar to Figure 1, but for initial perturbation (c).

The surface amplitude of v_r is 2.1 m s^{-1}, and $v_r \gg v_t$ ($v_t = 3.2 \times 10^{-2}$ m s^{-1}). Again $T = 131^m$. A similar behaviour is found to result from perturbations of type (c) too.

We have found that this behaviour is exhibited also by all the solar models we have considered (excluding the unstable $n = 0$ polytrope, for which we obtained a period of $187^m.2$, which is equal to that of the Kelvin mode of an incompressible fluid sphere (Thomson, 1863), and a characteristic growth time of $31^m.7$ for the g_1 mode, agreeing with that obtained by Pekeris (1938) for the compressible sphere). Oscillation periods vary from 114^m for the Schatzman and Maeder model and 117^m for model H of Christensen-Dalsgaard and Gough (1980), up to 131^m for the $n = 3$ polytrope.

By considering the response to perturbations of types (a), (b), and (c) we do not find a dominant component of the resultant motion with a period of 160^m. Moreover, it is always the case that $v_r \gg v_t$ at the surface (except in the case of the polytrope with $n = 0$). In the terminology of eigenmodes, we can say that at the surface of the Sun p modes are dominant for the perturbations we have considered. We infer that one should observe mainly radial oscillations of the solar surface. That is indeed just what is observed (our attempts to identify transverse modes have not yet succeeded).

4. Discussion

The foregoing analysis raises the question of whether we are likely to detect low-frequency g modes at the surface in general? In all the calculations considered above, low-frequency g modes are confined to the central regions of the Sun, and at the surface the motion is dominated by p waves and their wakes (see Kosovichev and Severny, 1981). Whether g modes can appear at the surface is a question that should be considered.

To answer this question we calculated $l = 2$ g modes for the $n = 3$ polytrope, paying particular attention to g_4, whose period $T = 148^m.6$ is the nearest to 160^m. (We found that our time-dependent programme reproduces all the g modes perfectly, in complete agreement with earlier calculations by Vandakurov (1967) and Robe (1968).) Then we considered the following problem: assuming that this g_4 mode is present initially with $v_r = 1$ m s^{-1} at $r = R_\odot$, can it dominate at the surface after the introduction of some additional perturbation? The answer, as our calculations show, is pessimistic: namely, the g_4 mode contributes substantially to v_r only if a central perturbation of type (a),

$$\frac{\rho'}{\rho_0} = \varepsilon \left(\frac{\rho'}{\rho_0}\right)_{g_4} \left(\frac{r}{a}\right)^2 \exp\left(-0.5\,\frac{r^2}{a^2}\right) P_2(\cos\theta)\,, \tag{8}$$

has a relative amplitude ε smaller than 0.1%. For the perturbation near the boundary of type (b), ε must be less than 1%.

Figure 4 illustrates the result for case (a). The pattern of the radial component of velocity consists of bursts of oscillations with period 131^m, with large peak amplitude $v_{r,p}$, superposed on the slower $148^m.6$ oscillation associated with the original g_4 mode. The latter has amplitude $v_{r,g}$ comparable with $v_{t,g}$, which is only about one tenth of $v_{r,p}$.

Our computations do not explain how the 160^m period can appear, because all the

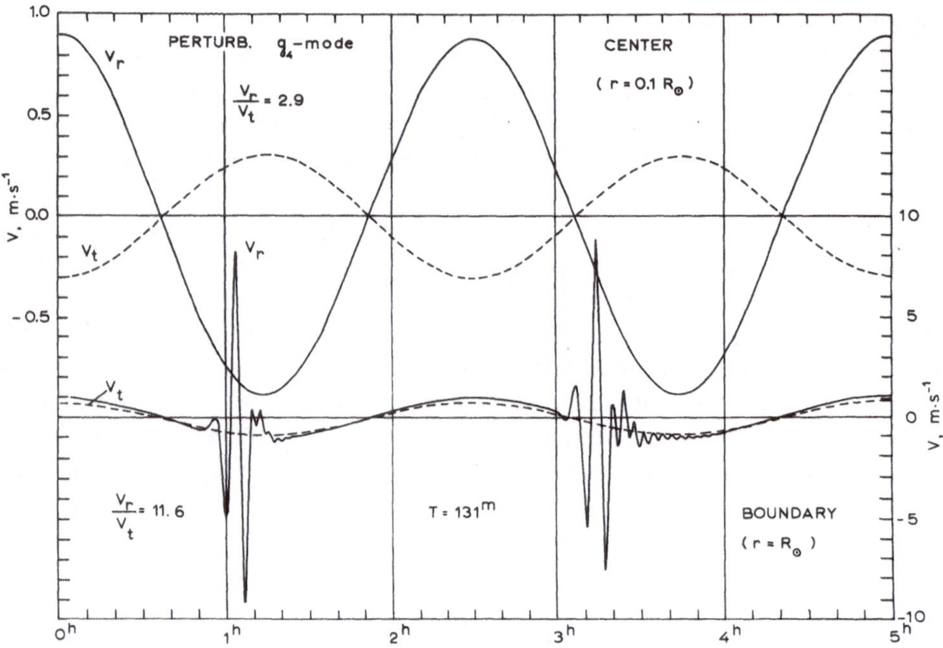

Fig. 4. Similar to Figure 1, but with g_4 (l = 2) present initially together with perturbation (8) with $\varepsilon = 0.1\%$.

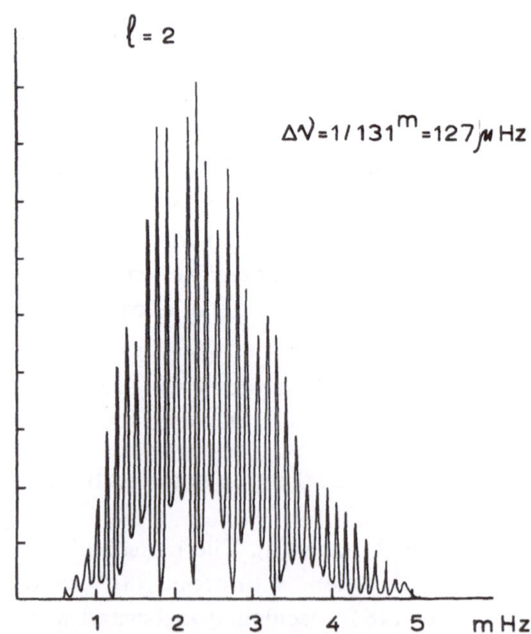

Fig. 5. Power spectrum near 3 mHz of quadrupole oscillations of the surface of the n = 3 polytrope following an initial perturbation of type (a).

perturbations we considered lead to the appearence of oscillations like p modes with periods $\lesssim 131^m$. The 160^m oscillations that are evident at the surface of the Sun indicate that perturbations inside the Sun of the type we have considered are not very strong. Otherwise p modes would predominate, and we would see only the higher-frequency oscillations.

The power spectrum of solar oscillations at the surface for each l and $n = 3$ has, besides the peak corresponding to $T = 131^m$, a set of discrete equidistant peaks in the region around 3 mHz, with a spacing $\Delta \nu = 1/131^m = 127$ μHz (see Figure 5). This power spectrum is qualitatively very similar to those found from whole-disk measurements (see Claverie *et al.*, 1979; Grec *et al.*, 1980).

Acknowledgements

We are gratefull to Dr D. O. Gough for reading and considerable improvement of the manuscript.

References

Christensen-Dalsgaard, J. and Gough, D. O.: 1976, *Nature* **259**, 89.
Christensen-Dalsgaard, J. and Gough, D. O.: 1980, *Nature* **288**, 544.
Christensen-Dalsgaard, J., Dilke, F. W. W., and Gough, D. O.: 1974, *Monthly Notices Roy. Astron. Soc.* **169**, 429.
Claverie, A., Isaak, G. R., McLeod, C. P., van der Raay, H. B., and Roca Cortes, T.: 1979, *Nature* **282**, 591.
Dziembowski, W. and Pamyatnykh, A. A.: 1978, in J. Rösch (ed.), *Plein feux sur la physique solaire*, CNRS, Paris, p. 135.
Grec, G., Fossat, E., and Pomerantz, M.: 1980, *Nature* **288**, 541.
Kosovichev, A. G. and Severny, A. B.: 1981, *Astron. Zh. Pisma* **7**, 34.
Iben, I. and Mahaffy, J.: 1976, *Astrophys. J.* **209**, L39.
Pekeris, C. L.: 1938, *Astrophys. J.* **88**, 189.
Robe, H.: 1968, *Ann. Astrophys.* **31**, 475.
Schatzman, E., Maeder, A., Angrand, F., and Glowinski, R.: 1981, *Astron. Astrophys.* **96**, 1.
Thomson, W.: 1863, *Phil. Trans. Roy. Soc. London* **153**, 583.
Vandakurov, Yu. V.: 1967, *Astron. Zh.* **44**, 786.
Vorontsov, S. V. and Zharkov, V. N.: 1968, *Astron. Zh.* **55**, 84.

A CONVENIENT METHOD TO OBTAIN STELLAR EIGENFREQUENCIES* **

M. KNÖLKER and M. STIX

Kiepenheuer-Institut für Sonnenphysik, Schöneckstr. 6, D-7800 Freiburg i.Br., F.R.G.

Abstract. The differential equations describing stellar oscillations are transformed into an algebraic eigenvalue problem. Frequencies of adiabatic oscillations are obtained as the eigenvalues of a banded real symmetric matrix. We employ the Cowling-approximation, i.e. neglect the Eulerian perturbation of the gravitational potential, and, in order to preserve selfadjointness, require that the Eulerian pressure perturbation vanishes at the outer boundary. For a solar model, comparison of first results with results obtained from a Henyey method shows that the matrix method is convenient, accurate, and fast.

1. Introduction

Stellar oscillations can be described as solutions of linear equations which are obtained from a perturbation of the equations of the internal constitution of stars. The Henyey method (Henyey *et al.*, 1964) which has been employed in most cases to solve these equations, requires an explicit first guess at the eigenfrequency, which then is improved by iteration. However, it often turns out that the desired eigenvalue is not found, inspite of a rather close guess. In the present contribution we therefore use a different method. We transform the system of equations (Section 2A) into an algebraic eigenvalue problem (Section 2B). Castor (1971) has described such a method for radial oscillations. Here we consider the general non-radial case; but unlike Sobouti (1977), who expanded the solutions in terms of complete sets of basis functions, and then obtained the algebraic eigenvalue problem via the Rayleigh–Ritz variational scheme, we employ here only the usual expansion into spherical harmonics and use a difference scheme for the radial part. Since we restrict ourselves to the adiabatic case, our matrix will be real and symmetric. A special consideration will be devoted to the choice of boundary conditions (Section 2C), which must not be in conflict with the self-adjointness, i.e. energy-conservation, of the system.

Standard routines can be employed to obtain all n eigenvalues of a n^2 matrix. Although these also use iteration techniques, they do not in general depend on explicit first guesses, and, in addition, are extremely rapid, in particular if – as in our case – the matrix has a band structure. We report first results in Section 3, and compare them to results obtained with the Henyey method.

* Proceedings of the 66th IAU Colloquium: *Problems in Solar and Stellar Oscillations*, held at the Crimean Astrophysical Observatory, U.S.S.R., 1–5 September, 1981.

** Mitteilungen aus dem Kiepenheuer-Institut Nr. 213.

Solar Physics **82** (1983) 331–341. 0038–0938/83/0822–0331$01.65.

2. The Eigenvalue Problem

A. Differential equations

We consider Lagrangian perturbations $\delta\rho$, $\delta p\ldots$, and Eulerian perturbations ρ_1, $p_1\ldots$, both proportional to $\exp(i\omega t)$. Adiabatic oscillations of small amplitude are described by

$$\frac{\delta T}{T_0} = \nabla_{\text{ad}} \frac{\delta p}{P_0} \tag{1}$$

and by the conservation of momentum and mass,

$$\omega^2 \delta \mathbf{r} = -\frac{\rho_1}{\rho_0^2} \nabla P_0 + \frac{1}{\rho_0} \nabla p_1 , \tag{2}$$

$$\frac{\delta\rho}{\rho_0} = -\nabla \cdot \delta\mathbf{r} \tag{3}$$

(e.g. Ledoux and Walraven, 1958), where T_0, $P_0\ldots$ are equilibrium quantities, and the Eulerian perturbation, Φ_1, of the gravitational potential has been neglected (Cowling, 1941). The latter approximation was found to change the frequencies of low order radial p-modes by up to 3%, while the effect on radial modes with periods around 5 min was only ~0.2% (Knölker, 1978). We anticipate that the non-radial oscillations, in particular at large l, will be affected still less (e.g. Cox, 1980, p. 248).

Equations (1) and (3) are supplemented by the equation of state

$$P = \frac{\rho \mathcal{R} T}{\mu} + \frac{aT^4}{3} = \frac{\rho \mathcal{R} T}{\beta\mu} , \tag{4}$$

where $\mathcal{R}$ is the gas constant, μ the mean molecular weight, a the radiation density constant, and β the ratio gas pressure/total pressure. We expand the unknown functions in terms of spherical harmonics, $Y_l^m(\theta, \varphi)$, linearize (4), eliminate the horizontal components of the displacement, $\delta\mathbf{r}$, and the perturbations of density and temperature, and introduce a non-dimensional radial displacement, $x = \delta r/r_0$, (Lagrangian) pressure perturbation, $p = \delta p/P_0$, and frequency, $\sigma = \omega(4\pi G\bar{\rho}_0)^{-1/2}$. The resulting two equations for x and p are written in such a way that the equilibrium quantities, c_i, of Baker and Kippenhahn (1962, Equations (27)) can be used (we have, in addition, $c_{20} = c_3\rho_0/\bar{\rho}_0$):

$$\frac{1}{c_4} \frac{\mathrm{d}x}{\mathrm{d}\ln P_0} = \left(3 - \frac{l(l+1)}{c_3\sigma^2}\right) x + \left(\Gamma - c_4 \frac{l(l+1)}{c_3\sigma^2}\right) p , \tag{5}$$

$$\frac{\mathrm{d}p}{\mathrm{d}\ln P_0} = -\left(c_3\sigma^2 + 4 - \frac{l(l+1)}{c_3\sigma^2} - c_{20}\right)x + \left(c_4\frac{l(l+1)}{c_3\sigma^2} - 1\right)p\,, \tag{6}$$

where $\Gamma = c_5 - c_6/c_2 = \Gamma_1^{-1}$, and $\Gamma_1 = (\mathrm{d}\ln P_0/\mathrm{d}\ln \rho_0)_{\mathrm{ad}}$.

B. Transformation to an Algebraic Eigenvalue Problem

In order to solve Equations (5) and (6), we have so far used a Henyey type code, originally written by Baker and Kippenhahn (1962), and extended to the non-radial case by Knölker (1978). In the present contribution the equations are however transformed into the problem of finding the eigenvalues of a real symmetric matrix. We first introduce a new variable

$$y = x'/c_4 - 3x - \Gamma p\,, \tag{7}$$

where the prime denotes a derivative with respect to $\ln P_0$. We eliminate p from (5) and (6) and obtain

$$\sigma^2 x = -Ax'' + Bx' + Cx + \frac{1}{c_3\Gamma}\left(1 - \Gamma - \frac{\Gamma'}{\Gamma}\right)y + \frac{y'}{c_3\Gamma}\,, \tag{8}$$

$$\sigma^2 y = \frac{l(l+1)}{c_3\Gamma}(c_4 y - x' + (3c_4 - \Gamma)x)\,, \tag{9}$$

where

$$A = (c_3 c_4 \Gamma)^{-1}\,,$$

$$B = A(\Gamma'/\Gamma + c_4'/c_4 + 3c_4 - 1)\,,$$

$$C = \frac{1}{c_3\Gamma}(c_{20}\Gamma - 4\Gamma - 3\Gamma'/\Gamma + 3)\,.$$

The next step is the transformation

$$x = au\,, \qquad y = a\sqrt{l(l+1)}\,v\,, \tag{10}$$

where the condition $a'/a = B/2A$ would delete the u'-term in the differential equation for u. Here we use instead $a'/a = (B + A')/2A$. This ensures that the symmetry of our matrix will be conseved at the same time. The resulting equations for u and v can then be written in the form

$$\sigma^2 u = -\tfrac{1}{2}((Au)'' + Au'') + Du + \tfrac{1}{2}((Gv)' + Gv') + Fv\,, \tag{11}$$

$$\sigma^2 v = Ev - \tfrac{1}{2}((Gu)' + Gu') + Fu\,, \tag{12}$$

where

$$D = C + A''/2 + Ba'/a - Aa''/a\,,$$

$$E = c_4 l(l+1)/(c_3\Gamma)\,,$$
$$F = G(3c_4 - \Gamma - a'/a) + G'/2\,,$$
$$G = \sqrt{l(l+1)}/(c_3\Gamma)\,.$$

We now introduce a 'staggered mesh' (e.g. Williams, 1969): We consider the values v_i of v at equidistant levels of $\ln P_0$, and the values $u_{i+1/2}$ at intermediate levels, and replace the derivatives by centered differences. The unknowns are then arranged into a vector $\mathbf{z} = (v_1, u_{3/2}, v_2, u_{5/2} \ldots)$, and Equations (11) and (12) are combined into the system

$$\sigma^2 \mathbf{z} = \mathbf{N}\mathbf{z}\,, \tag{13}$$

where the special combinations of the derivatives appearing in (11) and (12) guarantee the symmetry of the matrix $\mathbf{N}$. In order to limit the bandwidth of $\mathbf{N}$ to 5, we have replaced $(Fv)_{i+1/2}$ by $(F_{i+1}v_{i+1} + F_i v_i)/2$ and u_i by $(u_{i+1/2} + u_{i-1/2})/2$ and all coefficients $A_{i+1/2}, G_{i+1/2} \ldots$ are eliminated by the proper arithmetic means of such quantities at neighbouring levels. Errors arising from these operations are only of order h^2, where $h = (\ln P_0)_{i+1} - (\ln P_0)_i$, consistent with the difference scheme used. – As an alternative we have arranged the unknowns u_i and v_i at *all* levels into an eigenvector. Direct use of (11) and (12), again with a scheme of centered differences, then also yields a problem of type (13) with a symmetric matrix, but in this case the bandwidth is 7.

C. Boundary conditions

We place the outer boundary for our oscillating star at a level where the optical depth is small ($\tau = \tau_B = 2.5 \times 10^{-4}$). The equilibrium atmosphere outside this level is adapted to the atmosphere of Vernazza *et al.* (1976), and P_0 and ρ_0 are finite at $\tau = \tau_B$. The 'zero-'boundary conditions, used by Chandrasekhar (1964) and Unno *et al.* (1979, p. 85), are therefore not satisfied. In order to keep the problem self-adjoint we require that the expression $p_1 + \rho_0 \Phi_1$ vanishes at the boundary. Due to the neglect of Φ_1 this means that the Eulerian perturbation, p_1, of the pressure vanishes. In terms of the variables u and v this condition is $Au' = Gv - Au(\Gamma + a'/a - 3c_4)$. Using Equations (10) and (11) we can calculate the expression

$$\int [(11)u^* + (12)v^* - (11)^*u - (12)^*v]\, \mathrm{d}\ln P_0\,,$$

where the asterisk denotes the complex conjugate, and thus show that

$$(\sigma^2 - \sigma^{*2}) \int (uu^* + vv^*)\, \mathrm{d}\ln P_0 = [u(Au' - Gv)^* - u^*(Au' - Gv)]_o^i\,. \tag{14}$$

This expression vanishes at the inner boundary (i), where we require $u = 0$, and at the outer boundary (o), due to the above condition and the reality of $A, G \ldots$.

From Equations (7) and (9) we see that, at the outer boundary,

$$\sigma^2 y = -\frac{l(l+1)}{c_3}(x + c_4 p) = -\frac{l(l+1)c_4}{c_3 P_0}\, p_1 = 0\,.$$

Hence, if $\sigma^2 \neq 0$, we have $y = 0$ and, therefore, $v = 0$. The variable v_1 can thus be dropped from the list of our unknowns. At the same time, we find from (9) that $x' = (3c_4 - \Gamma)x$ at the outer boundary, and, by (10), $u' = bu$, where $b = 3c_4 - \Gamma - a'/a$. In the form $(u_{3/2} - u_{1/2})/h = b(u_{3/2} + u_{1/2})/2$, this condition is used in order to eliminate $u_{1/2}$ from Equation (11) for $u_{3/2}$. The outer boundary condition is thus incorporated into the first line of our matrix, and the symmetry is preserved.

In the present contribution we exclude from our model the central region of the Sun. We set $u = 0$ at, say, $r = 0.2\, r_\odot$, and count our levels so that $U_{N+1/2} = 0$. We drop this variable from the list of our unknowns. Hence, the eigenvector is $\mathbf{z} = (u_{3/2}, v_2, \ldots u_{N-1/2}, v_N)$, and the size of our matrix is $[2(N-1)]^2$.

After the evaluation of $\mathbf{z}$, all operations described in this section can be reversed in order to recover the variables x and p at all levels. We normalize the solution such that $x = 1$ at the outer boundary.

The condition $u' = bu$ can also be incorporated into the direct difference scheme. An asymmetry is then obtained in the first line of the matrix of bandwidth 7, and a Jacobian transformation, leaving the eigenvalues unchanged, must be applied to the vector $\mathbf{z}$ in order to restore the symmetry (e.g. Acton, 1970, p. 316).

Instead of the condition $p_1 = 0$ at the outer boundary, Ando and Osaki (1975) have employed the condition of complete reflexion of the stellar oscillations at the surface. This is possible only for those frequencies which fall into the range of evanescent atmospheric waves. In the case of radial pulsations ($l = 0$), and frequencies small compared to the atmospheric cut-off, the condition has the form (Baker and Kippenhahn, 1965)

$$p = -(4 + 3\sigma^2)x\,,$$

or

$$\sigma^2 u = -Au' + A\left(3c_4 - 4\Gamma c_4 - \frac{a'}{a}\right)u\,. \tag{15}$$

The surface term on the right-hand side of (14) is then $-(\sigma^2 - \sigma^{*2})uu^*$. In combination with the left-hand side this means that the eigenvalue must be real. This must not, however, be understood as a proof of selfadjointness since, by assumption, we have restricted ourselves to the frequencies of evanescent waves in the atmosphere.

Equation (15) can be incorporated into the algebraic system (13). The ensuing asymmetry of the matrix N again can be removed by a Jacobian transformation.

3. Results

We report here only a few results in order to demonstrate how the method outlined in the preceding section works. For the equilibrium we use a model computed with a stellar envelope program similar to that of Baker and Temesváry (1966), with $T_{\text{eff}} = 5770$ K, and extending in radius down to $r = 0.2\, r_\odot$. Böhm-Vitense's (1958) formulation of the

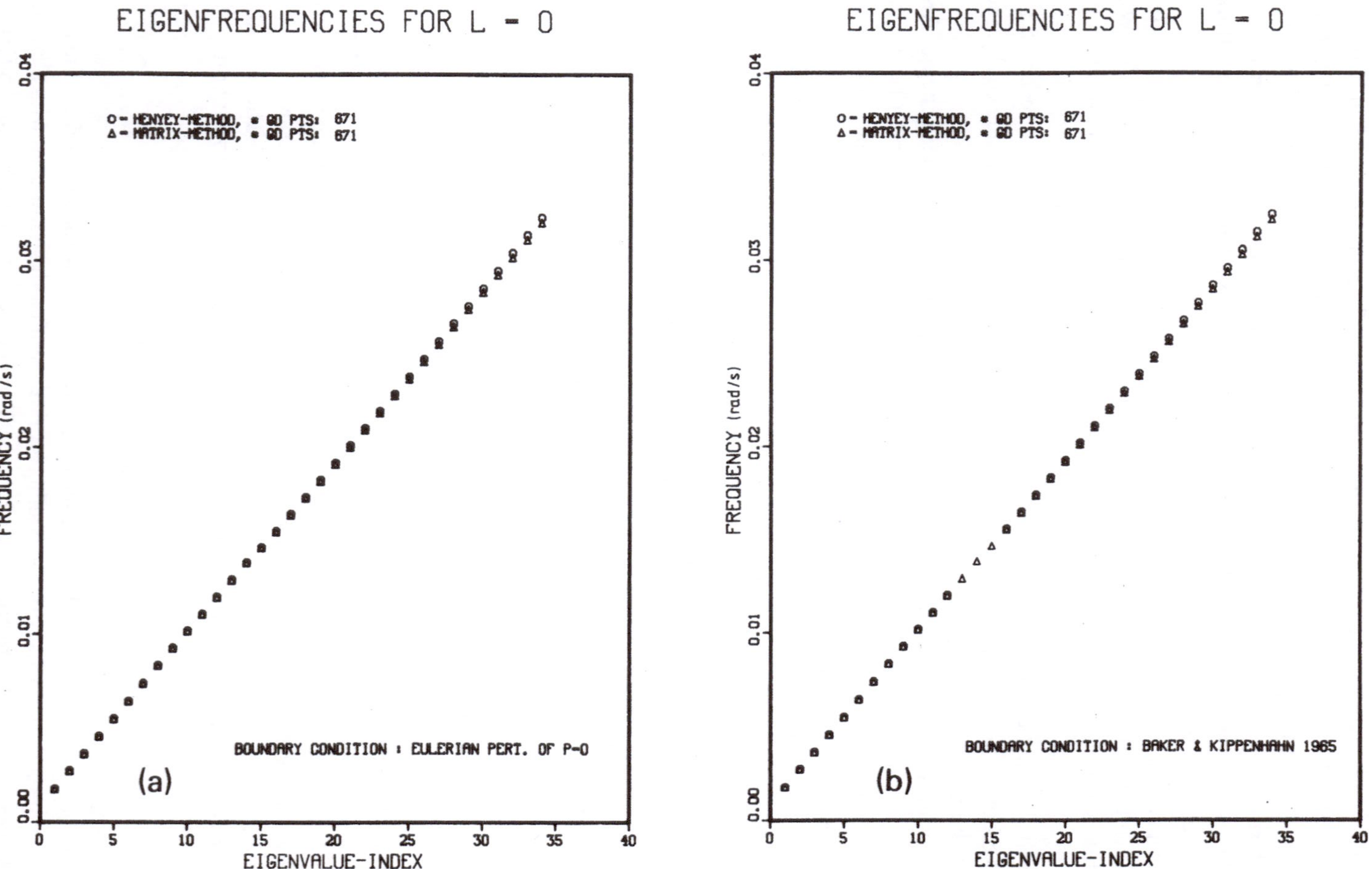

Fig. 1. Eigenfrequencies, ω, of the first 34 radial pulsations, computed with the Henyey and matrix methods: (a) with zero Eulerian pressure perturbation, (b) with complete reflexion at the outer boundary.

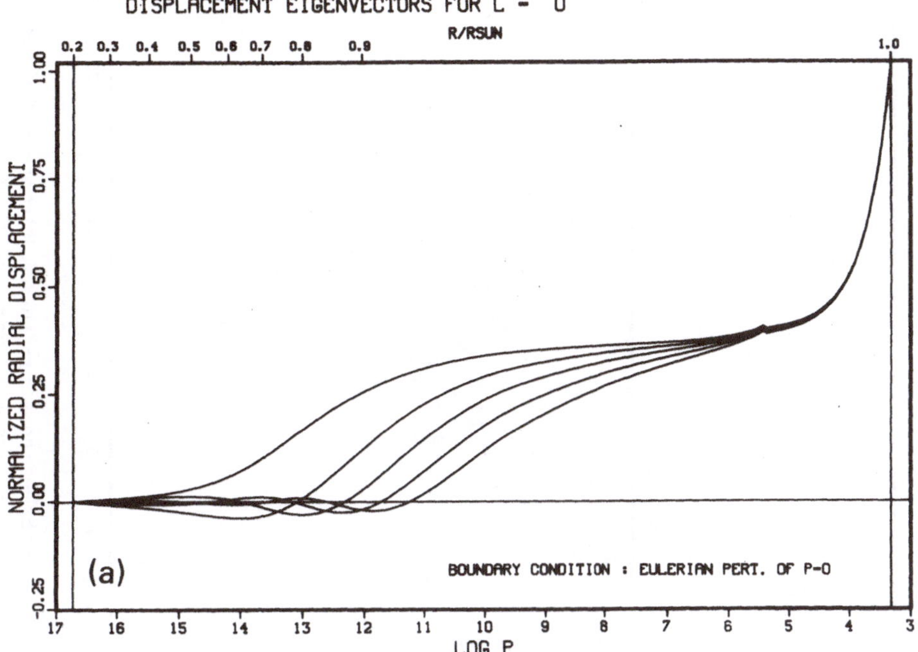

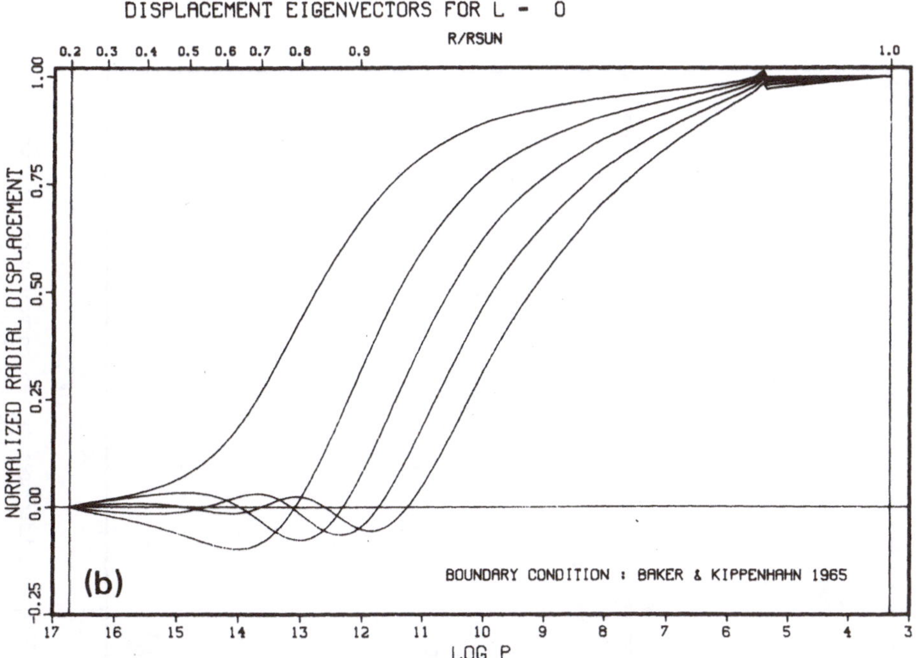

Fig. 2. Displacement eigenfunctions, $\delta r/r_0$, for the first five radial pulsations: (a) zero Eulerian pressure perturbation, (b) complete reflexion at the outer boundary.

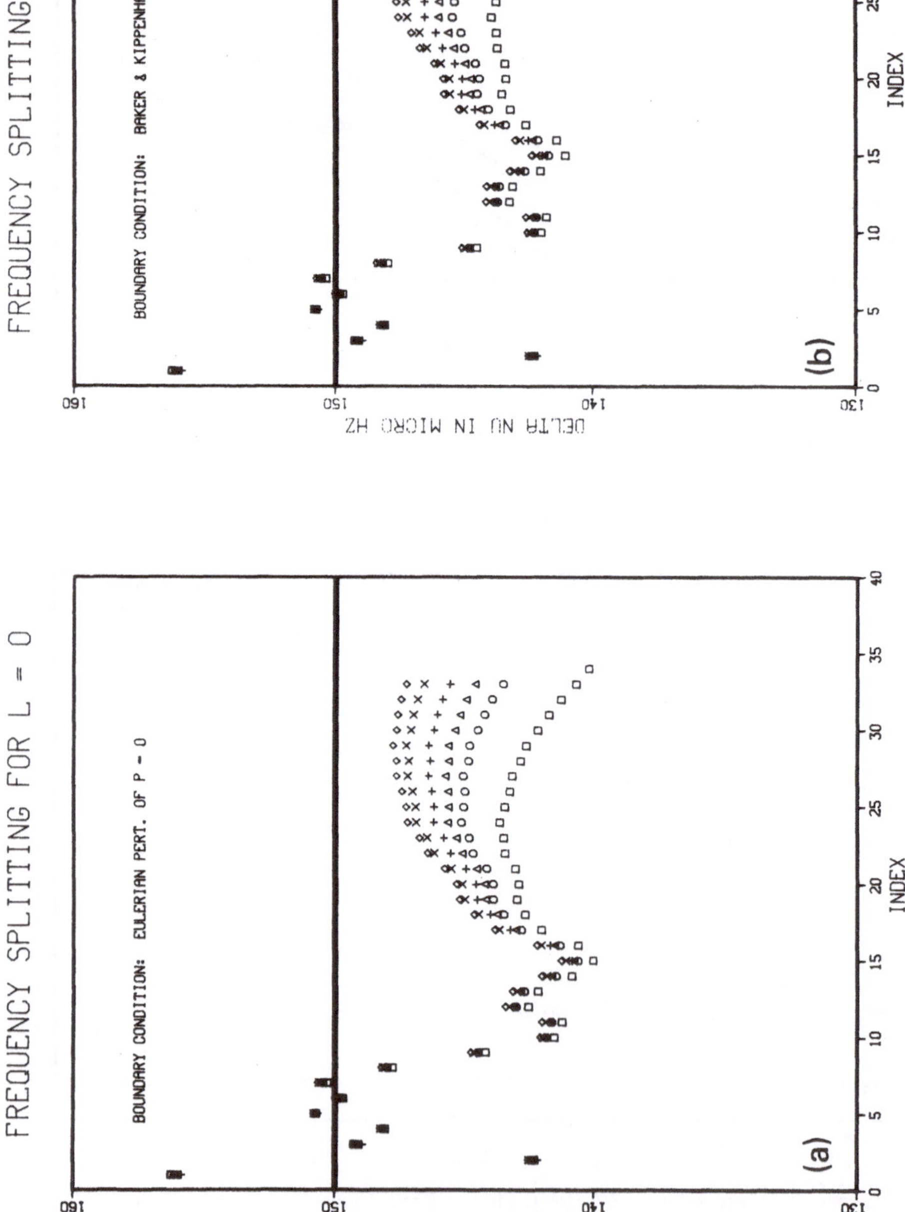

Fig. 3. Frequency differences, $\Delta\nu$, between two consecutive radial pulsation eigenfrequencies: (a) zero Eulerian pressure perturbation, (b) complete reflexion at the outer boundary. The various symbols mark the cases of 269, 336, 384, 448, 537, and 671 grid points, from below in this order; the horizontal line at 150 μHz is the asymptotic value $(2 \int c^{-1}\,dr)^{-1}$.

mixing length theory is used, with α = mixing length/scale height = 1. For the opacity we use the tables of Cox and Tabor (1976), and the abundances of H, He, and heavier elements are specified by $X = 0.7$, $Y = 0.28$, and $Z = 0.02$.

Eigenfrequencies, ω, of radial pulsations ($l = 0$), are shown in Figures 1a, b, up to the 34th mode. The two choices of boundary conditions discussed above virtually lead to the same frequencies, although the corresponding eigenvectors, depicted in Figures 2a, b, exhibit substantial differences near the outer boundary. The results obtained from the Henyey and matrix methods differ at high frequencies only. The reason is probably the intrinsic error of the finite difference scheme: The matrix method, applied to $N = 336$ and $N = 671$, produces frequency differences of the same order as those shown for the two methods in Figure 1. The frequencies found by the matrix method have been used as first guesses for the Henyey-frequencies shown in Figure 1. In three cases (Figure 1b, modes 13 to 15) the iteration was still unsuccessful. This illustrates our point that the first guesses for the Henyey method must be very close indeed.

The displacement eigenvectors shown in Figures 2a, b are normalized to 1 at the outer boundary. This obscures somewhat the close similarity of the cases *a* and *b* in the deeper part: all the radial nodes lie at the same levels. The small hump at $\log P_0 = 5.4$ is caused by a numerical differentiation of the temperature gradient ∇ which we need in order to evaluate the coefficient D of Equation (11) above; ∇ has a sharp maximum at this level, cf. e.g. Figure 2 of Ando and Osaki (1975). We hope to smooth the hump through the use of analytical derivatives for all coefficients needed in the program.

The convergence of the eigenfrequencies with increasing number, N, of grid points is demonstrated in Figures 3a, b. The frequency splitting, $\Delta\nu$, between two neighbouring modes converges rapidly for the lower modes, and more slowly for the higher ones. In the range arround the 5-min oscillations (modes 20 to 25, say) the difference in $\Delta\nu$ between the cases $N = 537$ and $N = 671$ is less than 0.5 μHz. We made however no effort to adjust our equilibrium model so that the observed value for $\Delta\nu$ around 136 μHz (Grec *et al.*, 1980) would be obtained. Our $\Delta\nu$ values lie consistently higher, as does the asymptotic value $(2\int c^{-1}\,\mathrm{d}r)^{-1}$, represented by the horizontal line at 150 μHz in Figures 3a, b.

First results of a non-radial case, $l = 100$, are shown in Figure 4. The frequencies of the first 13 p-modes were computed using both the matrix and the Henyey methods. For comparison, the frequencies computed by Ando and Osaki (1975) agree very well with our results. In the intermediate range (modes 5 to 8), where we obtain slightly different frequencies from the two methods, their results lie almost exactly in the middle between our two values.

We hope to report results for more l-values, and also to show eigenvectors for non-radial oscillations in the near future. In particular, we hope that our method permits a convenient access to the eigenfrequencies of g-modes as well. To this end, we plan to replace our present equilibrium envelope by a complete stellar model.

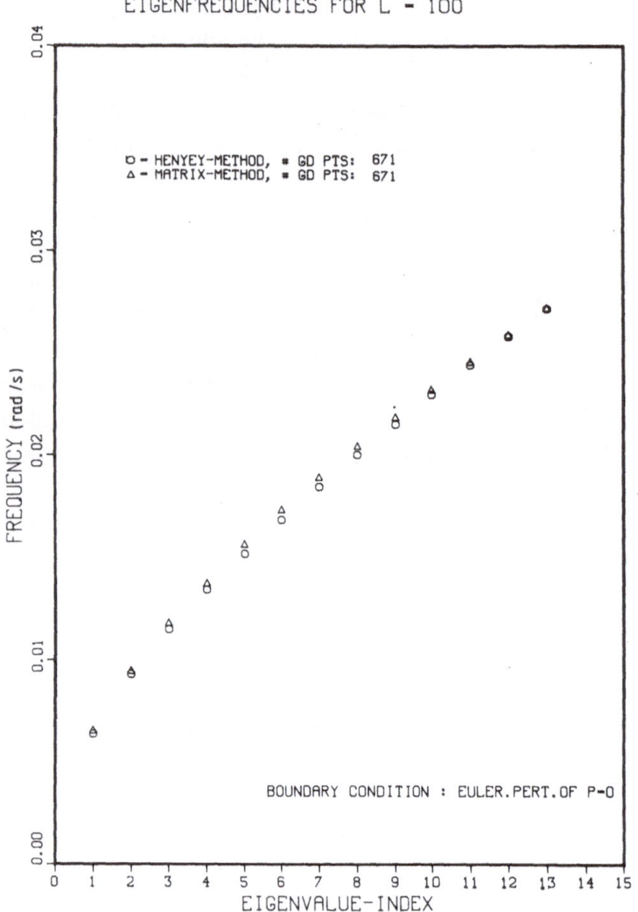

Fig. 4. Eigenfrequencies, ω, of the first 13 non-radial pulsations for $l = 100$, computed with the Henyey and matrix methods, with the condition of zero Eulerian pressure perturbation at the outer boundary.

Acknowledgements

We thank the Deutsche Forschungsgemeinschaft for financial support. The numerical calculations were carried out at the UNIVAC 1100/81, Rechenzentrum der Universität Freiburg.

References

Acton, F. S.: 1970, *Numerical Methods That Work*, Harper and Row, New York.
Ando, H. and Osaki, Y.: 1975, *Publ. Astron. Soc. Japan* **27**, 581.
Baker, N. and Kippenhahn, R.: 1962, *Z. Astrophys.* **54**, 114.
Baker, N. and Kippenhahn, R.: 1965, *Astrophys. J.* **142**, 868.
Baker, N. and Temesváry, S.: 1966, *Tables of Convective Stellar Envelope Models*, 2nd ed., NASA, New York.

Böhm-Vitense, E.: 1958, *Z. Astrophys.* **46**, 108.
Castor, J. I.: 1971, *Astrophys. J.* **166**, 109.
Chandrasekhar, S.: 1964, *Astrophys. J.* **139**, 664.
Cox, A. N. and Tabor, J. E.: 1976, *Astrophys. J. Suppl.* **31**, 271.
Cox, J. P.: 1980, *Theory of Stellar Pulsation*, Princeton Univ. Press.
Cowling, T. G.: 1941, *Monthly Notices Roy. Astron. Soc.* **101**, 367.
Grec, G., Fossat, E., and Pomerantz, M.: 1980, *Nature* **288**, 541.
Henyey, L. G., Forbes, J. E., and Gould, N. L.: 1964, *Astrophys. J.* **139**, 306.
Knölker, M.: 1978, Diplomarbeit, Univ. Göttingen.
Ledoux, P. and Walraven, Th.: 1958, in S. Flügge (ed.), *Handbuch der Physik*, Springer, Berlin, Vol. 51, p. 353.
Sobouti, Y.: 1977, *Astron. Astrophys.* **55**, 327.
Unno, W., Osaki, Y., Ando, H., and Shibahashi, H.: 1979, *Nonradial Oscillations of Stars*, Univ. Tokyo Press.
Vernazza, J. E., Avrett, E. H., and Loeser, R.: 1976, *Astrophys. J. Suppl.* **30**, 1.
Williams, G. P.: 1969, *J. Fluid Mech.* **37**, 727.

ON THE INFLUENCE OF NONLINEARITIES ON THE EIGENFREQUENCIES OF FIVE-MINUTE OSCILLATIONS OF THE SUN*

GAETANO BELVEDERE

Istituto di Astronomia, Università di Catania, Italy

DOUGLAS GOUGH

Institute of Astronomy, and Department of Applied Mathematics and Theoretical Physics, University of Cambridge, England

and

LUCIO PATERNÒ

Osservatorio Astrofisico di Catania and Istituto di Fisica, Facoltà di Ingegneria, Università di Catania, Italy

Abstract. Fitting the results of linear normal-mode analysis of the solar five-minute oscillations to the observed $k - \omega$ diagram selects a class of models of the Sun's envelope. It is a property of all the models in this class that their convection zones are too deep to permit substantial transmission of internal g modes of degree 20 or more. This is in apparent conflict with Hill and Caudell's (1979) claim to have detected such modes in the photosphere.

A proposal to resolve the conflict was made by Rosenwald and Hill (1980). They pointed out that despite the impressive agreement between linearized theory and observation, nonlinear phenomena in the solar atmosphere might influence the eigenfrequencies considerably. In particular, they suggested that a correct nonlinear analysis could predict a shallow convection zone. This paper is an enquiry into whether their hypothesis is plausible.

We construct $k - \omega$ diagrams assuming that the modes suffer local nonlinear distortions in the atmosphere that are insensitive to the amplitude of oscillation over the range of amplitudes that are observed. The effect of the nonlinearities on the eigenfrequencies is parameterized in a simple way. Taking a class of simple analytical models of the Sun's envelope, we compute the linear eigenfrequencies of one model and show that no other model can be found whose nonlinear eigenfrequencies agree with them. We show also that the nonlinear eigenfrequencies of a particular solar model with a shallow convective zone, computed with more realistic physics, cannot be made to agree with observation. We conclude, therefore, that the hypothesis of Rosenwald and Hill is unlikely to be correct.

1. Introduction

In the computation of solar five-minute acoustic oscillations it is usual to apply linearized theory out to some level high in the atmosphere. There boundary conditions are applied. The dynamical boundary condition normally chosen is either that the Lagrangian pressure perturbation vanishes or that the solution matches onto a causal linear eigenfunction of a plane parallel atmosphere. The causal eigenfunctions are those that

* Proceedings of the 66th IAU Colloquium: *Problems in Solar and Stellar Oscillations*, held at the Crimean Astrophysical Observatory, U.S.S.R., 1–5 September, 1981.

Solar Physics **82** (1983) 343–354. 0038–0938/83/0822–0343$01.80.

correspond to forcing from below; they exclude the possibility of incoming waves from infinity or evanescent motion produced by pressure perturbations from above.

The eigenfrequencies of the five-minute oscillations, excepting the chromospheric modes, hardly depend on which of those boundary conditions is chosen. This is so even if the boundary condition is applied no higher in the atmosphere than at the temperature minimum, as has been demonstrated for the modes of high degree by Berthomieu *et al.* (1980) and for the modes of low degree by Christensen-Dalsgaard and Gough (1981a). The reason is clear, and is almost taken for granted in modern discussions of stellar pulsation (e.g., Unno *et al.*, 1979; Cox, 1980): provided its eigenfrequency is substantially below the value of Lamb's acoustical cut-off frequency characteristic of the photospheric regions, a mode cannot propagate through the atmosphere, and the motion in the subphotospheric zone of propagation, where the oscillation is mainly controlled, is essentially decoupled from conditions high in the atmosphere. Of course, any mode whose frequency is great enough to permit trapping in the chromosphere is liable to be influenced by the atmospheric boundary conditions. We exclude these from our discussion.

An alternative argument is the following: owing to the rapid decrease of density and pressure with height, the normal-mode equations resemble equations that have a singularity just above the level at which the boundary conditions are applied. The adiabatic wave equation, for example, admits only one regular solution at that singularity, as can easily be seen if one approximates the undisturbed solar envelope by a polytrope (Lamb, 1932). Even though pressure and density do not actually vanish in the true atmosphere, the eigenfunctions are qualitatively similar to their polytropic counterparts, and any boudary condition of the type mentioned above combines comparable amounts of the singular and regular solutions. In the propagation zone beneath the photosphere the singular solution has declined almost to zero. Thus in the region where the dynamics is controlled, the eigenfunction is essentially independent of the amount of singular solution admitted by the boundary conditions, and is therefore indistinguishable from the regular solution. For this reason it is common practice in stellar pulsation theory to dispense with the upper atmosphere, and simply select what is apparently the regular solution.

These arguments rely on the validity of the linearization of the equations of motion. The amplitudes of the solar modes are low enough that linearized theory should be a good approximation beneath the photosphere. However, nonlinear processes are unlikely to be negligible in the atmosphere. Hill *et al.* (1978) have argued that nonlinear distortions of the eigenfunctions provide a substantial contribution to the oscillatory signal in the SCLERA diameter measurements, and Stebbins *et al.* (1980) have found that the height-dependence of the radial velocity amplitude deviates substantially from the prediction of linear theory (though in the opposite sense to that inferred by Hill *et al.*, 1978). What has not been established, however, is whether such distortions in the evanescent region are sufficient to influence the eigenfrequencies.

Comparisons of the eigenfrequencies of linear theory with observations of five-minute oscillations are leading us to a model of the solar interior (Berthomieu *et al.*, 1980;

Lubow *et al.*, 1980; Christensen-Dalsgaard and Gough, 1981a, b). An important feature of this model is that the convection zone extends through some thirty percent, by radius. This is roughly equal to, though somewhat deeper than, what had commonly been inferred from stellar evolution calculations. However, it presents an embarassment to an interpretation of some of the SCLERA diameter measurements, and to that we now turn our attention.

Hill (1978) and Hill and Caudell (1979) have claimed to have detected oscillation modes of degree between 20 and 40 with periods of 45 min and 66 min. At first sight the most obvious interpretation of such oscillations is that they are *g* modes trapped in the radiative interior. However, if the convection zone really is a deep as the five-minute oscillations indicate, it would attenuate high-degree *g* modes so severely that the observed amplitudes in the atmosphere would imply apparently implausible values in the radiative interior (Dziembowski and Pamjatnykh, 1978). If, on the other hand, one were to adopt a model with a shallow convection zone, one would not be faced with this difficulty. Moreover, one might even predict a neutrino flux in agreement with observation (Christensen-Dalsgaard *et al.*, 1979). However, it would still not be easy to explain the precise values of the frequencies of the oscillations detected by Hill and Caudell (1979) using linearized normal-mode analysis (Christensen-Dalsgaard *et al.*, 1980); and furthermore, the agreement between theory and observation would be destroyed for the five-minute oscillations.

It has been postulated by Rosenwald and Hill (1980) that the root of the problem lies in the linearization of the dynamics. If nonlinear processes are important in the upper part of the atmosphere it is not meaningful to apply boundary conditions to linear eigenfunctions there, however plausible the boundary conditions themselves might be. Consequently it is not out of the question that the ability of linearized theory to reproduce the observations of the five-minute oscillations is merely fortuitous. If that were the case we would have to rescind our conclusions about the structure of the Sun. It is therefore important to assess the credibility of the postulate. That is the purpose of the investigation reported in this paper.

2. Formulation of the Problem

We first assume that linearized theory is a good representation of the oscillations beneath the photosphere, as did Rosenwald and Hill (1980). We also adopt the adiabatic approximation; the computations of Berthomieu *et al.* (1980) provide adequate justification for that. We confine attention to modes of high degree l. These decay rapidly with depth, provided they are taken to be regular at the centre of the Sun. Therefore it suffices to consider just the outer part of the envelope of the Sun, extending to a depth no greater than half the solar radius. Moreover, perturbations to the gravitational potential can be neglected.

Subject to these approximations the linearized normal mode equations reduce to a second-order ordinary differential system (e.g. Unno *et al.*, 1979; Cox, 1980). Normally, one homogeneous boundary condition is applied at the base of the envelope, and another at the surface. Most commonly, the displacement eigenfunction is taken to vanish at the base of the envelope, though sometimes the solution is matched onto an asymptotic representation of the solution that decays with depth. We shall not address here the issue of whether the solution that is singular at the centre of the Sun can also be present.

It is convenient for our discussion to take the outer boundary as the photosphere. In purely linear theory the boundary condition that must be applied there is simply that the envelope eigenfunction matches smoothly onto the corresponding causal eigenfunction in the atmosphere. But if nonlinearities are important in the atmosphere, this condition must be rejected.

If that were all there is to say, a solution beneath the photosphere could be found for any frequency ω, and the dispersion relation measured by Deubner *et al.* (1979), for example, could be rationalized with any model of the Sun. Therefore, if the class of acceptable solar models is to be restricted, the solutions of the oscillation equations must be constrained further.

Beneath the photosphere ($r < R$) we represent the perturbation by

$$\mathbf{y}(\mathbf{r}, t) = (y_1, y_2) \equiv \mathrm{Re}\,\{[r^{-1}\xi(r), (g\rho r)^{-1}\delta p(r)]S_{lm}(\theta, \phi)e^{i\omega t}\} \tag{2.1}$$

with respect to spherical polar co-ordinates (r, θ, ϕ), where $g(r)$ is the gravitational acceleration, $\rho(r)$ is the undisturbed density, ξ and δp are the oscillation eigenfunctions representing vertical displacement and Lagrangian pressure perturbation, S_{lm} is a tesseral harmonic of degree l and order m, and t is time. Above the photosphere we represent the two linearly independent solutions by

$$\mathbf{y}^- = (y_1^-, y_2^-), \qquad \mathbf{y}^+ = (y_1^+, y_2^+)\,. \tag{2.2}$$

These are what Hill *et al.* (1978) call β_- and β_+, and are respectively the causal and noncausal linear eigenfunctions. We suppose them to be normalized such that $y_1^-(R) = 1$ and $y_1^+(R) = 1$. In the spirit of Rosenwald and Hill (1980) we now introduce a parameter λ such that the actual nonlinear atmospheric solution evaluated at the photosphere ($r = R$) is

$$\mathbf{y}_\lambda = y_1(R)[(1 - \lambda)\mathbf{y}^- + \lambda\mathbf{y}^+]\,. \tag{2.3}$$

Continuity of displacement and Lagrangian pressure perturbation at the photosphere is then given by

$$\mathbf{y} = \mathbf{y}_\lambda \quad \text{at} \quad r = R\,. \tag{2.4}$$

Notice that λ merely parameterizes the result of performing an appropriate nonlinear analysis of the atmosphere, the details of which we do not specify. The ratio $\lambda/(1 - \lambda)$ of the contributions from $\mathbf{y}^+$ and $\mathbf{y}^-$ at the photosphere can thus be regarded as a formal measure of the degree of nonlinearity. Clearly a value of λ exists for any frequency ω. Linear theory results when $\lambda = 0$. Our task now is to determine what reasonable restrictions on λ should be imposed.

It is certainly plausible that nonlinear processes in the atmosphere should depend on the frequency ω, the horizontal wavenumber $k = [l(l+1)]^{1/2}R^{-1}$ and, of course, the amplitude A of the oscillations. However, since the scale height of the atmosphere is much less than k^{-1}, any localized nonlinearity cannot depend on k; in the atmosphere all p modes resemble radial oscillations. Moreover, the observations of Deubner *et al.* (1979) of five-minute oscillations of high degree, and the observations reported by Grec *et al.* (1980) of five-minute oscillations of low degree (Fossat and Grec, private communication) show no systematic frequency variation with changing amplitude*. Therefore, provided we confine our attention to those modes whose amplitudes are large enough to have been detected, we may safely infer that λ is a function of ω alone. Notice that the assumption about the dependence of λ on A is not that λ is independent of A for all A, but simply that its value does not vary substantially in the range of interest. In the analysis that follows, that value is permitted to be quite different from the value at $A = 0$.

We recall that the observed diagnostic $(k - \omega)$ diagram can be reproduced by the linear eigenfrequencies of a solar model with a substantial convection zone (Berthomieu *et al.*, 1980; Lubow *et al.*, 1980). This model we call model A. The question we must answer is whether for a model with a shallow convection zone (which we call model C), whose linear eigenfrequencies do not agree with observation, a function $\lambda(\omega)$ exists such that

$$\omega_n(k, 0; \mathrm{A}) = \omega_n(k, \lambda; \mathrm{C}) \tag{2.5}$$

for all n and k, where $\omega_n(k, \lambda; \mathrm{M})$ denotes the f-mode eigenfrequency ($n = 0$) or p-mode eigenfrequency of order $n > 0$ of oscillations of solar model M, having horizontal wavenumber k in the photosphere and computed with linearized theory beneath the photosphere subject to the boundary condition (2.4). If such a function were to exist, it would not be implausible that model C is a good model of the Sun; then the postulate of Rosenwald and Hill would deserve further consideration.

3. A Simple Illustrative Example

We first approximate the outer layers of the Sun by an isothermal atmosphere of perfect gas supported by a polytrope of index μ. Because we are restricting attention to modes of high degree, we also make the plane-parallel approximation, and take the gravitational acceleration g to be constant. With respect to Cartesian co-ordinates $\mathbf{x} \equiv (x, y, z)$, with z increasing downwards, the equilibrium state is defined by:

$$p = p_0 e^{z/H}, \qquad \rho = \rho_0 e^{z/H} \tag{3.1}$$

when $z < 0$, and

$$p = p_0(1 + z/z_0)^{\mu+1}, \qquad \rho = (\mu + 1)(g z_0)^{-1} p_0 (1 + z/z_0)^{\mu} \tag{3.2}$$

* We recognise that some of the apparent amplitude variation observed is a product of interference between modes with similar frequencies. That does not alter our conclusion.

when $z > 0$, where p and ρ are pressure and density, $H = p_0/g\rho_0$ is the scale height of the isothermal atmosphere, and p_0, ρ_0 and z_0 are constants.

The constant z_0 is a measure of the depth of the transition between the isothermal and polytropic regions of the model; if the atmosphere were absent, p and ρ would vanish at $z = -z_0$. Notice that we have not imposed the condition $(\mu + 1)p_0 = g\rho_0 z_0$, which would imply continuity of density, and hence of temperature, at $z = 0$. We have in mind treating the superadiabatic boundary layer at the top of the solar convection zone simply as a temperature discontinuity. Our justification for so doing is the insensitivity of p-mode eigenfrequencies to details of the structure of that boundary layer (Berthomieu *et al.*, 1980); it is upon the temperature jump across that boundary layer that the eigenfrequencies depend. Thus we assume that the polytrope is adiabatically stratified, taking $\mu = (\gamma - 1)^{-1}$, where γ is the adiabatic exponent $(\partial \ln p/\partial \ln \rho)_{\text{ad}}$ of the gas. Notice that $z_0 = (\mu + 1)(T_0/T_a)H$, where T_0 is the limit of the temperature in the polytrope as $z \to 0$, and T_a is the temperature of the isothermal atmosphere. Thus z_0 and H have similar magnitudes.

The polytropic layer is supposed to represent that part of the convection zone within which the p modes are trapped. In reality μ varies, due to ionization, and the value we adopt is intended to be representative of the trapping region. Changing the structure of the superadiabatic boundary layer in a realistic solar model, which can be achieved by changing the mixing length, for example, changes both the temperature jump across that boundary layer and the temperature stratification beneath. This changes the depth of the convection zone. It also moves the ionization zones, and so changes μ. Hence, in general, two distinct polytropic models should differ in both z_0 and μ. Nevertheless, to keep the analysis simple, we consider explicitly only the effect of changing z_0. We do, however, allow for the possibility that γ is different in the atmosphere and the convection zone. Thus we use γ to specify the constant adiabatic exponent in the isothermal atmosphere, and use only μ in the adiabatically stratified polytrope beneath.

We next calculate the adiabatic normal modes of oscillation of the model. We adopt the divergence of the velocity, χ, as dependent variable, and seek nontrivial separable solutions of the form

$$\chi(\mathbf{x}, t) = \text{Re}[X(z)e^{ikx + i\omega t}]. \tag{3.3}$$

In so doing we are ignoring the f modes. The vertical component of the associated displacement eigenfunction can be written

$$\xi(\mathbf{x}, t) = (gk^3)^{-1/2}\,\text{Re}[\Xi(z)e^{ikx + i\omega t}]. \tag{3.4}$$

We observe that the equation of continuity and the linearized adiabatic equation of state together imply $i\omega\delta p = -\gamma p\chi$. Thus continuity of δp and ξ at $z = 0$ is obtained by requiring that Ξ and X be continuous. Derivations of the equations of motion and discussions of some aspects of their solutions for isohermal and polytropic atmospheres are given by Lamb (1932). The problem treated here is just a straightforward generalization of Lamb's analysis.

In the isothermal atmosphere, two independent solutions are

$$X_+ = e^{\kappa_+ z/H}, \qquad X_- = e^{\kappa_- z/H}, \tag{3.5}$$

where

$$\kappa_\pm = -\tfrac{1}{2} [\![1 \pm \{1 - 4\gamma^{-1}[\sigma^2 - (\gamma - 1)\sigma^{-2}]Hk + 4H^2k^2\}^{1/2}]\!] \tag{3.6}$$

and $\sigma^2 = (gk)^{-1}\omega^2$. The solutions X_+ and X_- correspond respectively to the β_+ and β_- solutions of Hill *et al.* (1978). The associated displacement amplitudes are

$$\Xi_\pm = i\gamma\sigma(\sigma^2 - 1)^{-1}(\kappa_\pm + 1 + \sigma^{-2}Hk)e^{\kappa_\pm z/H}. \tag{3.7}$$

We confine attention to modes with $\omega < \omega_c$, where $\omega_c^2 = \gamma g/4H$ is the square of Lamb's critical cutoff frequency for radial modes in the atmosphere. Thus κ_+ and κ_- are real.

In the polytropic envelope the disturbance is given by

$$X = e^{-\zeta}U(-\alpha, \mu + 2, 2\zeta), \tag{3.8}$$

$$\Xi = \frac{i\gamma\sigma}{(\mu + 1)(\sigma^2 - 1)} \zeta e^{-\zeta}\left[\frac{\mathrm{d}U}{\mathrm{d}\zeta} + \left(\frac{\mu + 1}{\zeta} - \sigma^{-2} - 1\right)U\right], \tag{3.9}$$

where

$$2\alpha = \mu^{-1}\sigma^2 - (\mu + 2) \tag{3.10}$$

and $\zeta = k(z_0 + z)$. Here U is the confluent hypergeometric function that vanishes as $\zeta \to \infty$ (e.g. Abramowitz and Stegun, 1964).

The eigenvalues σ_n of σ are determined by demanding that at $\zeta = \zeta_0 \equiv kz_0$ the solutions (3.8) and (3.9) be continuous with an appropriate combination of (3.5) and (3.7). We take that combination as in (2.3). We recall that $kH \ll 1$ for modes with $\omega < \omega_c$ and that z_0 is comparable with H. Hence $\zeta_0 \ll 1$. We can therefore expand $\kappa_\pm$ in powers of kH, and U in powers of ζ_0, retaining only the leading significant terms. The result is

$$\sigma_n^2 \simeq s_n^2 - \frac{2\Gamma(\mu + n + 1)}{\mu\Gamma(\mu + 1)\Gamma(\mu + 2)\Gamma(n)} \frac{1 - K}{K}(2kz_0)^{\mu + 1}, \tag{3.11}$$

where

$$s_n^2 = 1 + 2n/\mu, \tag{3.12}$$

$$K(n, \lambda) = 1 - \lambda - [(1 - 2\lambda)\gamma^{-1}(s_n^2 - s_n^{-2}) + 2(\lambda - 1)s_n^{-2}]Hk, \tag{3.13}$$

n is a positive integer and Γ is the gamma function. The second term on the right-hand side of Equation (3.11) is the change in σ_n^2 produced by replacing a complete polytrope, extending to $z = -z_0$, with the composite model considered here. This change is small compared with σ_n^2.

We specify our reference model A by setting $z_0 = z_A$. Its dimensionless eigenfrequencies σ_{An} are given by Equations (3.11) and (3.12) with $\lambda = 0$, and we pretend that these agree with observation. We now consider model C, with $z_0 = z_C$. Its frequencies

σ_{Cn} are determined by choosing $\lambda(\omega)$ such that $\sigma_{CN} = \sigma_{AN}$ for some N, which yields, to leading order in Hk,

$$\lambda = -\tfrac{1}{4}[1 + (2\gamma - 1)s_N^{-4}]\left[1 - \left(\frac{z_A}{z_C}\right)^2\right]\left(\frac{\omega}{\omega_c}\right)^2 . \tag{3.14}$$

To assess how close the frequencies σ_{Cn} of the presumed nonlinear modes are to σ_{An}, we compare their differences with the corresponding differences $\sigma_{Cn}^{(L)} - \sigma_{An}$, where $\sigma_{Cn}^{(L)}$ are the linear eigenfrequencies of model C, computed with $\lambda = 0$. The result is

$$\frac{\sigma_{Cn} - \sigma_{An}}{\sigma_{Cn}^{(L)} - \sigma_{An}} \simeq \frac{(2\gamma - 1)(s_n^{-4} - s_N^{-4})}{1 + (2\gamma - 1)s_n^{-4}} . \tag{3.15}$$

Except when $n = N$, this is not zero. Thus the frequencies σ_{Cn} cannot be made to coincide with the linear eigenfrequencies of model A. Moreover, it is evident that the magnitude of the deviation is comparable with the difference between the linear eigenfrequencies of models A and C. This is true also if the polytropic index μ is permitted to be different in the two models. Moreover, it is not possible to vary both z_0 and μ in such a way as to make the frequencies of the two models agree for all values of n.

4. Analysis of a More Realistic Pair of Solar Models

Two solar envelope models were computed using the fast numerical programme described by Belvedere *et al.* (1980). The models were integrated inwards from the 'surface', where $r = R_s$ and $T = 4900$ K, to $r = R_0 \equiv R/2$, adjusting the conditions at $r = R_s$ to ensure that $r = R$ and $T = 5770$ K at an optical depth of $\frac{2}{3}$. A mesh spacing of 0.05 electron pressure scale heights was used. Model A was chosen to have abundances of helium and heavy elements: Y, $Z = 0.25, 0.02$, and a constant mixing length to pressure scale height ratio $\alpha = 1.65$ was adopted. Its convection zone was 220 Mm deep. Model C was computed with the same composition, but with $\alpha = 0.65$, and had a convection zone 40 Mm deep. The models resemble models A and C of Christensen-Dalsgaard *et al.* (1979), the main difference being in the chemical composition of model C beneath the convection zone.

The linearized adiabatic oscillation equations [Unno *et al.*, 1979, Equations (17.9), (17.10); Cox, 1980, Equations (17.50), (17.51)] were integrated outwards from $r = R_0$ using a fourth-order Runge–Kutta algorithm. At $r = R_s$ the solutions were matched onto the adiabatic solutions for a plane parallel isothermal atmosphere at $T = 4900$ K. For each integration ω was specified, and in cases where the boundary condition (2.4) had to be satisfied, the eigensolution was found by Newton–Raphson iteration on ω.

First, the eigenfrequencies ω_{An} of model A were computed with $\lambda = 0$. Then, for some value N of n, the frequencies ω_{Cn} of model C were forced to coincide with them. This, of course, required no iteration: the frequency ω_{CN} was chosen to be ω_{AN}, and subsequently $\lambda(\omega)$ was computed by insisting that the photospheric condition (2.4) be

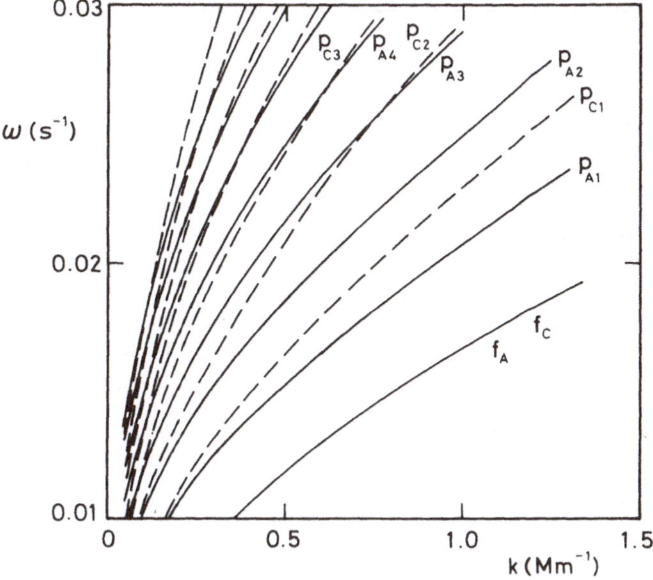

Fig. 1. Theoretical diagnostic diagrams for models A and C of the solar envelope; the frequencies ω are the eigenfrequencies of the linear normal mode analysis ($\lambda = 0$), and are expressed as continuous functions of the wave number k. Model A is our standard model; for the purposes of the argument presented in this paper its linear eigenfrequencies, indicated by the continuous curves, may be regarded as being in agreement with observation. Model C has a shallow convection zone; its eigenfrequencies are represented by the broken curves. In the frequency range plotted the f-mode frequencies of the two models differ by no more than a few parts per thousand.

satisfied. Finally, the remainder of the frequencies of model C were computed, using the same function $\lambda(\omega)$ in condition (2.4).

The linear eigenfrequencies of models A and C are shown in Figure 1. The frequencies of model A do not agree exactly with observation, due to the simplifying approximations used in the construction of the equilibrium model. Nevertheless, we shall pretend that they do. The linear p-mode eigenfrequencies of model C differ from observation substantially. The f-mode frequencies do agree, because these are essentially independent of stratification when $kR \gg 1$.

In Figure 2 we show for several orders N our measures $(1 - \lambda)/\lambda$ of the nonlinearity required to make the frequencies of model C coincide with the linear eigenfrequencies of model A. Infinities of $(1 - \lambda)/\lambda$ signify a pure β_- solution, and zeros signify a pure β_+ solution.

Finally we show in Figure 3 the complete diagnostic diagrams for model C constructed with the parameterized nonlinear boundary conditions (2.4) defined with the functions $\lambda(\omega)$ shown in Figure 2. Once again, the results are compared with the linear eigenfrequencies of model A, which we presume to be in agreement with observation. The deviations between the frequencies of the models are, in all cases, comparable with the deviations between the corresponding linear eigenfrequencies, as we found also for the simple model in Section 3.

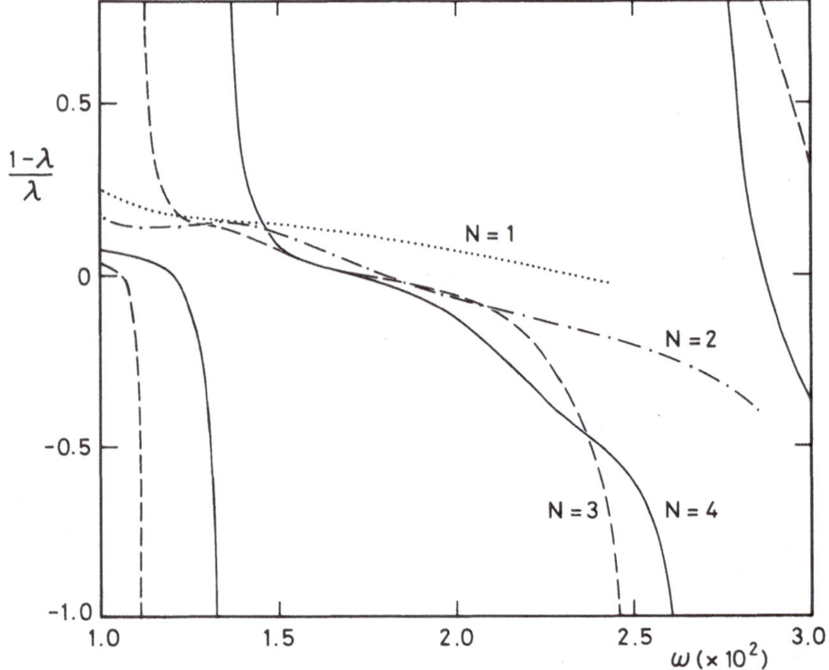

Fig. 2. Examples of the measures $(1 - \lambda)/\lambda$ of the nonlinearity that we have presumed to be present in the atmosphere of model C. They were constructed in such a way as to make the frequencies of oscillation of model C agree with the linear eigenfrequencies of model A for all p modes of order N as indicated. Infinities of $(1 - \lambda)/\lambda$ occur where the eigenfrequencies of models A and C intersect in Figure 1.

5. Summary and Conclusion

We have argued that the effects of atmospheric nonlinearities on the highest-amplitude five-minute oscillations depend only on frequency. Granted that this is so, we have shown that the oscillation frequencies of a particular solar model with a shallow convective zone, no matter what form the nonlinearities take, cannot be made to coincide with the linear eigenfrequencies of another model constructed to be in fair agreement with observation.

We have not demonstrated explicitly that this result is true for all solar models that differ substantially from our standard. Nevertheless, an analysis of a highly simplified model of the outer layers of the sun, consisting of an isothermal atmosphere on a polytropic envelope, suggests that this is indeed a general result. To find a nonlinear model whose frequencies agree with observation for n_0 values of the order n of the modes is likely to require the adjustment of at least n_0 parameters of the model to which the frequencies are sensitive, and possibly more. Sensitivity analyses of the linear eigenfrequencies by Berthomieu *et al.* (1980) and Lubow *et al.* (1980) suggest that such parameters do not exist. That is why we have not pursued the matter further.

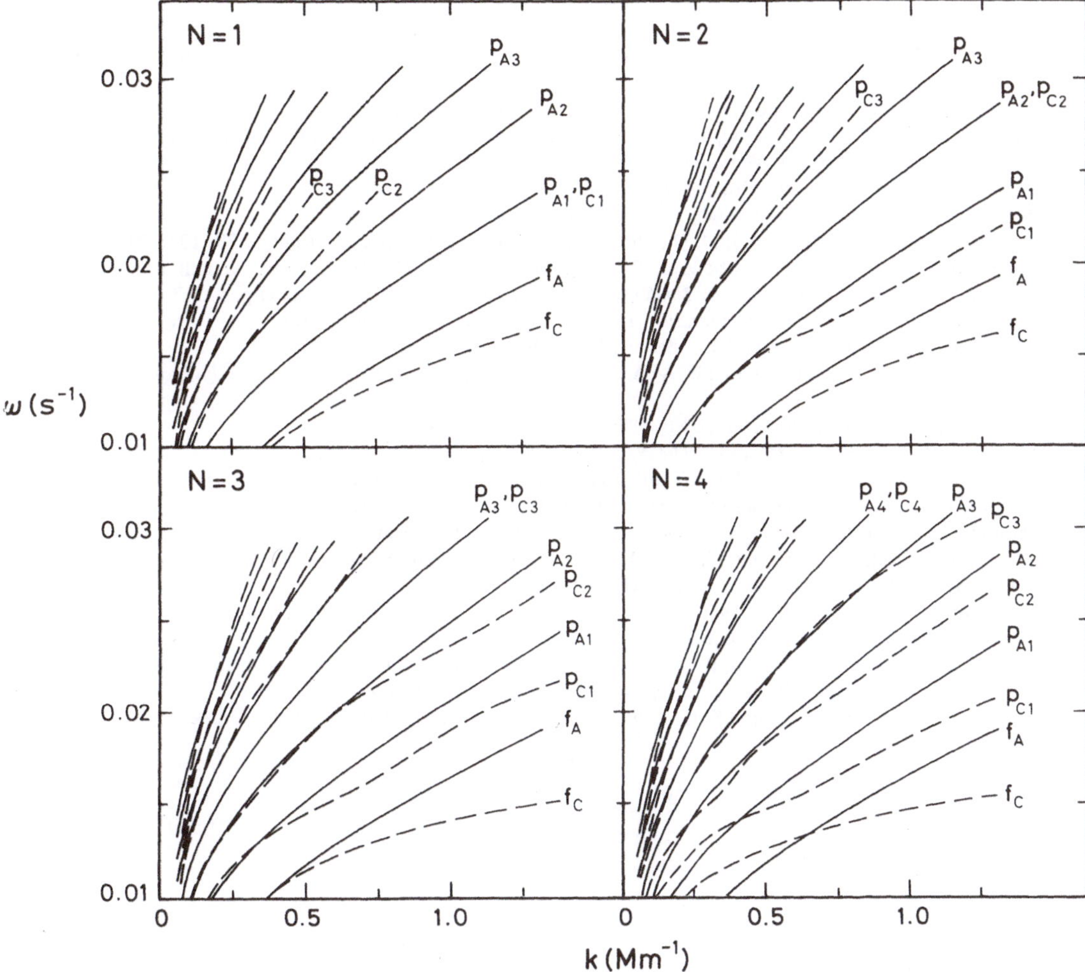

Fig. 3. Diagnostic diagrams for model C (broken curves) computed with the boundary condition (2.4) and the functions $\lambda(\omega)$ represented in Figure 2. The linear eigenfrequencies ($\lambda = 0$) of model A (continuous curves) are included for comparison.

We contend that the evidence we have provided renders it extremely unlikely that nonlinearities in the solar atmosphere can influence the five-minute oscillation eigenfrequencies to a noticeable extent, however much they may distort the eigenfunctions. We expect that this is so also for modes with longer periods, because for these the coupling between the atmosphere and the interior is even weaker that it is for the five-minute modes. Only the high-frequency modes that suffer trapping in the chromosphere are liable to be affected. Therefore we conclude that inferences concerning the internal structure of the Sun based on linear eigenfrequencies are probably correct.

References

Abramowitz, M. and Stegun, I. A.: 1964, *Handbook of Mathematical Functions*, US Gov. Printing Office, Washington, D.C.

Belvedere, G., Paternò, L., and Roxburgh, I.W.: 1980, *Astron. Astrophys.* **91**, 356.

Berthomieu, G., Cooper, A. J., Gough, D. O., Osaki, Y., Provost, J., and Rocca, R.: 1980, in H. A. Hill and W. A. Dziembowski (eds.), *Nonradial and Nonlinear Stellar Pulsation*, Springer, Heidelberg, p. 307.

Christensen-Dalsgaard, J. and Gough, D. O.: 1981a, *Nature* **288**, 544.

Christensen-Dalsgaard, J. and Gough, D. O.: 1981b, *Astron. Astrophys.* **104**, 173.

Christensen-Dalsgaard, J., Gough, D. O., and Morgan, J. G.: 1979, *Astron. Astrophys.* **73**, 121; **79**, 260.

Christensen-Dalsgaard, J., Dziembowski, W. A., and Gough, D. O.: 1980, in H. A. Hill and W. A. Dziembowski (eds.), *Nonradial and Nonlinear Stellar Pulsation*, Springer, Heidelberg, p. 313.

Cox, J. P.: 1980, *Theory of Stellar Pulsation*, Princeton University Press, Princeton, New Jersey.

Deubner, F.-L., Ulrich, R. K., and Rhodes, Jr, E. J.: 1979, *Astron. Astrophys.* **72**, 177.

Dziembowski, W. A. and Pamjatnykh, A. A.: 1978, in J. Rösch (ed.), *Plein Feux sur la Physique Solaire*, CNRS, Paris, p. 135.

Grec, G., Fossat, E., and Pomerantz, M.: 1980, *Nature* **288**, 541.

Hill, H. A.: 1978, in J. A. Eddy (ed.), *The New Solar Physics*, Westview Press, Boulder Colorado, p. 135.

Hill, H. A. and Caudell, T. P.: 1979, *Monthly Notices Roy. Astron. Soc.* **186**, 327.

Hill, H. A., Rosenwald, R. D., and Caudell, T. P.: 1978, *Astrophys. J.* **225**, 304.

Lamb, H.: 1932, *Hydrodynamics*, Cambridge University Press, Cambridge.

Lubow, S. H., Rhodes Jr, E. J., and Ulrich, R. K.: 1980, in H. A. Hill and W. A. Dziembowski (ed.), *Nonradial and Nonlinear Stellar Pulsation*, Springer, Heidelberg, p. 300.

Rosenwald, R. D. and Hill, H. A.: 1980, in H. A. Hill and W. A. Dziembowski (eds.), *Nonradial and Nonlinear Stellar Pulsation*, Springer, Heidelberg, p. 404.

Stebbins, R. T., Hill, H. A., Zanoni, R., and Davis, R. E.: 1980, in H. A. Hill and W. A. Dziembowski (eds.), *Nonradial and Nonlinear Stellar Pulsation*, Springer, Heidelberg, p. 381.

Unno, W., Osaki, Y., Ando, H., and Shibahashi, H.: 1979, *Nonradial Oscillations of Stars*, University of Tokyo Press, Tokyo.

ON WAVES IN NON-ISOTHERMAL, COMPRESSIBLE, IONIZED AND VISCOUS ATMOSPHERES*

L. M. B. C. CAMPOS

Instituto Superior Técnico, 1096 Lisboa, Portugal

Abstract. A review is given of the properties of waves in atmospheres, with particular emphasis on (Section 1) the variation of amplitude and phase with altitude for propagating waves (Figures 1 to 4) and the waveforms of standing modes (Figure 5). The cases dealt with concern waves under the combined influences of gravity and compressibility, and examine the effects of: (Section 2) temperature gradients in a non-isothermal atmospheric model; (Section 3) external magnetic field, either vertical or horizontal; (Section 4) dissipation by viscosity and electrical resistance. The results are relevant to (Section 5) the assessment of atmospheric wave growth and shock formation, and to the calculation of heating functions describing the deposition of wave energy.

1. Approximate and Exact Theories of Atmospheric Waves

The study of waves in atmospheres has been strongly influenced by the basic case (Rayleigh, 1890; Lamb, 1932 § 309) of acoustic-gravity waves in isothermal atmospheres, which exhibit for propagating modes an amplitude growing exponentially on twice the scale height and a phase increasing linearly with altitude. The aim of the present communication is to indicate the extent to which these properties of atmospheric waves are modified by the presence of: (Section 2) temperature gradients; (Section 3) external magnetic fields; (Section 4) viscous or resistive dissipation.

In all these cases the linear wave equations describing small amplitude waves have variable coefficients, due to atmospheric stratification, viz., variation of density, temperature, wave speed or damping with altitude. Taking these coefficients approximately constant leads to the W.K.B.J. approximation (Brekhovskikh, 1961; Moore and Spiegel, 1964; Lighthill, 1967; McLellan and Winterberg, 1968; Yeh and Liu, 1974; Campos, 1982); this approximation leads necessarily to sinusoidal waves, which can be described in terms of dispersion relation, phase speed, group velocity, etc. The W.K.B.J. approximation applies only to high-frequency waves over short distances, i.e., is *invalid* for: (i) wavelengths comparable to or larger than the scale height, i.e., the main part of the wave spectrum in the solar photo- and chromospheres (Bray and Loughhead, 1974; Athay, 1976); (ii) asymptoticaly for all frequencies at large distances compared with the wavelength, e.g., for waves generated in the photosphere and propagating in the high corona.

In order to describe the wave fields for all wavelengths (including those comparable to the scale height), and all distances (including asymptotically), the wave equations must be solved *exactly*, i.e., taking into account the dependence of the coefficients on

* Proceedings of the 66th IAU Colloquium: *Problems in Solar and Stellar Oscillations*, held at the Crimean Astrophysical Observatory, U.S.S.R., 1–5 September, 1981.

Solar Physics **82** (1983) 355–368. 0038–0938/83/0822–0355$02.10.

altitude (Lamb, 1908; Hide, 1956; Yanowitch, 1967; Thorpe, 1968; Zhugzhda, 1971; Nye and Thomas, 1976; Thomas, 1978; Leroy, 1982). Although the frequency is conserved (if atmospheric properties do not depend on time), there is no wavenumber since the waveform is generally not sinusoidal in altitude, and dispersion relation, group velocity, etc... do not exist. The wave equation can be solved exactly for some forms of the variable coefficients, corresponding to specific atmospheric models, for which the wavefields are expressed at all altitudes in terms of special functions, e.g., Bessel or hypergeometric. Besides, it is possible to determine asymptotic properties for a wide range of atmospheric models (Campos, 1983a, b, c).

Thus we are concerned with wave fields which are given: (i) exactly at all altitudes and frequencies for specific (isothermal or non-isothermal) atmospheric models; (ii) asymptoticaly at high altitude, for all frequencies, for any atmosphere with bounded temperature and vanishing density. The asymptotic form of the exact wave field for specific atmospheric models (i) can be used to check the general asymptotic laws for waves (ii).

2. Acoustic-Gravity Waves in a Non-Isothermal Atmosphere

Waves under the combined influences of compressibility and gravity have been studied extensively in connection with the Earth's atmosphere (Yih, 1965; Beer, 1974; Gossard and Hooke, 1975; Lighthill, 1978), besides solar applications (e.g., Biermann, 1948; Mein, 1978, 1980). We consider three-dimensional acoustic-gravity waves in a model atmosphere with the temperature profile

$$T(z) = T_\infty + (T_0 - T_\infty)e^{-\theta z}, \tag{1}$$

where we may choose at will: (i) the initial T_0 and asymptotic T_∞ temperature, and thus the degree of heating or cooling with altitude; (ii) the steepness parameter θ, allowing larger or smaller values of the maximum temperature gradient $\theta(T_\infty - T_0)$ or conductive heat flux. Some of the temperature profiles in the family (1) are illustrated in Figure 1, where we plot the ratio of temperatures $T(z)/T_0$ (or its logarithm) against altitude z (made dimensionless dividing by the asymptotic scale height $L \equiv RT_\infty/g$, where R is the gas constant and g the acceleration of gravity): (top) the steepness parameter is kept fixed at $\theta L = 1$, and the ratio of asymptotic to initial temperature is given four values $T_\infty/T_0 = 0.5, 1, 10, 100$ corresponding to cooling with altitude, the isothermal case, and moderate or intense heating with alititude; (bottom) the ratio of temperatures is kept fixed at $T_\infty/T_0 = 10$, and the steepness parameter is given four values $\theta L = 0.5, 1, 2, 4$ corresponding to an increasingly steeper approach to the asymptotic temperature, and larger maximum temperature gradient.

It can be shown (Campos, 1983a) that the vertical velocity W at altitude z, for a wave of frequency ω and horizontal wavevector $\mathbf{k}$, satisfies

$$L^2(1 - \beta e^{-\theta z})W'' - LW' - \{k^2L^2(1 - \beta e^{-\theta z}) + \\ + (\omega/2\omega_2)^2 + (k\omega_1/\omega)^2\} W = 0, \tag{2}$$

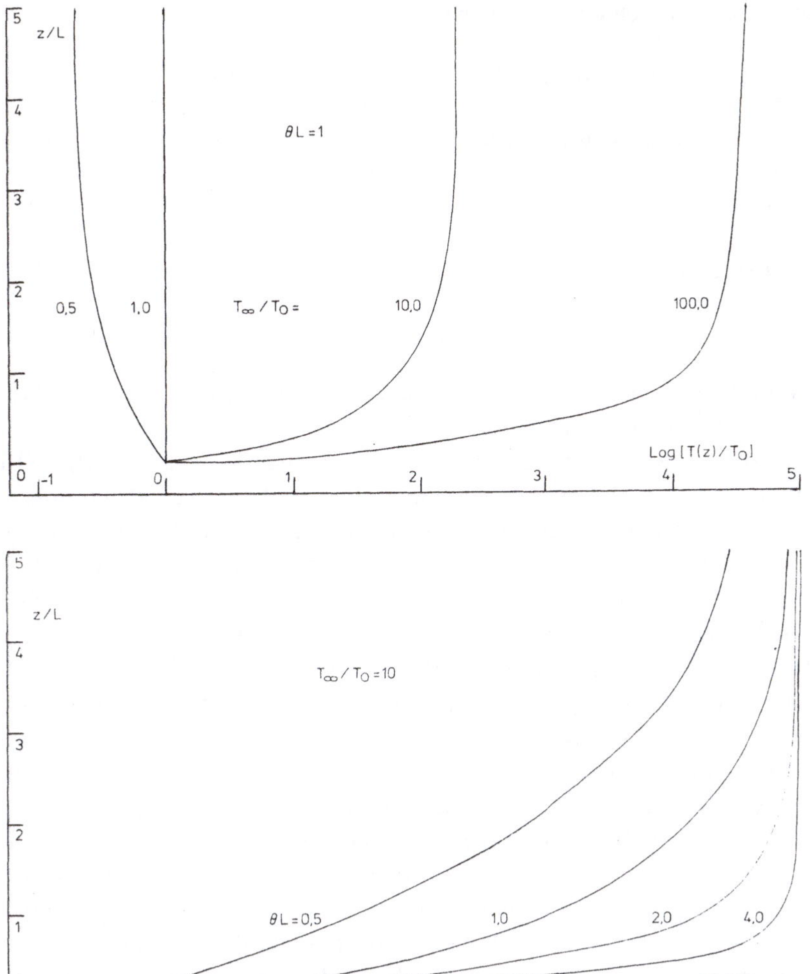

Fig. 1. Temperature profiles for a range of atmospheric models (1) with four ratios of asymptotic to intitial temperature (top) and four values of steepness parameter (bottom).

where $L = RT_\infty/g$ is the asymptotic scale height, $\beta = 1 - T_0/T_\infty$ a measure of the non-isothermality of the temperature profile, and ω_1, ω_2 denote the cut-off frequencies:

$$\omega_1 \equiv (g/c_\infty)\sqrt{(\gamma - 1)}\,, \qquad \omega_2 \equiv c_\infty/2L\,, \tag{3a, b}$$

where $c_\infty \equiv \sqrt{(\gamma R T_\infty)}$ is the asymptotic sound speed and γ the ratio of specific heats. Propagation is only possible below ω_1 (gravity mode) or above ω_2 (acoustic mode), since then the vertical wavenumber

$$K \equiv \{(\omega^2/\omega_2^2 - 1)/4L^2 + k^2(\omega_1^2/\omega^2 - 1)\}^{1/2} \tag{4}$$

is real, and the wave velocity field is given by

$$W(z; \mathbf{k}, \omega) = W(0; \mathbf{k}, \omega) e^{z/2L} e^{iKz} \{E(z; \mathbf{k}, \omega)/E(0; \mathbf{k}, \omega)\} , \tag{5a}$$

where: (i) the first three factors are the initial wave velocity, the exponential amplitude growth and linear phase increase as for a Lamb wave in an isothermal atmosphere; (ii) the factor in curly brackets, which involves the hypergeometric function

$$E(z; \mathbf{k}, \omega) \equiv F(-1/(2\theta L) - k/\theta - iK/\theta, -1/(2\theta L) + k/\theta - iK/\theta; \quad 1 - 2iK/\theta; \beta e^{-\theta z}) , \tag{5b}$$

reduces to unity in the isothermal case $\beta = 0$, and otherwise concentrates all the effects of the non-uniform temperature profile.

It is clear from (5b) that the Lamb's wave (first three factors in (5a)) will be most modified in the region $e^{-\theta z} \sim 1$ of larger temperature gradient. At high altitude, as $z \to \infty$ and $e^{-\theta z} \to 0$, the hypergeometric function in (5b) tends to unity $E(\infty; \mathbf{k}, \omega) = 1$, so that Lamb's wave is regained (5a), with an extra constant factor $\{E(0; \mathbf{k}, \omega)\}^{-1}$, which corresponds to a constant amplitude factor $|E(0; \mathbf{k}, \omega)|^{-1}$ and a phase shift $-\arg\{E(0; \mathbf{k}, \omega)\}$. This is illustrated for vertical waves (with horizontal wavenumber $k = 0$) in Figure 2, where the logarithm of the ratio of the velocity spectrum at altitude z to the initial velocity $V \equiv \log\{W(z; 0, \omega)/W(0; 0, \omega)\}$ is plotted against dimensionless altitude z/L, for four values of the compactness parameter $\varepsilon \equiv kL = 0.5, 1, 2, 5$, corresponding to wavelengths $\lambda \equiv 2\pi L/\varepsilon$ larger than or comparable to the scale height: (left) the real part of V, which is the logarithm of the ratio of amplitudes, shows that waves grow exponentially in the asymptotic regime (straight lines for $z/L \to \infty$), but growth is faster than exponential in the region of larger temperature gradients ($z/L < 2$); (right) the imaginary part of V, which is the difference in phase (or argument) between the wave spectrum $W(z; 0, \omega)$ at altitude z and the initial value $W(0; 0, \omega)$, increases linearly with altitude in the asymptotic regime (like Kz for $z/L \to \infty$), but the phase increases faster in the region of larger temperature gradient. The faster than exponential amplitude growth and greater than linear phase increase with altitude are more noticeable when propagating through positive temperature gradients, for higher frequency waves (smaller kL); negative temperature gradients can be shown (Campos, 1983a) to have the reverse effect.

In general, for any atmosphere (isothermal or not) with bounded asymptotic temperature, acoustic-gravity waves grow exponentially in amplitude and increase linearly in phase in the asymptotic regime; the growth is faster in positive temperature gradients and smaller in negative ones, the effect being more noticeable for higher frequency waves. Thus the amplitude and phase as function of altitude are similar in the asymptotic regime for waves in isothermal or non-isothermal atmospheres with bounded temperature; the effect of temperature gradients at lower altitudes adds up to a constant amplitude factor and phase shift which will be larger for high frequency waves.

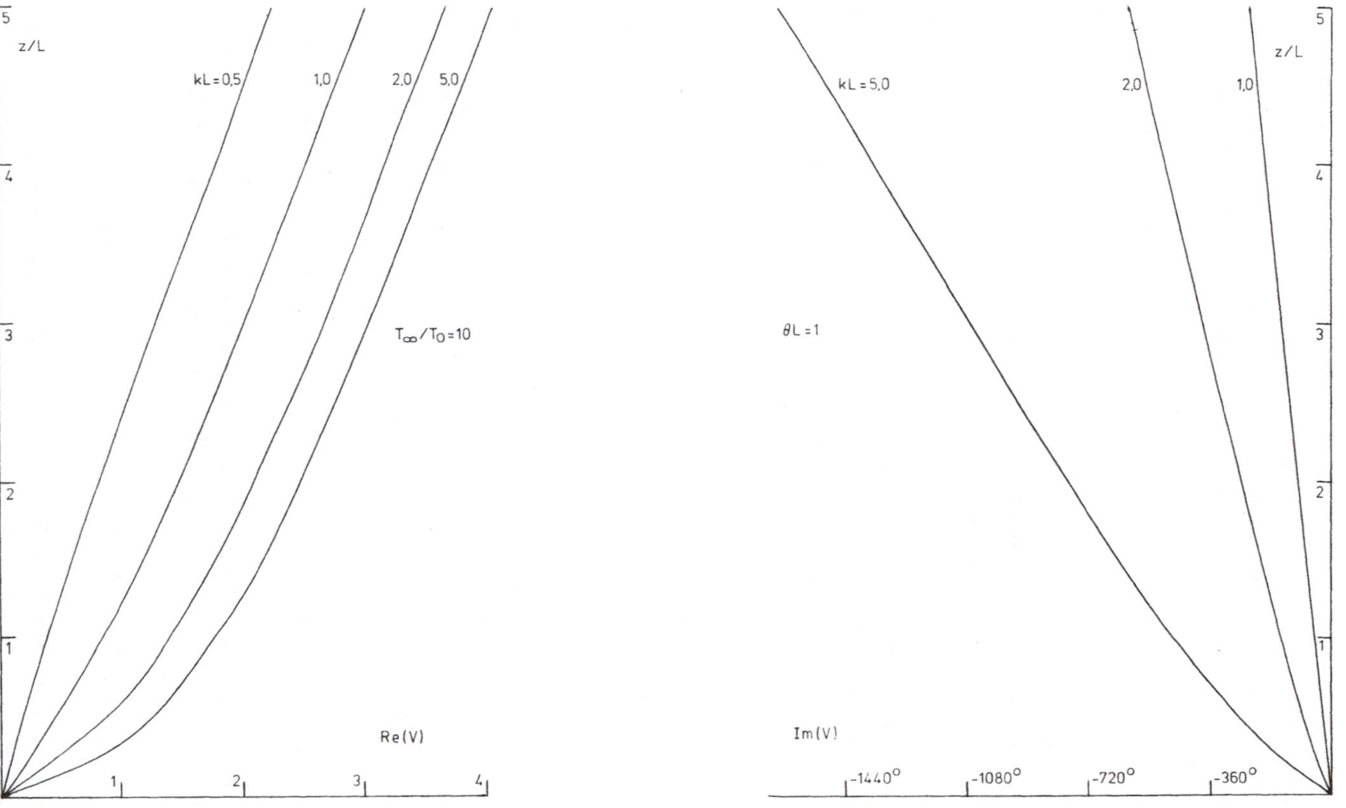

Fig. 2. Acoustic-gravity waves propagating (5a, b) vertically in the non-isothermal atmosphere of Figure 1. The logarithm of the ratio of amplitudes (left) and the phase shift (right) for the velocity perturbation at altitude z relative to initial value at altitude $z = 0$ are plotted against (dimensionless) altitude (divided by the scale height). Four values of the compactness parameter KL are considered.

3. Alfvén-Gravity and Magnetosonic-Gravity Waves

The effectsof an external magnetic field will be exhamined in the case of vertical waves, for which the wave fields depend only on altitude z and time t, but not on horizontal coordinate $\mathbf{x}$ (this is equivalent to vanishing horizontal wavevector $\mathbf{k} = 0$). We consider in the first instance a constant, vertical external magnetic field H_z, in which case (Campos, 1983b) the only propagating components of the velocity $\mathbf{v}$ and magnetic field $\mathbf{h}$ perturbations are horizontal and parallel, viz., along the x-axis, and satisfy

$$\partial^2 v_x/\partial t^2 - C_1^2 \partial^2 v_x/\partial z^2 = 0 = \partial^2 h_x/\partial t^2 - \partial\{C_1^2 \partial h_x/\partial z\}/\partial z\,, \tag{6a, b}$$

where C_1 denotes the Alfvén speed, which depends strongly on altitude, e.g., for an isothermal atmosphere under a constant magnetic field H it increases exponentially on twice the scale height for density:

$$C_1(z) \equiv \mu H/\sqrt{4\,\pi\rho(z)} = \{\mu H/\sqrt{4\,\pi\rho_0}\}\,\exp(z/2L)\,. \tag{7}$$

Thus, whereas in an homogeneous medium the Alfvén speed is constant, and the velocity and magnetic field perturbations satisfy the same equation (6a), in an atmosphere the Alfvén speed depends on altitude and the velocity and magnetic fields obey different laws, viz., (6b) has relative to (6a) an extra term $-2\,C_1\,\mathrm{d}C_1/\mathrm{d}z\;\partial h_x/\partial z$.

In any atmosphere for which the density vanishes at high altitude, viz. $\rho \to 0$ as $z \to \infty$, the Alfvén speed diverges $C_1 \to \infty$, and thus from (6a) $\partial^2 v_x/\partial z^2 \to 0$ and (6b) $\partial h_x/\partial z \to 0$, so that the velocity perturbation grows linearly and the magnetic field perturbation is asymptoticaly constant:

$$v_x(z;\,\omega) \sim \{a(\omega)z + b(\omega)\}\,v_x(0;\,\omega)\,,$$
$$h_x(z;\,\omega) \sim i(H_z/\omega)\,a(\omega)\,v_x(0;\,\omega)\,, \tag{8a, b}$$

for propagating Alfvén-gravity waves of frequency ω (we have used the induction equation $\partial h_x/\partial t + H_z\,\partial v_x/\partial z = 0$ to derive (8b) from (8a)). This result can be checked for an isothermal atmosphere (7), in which case (6a, b) can be solved exactly to yield the velocity and magnetic field perturbations at all altitudes (Campos, 1983b):

$$v_x(z;\,\omega) = v_x(0;\,\omega)\,\{H_0^{(2)}((2\omega L/c_1)e^{-z/2L})/H_0^{(2)}(2\omega L/c_1)\}\,, \tag{9a}$$

$$h_x(z;\,\omega) = i(H_z/c_1)\,v_x(0;\,\omega)e^{-z/2L}\,\{H_1^{(2)}((2\omega L/c_1)e^{-z/2L})/H_0^{(2)}(2\omega L/c_1)\}\,, \tag{9b}$$

where c_1 is the Alfvén speed at altitude $z = 0$, and $v_x(0;\,\omega)$ the initial velocity perturbation for a wave of frequency ω. It can be checked that the asymptotic forms of (9a, b) for large altitude z are (8a, b) with:

$$H_0^{(2)}(2\omega L/c_1) \times \{a(\omega), b(\omega)\} = 2i/\pi L,\ 1 - i2\phi/\pi - i(2/\pi)\log(2\pi L/c_1)\,, \tag{10a, b}$$

where ϕ is Euler's constant, and $H_n^{(2)}$ the Hankel function of second kind order n.

The boundary conditions used to determine the constants of integration in the solution of (6a) are: (i) the initial velocity perturbation $v_x(0;\,\omega)$ for a wave of frequency ω at altitude $z = 0$; (ii) the velocity perturbation at low altitude and high-frequency corresponds to an upward propagating wave, i.e., scales on $\exp\{i\omega(z/c_1 - t)\}$. The wave fields (9a, b) are thus determined from the condition, which appears relevant to the solar case, that we know the velocity perturbation of upward propagating waves of all frequencies at a given level $z = 0$, e.g., in the photosphere. The wave fields (9a, b) are plotted in dimensionless form:

$$V \equiv v_x(z;\,\omega)/v_x(0;\,\omega)\,, \qquad H \equiv c_1 h_x(z;\omega)/H_z v_x(0;\omega)\,, \tag{11a, b}$$

against dimensionless altitude z/L in Figures 3 and 4, the former for amplitudes (or modulus of (11a, b)) and the latter for phase shifts (arguments of (11a, b)). Four values of the compactness parameter $\varepsilon \equiv \omega L/c_1$ are considered, corresponding to wavelengths λ larger than or comparable to the scale height, and density changes by a factor of $\Delta = \exp(\lambda/L) = \exp(2\pi/\varepsilon)$ within a wavelength that are large to moderate. It is clear from Figures 3 and 4 that asymptoticaly as $z \to \infty$: (i) the velocity perturbation grows linearly, faster for higher frequencies; (ii) the magnetic field perturbation is constant asymptoticaly, and smaller for higher frequencies; (iii) the phase difference between the wave fields at altitude z and 0 is finite asymptoticaly as $z \to \infty$, and generally larger for high frequency waves.

Besides propagating waves, which can transport energy from one atmospheric region to another, we also consider standing modes, which appear as atmospheric oscillations. The boundary conditions for a wave trapped between the atmospheric layers $z = 0$ and $z = a$ are $v_x(0;\,\omega) = 0 = v_x(a;\,\omega)$; as $a \to \infty$ we find that the wave reflected from infinity cannot give a zero but only a finite amplitude, i.e., a 'node at infinity' corresponds to a finite, non-zero velocity perturbation. The solution of (6a) which vanishes at $z = 0$ and is finite at infinity corresponds to standing modes with frequencies and wavelengths:

$$\omega_n = c_1 j_n/2L\,, \qquad \lambda_n = 4\pi L/j_n\,, \tag{12a, b}$$

where j_n are the roots of the Bessel function $J_0(j_n) = 0$. The wave fields for the nth mode are given by the velocity and magnetic field perturbations:

$$v_x(z,t) = (\pi c_1/2L)\,\mathrm{Im}\,\{v_0(\omega_n)e^{-i\omega_n t}\}\,\{J_0(j_n e^{-z/2L})/J_1(j_n)\}\,, \tag{13a}$$

$$h_x(z,t) = (\pi H_z/2L)\,\mathrm{Re}\,\{v_0(\omega_n)e^{-i\omega_n t}\}\,\{J_1(j_n e^{-z/2L})/J_1(j_n)\}\,, \tag{13b}$$

where $v_0(\omega)$ is the spectrum at altitude $z = 0$ and frequency ω. The first four standing modes of vertical Alfvén-gravity waves in an isothermal atmosphere are illustrated in Figure 5, showing that asymptoticaly: (i) the velocity perturbation is finite but non-zero, increasing with the order of the mode; (ii) the magnetic field perturbation decays exponentially to zero.

Having considered two magneto-acoustic-gravity wave modes, namely, an acoustic-gravity wave (Section 2) and an Alfvén-gravity wave (Section 3), we now turn to a mode coupling compressibility and magnetism (besides gravity). This is a vertical magneto-sonic-gravity wave, corresponding to a constant, horizontal external magnetic field H_x, which propagates an horizontal, parallel magnetic field perturbation h_x and a vertical

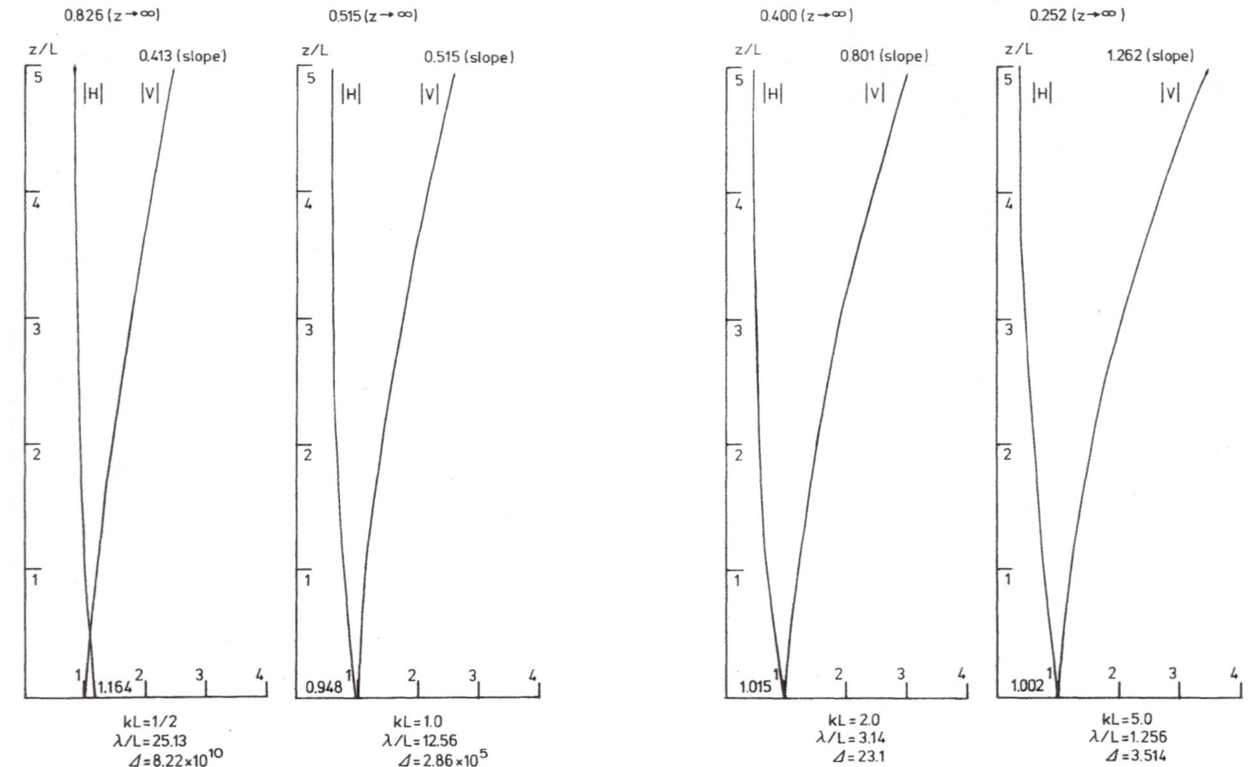

Fig. 3. Alfvén-gravity waves propagating vertically (9a, b) in an isothermal atmosphere: dimensionless (11a, b) amplitude of velocity (V) and magnetic field (H) perturbations versus dimensionless altitude, for four values of compactness parameter kL (or wavelength λ or density change Δ within a wavelength).

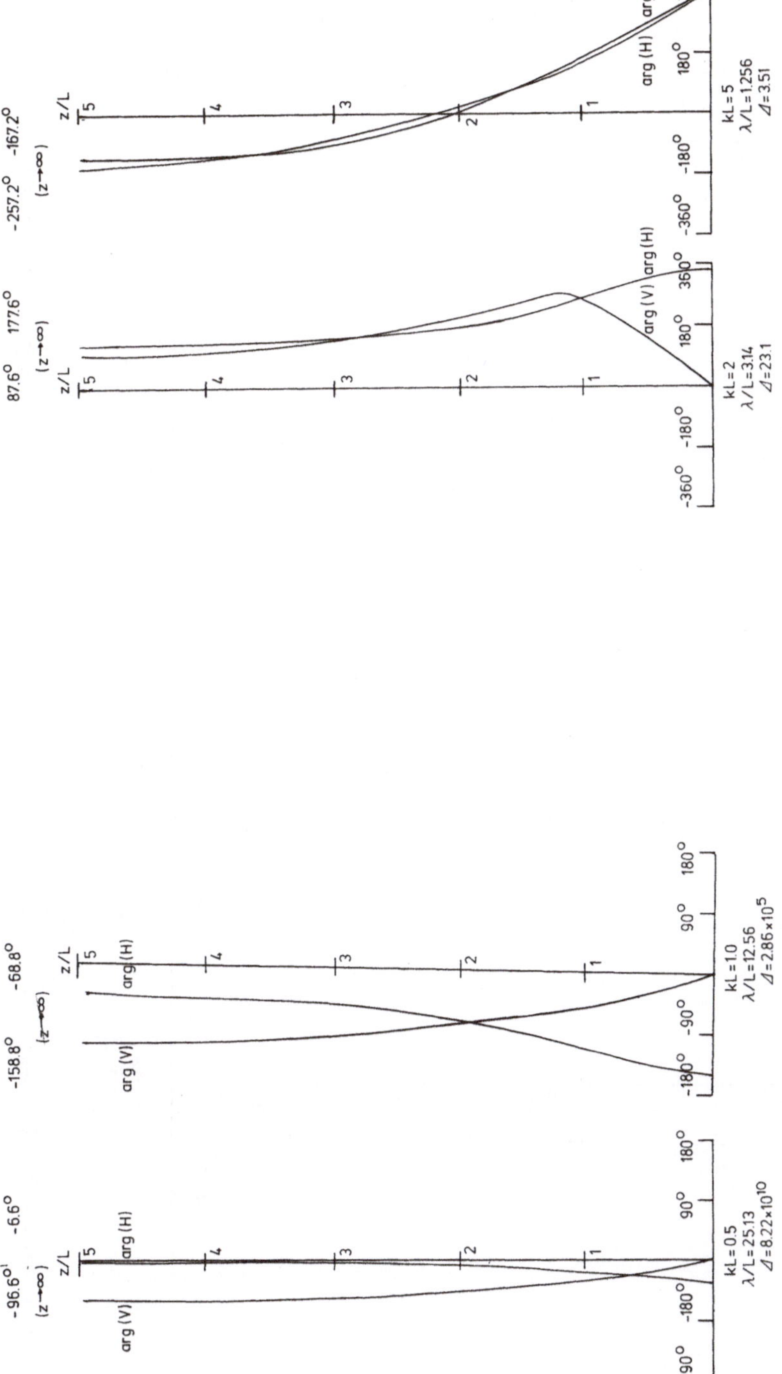

Fig. 4. As for Figure 3: difference in phases.

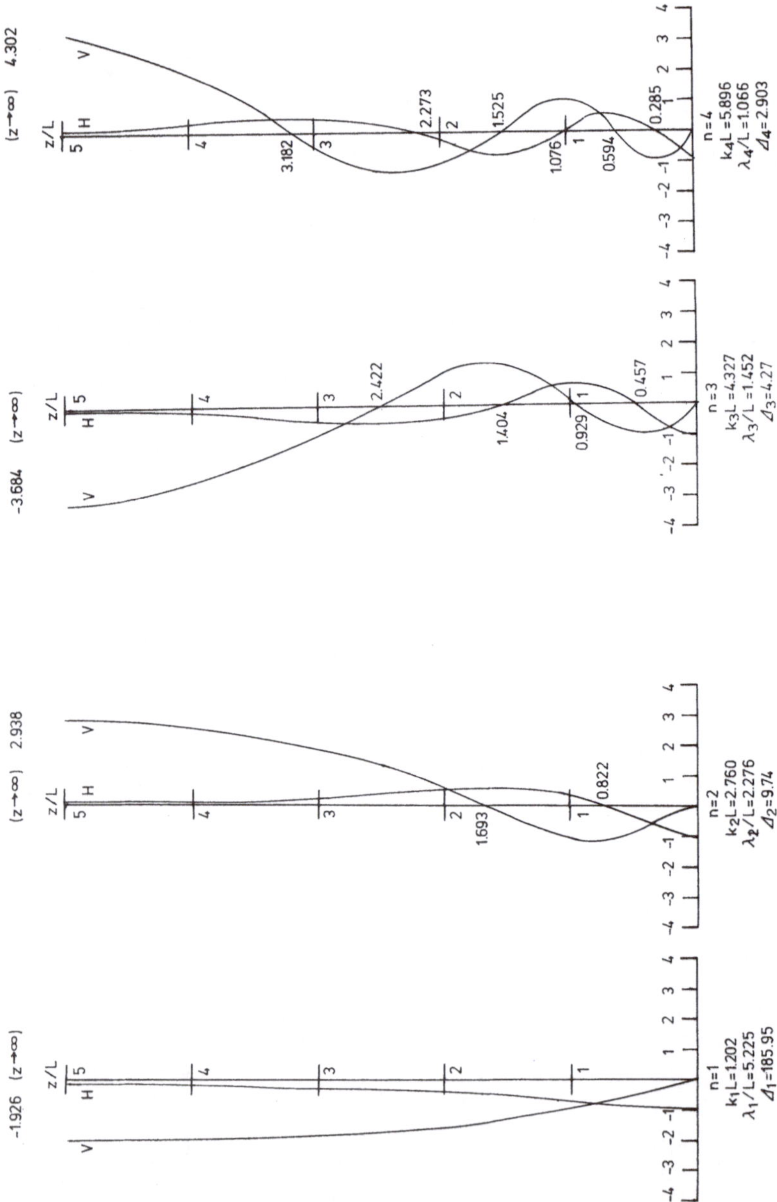

Fig. 5. Velocity (V) and magnetic field (H) waveforms for the first four standing modes (13a, b) of Alfvén-gravity waves in an isothermal atmosphere.

velocity perturbation v_z. The wave equations

$$\partial^2 v_z/\partial t^2 - (C_0^2 + C_1^2)\partial^2 v_z/\partial z^2 + \gamma g\, \partial v_z/\partial z = 0\,, \tag{14a}$$

$$\partial^2 h_x/\partial t^2 - \partial\{(C_0^2 + C_1^2)\partial h_x/\partial z\}/\partial z + \gamma g\, \partial h_x/\partial z = 0\,, \tag{14b}$$

where the sound $C_0(z)$ and Alfvén $C_1(z)$ speeds generally depend on altitude, can be used to derive asymptotic laws, and can be integrated exactly in terms of hypergeometric functions in the isothermal case. It can be shown (Campos, 1983b) that there exist two altitude ranges, separated by a transition layer, defined by the equality of sound and Alfvén speeds $C_0(z_*) = C_1(z_*)$: (i) below the transition layer the gas pressure predominates over the magnetic pressure, and the magnetosonic-gravity waves ressemble acoustic-gravity waves, which apart from modifications due to the magnetic field, exhibit exponential amplitude growth, whether standing or propagating; (ii) above the transition layer the magnetic pressure predominates, and the magnetosonic-gravity waves ressembles, apart from modifications due to compressibility, an Alfvén-gravity wave, exhibiting: (a) asymptoticaly finite velocity and decaying magnetic field for standing modes; (b) linearly diverging velocity and asymptoticaly constant magnetic field for propagating waves.

4. Dissipation by Viscosity or Electrical Resistance

As a first instance of dissipation of atmospheric waves we consider the effect of viscosity on a vertical acoustic-gravity wave. The velocity perturbation is vertical and satisfies the equation (Campos, 1983c)

$$\partial^2 v_z/\partial t^2 - C_0^2\, \partial^2 v_z/\partial z^2 + \gamma g\, \partial v_z/\partial z = \nu\, \partial^3 v_z/\partial z^2\, \partial t\,, \tag{15}$$

where the sound speed C_0 and kinematic viscosity ν depend generally on altitude. In an isothermal atmosphere the former is constant, and if the static viscosity is also constant, the kinematic viscosity varies inversely with density, i.e., grows exponentially with altitude, and (15) can be solved in terms of hypergeometric functions. The appearance of hypergeometric solutions once more (they also apply to magnetosonic-gravity waves) is associated with the fact that these are the simplest special functions with three singularities: (i) at $z = 0$, corresponding to the initial wave field; (ii) at $z = \infty$, corresponding to the asymptotic regime; (iii) at a transition or reflecting layer $z = z_*$ which separates two altitude ranges. In the case of viscous acoustic-gravity waves the reflecting layer is determined by the condition $\omega C_0(z) = \nu(z)$, so that it depends on frequency, and below it waves grow exponentially whereas above they tend to a finite asymptotic value.

For any atmosphere, isothermal or not, with asymptoticaly vanishing density $\rho \to 0$ for $z \to \infty$, the kinematic viscosity diverges $\nu \to \infty$ (if the static viscosity does not vanish), and it would appear from (15) that $\partial^2 v_z/\partial z^2 \to 0$, so that the velocity perturbation would grow linearly. However, this would correspond to $\partial v_z/\partial z$ asymptoticaly constant, so that the rate of dissipation of energy by viscosity, which scales on $(\partial v_z/\partial z)^2$ would be infinite when integrated over an atmospheric column from $z = 0$ to $z = \infty$, which is physically absurd since energy can be supplied to the waves only at finite rate.

The condition of finite dissipation rate (Yanowitch, 1967; Lyons and Yanowitch, 1974)

$$\int_0^\infty |\partial v_z(z;\omega)/\partial z|^2 \, dz < \infty , \tag{16}$$

implies that the velocity perturbation is asymptoticaly bounded for standing modes and propagating waves.

Having mentioned viscous dissipation of an hydrodynamic wave, we now turn to Ohmic resistive dissipation of Alfvén-gravity waves. In this case (16) is replaced by a condition of finite rate of dissipation by Joule effect, i.e., it holds with $\partial v_z(z;\omega)/\partial z$ replaced by the electric current $j(z;\omega)$ which scales on $\partial h_x(z;\omega)/\partial z$, where h_x is the magnetic field perturbation. The velocity v_x and the magnetic field h_x perturbations satisfy (Campos, 1983c):

$$\partial^2 v_x/\partial t^2 - C_1^2 \partial^2 v_x/\partial z^2 - C_1^2 \partial\{\chi \, \partial(C_1^{-2} \partial v_x/\partial t)/\partial z\}/\partial z = 0 , \tag{17a}$$

$$\partial^2 h_x/\partial t^2 - \partial(C_1^2 \partial h_x/\partial z)/\partial z - \chi \, \partial^3 h_x/\partial t \, \partial z^2 = 0 , \tag{17b}$$

which reduces to (6a, b) when the electrical diffusivity vanishes $\chi = 0$. The electrical diffusivity χ is approximately independent of density, and thus can be taken as a constant in an isothermal atmosphere, in which case the Alfvén speed is given by (7). The exact solution of (17a, b) appears again in terms of hypergeometric functions, so that there is a transition layer, specified by $\omega C_1(z) = \chi(z)$, separating: (i) a low altitude region where diffusion predominates, and the velocity and magnetic fields have a wavenumber k given by $k^2 = \omega/2\chi$; (ii) a high-altitude region where propagation predominates and the amplitude and phase laws are similar to those for non-dissipative hydromagnetic waves (Section 3).

It may be concluded that exponential amplitude growth is the exception rather than the rule for atmospheric waves, since it applies only to acoustic-gravity waves (standing or propagating), and is modified: (i) in the presence of an external magnetic field, with or without resistance, to linear growth for propagating waves and asymptoticaly finite,

TABLE I

Asymptotic laws[2] for the amplitude[1] of atmosphere waves

Type	Standing mode	Propagating wave
Acoustic-gravity	Exponential growth	Exponential growth
Alfvén-gravity	Finite, non-zero[3]	Linear growth[4]
Magnetosonic-gravity	Finite, non-zero[3]	Linear growth[4]
Viscous acoustic-gravity	Finite, non-zero	Finite, non-zero
Resistive Alfvén-gravity	Finite, non-zero[3]	Linear growth[4]

[1] The phase shift for propagating waves is linear in the acoustic-gravity case and asymptoticaly finite and non-zero in all other cases.
[2] These asymptotic laws apply to the velocity perturbation.
[3] The magnetic field perturbation decays exponentially.
[4] The magnetic field perturbation is asymptoticaly finite, non-zero.

non-zero amplitude for standing modes; (ii) in the presence of dissipation by viscosity to finite asymptotic amplitude for standing modes or propagating waves. The magnetic field perturbation (for magnetic and coupled modes) is asymptoticaly finite for propagating waves and decaying for standing modes. These results are summarized in the Table I, and apply to waves in isothermal and non-isothermal atmospheres with asymptoticaly bounded temperature and vanishing density. A consequence for all atmospheric magnetic or coupled modes is: (i) the kinetic energy $E = \rho v^2/2$ (per unit volume) tends asymptoticaly to zero, since the density decays (exponentially) faster than the square of velocity may grow (algebraically); (ii) the magnetic energy $G = \mu h^2/8\pi$ (per unit volume) decays to zero for some standing modes, but for propagating waves is asymptoticaly bounded. Thus a magnetic or coupled (magneto-acoustic) wave propagating upward in an atmosphere violates the equipartition of energy which holds in an homogeneous medium, since asymptoticaly all energy is magnetic.

5. Shock Fomation and Atmospheric Heating

As an indication of the relevance of these results to solar physics we recall that it is generally accepted that waves are generated in the solar photosphere; these consist (Campos, 1977) of three magneto-acoustic-gravity wave modes, and it is often argued that they grow into shocks during propagation upward into the chromosphere. If waves grow exponentially, then, even if the initial amplitude is small, after a few scale heights, the non-linear effects which lead to shock formation come into play. It should be borne in mind, however, that in the solar atmosphere both magnetic fields and dissipation mechanisms are present, and thus waves may grow only linearly or have bounded amplitude. In these cases it is necessary to calculate, for the particular wave mode and in the physical conditions considered, whether waves grow enough for shocks to form within the height range of interest or: (i) shock formation is delayed by the magnetic field until higher altitude; (ii) the dissipation mechanisms limit amplitude to a level at which non-linear effects are small and shocks do not form.

We have concerned ourselves with the properties of waves in non-isothermal, compressible, viscous and ionized atmospheres, and leave the detailed application to solar phenomena (Osterbrock, 1961; Uchida, 1968; Meyer, 1968; Hollweg, 1972; Mein and Mein, 1980) to subsequent works. We do however mention another consequence of these results, concerning the calculation of atmospheric heating functions, specifying the rate of deposition of wave energy with altitude. An accurate law of variation of wave amplitude and phase with altitude is necessary for the computation of the heating function, which in turn is critical for the establishment of solar atmospheric models. In this respect the W.K.B.J. approximation is of limited use, since it assumes sinusoidal waveforms and linear phases, and does not apply to wavelengths comparable to or larger that the scale height nor does it hold after several scale heights. Instead an exact theory should be used, applying to all frequencies and altitudes, yielding amplitude and phase laws from which the heating functions can be calculated as input to solar atmospheric models.

References

Athay, R. G.: 1976, *The Quiet Sun*, D. Reidel Publ. Co., Dordrecht, Holland.

Beer, T.: 1974, *Atmospheric Waves*, John Wiley.

Biermann, L.: 1948, *Z. Astrophys.* **25**, 161.

Brekhovskikh, L. M.: 1961, *Waves in Layered Media*, Academic Press, 2nd edition, New York, 1980.

Bray, R. J. and Loughhead, R. E.: 1974, *The Solar Chromosphere*, Chapman and Hall, London.

Campos, L. M. B. C.: 1977, *J. Fluid Mech* **81**, 529.

Campos, L. M. B. C.: 1982, 'On Anisotropic, Dispersive and Dissipative Wave Equations', to appear in *Port. Math.*

Campos, L. M. B. C.: 1983a, 'On Three-Dimensional Acoustic-Gravity Waves in Model Non-Isothermal Atmospheres', to appear in *Wave Motion* **5**.

Campos, L. M. B. C.: 1983b, 'On Magneto-Acoustic-Gravity Waves Propagating or Standing Vertically in an Atmosphere', to appear in *J. Phys. A*.

Campos, L. M. B. C.: 1983c, 'On Viscous and Resistive Dissipation of Atmospheric Waves', in preparation.

Gossard, E. and Hooke, W.: 1975, *Waves in the Atmosphere*, Elsevier, Amsterdam.

Hide, R.: 1956, *Proc. Roy. Soc.* **A223**, 276.

Hollweg, J. V.: 1972, *Cosmic Electr.* **2**, 423.

Lamb, H.: 1908, *Proc. Lond. Math. Soc.* **7**, 122

Lamb, H.: 1932, *Hydrodynamics*, Cambridge Univ. Press, London.

Leroy, B.: 1982, *Astron. Astrophys.* **97**, 245.

Lighthill, M. J.: 1967, *IAU Symp.* **28**, 429.

Lighthill, M. J.: 1978, *Waves in Fluids*, Cambridge Univ. Press, London.

Lyons, P. and Yanowitch, M.: 1974, *J. Fluid Mech.* **66**, 273.

Mein, N.: 1978, *Solar Phys.* **59**, 3.

Mein, N. and Mein, P.: 1980, *Astron. Astrophys.* **84**, 96.

Meyer, F.: 1968, in K. O. Kiepenheuer (ed.), 'Structure and Development of Solar Active Regions', *IAU Symp.* **35**, 487.

Moore, D. W. and Spiegel, E. A.: 1964, *Astrophys. J.* **139**, 48.

Nye, A. H. and Thomas, J.: 1976, *Astrophys. J.* **204**, 573.

Osterbrock, D. E.: 1961, *Astrophys. J.* **134**, 347.

Rayleigh, J. W. S.: 1890, *Phil. Mag.* **29**, 173.

Thomas, J.: 1976, *IAU Colloq.* **36**, 134.

Thorpe, S.: 1968, *Phil. Trans. Roy. Soc.* **263**, 563.

Uchida, Y.: 1968, *Solar Phys.* **4**, 30.

Yeh, C. S. and Liu, C. H.: 1974, *Rev. Geophys. Space Sci.* **12**, 193.

Yih, C. S.: 1965, *Non-Homogeneous Fluids*, MacMillan, New York.

Yanowitch, M.: 1967, *Can. J. Phys.* **45**, 2003.

Zhughzda, Y. D.: 1971, *Cosmic Elect.* **2**, 267.

SEISMOLOGY OF SUNSPOT ATMOSPHERES*

Y. D. ŽUGŽDA

Institute of Terrestrial Magnetism, Ionosphere and Radio Wave Propagation (IZMIRAN) of the USSR Academy of Sciences, USSR 142092 Moscow, U.S.S.R.

V. LOCĀNS

Radioastrophysical Observatory of the Latvian Academy of Sciences, 226524 Riga, U.S.S.R.

and

J. STAUDE

Zentralinstitut für solar-terrestrische Physik, Sonnenobservatorium Einsteinturm, DDR 1500 Potsdam, G.D.R.

Abstract. The present work deals with the theory of oscillations with periods of about 3 min observed in the chromosphere above sunspot umbrae. The model of these oscillations (slow mode magneto-acoustic waves trapped in a chromospheric resonant cavity) provides an independent method of checking empirical models of the chromosphere above sunspots.

Making use of this method, we investigate sunspot models which have been derived from spectroscopic data; the calculated periods of the oscillations fit well the observed periods.

1. Introduction

Empirical models of the thermodynamic structure of sunspot umbrae seem to be well established at photospheric levels. For lack of reliable observations and difficulties in their interpretation uncertainties arise, however, at larger heights starting from the temperature minimum ($T_{\min}$). Recently, the situation improved because EUV sunspot observations with high spatial and spectroscopic resolutions became available, including hydrogen Lα line contours from HRTS (Basri *et al.*, 1979) and OSO-8 (Kneer *et al.*, 1981); and a unified working model of sunspot structure from the subphotosphere (that is, the upper part of the convective zone) up to the base of the transition layer between chromosphere and corona has been derived using these data (Staude, 1981). Such models cannot be defined unambiguously, however, as long as the EUV data are available only for a small number of sunspots and spectral lines.

Oscillations in velocity and brightness observed at photospheric and chromospheric layers of sunspots could provide an independent method of investigating the atmospheric structure. In a recent paper, Žugžda and Locāns (1981) proposed such a type of sunspot seismology assuming a model for a chromospheric resonant cavity which is forced from below by acoustic noise producted in the convective zone.

In the following section we shall summarize observations and different efforts towards its interpretation. The subsequent sections will give a description of our models

* Proceedings of the 66th IAU Colloquium: *Problems in Solar and Stellar Oscillations*, held at the Crimean Astrophysical Observatory, U.S.S.R., 1–5 September, 1981.

Solar Physics **82** (1983) 369–378. 0038–0938/83/0822–0369$01.50.

of sunspot structure and chromospheric oscillations, followed by a discussion of the results and conclusions for further work.

2. Observations and Suggestions for the Interpretation

Intensity and velocity oscillations in chromospheric lines of sunspot umbrae are observed with periods P between about 2 and 3 min. Beckers and Tallant (1969) first discovered oscillatory-type phenomena in the Ca II H and K lines and the infrared triplet, which were called 'umbral flashes'. Havnes (1970) suggested that such flashes could be produced by compression waves. Later umbral velocity oscillations were also observed in Hα and other chromospheric lines, e.g. by Giovanelli (1972) and Giovanelli *et al.* (1978).

At photospheric levels of umbrae, oscillations occur in a much broader range of P between 2 and 8 min. Most of these observations concern velocity oscillations, but there are also reports on oscillations of the magnetic field vector (Mogilevsky *et al.*, 1973). A clear correlation with the chromospheric oscillations does not exist (Beckers and Schultz, 1972; Bhatnagar *et al.*, 1972). In the present paper we shall mainly deal with the chromospheric oscillations.

Recently umbral oscillations in velocity and partly also in brightness have been discovered, even in the transition region at a temperature of $T \approx 10^5$ K. These oscillations observed on the SMM spacecraft (Gurman *et al.*, 1982; Tandberg-Hanssen *et al.*, 1981) show periods of $129 \lesssim P \lesssim 173$ s similar to those in the chromosphere; they seem to represent upward propagating compression waves. Clear correlations between the parameters P, oscillation amplitude, umbral area, and the magnetic field **B** do not exist.

In theoretical efforts to explain the umbral oscillations, most authors looked for the resonant response of the umbral atmosphere to forcing from below. Different wave modes have already been considered:

Uchida and Sakurai (1975) assumed Alfvén waves already being downward reflected at the layers around T_{min} by the strong increase of the Alfvén speed v_A with increasing height and decreasing mass density ρ (see Figures 3 and 4). However, there is no effective upward reflection of downward propagating waves from the subphotosphere (Thomas, 1978; Žugžda and Locāns, 1980, 1982), hence, a resonant cavity cannot form.

'Fast' mode magneto-atmospheric waves were considered by Antia and Chitre (1979) as well as by Scheuer and Thomas (1981), and Thomas and Scheuer (1982). Downward reflection of these waves can occur due to the increase of v_A, as in the case of the Alfvén mode, but now also an upward reflection from the subphotosphere is possible due to the increase of the sound speed c_S, and a photospheric resonant cavity will form. This mode could explain the umbral oscillations observed in the photosphere. At higher levels, a pure acoustic wave will propagate along the magnetic field; its energy is much smaller than that of the mainly horizontal oscillations in the photosphere, but the amplitude is large enough to explain observed chromospheric oscillations. However, in this model the oscillations in the chromosphere should show a clear correlation to those in the photosphere which contradicts existing observations. Moreover, of course, the model cannot explain the oscillations observed in the transition region if horizontal and vertical motions are assumed to be zero there.

Žugžda and Locāns (1981) proposed a model for slow-mode magneto-acoustic waves which will be described in more detail in the following section. In the limiting case of a strong magnetic field, hence $v_A^2 \gg c_S^2$, the mode degenerates to a sound wave travelling along the magnetic field. Downward reflection occurs at the transition region where c_S rapidly increases, while a lower boundary exists at T_{min}: a chromospheric resonant cavity is formed due to the reflection of propagating waves at the frequency $\omega_0 = \gamma g/2c_S$, where $\gamma = c_P/c_V$ is the ratio of specific heats and g is the gravity. For frequencies $\omega < \omega_0$ the transmission of the layers around T_{min} is strongly reduced, while for $\omega > \omega_0$ we have only little reflection and therefore a lot of leakage from the cavity above T_{min}.

Hollweg and Roberts (1981) studied the resonance of 'tube waves' in a strong magnetic field decreasing with height. Till now only a rough model for the atmosphere above the sunspot was assumed; the consideration of a sufficiently shallow sunspot chromosphere would hardly result in a strong decrease of the magnetic field with height. The model is likely to fit magnetic tubes at the supergranular boundaries rather than sunspots.

Most of the hitherto existing models used very simple and schematic assumptions concerning the atmospheric structure in the umbra, e.g., a constant $\beta = -\partial T/\partial z$ (z is the vertical coordinate increasing upward, i.e., in the opposite direction to the gravity $\mathbf{g}$) or an isothermal layer limited by a subphotosphere with constant gradient β and an infinite gradient at the transition region were assumed. However, a detailed comparison with observations requires the use of a more realistic model of the umbral atmosphere which will be described in the following section.

3. Physical Assumptions and Methods of the Analysis

3.1. Models for the Atmospheric Structure of Umbrae

Details of our methods to derive a sunspot working model were published in a recent paper (Staude, 1981), therefore only a short summary will be given here. From the literature we depended as much as possible upon observed spectroscopic data and as little as necessary upon model parameters. The set of model parameters was completed by our own procedures to obtain a rather self-consistent model with uniform assumptions concerning the chemical composition, the methods to solve the equations of state, of hydrostatic equilibrium, of deviations from LTE, to calculate the absorption coefficients, the conversion of different height scales, etc.

The procedures are partly based on the LINEAR code by Auer *et al.* (1972), hence the states H, H^-, H^+, H_2, H_2^+ are considered for hydrogen, moreover, He and nine other elements are taken into account with two states of ionization, and the ground state of H is permitted to depart from LTE. All thermodynamic quantities for the analysis in the following subsection were calculated using the procedures mentioned above; this concerns, e.g., the effective molecular weight μ, γ, ρ, c_S, and v_A.

The resulting photospheric part of our model 1 is similar to that of Stellmacher and Wiehr (1975), the lower chromosphere for $T < 11000$ K is similar to the model of Teplitskaya *et al.* (1978), while the further extrapolation up to $T = 40000$ K was carried out to explain EUV observations such as the Lα data mentioned in Section 1. In addition to this sunspot model 1 (Staude, 1981) two sunspot models with steeper gradients β, and therefore larger densities in the chromosphere, were calculated for comparison: the extent of the chromosphere between $T_{\rm min} = 3000$ K and $T = 22000$ K is 1400 km for model 1, but 1050 and 700 km for models 2 and 3, respectively. At a temperature of $T = 20000$ K where the Lα line core is formed, the related electron densities n_e are 4.9×10^{10}, 1.9×10^{11}, and 7.8×10^{11} cm^{-3} for spots 1, 2, and 3, respectively. For comparison, models of the mean undisturbed solar atmosphere and a plage were derived from the data of Basri *et al.* (1979); at 20000 K the corresponding n_e values are 2.2×10^{10} and 4.0×10^{11} cm^{-3}. These values should be compared with observed data, e.g., with the contrast φ, that is the intensity ratio in the Lα line core between spot and undisturbed Sun: Kneer *et al.* (1981) measured $\varphi \approx 3$, while Basri *et al.* (1979) obtained $1.7 \leq \varphi \leq 6.9$ and an average value of 2.5. Following Kneer *et al.* (1981) we roughly estimate φ from the n_e ratios of the models assuming the limiting cases of optically thick and thin radiations: φ values between 1.5 and 5, 3 and 80, and 6 and 1300 were obtained for spots 1, 2, and 3, respectively. Hence, all three models are compatible with the observed data in the physically more probable case of optically thick radiation. Considering further spectroscopic data, however, spot model 1 as derived by Staude (1981) has evidently the most realistic structure while spot 3 seems unlikely. Figures 1 to 3 show the z-dependence of T, ρ, and n_e for the spot models 1 and 2 as well as for the quiet Sun and a plage for comparison. Figure 4 gives c_S and $v_{\rm A}$ for spots 1 and 2 , and Figure 6 includes $T(z)$ for the chromospheres of all three spot models.

3.2. Model for the oscillations

The model of a chromospheric resonant cavity for slow mode magneto-acoustic waves was proposed by Žugžda and Locāns (1981) and outlined already in Section 2 above. The theory is based on an analytic solution of the equations of ideal magnetohydrodynamics derived by Syrovatsky and Žugžda (1967) with the aim of studying oscillatory convection in a sunspot. Additional assumptions are: a compressible stratified medium with constant values of β, γ, $\mathbf{B}$, and $\mathbf{g}$, parallel directions of magnetic field and gravity, and a strong magnetic field satisfying $c_S^2 \ll v_{\rm A}^2$. Linearizing the equations about the basic state gives the following solution for the vertical velocity:

$$v_z = \left(\frac{\rho}{\rho_0}\right)^{-1/2} Z_{\pm n}(\xi) \exp(-i\omega t + ik_\perp r_\perp),$$

$$\xi = \frac{2\omega}{|\beta|}\left(\frac{\mu T}{\gamma R}\right)^{1/2}, \qquad n = \frac{\mu g}{\beta R} - 1, \qquad \omega = 2\pi/P, \tag{1}$$

where $Z_{\pm n}$ are the general solutions of Bessel's equation, $k_\perp$ and $r_\perp$ are the horizontal wave and radius vectors, respectively, and R is the gas constant.

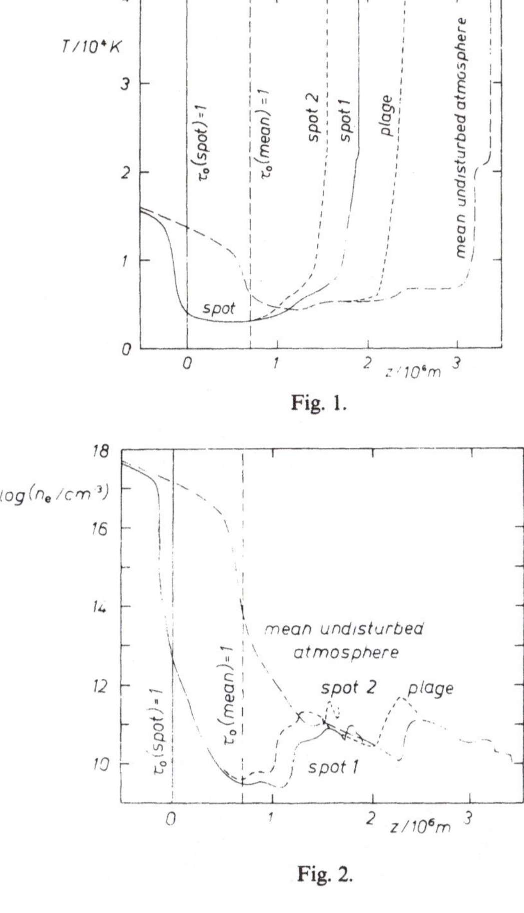

Fig. 1.

Fig. 2.

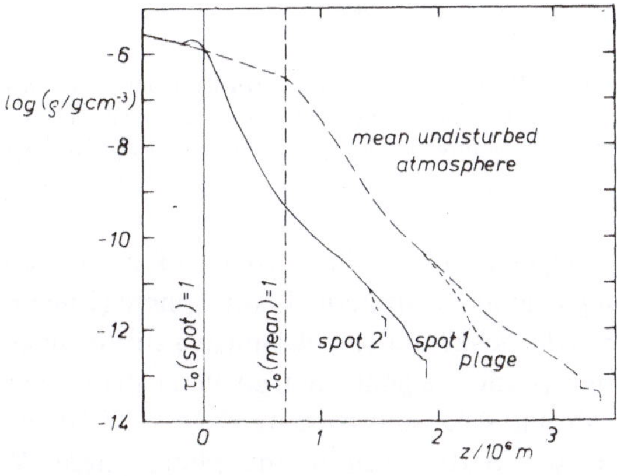

Fig. 3.

Fig. 1, 2, and 3. Temperature T, mass density ρ, and electron density n_e versus the geometrical height z. The sunspot models 1 and 2 as well as models for the mean undisturbed Sun and a plage are plotted for comparison. $z = 0$ corresponds to an optical depth of $\tau_0 = 1$ for $\lambda_0 = 5000$ Å in the sunspots.

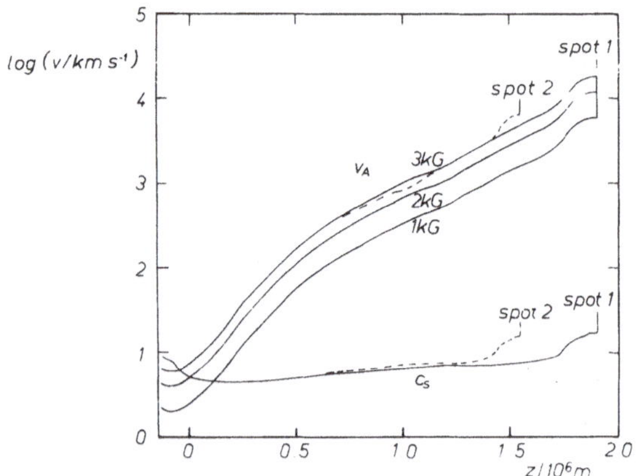

Fig. 4. Alfvén speed v_A and sound speed c_S versus z for the sunspot models 1 and 2, v_A for three different values of B. T_{min} corresponds to a height of z = 500 km.

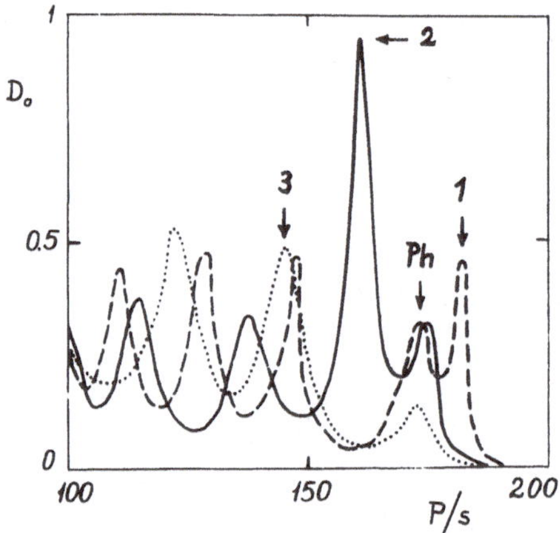

Fig. 5. Transmission D_0 versus period P for waves travelling vertically from the lower photosphere to the upper chromosphere of the three sunspot models. Arrows with numbers indicate the fundamental resonant peaks of the chromospheric cavity in the three models, while Ph marks the photospheric resonance.

The solution (1) does not depend on $k_\perp$. The existence of such a solution is evident from the solution for a strong field in an isothermal atmosphere (Žugžda and Džalilov, 1981); the peculiarity is due to the behaviour of a plasma in a strong magnetic field where motions along the field cannot produce significant transverse displacements ($v_\perp \ll v_z$). Figure 4 shows that the strong field approximation $c_S^2 \ll v_A^2$ is well satisfied in the sunspot chromosphere and partly even in the photosphere. We assume the reflection of waves from the lateral boundaries of the resonator resulting in a standing wave in a horizontal direction. Therefore, we consider only vertically-propagating waves.

In order to investigate the effectiveness of the resonant cavity, the transmission D_0

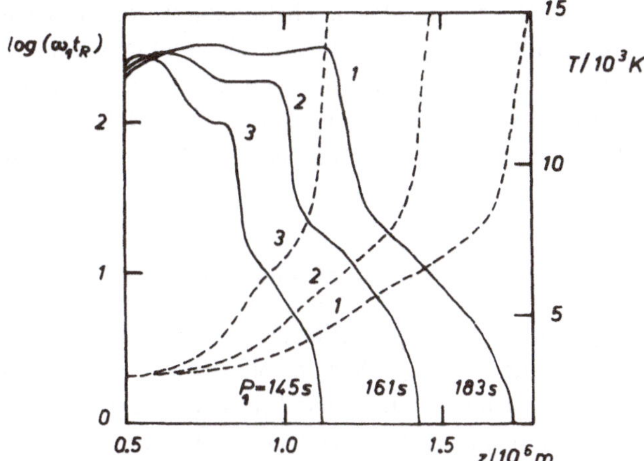

Fig. 6. Height dependence of the radiative decay time t_R of waves normalized to the period $P_1 = 2\pi/\omega_1$ of the fundamental resonant mode of the corresponding chromospheres for the three sunspot models (full lines). For comparison the $T(z)$ functions are also plotted (dashed lines).

for waves propagating through photosphere and chromosphere is calculated. The procedure is similar to that of Uchida (1967), and the algorithm agrees with that which has been used by Žugžda and Locāns (1980, 1982) in a study of Alfvén wave propagation. Such a model in terms of resonant transmission, that is, a filtering of a broad band, incident flux, differs from the approach of some other authors who assumed fixed boundary conditions and looked for the eigenvalues of the problem. Our model atmospheres with given $T(z)$ are subdivided into some 10 layers each of which can be represented by a constant β; the solutions of the equations for neighbouring layers are connected to each other by assuming boundary conditions which guarantee the conservation of mass and momentum.

The eigenvalue approach is successful for complete wave reflection from the lower and upper boundaries of a resonant cavity. For instance, the quiet Sun, photospheric oscillations (300 s) become non-propagating waves above and below the resonance layer in the convective zone; these are trapped waves. For incomplete reflection from the boundaries of such a layer we get complex eigenvalues due to the emission of waves from the resonator; the imaginary part (determining the Q-(quality) factor of the resonator) describes the damping of waves in the resonator due to the leakage of energy across the boundaries. These waves could be called 'semi-trapped waves'. If the characteristic time of wave damping is smaller than the oscillatory period, then we have no real resonance. We get accurate results by means of the eigenvalue approach if there are sharp, strongly reflecting boundaries of the considered layer. For gradual boundaries with a weak reflection we give preference to our method for studying the transfer of waves through the considered atmospheric layer. Here the source of the waves is assumed to be placed far enough from the considered region; no artificial layer of wave reflection is introduced at the starting level. The maximum of the coefficient of wave energy transmission determines the real part of the eigenvalue, while its imaginary part or the quality of the resonance is determined by the halfwidth of the resonant peak. In this way, the calculation of the transmission renders accessible the study of both the resonant filtering of oscillations and the response of the resonator to forcing from oscillatory sources inside the resonant layer.

4. Discussion of the Results

Figure 5 shows the dependence of the transition D_0 on the period P for waves travelling through the photosphere and chromosphere of our three sunspot models. The fundamental resonant modes for models 1, 2, and 3 (indicated by arrows) have values of P_1 = 183, 161, and 145 s, respectively, which are within the range of the observed periods of chromospheric oscillations. The periods of the resonant oscillations decrease with decreasing extent of the chromosphere; this applies also to the other resonant peaks at higher frequencies (harmonics) which are also sometimes detected. But larger values of P are more frequently observed which with the spectroscopic data supports model 1 and perhaps 2 more than 3. Unfortunately no simultaneous observations of EUV line profiles and oscillations in the chromosphere of sunspots are available, therefore our discussion must be limited to such vague statements.

The present model is also compatible with the observed oscillations at $T \approx 10^5$ K because the free boundary conditions in our treatment permit an oscillation of the thin transition layer as a whole. Sometimes several periods are observed simultaneously in one spot which could be explained by the existence of harmonics of low order (without integer multiples of P_1, due to the inhomogeneity of the resonator) in our model, but a detailed comparison with observations requires a more exact consideration of additional effects such as those discussed below.

Another feature (marked by Ph) in Figure 5 is still worth mentioning: A maximum of $P \approx 173$ s for all three sunspot models is clearly due to a photospheric resonant cavity, because the photospheric part is the same in our three models. Note that this period is similar to those in the models of a photospheric resonant cavity for the fast mode (see Section 2) and it could explain the observed photospheric oscillations as well.

Dissipative processes have been neglected in the present analysis. In order to get some information on the possible effect of radiative relaxation we calculated the radiative decay time t_R for continuum radiation in an optically thin medium,

$$t_R = \frac{\rho c_V}{16\,\sigma \varkappa_0 T^3}\,,$$

where σ is the Stefan–Boltzmann constant and $\varkappa_0$ is the absorption coefficient per unit length at a wavelength of λ_0 = 5000 Å. Figure 6 shows the resulting values of t_R related to the fundamental resonant periods for our three chromospheric sunspot models. It is clearly demonstrated that $\omega_1 t_R \gg 1$ for the greatest part of the chromospheres, therefore the waves correspond to almost adiabatic perturbations and radiative damping can be neglected in this rough estimate. For the upper chromosphere and transition region, however, a more exact treatment is required, and line radiative losses should be considered also in the lower chromosphere as it has been done by Giovanelli (1978, 1979).

5. Conclusions

Our model of a chromospheric resonant cavity for slow-mode magneto-acoustic waves is compatible with the properties of observed oscillations in sunspot chromospheres and even in the transition layer. This refers to the periods, but also to the observed independence of the period on the magnetic field strength. The oscillations are clearly determined by the atmospheric structure, that is to say, mainly by the temperature gradient or the extent of the chromosphere, thereby providing a method to test sunspot model atmospheres. The proposed sunspot models, especially model 1, result in resonant oscillations the periods of which agree with those from observations. It would be highly desirable to get simultaneous observations of oscillations and of EUV line profiles, e.g., of Lα, from the same sunspots in order to check the proposed models by more reliable data.

At photospheric levels, our model provides a resonant cavity as well, producing periods which are similar to those of a model for the fast mode (Scheuer and Thomas, 1981). This result points out the necessity to consider the linear interaction of waves in a strong magnetic field in order to investigate the photospheric resonator (Žugžda and Džalilov, 1981).

The photosphere is likely to build up a mixed resonance from two oscillatory modes because the linear interaction of the slow and fast modes is very strong there. In our opinion, a better diagnosis of sunspot atmospheres by means of oscillations could be achieved by considering the full system of wave equations, as has been done by Thomas and Scheuer (1982), but by using the method of Žugžda and Locāns (1982). Thomas and Scheuer (1982) replaced the reflection from a flat gradient of temperature by reflection form a sharp gradient; this procedure could result in an inaccurate determination of the resonant frequencies if there would be only a small real reflection of waves from the lower photosphere and convective zone (see Section 3.2).

A more exact analysis should consider the effects of wave absorption which could strongly influence the quality of atmospheric resonances; for Alfvén waves such effects have been demonstrated by Žugžda and Locāns (1982).

The treatment of radiative relaxation has to take into account the influence of lines including non-LTE effects in an optically thick medium. In the transition layer the validity of ionization equilibrium could perhaps be called in question. In order to improve the diagnostic means the heights of formation of the observed lines should be calculated, and the influence of the perturbations introduced by the waves in the thermodynamic quantities and thereby in the line profiles should be considered. Some of these improvements to our model are being prepared.

Acknowledgements

The authors would like to thank Dr John W. Leibacher who brought the unpublished work of Gurman *et al.* (1981) to our attention and gave encouraging and helpful comments as well as suggestions resulting in an improved representation of our paper. This work has benefitted considerably from discussions at the IAU Colloquium No. 66 on 'Problems in Solar and Stellar Oscillations' at the Crimean Astrophysical Observatory of the U.S.S.R. Academy of Sciences.

References

Antia, H. M. and Chitre, S. M.: 1979, *Solar Phys.* **63**, 67.
Auer, L. H., Heasley, J. N., and Milkey, R. W.: 1972, Kitt Peak National Observ. Contrib. No. 555.
Basri, G. S., Linsky, J. L., Bartoe, J.-D., F., Brueckner, G., and Van Hoosier, M. E.: 1979, *Astrophys. J.* **230**, 924.
Beckers, J. M. and Schultz, R. B.: 1972, *Solar Phys.* **27**, 61.
Beckers, J. M. and Tallant, P. E.: 1969, *Solar Phys.* **7**, 351.
Bhatnagar, A., Livingston, W. C., and Harvey, J. W.: 1972, *Solar Phys.* **27**, 80.
Giovanelli, R. G.: 1972, *Solar Phys.* **27**, 71.
Giovanelli, R. G.: 1978, *Solar Phys.* **59**, 293.
Giovanelli, R. G.: 1979, *Solar Phys.* **62**, 253.
Giovanelli, R. G., Harvey, J. W., and Livingston, W. C.: 1978, *Solar Phys.* **58**, 347.
Gurman, J. B., Leibacher, J. W., Shine, R. A., Woodgate, B. E., and Henze, W.: 1982, *Astrophys. J.* **253**, 939.
Havnes, O.: 1970, *Solar Phys.* **13**, 323.
Hollweg, J. V. and Roberts, B.: 1981, *Astrophys. J.* **250**, 398.
Kneer, F., Scharmer, G., Mattig, W., Wyller, A., Artzner, G., Lemaire, P., and Vial, J. C.: 1981, *Solar Phys.* **69**, 289.
Mogilevsky, E. I., Obridko, V. N., and Shelting, V. D.: 1973, *Radiofizika* **16**, 1357.
Scheuer, M. A. and Thomas, J. H.: 1981, *Solar Phys.* **71**, 21.
Staude, J.: 1981, *Astron. Astrophys.* **100**, 284.
Stellmacher, G. and Wiehr, E.: 1975, *Astron. Astrophys.* **45**, 69.
Syrovatsky, S. I. and Žugžda, Yu. D.: 1967, *Astron. Zh.* **44**, 1180.
Tandberg-Hanssen, E., *et al.*: 1981, *Astrophys. J.* **244**, L127.
Teplitskaya, R. B., Grigoryeva (Efendieva), S. A., and Skochilov, V. G.: 1978, *Solar Phys.* **56**, 293.
Thomas, J. H.: 1978, *Astrophys. J.* **225**, 275.
Thomas, J. H. and Scheuer, M. A.: 1982, *Solar Phys.* **79**, 19.
Uchida, Y.: 1967, *Astrophys. J.* **147**, 181.
Uchida, Y. and Sakurai, T.: 1975, *Publ. Astron. Soc. Japan* **27**, 259.
Žugžda, Y. D. and Džalilov, N. S.: 1981, *Astron. Zh.* **58**, 838.
Žugžda, Y. D. and Locāns, V.: 1980, *Investigations of the Sun and Red Stars (Zinātne, Riga)* **11**, 62.
Žugžda, Y. D. and Locāns, V.: 1981, *Pis'ma v Astron. Zh. (Soviet Astron. J. Letters)* **7**, 44.
Žugžda, Y. D. and Locāns, V.: 1982, *Solar Phys.* **76**, 77.

ADIABATIC OSCILLATIONS OF A DIFFERENTIALLY-ROTATING STAR*

Second-Order Perturbation Theory

S. V. VORONTSOV

Department of Theoretical Physics, Institute of Physics of the Earth, Academy of Sciences of the U.S.S.R., Moscow 123810, U.S.S.R.

Abstract. Perturbation theory is developed for calculating the influence of slow differential rotation on the adiabatic nonradial modes of stellar oscillations. The effects of Coriolis forces and ellipticity are analysed simultaneously using the perturbation technique for Hermitian operators which is developed up to second order in eigenvalues and to first order in eigenvectors.

1. Introduction

The effect of slow differential rotation on linear adiabatic oscillations of a star was first analysed by Hansen *et al.* (1977) using a variational-type perturbation technique. A general expression for first-order corrections to eigenfrequencies was obtained and reduced to a form convenient for computation of a particular type of differential rotation distributions. For a wider class of angular velocity distributions, the reduction was done by Cuypers (1980).

In the present paper the perturbation theory is developed up to second order in the eigenfrequency and first-order corrections to the eigenfunctions are also determined. In addition to the Coriolis effects, the rotational distortion of the stellar configuration and centrifugal forces are taken into account.

2. Perturbation Theory

Linear adiabatic oscillations of a differentially-rotating star are analysed in an inertial frame. The distribution of the velocity of rotation is assumed to be stationary and φ-independent in a spherical coordinate system (r, θ, φ). General vector equations – the moment equation, the Poisson equation for gravitational perturbations and the continuity equation – may be written as

$$\begin{aligned}\rho_0[-\omega^2\mathbf{u} + \Omega(2i\omega\hat{z} \times \mathbf{u} - 2m\omega\mathbf{u}) + \Omega^2(2im\hat{z} \times \mathbf{u} - m^2\mathbf{u})] &= \\ = \nabla(K\nabla\cdot\mathbf{u}) - \nabla[\mathbf{u}\cdot\rho_0(\nabla\psi_0 - \hat{r}_c r_c\Omega^2)] - \rho'(\nabla\psi_0 - \hat{r}_c r_c\Omega^2) &- \\ - \rho_0\nabla\psi' - \rho_0\hat{r}_c r_c(\mathbf{u}\cdot\nabla\Omega^2)\,, &\end{aligned} \tag{1}$$

* Proceedings of the 66th IAU Colloquium: *Problems in Solar and Stellar Oscillations*, held at the Crimean Astrophysical Observatory, U.S.S.R., 1–5 September, 1981.

Solar Physics **82** (1983) 379–382. 0038–0938/83/0822–0379$00.60.

$$\nabla^2\psi' = 4\pi G\rho' , \tag{2}$$

$$\rho' = -\nabla\cdot(\rho_0\mathbf{u}) . \tag{3}$$

Here ω is the angular frequency of oscillations, $\mathbf{u}$ is the vector displacement field, K is the adiabatic compressibility ($K = \Gamma_1 p_0$, where p_0 is the unperturbed pressure), ρ_0 and ψ_0 are the unperturbed density and gravitational potential, ρ' and ψ' are their Eulerian perturbations, r_c is the distance from the axis of rotation (z), $\hat{r}_c$ and $\hat{z}$ are the unit vectors in r_c, z-directions. The solutions of Equations (1)–(3) must satisfy the usual free-boundary conditions.

Equation (1) may be written in operator form

$$-\omega^2\mathbf{u} + \Omega(2i\omega\hat{z}\times\mathbf{u} - 2m\omega\mathbf{u}) + \Omega^2(2im\hat{z}\times\mathbf{u} - m^2\mathbf{u}) = = -H_0\mathbf{u} - \Omega_0^2(\Psi + E)\mathbf{u} . \tag{4}$$

Here H_0 is an integro-differential Hermitian operator corresponding to the zero-order boundary-value problem (a nonrotating, spherically symmetric star):

$$H_0\mathbf{u}_0 = \omega_0^2\mathbf{u}_0 . \tag{5}$$

Operators Ψ and E correspond to the influence of centrifugal forces and ellipticity. They are determined by equivalency of Equations (1), (4); an explicit form of these operators is not necessary because only their matrix elements, calculated by a variational technique, will be needed.

In Equation (4), Ω_0 denotes the average angular velocity of rotation, so that

$$\Omega(r, \theta) = \Omega_0\Omega_d(r, \theta) . \tag{6}$$

A small parameter $\lambda = \Omega_0/\omega_0$ is then introduced and the solutions of Equation (4) are found in the form

$$\omega = \omega_0(1 + \sigma_1\lambda + \sigma_2\lambda^2 + \ldots) , \tag{7}$$

$$\mathbf{u} = \mathbf{u}_0 + \mathbf{u}_1\lambda + \mathbf{u}_2\lambda^2 + \ldots . \tag{8}$$

Substitution of expansions (7), (8) into Equation (4) leads to the system of equations of perturbation theory

$$[I - \omega_0^{-2}H_0]\mathbf{u}_0 = 0 , \tag{9}$$

$$[I - \omega_0^{-2}H_0]\mathbf{u}_1 = -2\sigma_1\mathbf{u}_0 - \Omega_d[2m - 2i\hat{z}\times]\mathbf{u}_0 , \tag{10}$$

$$\begin{aligned}[I - \omega_0^{-2}H_0]\mathbf{u}_2 = {} & -2\sigma_1\mathbf{u}_1 - (\sigma_1^2 + 2\sigma_2)\mathbf{u}_0 - \Omega_d[2m - 2i\hat{z}\times]\mathbf{u}_1 - \\ & - \sigma_1\Omega_d[2m - 2i\hat{z}\times]\mathbf{u}_0 - \\ & - \Omega_d^2[m^2 - 2im\hat{z}\times]\mathbf{u}_0 + [\Psi + E]\mathbf{u}_0 ,\end{aligned} \tag{11}$$

where I is the identity operator. Equation (9) represents the zero-order problem; its solutions are orthogonal and assumed to be normalized in the sense of a scalar product defined by

$$(\mathbf{u}, \mathbf{u}') = \int_V \rho_0 \mathbf{u}^* \cdot \mathbf{u}' \, dv \,, \tag{12}$$

where V is the volume occupied by the star. Scalar multiplication of Equation (10) onto $\mathbf{u}_0$ gives the first-order correction to eigenfrequency

$$\sigma_1 = -(\mathbf{u}_0, \Omega_d[m - i\hat{z} \times]\mathbf{u}_0) \,, \tag{13}$$

which is the result obtained by Hansen *et al.* (1977). The first-order correction to the eigenfunctions is also found from Equation (10):

$$\mathbf{u}_1 = \sum_{l',n' \neq l,n} a^{l',m,n'}_{l,m,n,1} \mathbf{u}_0' + \{-\Omega_d[2m - 2i\hat{z} \times]\mathbf{u}_0\}_T \,, \tag{14}$$

$$\frac{\omega_0^2 - (\omega_0')^2}{\omega_0^2} a^{l',m,n'}_{l,m,n,1} = (-\mathbf{u}_0', \Omega_d[2m - 2i\hat{z} \times]\mathbf{u}_0) \,, \tag{15}$$

where index T denotes orthogonal projection onto the field of toroidal vector spherical harmonics and the corresponding term in Equation (14) represents a torsional-type correction to the eigenfunctions. The second-order correction to the eigenfrequency is found from Equation (11):

$$2\sigma_2 - \sigma_1^2 = -(\Omega_d[2m - 2i\hat{z} \times]\mathbf{u}_0, \mathbf{u}_1) - m(\mathbf{u}_0, \Omega_d^2[m - 2i\hat{z} \times]\mathbf{u}_0) + \\ + (\mathbf{u}_0, [\Psi + E]\mathbf{u}_0) \,. \tag{16}$$

The reduction of the scalar products in (13)–(16) to a form convenient for computation (unidimensional integrals containing radial eigenfunctions) is given by Vorontsov (1981). The angular velocity distribution is represented by a finite set $\Omega = \Omega_0 \sum_l \Omega_{dl}(r) Y_{l_0}(\theta, \varphi)$ and the angular dependences are reduced to a computation of angular integrals containing three vector spherical harmonics, which is simply performed in terms of Wigner's 3-j symbols.

Computationally-convenient formulae (Vorontsov, 1981) are given for the case $\partial\Omega/\partial z = 0$. The simplifying assumption was used that the effect of differentiality of rotation on the stellar configuration is relatively small, i.e., the theory of rotational distortion of a star for the case of rigid rotation is appropriate. A corresponding expression for the matrix elements of the operator $\Psi + E$ is given by Vorontsov and Zharkov (1981), taking into account possible discontinuities in density distribution in a star.

Acknowledgement

The author thanks Prof. V. N. Zharkov for useful discussions.

References

Cuypers, J.: 1980, *Astron. Astrophys.* **89**, 207.
Hansen, C. J., Cox, J. P., and Van Horn, H. M.: 1977, *Astrophys. J.* **217**, 151.
Vorontsov, S. V.: 1981, *Soviet Astron.-AJ* **58**, 1275.
Vorontsov, S. V. and Zharkov, V. N.: 1981, *Soviet Astron.-AJ* **58**, 1101.

EXCITATION OF THE SOLAR OSCILLATIONS BY OBJECTS CONSISTING OF y-MATTER*

S. I. BLINNIKOV

Institute of Theoretical and Experimental Physics, Moscow, U.S.S.R.

and

M. YU. KHLOPOV

Institute of Applied Mathematics, Academy of Sciences, Moscow, U.S.S.R.

Abstract. Modern development of particle physics makes it probable that new, still undiscovered, particles exist interacting with the ordinary matter by means of gravitation only. Okun suggested to call such particles the y-particles and to call the matter consisting of them the y-matter. We show that planet-like object orbiting inside the Sun and consisting of y-matter may explain 160 min nonradial solar oscillations.

The difficulties of explanation within the frame of standard solar model of 160-min oscillations observed in Crimean Observatory (Severny *et al.*, 1976, 1979) are well known (Vorontsov and Zharkov, 1981). In fact these oscillations perhaps are not free, but forced. In the paper by Severny *et al.* (1979) a possibility of excitation of the oscillations by a small black hole orbiting the Sun at the depth 2×10^4 km below the photosphere was mentioned (D. O. Gough in his remark at Toulouse conference 1978 on solar physics was apparently the first to suggest such a possibility). Modern particle physics may offer another possibility, to the same extent exotic, but perhaps free of some difficulties of the black hole model.

Recently Okun (1980) suggested that there may exist so called y-particles having the only mutual interaction with the ordinary matter – the gravitational one (the history of this idea may be traced back the papers by Lee and Yang (1956) and Kobzarev *et al.* (1966)). These y-particles may have their own interactions, similar to the weak, strong and electromagnetic interactions of ordinary particles. The existence of such y-interactions, not acting on the ordinary matter, may lead as a result of the cosmological evolution to the formation by the y-matter of compact astronomical objects (y-stars and y-planets) which may be discovered only by their gravitational effect on the ordinary matter.

Oscillations of celestial bodies might be an example of y-matter manifestations. Unlike the Okun's (1980) suggestion to search for terrestrial oscillations, we draw attention to the solar pulsations with the period 160 min. According to Severny *et al.* (1979) the parameters of these pulsations are: the period $P = 160^{\mathrm{m}}.010 \pm 0^{\mathrm{m}}.004$, the velocity amplitude 0.5–1 m s^{-1}.

* Proceedings of the 66th IAU Colloquium: *Problems in Solar and Stellar Oscillations*, held at the Crimean Astrophysical Observatory, U.S.S.R., 1–5 September, 1981.

Solar Physics **82** (1983) 383–385. 0038–0938/83/0822–0383$00.45.

In the approximation of hydrostatic adiabatic tide (Zahn, 1966) the body of the mass m at the depth h under the surface of the star of mass M and radius R displaces the surface for the distance of order

$$\Delta R \simeq \frac{R}{h}\frac{m}{M}R\,. \tag{1}$$

The estimate (1) is valid if $h/R \ll 1$, but $h^2/R^2 \gg m/M$. The velocity amplitude 1 m s^{-1} corresponds to the total displacement of the surface ΔR (the double amplitude) of about 3 km. Substituting $h = 2 \times 10^4$ km, $R = R_\odot$, $M = M_\odot$ into (1) we obtain $m = 10^{-7} M_\odot$. This estimate is not very reliable. If we take into account that the observed ΔR may be underestimated because of the averaging over the large area of the solar disk $\sim \pi R^2/2$ (and in Equation (1) ΔR corresponds to the area of order πh^2), then m may increase substantially. On the other hand, the estimate of m may strongly decrease due to the resonance with a solar g-mode, so we prefer to use the value $m = 10^{-7} M_\odot = 2 \times 10^{26}$ g (the exact value is necessary for the model with a black hole, in the case of the y-planet the crude estimate is sufficient).

According to the standard solar model (Allen, 1977) at the depth $h = 2 \times 10^4$ km the density $\rho = 2 \times 10^{-4}$ g cm^{-3}, the temperature $\simeq 10^5$ K, so the sound speed is $v_s = 3 \times 10^6$ cm s^{-1}. The velocity of the body is $v = 4.4 \times 10^7$ cm s^{-1}, i.e. the accretion is supersonic. For $m = 10^{-7} M_\odot$ we find the accretion radius

$$R_\mathrm{A} = 2Gm/v^2 \simeq 10^4 \text{ cm}\,. \tag{2}$$

By the standard formula for the accretion rate we obtain

$$\dot{m} = \pi R_\mathrm{A}^2 \rho v = 10^{12} \text{ g s}^{-1}\,. \tag{3}$$

Such an accretion rate onto the black hole with account for the magnetic field results in the energy release (Shvartsman, 1971; Bisnovatyi-Kogan and Ruzmaikin, 1974) of order $0.1\dot{m}c^2 \simeq 10^{32}$ erg s^{-1}. This is comparable to the solar luminosity and so absolutely unacceptable. The account for the radiation pressure decreases the luminosity of the hole down to the Eddington limit $\simeq 10^{31}$ erg s^{-1} – being unacceptable either, so black hole must excite the oscillations by its thermal effect and not by gravity.

Accretion onto y-planet differs substantially from the accretion onto a black hole. If radius of y-planet, $r > R_\mathrm{A}$, then the solar matter is not captured by the y-planet. If $r \sim 10^8$ cm, as it is in the case of ordinary planets, i.e. $r/R_\mathrm{A} \gg v/v_s$, the velocity perturbations of the infalling gas Δv turn to be everywhere less than v_s: $\Delta v \lesssim vR_\mathrm{A}/r$. It means that if the accretion shock wave is formed, then the shock is weak and maximum heating of the matter is low, being of order $\dot{m}(\Delta v)^3/v_s$, where in contrast to Equation (3) $\dot{m} \sim \pi r^2 \rho v$ is the mass perturbed (but not captured!) by the y-planet in unit time. So the heating is less than 10^{24} erg s^{-1}. Thus the y-planet induces inside the Sun only gravitational and weak acoustic effects. The same is true even if our evaluation of m is underestimated by orders of magnitude (it is surely true if m is overestimated). The mass m must not be too large. 'Spreading' m along the ring of radius $R_\odot - h \simeq R_\odot$ (which is reasonable since $P = 160^{m} = \frac{1}{9}$ day $\ll 88$ days $= P_{♀}$ – the period of Mercurian revolution) we obtain the perturbation of the perihelion precession of the Mercury (Landau and Lifshitz, 1965) per one revolution

$$\delta\varphi = \frac{6\pi m R_{\odot}^2}{4 M_{\odot} d_{♀}^2}, \tag{4}$$

where $d_{♀}$ is the mean distance of the Mercury from the Sun. For $m = 10^{-7} M_{\odot}$ Equation (4) gives the precession of the perihelion of Mercury $\sim 0\overset{''}{.}02$ per century. However, if m is higher by 2–3 orders of magnitude then the value of $\delta\varphi$ will contradict the predictions of general relativity. Besides that the oscillations of the whole figure of the Sun will be large, since the centre of mass of the system is to be at rest.

Leaving aside the question of origin of the y-planet inside the Sun, we note that the value $m = 10^{26}$ g corresponds to 1–10% of the amount of y-matter which might be accreted by the Sun during its lifetime if the parameters of the interstellar y-matter in the Galaxy are the same as for ordinary matter.

The presence of a body inside the Sun may be in principle discovered by its gravitational effect on the solar probe approaching the Sun for a distance of few $R_{\odot}$. For the test of this opportunity the delaited analysis is needed of possible noncircular (and non-elleptic!) orbits inside the Sun.

Summarizing we conclude that the presence of the y-matter inside the Sun seems now not excluded. More serious consideration of the y-planet inside the Sun (e.g. the problems of its formation and evolution due to the tidal friction) would be necessary if the existence of 160 min oscillations becomes generally accepted, if phase of the oscillations and their period are stable, and if alternative theoretical models fail to explain this phenomenon.

Acknowledgement

We are grateful to L. B. Okun for valuable discussions which stimulated the writing of this paper.

References

Allen, C. W.: 1977, *Astrophysical Quantities*, Moscow, Mir, p. 234.
Bisnovatyi-Kogan, G. S. and Ruzmaikin, A. A.: 1974, *Astrophys. Space Sci.* **28**, 45.
Kobzarev, I. Yu., Okun, L. B., and Pomeranchuk, I. Ya.: 1966, *J. Nucl. Phys. USSR* **3**, 1154.
Landau, L. D. and Lifshitz, E. M.: 1965, *Mechanics*, Moscow, Nauka, p. 55.
Lee, T. D. and Yang, C. N.: 1956, *Phys. Rev.* **104**, 254.
Okun, L. B.: 1980, *Zh. Eksper. Teor. Fiz. (JETP)* **79**, 694.
Severny, A. B., Kotov, V. A., and Tsap, T. T.: 1976, *Nature* **259**, 89.
Severny, A. B., Kotov, V. A., and Tsap, T. T.: 1979, *Astron. Zh.* **56**, 1137.
Shvartsman, V. F.: 1971, *Astron. Zh.* **48**, 479.
Vorontzov, S. V. and Zharkov, V. N.: 1981, *Uspekhi Fiz. Nauk* **134**, 675.
Zahn, J.-P.: 1966, *Ann. Astrophys.* **29**, 313.

NONLINEAR ANELASTIC MODAL THEORY FOR SOLAR CONVECTION*

(Invited Review)

JEAN LATOUR

*Joint Institute for Laboratory Astrophysics**, University of Colorado, and Observatoire de Nice, BP252, F06007 Nice CEDEX, France*‡

JURI TOOMRE

Department of Astrophysical, Planetary and Atmospheric Sciences†*, and Joint Institute for Laboratory Astrophysics**, University of Colorado, Boulder, CO 80309, U.S.A.*

and

JEAN-PAUL ZAHN

Observatoire du Pic-du-Midi et de Toulouse, F65200 Bagnères de Bigorre, France

Abstract. Preliminary solar envelope models have been computed using the single-mode anelastic equations as a description of turbulent convection. This approach provides estimates for the variation with depth of the largest convective cellular flows, akin to giant cells, with horizontal sizes comparable to the total depth of the convection zone. These modal nonlinear treatments are capable of describing compressible motions occurring over many density scale heights. Single-mode anelastic solutions have been constructed for a solar envelope whose mean stratification is nearly adiabatic over most of its vertical extent because of the enthalpy (or convective) flux explicitly carried by the big cell; a sub-grid scale representation of turbulent heat transport is incorporated into the treatment near the surface. The single-mode equations admit two solutions for the same horizontal wavelength, and these are distinguished by the sense of the vertical velocity at the center of the three-dimensional cell. It is striking that the upward directed flows experience large pressure effects when they penetrate into regions where the vertical scale height has become small compared to their horizontal scale. The fluctuating pressure can modify the density fluctuations so that the sense of the buoyancy force is changed, with buoyancy braking actually achieved near the top of the convection zone. The pressure and buoyancy work in the shallow but unstable H^+ and He^+ ionization regions can serve to decelerate the vertical motions and deflect them laterally, leading to strong horizontal shearing motions. It appears that such dynamical processes may explain why the amplitudes of flows related to the largest scales of convection are so feeble in the solar atmosphere.

1. Introduction

The structure of the solar atmosphere is determined largely by the convection just below the surface and the waves that it can generate. The coupling of these turbulent motions with magnetic fields must cause most of what is observed on the Sun. However, theoretical understanding of the dynamics of the solar convection zone and associated motions in the atmosphere is still very incomplete. For instance, no detailed theoretical

* Proceedings of the 66th IAU Colloquium: *Problems in Solar and Stellar Oscillations*, held at the Crimean Astrophysical Observatory, U.S.S.R., 1–5 September, 1981.

** JILA is operated jointly by the University of Colorado and the National Bureau of Standards.

† Formerly Department of Astro-Geophysics.

‡ Now at Observatoire du Pic-du-Midi et de Toulouse.

Solar Physics **82** (1983) 387–400. 0038–0938/83/0822–0387$02.10.

explanations are available for the observed discrete scales of motion seen as granulation, mesogranulation and supergranulation. Nor are there reliable predictions for differential rotation with both latitude and depth, nor for the magnetic dynamo action as evidenced by global field reversals. Nor is it clear why giant cells, or convective flows with horizontal scales comparable to the total depth of the convection zone, are essentially undetectable at the surface of the Sun. What are required are far more explicit dynamical theories to try to resolve these issues, though all such analyses of turbulent flows are quite formidable tasks. Certainly strides have been made recently through nonlinear studies of incompressible convection in rotating spherical shells (e.g. Gilman, 1978, 1980) and of dynamo action therein (Gilman and Miller, 1981), through linear instability and nonlinear modal analysis of compressible fluids in similar configurations (e.g. Glatzmaier and Gilman, 1981; Marcus, 1980), and by nonlinear simulations of fully compressible convection in planar geometries (Hurlburt *et al.*, 1982). Further, recent dynamical models of granulation have included fairly realistic equations of state and effects of radiative transfer in the optically thin atmosphere (Nelson and Musman, 1977, 1978; Dravins *et al.*, 1981; Nordlund, 1980, 1982). Still, although progress has been made in describing the dynamics of convection in a more detailed (and thus presumably more reliable) manner than that of the mixing-length approach, many of the basic consequences of convection in the Sun have yet to be explained.

We have been developing theoretical descriptions for solar and stellar convection that use anelastic modal equations. These result after making two approximations to the full dynamical equations. First, an anelastic approximation is used to filter out the high frequency acoustic waves that might be present in the compressible motions but which would not normally contribute much to the convective transport. Second, we treat the convection as if it had only a discrete spectrum of horizontal scales. The horizontal structure is expanded in a finite number of horizontal planforms or modes. Such a truncation is used to make the problem tractable, choosing to emphasize accurate representation of the vertical and temporal structure at the expense of the horizontal. This truncated modal approach is motivated by the seemingly cellular character of turbulent convection, and similar methods have worked reasonably well in theoretical descriptions of laboratory convection (Toomre *et al.*, 1977, 1982). The resulting nonlinear anelastic modal equations are capable of describing compressible convection over multiple scale heights and with all the complexities arising from realistic equations of state in a star. The equations have a spatial differential order, $(6n + 3)$, with respect to the vertical coordinate z which depends upon the number of horizontal modes, n, retained in the analysis (Latour *et al.*, 1976, hereafter called as Paper I). Although the differential order is high and their solution difficult, these equations have the distinct advantage that they can be generalized to describe the coupling of convection with rotation, magnetic fields and pulsation in the star.

The anelastic modal procedure has been used to study convection both in A-type stars and in the Sun. Although the stellar applications have been carried out so far mainly with the single-mode equations, the results obtained for convection in the outer envelope of an A star have turned out to be very instructive. The A stars have proved to be a useful framework of developing and refining the modal anelastic procedure, for here, unlike in the Sun, the mean structure is not overly sensitive to the details of the convection. The nonlinear modal studies have revealed that the two convection zones in A stars are dynamically coupled by the convective motions penetrating through the intervening stable material (Toomre *et al.*, 1976, Paper II; Toomre, 1980; Zahn, 1980; Nelson, 1980; Latour *et al.*, 1981, Paper III). The convection associated with cells of large horizontal scale is found to be driven principally by buoyancy forces in the deeper He^{++} unstable zone. It is striking that these motions of supergranular scale are able to penetrate upward all the way to the surface of the star, contrary to mixing-length predictions. Thus the convection is not simply confined to the unstable regions, and for A stars this means that diffusive gravitational separation of elements cannot be occurring in what previously was supposed to be a quiescent region between the H^+ and He^{++} convection zones. Another noteworthy result concerns the nature of the supergranular scale flow in the shallow H^+ zone. Analysis of the buoyancy and pressure work terms reveals that pressure effects dominate in the upper zone. The predominately vertical motions in the convection zone deeper down are turned into strong horizontal shear flows in the upper zone, largely as a result of the strong braking of vertical momentum in this region. This serves to diminish the vertical velocity amplitudes that are actually visible in the atmosphere. The convection of small horizontal scales, like granulation, can experience buoyancy driving in the H^+ zone, while supergranular scales are strongly braked. In the latter, significant pressure fluctuations can serve to change the sense of the density fluctuations, so that a rising fluid element near the top of this unstable zone is heavier than its surroundings and experiences buoyancy braking. Net work is extracted by both scales of convection from this highly unstable zone, though the effects are very different.

The work with A stars led to the suggestion that supergranulation in the Sun may well possess strong horizontal flows in the H^+ zone just below the surface (e.g. Toomre, 1980). Thus a shallow but highly unstable region may be able to effectively prevent large-scale cellular motions from getting through into the atmosphere with any significant portion of their original momentum. The deflection of large-scale flows appears to be a consequence of the rapidly decreasing scale height just below the surface of the star; similar behavior is also seen in modal analysis of convection in polytropes when the effects of stratification are significant (Massaguer and Zahn, 1980; Latour *et al.*, 1982). If such predictions continue to be borne out by modal solutions for solar convection, then this may explain why the observed vertical velocity amplitudes in supergranular flows in the atmosphere are so small. This may also explain why the giant cells of global scale, suggested by the magnetic field patterns, are below the present level of detection for velocities in the atmosphere. These giant cell flows may well be deflected by the shrinking scale height at the depth of the He^+ ionization in the Sun.

The basic problem in dealing with the solar convection zone is that the overall mean structure of the outer envelope is very sensitive to the detailed treatment of the convective motions just below the surface. Since the convection is responsible for almost all the transport, the mean structure is close to adiabatic throughout most of the convection zone. However, which particular adiabat is chosen depends sensitively upon the very rapid transition from radiative to adiabatic conditions just below the surface. Thus the details of the convection at the very top of the zone strongly influence the choice of adiabat, which in turn controls the depth of the convection zone in such envelope models. This means that a single-mode anelastic description of the convection must be supplemented by some representation of the heat flux being carried by other scales of motions. Certainly the longer term goal will be to explicitly include a sufficient number of horizontal modes so that the dominant scales of convection are being computed in a self-consistent manner. However, that poses a formidable task mathematically, for one may expect that at the very least about four modes will be required in the analysis. These four horizontal modes probably need to be of disparate scale, much like granulation, mesogranulation, supergranulation and giant cells.

We sought to minimize our computational difficulties at first by simply using a mean structure for the Sun constructed from mixing-length models to test the gross behavior of the modal convection as the horizontal scale is varied. The use of these highly simplified models, where the feedback between the convective flux and the mean structure is largely severed, has suggested that convection cells with the large horizontal scales of supergranulation are driven mainly by He^{++} and display strong horizontal shear layers just below the surface. Modes of intermediate scale like that of mesogranulation primarily feel the effects of He^{+}, while only cells with the small scale of granulation get much buoyancy driving in the H^{+} zone. We have now advanced to considerably more realistic descriptions in which the nonlinear feedback of the single anelastic mode upon the mean structure is fully implemented. We have accomplished this by introducing the effects of unresolved small scales of convection and turbulence as a diffusion of mean entropy, while treating the dynamics of the large-scale convection cell explicitly by solving the full anelastic equations. Such a procedure has worked out quite well in building preliminary models of supergranulation and giant cells: with the diffusive scale of such an eddy process restricted to be less than 1 Mm, we find that a single large-scale mode can transport nearly the full solar flux over most of the convection zone without developing any noticeable pathologies in the mean stratification. Further, we note that the cellular motions in our solar model extend over multiple density scale heights, much as we anticipated from our work with A stars.

2. Formulating the Problem

In this preliminary study of the hydrodynamics of the solar convection zone, we shall use the anelastic modal equations in their simplest form by retaining only a single mode. The class of solutions investigated has an upwelling flow at the center of the three-dimensional hexagonal convection cell. The computations are relevant only to the largest convective cellular flows in the Sun, with their horizontal scales being comparable to the overall depth of the convection zone. Our notation and formulation of the equations will be identical to that used in Paper II, though we will introduce additional terms to represent the transport of heat and the diffusion of momentum by turbulent motions with scales much smaller than the explicit cellular mode.

A. Single-mode anelastic approach

We shall here briefly recall some of our notation for the single-mode anelastic representation of the convection, but for the sake of brevity ask the reader to refer to Paper III for a detailed exposition of the equations. Each thermodynamic variable is separated into its horizontal mean, which depends only on the vertical coordinate z and time t, and its fluctuations relative to that mean value. The coordinate z is taken as the depth below the surface of the Sun and thus increases downwards. The fluctuations will be factorized so as to separate their amplitude functions from their specified horizontal planform function $f(x, y)$. The temperature field will thus be represented as

$$T(x, y, z, t) = \overline{T}(z, t)[1 + f(x, y)\Theta(z, t)] , \tag{2.1}$$

where $\overline{T}$ is the mean temperature and θ the amplitude function of its relative fluctuations. The density field is similarly described by its mean value $\overline{\rho}$ and the amplitude function Λ of its relative fluctuation, and so too pressure in terms of $\overline{P}$ and Π. Likewise enthalpy, entropy and thermal conductivity, with $\overline{H}$, $\overline{S}$, $\overline{K}$ being their mean values and H, S, K their absolute fluctuation functions. The momentum vector $m_i = \rho v_i$ is divergence free within the anelastic approximation if we ignore the explicit time dependence of $\overline{\rho}$ associated with possible radial pulsations. Thus the momentum vector can be represented as

$$m_i = \left\{a^{-2}\frac{\partial f}{\partial x}DW,\ \ a^{-2}\frac{\partial f}{\partial y}DW,\ \ fW\right\}, \tag{2.2}$$

assuming that the vertical component of vorticity is negligible and denoting $D \equiv \partial/\partial z$. The horizontal planform fluctuations satisfy

$$\left(\frac{\partial^2}{\partial x^2} + \frac{\partial^2}{\partial y^2}\right)f + a^2 f = 0, \qquad \overline{(f)^2} = 1 , \tag{2.3}$$

where the overbar denotes horizontal averaging. The planform functions within a single-mode representation are characterized by their horizontal wavenumber a and their self-interaction constant $C = \frac{1}{2}\overline{(f)^3}$, which is zero for two-dimensional rolls and $6^{-1/2}$ for the three-dimensional cells with a hexagonal planform considered here.

The use of such mean and fluctuating variables yields a system of nonlinear partial differential equations with z and t as independent variables. The single-mode anelastic equations are stated as Equations (2.7) to (2.12) in Paper III, and involve a fluctuating horizontal vorticity equation and a fluctuating heat equation for the particular mode under consideration, plus a mean-momentum equation and a mean-heat equation which are coupled by various nonlinear terms to the fluctuations. The single-mode system of equations is of 3rd order in time derivatives, and of 9th order in spatial derivatives. The nonlinear differential system will be solved by finite differences, the details of which are also provided in Paper III.

B. Mean model and boundary conditions

Our computational domain in the vertical extends over much of the depth of the solar convection zone, with the lower boundary placed at a depth of about 256 Mm below the solar surface so that the effects of motions in the stable stratification below the convection zone can be resolved. The placement of the upper boundary of the computational domain presents greater difficulties. It would be most appropriate to extend the domain well into the solar chromosphere in order to study the possible upward penetration of these largest scales of convection into the stable atmosphere. However, severe computational difficulties arise even when dealing with the full complexities of the hydrogen ionization zone just below the surface: the pressure scale height varies by a factor of about 20 in going from a depth of 7 Mm (where H is fully ionized) to the photosphere, and the conductivity $\overline{K}$ varies by several orders of magnitude as a consequence of hydrogen ionization. Indeed, the pressure scale height ranges from $H_p \simeq 0.1$ Mm in the photosphere to $H_p \simeq 50$ Mm near the bottom of the convection zone (at a depth of about 200 Mm), thereby changing by a factor of about 500. The numerical difficulties in describing compressible flows over such a range of properties are most formidable, and we preferred in these preliminary calculations to avoid some of them by placing our upper boundary at a depth where hydrogen is already largely (80%) ionized. Thus our computational domain starts at a depth of 5.7 Mm and extends downwards a further 250 Mm, with the scale height H_p varying by a factor of about 25 between the top and bottom of the domain.

We have adopted a fairly standard chemical composition of $X = 0.73$, $Y = 0.25$, and $Z = 0.02$ for our solar model, used the Cox and Stewart (1965) opacity tables, and have imposed the mean temperature and density at our upper boundary at $z = 5.7$ Mm to have the values $\overline{T}_0 = 3.774 \times 10^4$ K and $\bar{\rho}_0 = 1.880 \times 10^{-4}$ g cm^{-3} in keeping with standard calibrated mixing-length envelope models (cf. Gough and Weiss, 1976). At the lower boundary we assert that the energy flux is purely radiative and thus specify the mean temperature gradient there. We require the vertical momentum W and the temperature fluctuation θ to vanish at the upper and lower boundaries, which are otherwise stress-free. Such an imposition of conditions at the upper artificial boundary is certainly arbitrary and we have tried others, but the choice made here has the modest advantage of being the simplest.

C. Representation of small-scale turbulence

The single mode of large horizontal scale is likely to be able to transport a significant fraction of the solar flux over only the deeper portions of the convection zone, with smaller cellular scales having roles at shallower depths. Further, a cascade process must be present that transfers energy from the largest cellular scales being described by the modal equations to the smaller scales at which viscous dissipation occurs. We introduced in Paper III a representation for the energy cascade to smaller scales by means of a turbulent viscosity $\bar{\nu}_T$, with the latter a function of the shear being experienced by the modal flow. The turbulent viscosity serves to parameterize sub-grid shearing instabilities and consequent diffusion of vorticity by a succession of small-scale eddies.

We must here also introduce a representation for the turbulent heat transport associated with the much smaller scales of motion. Our imposition of an artificial upper boundary means that transport by the large mode becomes ineffective in a narrow region below this boundary. If radiation alone were required to carry all the flux there, then the temperature gradient that would result would be so steep that the whole structure of the convection zone would be drastically changed, making it much shallower than acceptable. Such a thermal boundary layer can be avoided by allowing the flux to be carried by the much smaller turbulent scales that must be present there, for the mean stratification is still highly unstable to convection and the pressure scale height is small. We therefore incorporate a turbulent transport term into the mean heat equation that serves to diffuse the mean entropy field should its gradient attempt to become very steep. Such turbulent heat transport is analogous to what would result from a local mixing-length approximation, though by limiting the length scale of mixing to be less than about 1 Mm, we can confine these processes to the vicinity of the upper boundary.

The turbulent heat transport by small scales can be accommodated by introducing the additional term

$$-D(\bar{\rho}\bar{T}\chi_T D\bar{S}) \tag{2.4}$$

to the left-hand side of the mean-heat equation (2.12) in Paper III, with this term representing the divergence of the turbulent heat flux. Here χ_T is a turbulent diffusivity $u_T l$, with l the mixing length linked to the local pressure scale height but possessing an imposed upper bound. The effective turbulent velocity u_T is estimated, as in the mixing-length approach, from the kinetic energy acquired by a small adiabatic parcel of fluid traveling a distance l under the acceleration of buoyancy, thus obtaining

$$\chi_T = l^2\left[-g\left(\frac{\partial\bar{S}}{\partial\ln\bar{\rho}}\right)^{-1} D\bar{S}\right]^{1/2}. \tag{2.5}$$

The modified mean-heat equations (2.12) from Paper III may then be restated in conservative form to clearly identify the sources and sinks, with the equation becoming

$$\bar{\rho}\bar{T}\frac{\partial\bar{S}}{\partial t} + D\left[WH - W\frac{P}{\bar{\rho}} - \bar{K}D\bar{T} - \bar{\rho}\bar{T}\chi_T D\bar{S}\right] = gW\Lambda - WPD\left(\frac{1}{\bar{\rho}}\right) + \\ + \frac{(\bar{\nu}+\bar{\nu}_T)}{\bar{\rho}}\left[a^{-2}(\phi + 2a^2W)^2 + 4(DW)^2\right], \tag{2.6}$$

with all the notation the same as that introduced in Paper III.

Similarly, we modify the fluctuating heat equation (2.9) in Paper III by introducing a term to its right-hand side which represents turbulent diffusion acting on the fluctuating entropy, with the expression being

$$+\bar{\rho}\bar{T}\chi_T(D^2 - a^2)S\,. \tag{2.7}$$

This term is just the fluctuating analog of expression (2.4) to leading order.

We will find that the turbulent heat transport (2.4) is of importance only near our artificially imposed upper boundary and negligible over most of the computational domain. By adjusting the value of l in that term we can adjust the depth of the unstable zone so that it has a reasonable value as judged by properly evolved full solar models (e.g., Gough and Weiss, 1976). Thus the term is of key importance in selecting the adiabat for the convection zone, but does not otherwise control the dynamics of the large-scale mode which will serve to keep the stratification in the rest of the zone adiabatic.

3. Results and Discussions

We have obtained steady solutions for a number of single-mode convective flows with a hexagonal planform, and in all these there has been a full feedback of the nonlinear convection upon the mean stratification. We will concentrate here on discussing the properties of one of these solutions of large horizontal scale, for it will serve to show what is characteristic of all these anelastic modal solutions for solar convection. The steady states are attained by time evolving the solutions from various different initial conditions, sometimes from an initial mean structure based on the usual mixing-length description and with the modal perturbation fields of very small amplitude, but more often from other evolved modal solutions nearby in parameter space.

A. Flow momenta and thermodynamic fluctuations

Figure 1a presents the variation of the vertical and horizontal momentum amplitude functions, W and U, with depth z over our computational domain, with the solution displayed in equal increments of $\log\bar{P}$ so that the structures near the top, where the scale height gets small, are readily visible. The actual numerical solutions were constructed on a highly stretched grid so adjusted that all boundary-layer features are adequately resolved, with typically 300 mesh points being used for each variable in the vertical. The horizontal wavenumber of this particular solution is $a = 15$ when based on the 250 Mm depth of our computational domain, thus yielding a horizontal wavelength for this hexagonal cell of 140 Mm. The vertical momentum W in this solution peaks at $z \approx 60$ Mm and decreases sharply as the bottom of the convection zone is reached at $z \approx 200$ Mm. The vertical momentum also decreases steadily as the upper boundary is approached. Although this is partly the result of our boundary condition there, the decrease is primarily attributable to buoyancy and pressure forces, triggered by the shrinking scale height, that deflect the upward directed momentum at cell center into a strong horizontal shearing flow. The plot of horizontal momentum U in Figure 1a

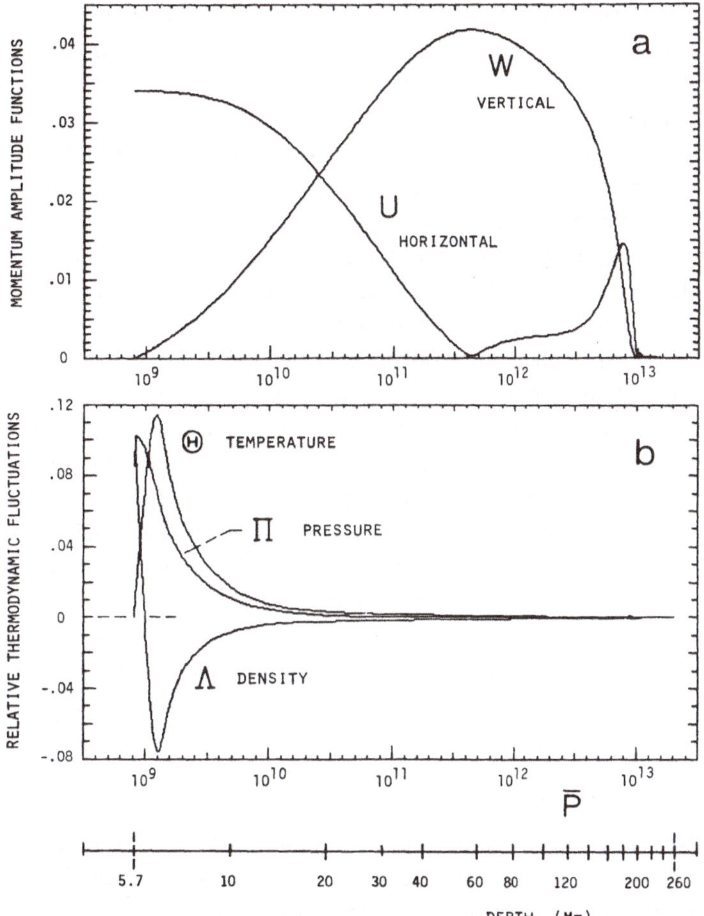

Fig. 1. Variation with depth (or $\log \bar{P}$) in the solar envelope of a typical single-mode nonlinear convection solution. The horizontal wavenumber for this hexagonal cell of large horizontal scale is $a = 14.0$, corresponding to a size of 140 Mm, and the flow is predominantly upward at cell center. Shown in (a) are the amplitude functions of the vertical momentum W and horizontal momentum U in dimensional units of (g cm^{-2} s^{-1}), and in (b) the amplitude functions of the relative thermodynamic fluctuations for temperature, θ, density, Λ, and pressure, Π.

shows that this component is especially prominent from the upper boundary to a depth of about 30 Mm, with another shearing flow also present near the bottom of the convection zone. Thus this large-scale cellular flow occupies the full extent of the unstable domain here, unlike what might have been expected from mixing-length predictions; there is little to confine the motions to a mere scale height or so. Certainly some of the variations seen in the solution near the top of the computational domain will be artifacts arising from the placement of our arbitrary upper boundary, though the striking tendency for the cellular convection to extend over many vertical scale heights appears to be a robust result.

The amplitude functions for the relative thermodynamic fluctuations of temperature, density and pressure are shown as θ, Λ, and Π in Figure 1b. They are of the same order throughout the zone and peak near the middle of the He^+ ionization region at a depth of about 7 Mm (see also Figure 2b). The amplitudes of these relative thermodynamic fluctuations are small: about 0.12 for the maximum of θ, but only of order 0.01 or less over most of the convection zone. The linearized relation $\Lambda = a\Pi - b\theta$ that links the relative thermodynamic fluctuations within the anelastic approximation is therefore justified. (The positive coefficients a, b depend upon $\overline{T}$ and $\overline{\rho}$ and the equation of state, and would just be unity if the gas were perfect.) When the pressure fluctuation Π is negligible, as within the Boussinesq approximation, then the density fluctuation Λ is just the mirror image of the temperature fluctuation θ. However when the stratification effects are significant, as in the upper portions of the solar convection zone, Π may be of the same order as θ and then Λ responds to both fluctuations. This may lead to a rising fluid element actually experiencing buoyancy braking near the top of an unstable zone, largely because Π can even change the sign of Λ. Although there is a hint of this near the upper boundary in Figure 1a, we do not place much credence in this because of our arbitrary boundary conditions there. However, such buoyancy braking arising from the effect of Π on Λ has a major role in the nonlinear modal solutions for convection studied in polytropes by Massaguer and Zahn (1980) and in the A-star convective envelopes of Paper III. Thus buoyancy braking caused by effects of compressibility where the scale height is small may also turn out to be an important process in the Sun, though we would have to extend the computational zone into the atmosphere to be sure of this.

B. What drives the motions

In order to understand the dynamics of such large-scale convection, it is useful to examine the rates of working by buoyancy and pressure forces, for these will largely control the power integrals for these flows. The expression $E_B = gW\Lambda$ represents work done by the mean pressure, or thus by buoyancy, while that by fluctuating pressure is $E_P = W\Pi\overline{P}D(1/\overline{\rho})$. In addition, a term E_V would represent the work done by viscous forces, but will not be shown explicitly. We should note that E_P represents the work done to modify the volume of a fluid parcel as it moves vertically, with this term vanishing in the abscence of density stratification, either in Boussinesq convection or in the comparable assumptions made in mixing-length treatments. The notation for E_P and E_B is such that positive terms increase the local kinetic energy in the flow. The net work produced by fluctuating pressure, or thus the integral in the vertical of E_P, plus that done by buoyancy, is converted into heat by viscous dissipation. These processes are however usually not in local balance, and thus flux terms serve to redistribute the energy vertically across the layer.

Figure 2a shows the variation with depth of E_B and E_P in the representative solution. We note that rates of working by buoyancy and pressure possess a relatively narrow peak at a depth of about 7 Mm in the middle of the He^+ ionization region (see Figure 2b). The buoyancy term E_B is consistently of greater amplitude than that of the pressure term E_P, and near the upper boundary a small region of buoyancy braking is

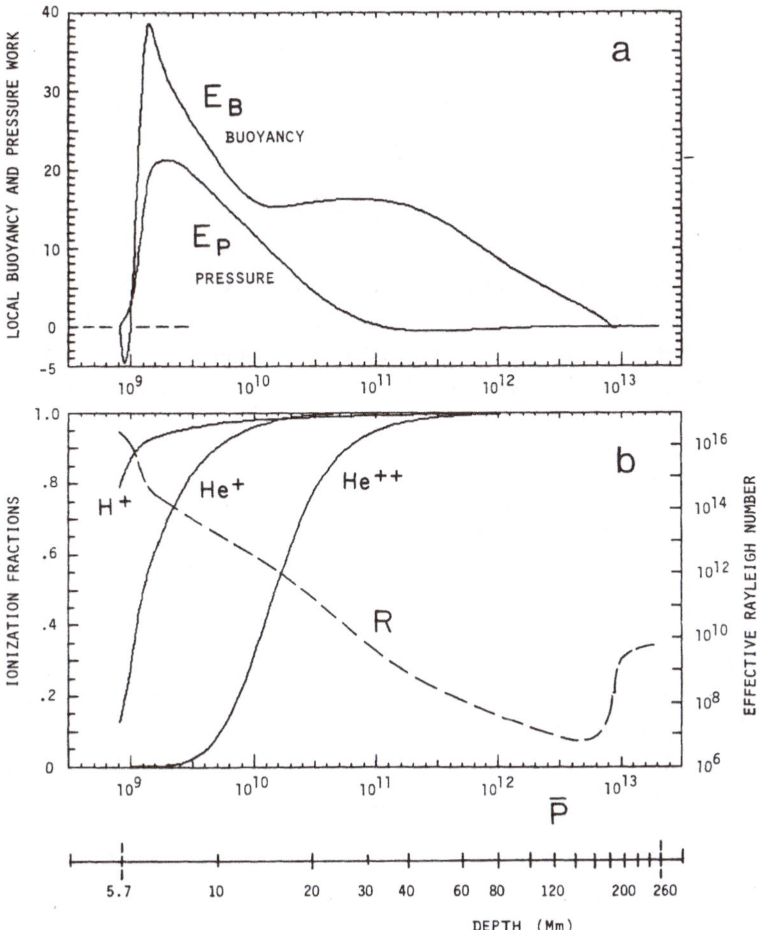

Fig. 2. Further properties of the nonlinear convection in the large-scale hexagon solution shown in Figure 1. (a) Variation with depth (or $\log \bar{P}$) of the local work done by the buoyancy (E_B) and by the fluctuating pressure (E_P). (b) Ionization fractions for H^+, He^+, and He^{++} with depth in this solar envelope, and the variation of the local effective Rayleigh number R. E_B and E_P are shown in the same arbitrary units.

evident. Further, the E_B profile in Figure 2a displays a plateau with a faint secondary maximum in the region where the second helium ionization is more than half completed, as seen in the curve of the ionization fraction of He^{++} in Figure 2b. Such a profile for E_B suggests that the buoyancy work used to drive the convective motions is largely concentrated in the ionization regions, with both the He^+ and He^{++} processes having a role in such driving. We should emphasize that the rate of working by the pressure terms through E_P has a significant effect on the flow, though the overall integral of E_B dominates. Closer inspection of the individual components that go into E_P reveal that pressure forces at depths shallower than about 20 Mm serve to decelerate the upward directed flow at the center of the cell and drive the consequent strong horizontal motions. The overall contributions by E_P in these regions is positive, meaning that net work is extracted by pressure terms from the unstable stratification, and so too by the buoyancy terms, but the pressure field has the key role in the initial deflection of the flow.

Figure 2b presents the degree of ionization of H and He with depth, and thus serves to quantify the locations of the ionization regions. Shown also is the variation of local Rayleigh number, based on the local degree of superadiabaticity, pressure scale height and enhanced turbulent viscosity. This Rayleigh number R ranges from about 10^{16} at the top to about 10^7 near the bottom of the convection zone, indicating that we have increased the viscosity typically by a factor of about 10^{12} over that of the natural molecular viscosity of the solar plasma. Such an enhancement of the viscosity, primarily by the turbulent viscosity $\bar{\nu}_T$, has made the numerical calculations tractable, and we have not found the resulting solutions to be overly sensitive to the functional forms used for $\bar{\nu}_T$.

C. The energy transports

Figure 3 displays how the various fluxes of energy in such modal convection vary with depth. Shown are the enthalpy or convective heat flux F_C, the radiative flux F_R, the viscous shear flux E_V, and the turbulent eddy heat flux F_E. These fluxes have been defined in Paper III, but for F_E, the divergence of which is introduced in Equation (2.4); all the fluxes have been normalized by the total flux. The conservation of energy in a steady state requires that the sum of these fluxes, together with the kinetic energy flux F_K not shown here, must be constant throughout the envelope. An interesting result is that F_C is dominant over most of the convection zone and is close to unity. Thus the large-scale convection mode is very efficient in transporting most of the solar flux. Near

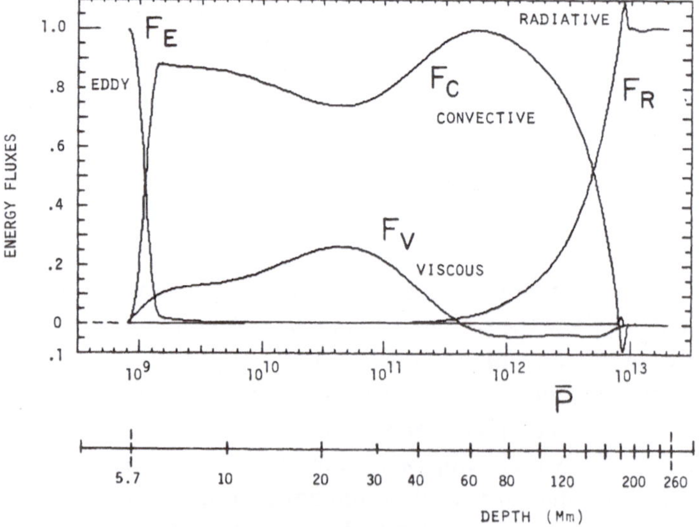

Fig. 3. Variation with depth (or $\log\bar{P}$) in the hexagon solution shown in Figure 1 of the enthalpy or convective heat flux F_C, the radiative flux F_R, the viscous shear flux F_V, and the small-scale turbulent eddy flux F_E. All fluxes have been scaled by the upward total flux.

the bottom of the zone, F_C is naturally supplanted by the radiative flux F_R, and even becomes slightly negative in a narrow overshooting region, forcing F_R to be larger than unity. Near our artificial upper boundary, the heat transport F_E by small-scale eddies takes over, thereby preventing the formation of a thermal boundary layer there since we force F_C to vanish. The viscous shear flux F_V represents a net kinetic energy transport by viscous stresses arising from our enhanced turbulent viscosity. This flux is not negligible over a significant portion of the zone, with F_C and F_V together accounting for most of the total flux over a large depth range. Figure 3 thus serves to emphasize that even a single large-scale mode, akin to possibly a giant cell, can make the mean stratification nearly adiabatic over most of the vertical extent of the convection zone.

4. Conclusions

The single-mode nonlinear solutions that we have studied all appear to suggest that the largest scales of convection in the Sun are prevented from getting through into the atmosphere with any significant fraction of their primary momentum. The shrinking scale height as the surface of the Sun is approached from below can produce major compressibility effects in the dynamics of large-scale convection that serve to deflect the cellular motions laterally. Thus fairly strong horizontal shearing flows may be present below the surface for scales of convection comparable to giant cells, and even some signature of them may extend into the atmosphere due to viscous stresses, whereas the vertical component of motion getting through into the photosphere is feeble at best. Certainly the preliminary anelastic modal solutions that we have considered so far only suggest such behavior, for we have imposed an artificial upper boundary in order to simplify the computational tasks, thereby ignoring much of what goes on explicitly in the hydrogen ionization zone. Also, our results are influenced by the enhanced turbulent viscosity that we have introduced to effectively reduce the local Rayleigh number and make the computations tractable. In particular, the vertical and horizontal momentum amplitude functions shown in Figure 1a translate into flow fields in which the maximum vertical velocity of about 10 m s^{-1} is attained at a depth of about 12 Mm and a peak horizontal velocity of about 100 m s^{-1} occurs at a depth of 7 Mm. These values can be changed by a factor of about 10 by changing the turbulent viscosity $\bar{\nu}_T$ by a factor of 10^5, thereby emphasizing that we are presently uncertain about the flow amplitudes that will be attained just below the surface. These modifications in $\bar{\nu}_T$ however appear to have little impact on the overall structure of the convective mode over the bulk of the zone where its convective heat transport is sufficient to account for most of the solar flux.

Future improvements to these preliminary anelastic models of solar convection will require a better representation of the small-scale turbulence and the inclusion of additional horizontal modes in the analysis. These modes will need to encompass horizontal scales of convection comparable to granulation, mesogranulation and supergranulation, in addition to those of giant cells, and it is not readily apparent just how sensitive the results will be to the number and choice of modes in such a representation of compressible turbulence. Clearly the analysis of the convective flows must also be extended into the atmosphere and radiative transfer effects there taken into account. We recognize that such steps are necessary before we can have any sense of comfort about the predictions of anelastic convection theory applied to the Sun, and we are thus engaged in implementing such improvements. What these preliminary single-mode solutions presently do provide is a sequence of nonlinear numerical experiments to help develop our intuition about highly compressible flows. Certainly many of the results are at variance with what is assumed in the effectively Boussinesq mixing-length approaches, for the convection here readily extends over multiple scale heights and is thereby decidedly nonlocal in character. The preliminary results to date suggest that prominent horizontal flows may be associated with the largest scales of convection at reasonably shallow depths below the surface. These may be capable of being sampled by the use of the five-minute oscillations as probes of such convective structures.

Acknowledgements

We thank Dr Douglas Gough for scientific advice about evolved and calibrated solar models and their sensitivities to changes in parameters. This continuing research was supported in part by the National Aeronautics and Space Administration through grants NSG–7511 and NAGW–91, by the Air Force Geophysics Laboratory through contract F19628–77–C–0104, by the National Science Foundation through grant ATM 80–20426, by travel grants from the Ministere des Affaires Etrangers, Paris, and by the Centre National de la Recherche Scientifique. The numerical calculations were partly carried out at Vaxlab at the University of Colorado and at the National Center for Atmospheric Research in Boulder, the latter sponsored by the National Science Foundation.

References

Cox, A. N. and Stewart, J. N.: 1965, *Astrophys. J. Suppl.* **11**, 94.
Dravins, D., Lindegren, L., and Nordlund, A.: 1981, *Astron. Astrophys.* **96**, 345.
Gilman, P. A.: 1978, *Geophys. Astrophys. Fluid Dynamics* **11**, 157.
Gilman, P. A.: 1980, *Highlights Astron.* **5**, 91.
Gilman, P. A. and Miller, J.: 1981, *Astrophys. J. Suppl.* **46**, 211.
Glatzmaier, G. and Gilman, P. A.: 1981, *Astrophys. J. Suppl.* **45**, 351.
Gough, D. O. and Weiss, N. O.: 1976, *Monthly Notices Roy. Astron. Soc.* **176**, 589.
Hurlburt, N. E., Toomre, J., Massaguer, J. M., and Graham, E.: 1982, *Astron. Astrophys.*, submitted.
Latour, J., Spiegel, E. A., Toomre, J., and Zahn, J.-P.: 1976, *Astrophys. J.* **207**, 233 (Paper I).
Latour, J., Toomre, J., and Zahn, J.-P.: 1981, *Astrophys. J.* **248**, 1081 (Paper III).
Latour, J., Massaguer, J. M., Toomre, J., and Zahn, J.-P.: 1982, *Astron. Astrophys.*, submitted.
Marcus, P. S.: 1980, *Astrophys. J.* **240**, 203.
Massaguer, J. M. and Zahn, J.-P.: 1980, *Astron. Astrophys.* **87**, 315.
Nelson, G. D.: 1980, *Astrophys. J.* **238**, 659.
Nelson, G. D. and Musman, S.: 1977, *Astrophys. J.* **214**, 912.
Nelson, G. D. and Musman, S.: 1978, *Astrophys. J. Letters* **222**, L69.
Nordlund, A.: 1980, in D. F. Gray and J. L. Linsky (eds.), 'Stellar Turbulence', *IAU Colloq.* **51**, 213.
Nordlund, A.: 1982, *Astron. Astrophys.*, submitted.
Toomre, J.: 1980, *Highlights Astron.* **5**, 571.
Toomre, J., Zahn, J.-P., Latour, J., and Spiegel, E. A.: 1976, *Astrophys. J.* **207**, 545.
Toomre, J., Gough, D. O., and Spiegel, E. A.: 1977, *J. Fluid Mech.* **79**, 1.
Toomre, J., Gough, D. O., and Spiegel, E. A.: 1982, *J. Fluid Mech.*, in press.
Zahn, J.-P.: 1980, in D. F. Gray and J. L. Linsky (eds.), 'Stellar Turbulence', *IAU Colloq.* **51**, 1.

ON THE DETECTION OF SUBPHOTOSPHERIC CONVECTIVE VELOCITIES AND TEMPERATURE FLUCTUATIONS*

D. O. GOUGH

Institute of Astronomy, and Department of Applied Mathematics and Theoretical Physics, University of Cambridge

and

J. TOOMRE

Joint Institute for Laboratory Astrophysics, and Department of Astrophysical, Planetary and Atmospheric Sciences, University of Colorado, Boulder, Colo., U.S.A.

Abstract. A procedure is outlined for estimating the influence of large-scale convective eddies on the wave patterns of five-minute oscillations of high degree. The method is applied to adiabatic oscillations, with frequency ω and wave number k, of a plane-parallel polytropic layer upon which is imposed a low-amplitude convective flow. The distortion to the $k - \omega$ relation has two constituents: one depends on the horizontal component of the convective velocity and has a sign which depends on the sign of ω/k; the other depends on temperature fluctuations and is independent of the sign of ω/k. The magnitude of the distortion is just at the limit of present observational sensitivity. Thus there is reasonable hope that it will be possible to reveal some aspects of the large-scale flow in the solar convection zone.

1. Introduction

The structure of the waves that produce the five-minute oscillations of the Sun depend principally on the mean vertical stratification of temperature. That stratification determines a fairly well defined sequence of relations between the frequency and the horizontal wave number of the oscillations. This has been exhibited for the real Sun in power spectra of Doppler measurements (e.g., Deubner *et al.*, 1979), and compared with theory to determine the adiabat in the isentropic part of the convection zone (e.g., Berthomieu *et al.*, 1980; Lubow *et al.*, 1980). However, these relations are not perfectly maintained: frequency is not precisely determined because the modes do not persist indefinitely, and the wave patterns are distorted by rotation and the inhomogeneities associated with convection. It is the purpose of this paper to report a preliminary theoretical assessment of the magnitude of the distortions, with a view to the eventual measurement of the velocity and temperature fluctuations in the convection zone. In an accompanying paper (Hill *et al.*, 1983) observational evidence for such distortions is presented.

We restrict our attention to modes of high degree. In that case the oscillations are trapped in a shallow wave guide just beneath the photosphere: the base of the trapping region is at a depth of about $nl^{-1}R$, where n and l are the order and degree of the mode

* Proceedings of the 66th IAU Colloquium: *Problems in Solar and Stellar Oscillations*, held at the Crimean Astrophysical Observatory, U.S.S.R., 1–5 September, 1981.

Solar Physics **82** (1983) 401–410. 0038–0938/83/0822–0401$01.50.

and R is the radius of the Sun. Because the wave guide is shallow we may neglect the curvature of the Sun and consider the dynamics of a plane atmosphere under a constant gravitational acceleration g. Moreover, we may associate with the mode a constant horizontal wave number $k \simeq lR^{-1}$.

The solar envelope is considered to be in a state of convection. Convective fluctuations in the thermodynamic properties of the envelope are regarded as being small compared with the corresponding horizontally averaged values, so that linearization about the mean state is valid. Furthermore, we consider the influence on the waves of only the largest scales of convective motion – the giant cells, perhaps – so that the temporal and horizontal spatial variations of the convective flow can be ignored compared with those of the waves. Essentially we are retaining just the leading terms in a JWKB expansion in time and horizontal Cartesian co-ordinates. Thus, we compute locally the perturbation to the dispersion relation between the frequency ω and the horizontal wave number k.

We impose no restriction on the vertical scale of variation. Convection has a tendency to form thin horizontal boundary layers, and we wish to allow for the possibility that these may be no thicker than the vertical scale of the waves. Thus we have in mind, for example, the problem of determining whether there are shear layers in the giant cells, such as have been found theoretically by Latour *et al.* (1983) with the single-mode representation of convection.

2. Perturbations to the Dispersion Relation

We first ignore the large-scale convective motion. The stratification of the envelope can then be specified by any two thermodynamic state variables, which are functions of just the vertical co-ordinate z. Here we choose the adiabatic sound speed $c(z)$ and the adiabatic exponent $\gamma(z) \equiv (\partial \ln p/\partial \ln \rho)_s$, where p and ρ are pressure and density and the derivative is taken at constant specific entropy s. Any linear normal mode of oscillation can be represented as a superposition of waves with the same horizontal wave number k. We assume the wave motion to be adiabatic. A wave variable χ, say, associated with such a wave can then be written in the separated form

$$\chi(\mathbf{x}, t) = \mathrm{Re}[X(z)e^{ikx + i\omega t}] \tag{2.1}$$

with respect to suitably orientated Cartesian co-ordinates $\mathbf{x} \equiv (x, y, z)$ and time t; the frequency ω is real (which, without loss of generality, we take to be positive) and depends on $|k|$ and the structure of the unperturbed atmosphere. It may be written

$$\omega^2 = \int f(n, |k|; c^2, \gamma)\, \mathrm{d}z\,. \tag{2.2}$$

Note that ω is independent of the horizontal direction of propagation of the wave.

The effect of the rotation of the Sun is simply to translate the wave pattern horizontally. We assume that this has been accounted for, and concentrate on the convective eddies. These influence the wave via a combination of advection and a modification to the function f resulting from variations in c^2 and γ. We ignore the influence of the vertical component of the convective velocity, for in the high wave number limit considered here we expect it to be much smaller than the other contributions to the distortion of the wave pattern.

It is evident that only the x component of the velocity of advection of the wave pattern represented by (2.1) has any significance. Let this velocity be $\tilde{U}$. Then the frequency ω is simply augmented by $k\tilde{U}$. Clearly, if the horizontal component of the convective velocity is independent of z throughout the region within which the wave is trapped, $\tilde{U}$ is simply the x component of that velocity. If the convective velocity varies with z, then once again $\tilde{U}$ depends only on its x component $U(z)$; it is simply an average of U weighted with the kinetic energy density of the wave (cf. Gough, 1978):

$$\tilde{U} = \int B(z)\, U(z)\, \mathrm{d}z\,, \tag{2.3}$$

where

$$B(z) \equiv \frac{\rho \mathbf{u}\cdot\mathbf{u}}{\int \rho \mathbf{u}\cdot\mathbf{u}\, \mathrm{d}z}\,, \tag{2.4}$$

and is independent of the sign of k; $\mathbf{u}(z)$ is the velocity amplitude of the wave and ρ is the density of the unperturbed envelope. The integrals are presumed to be taken over the entire extent of the envelope. Since horizontal density variations induced by the convection are presumed to be small, ρ may be replaced by the horizontally averaged density $\bar{\rho}$, which is a function of z alone.

The changes to the dispersion relation arising from the slow horizontal variation in the thermodynamic state of the gas can be obtained simply by expanding c^2 and γ about their horizontally averaged values. Setting

$$\begin{aligned} c^2(\mathbf{x}, t) &= \overline{c^2}(z) + \delta c^2(\mathbf{x}, t)\,, \\ \gamma(\mathbf{x}, t) &= \bar{\gamma}(z) + \delta\gamma(\mathbf{x}, t)\,, \end{aligned} \tag{2.5}$$

one obtains formally for the linearized change $\delta\omega^2$ in ω^2:

$$\delta\omega^2 = \int \overline{c^2}\, \frac{\delta f}{\delta c^2}\, \frac{\delta c^2}{\overline{c^2}}\, \mathrm{d}z + \int \bar{\gamma}\, \frac{\delta f}{\delta \gamma}\, \frac{\delta\gamma}{\bar{\gamma}}\, \mathrm{d}z\,, \tag{2.6}$$

where $\delta f/\delta c^2$ and $\delta f/\delta\gamma$ are the variational derivatives of f with respect to c^2 and γ.

The results can be combined to yield an expression for the total local variation $\Delta\omega$ in the frequency ω of the component of the wave pattern whose local horizontal wave number is k. This may be written in the form

$$\frac{\Delta\omega}{\omega} = \frac{k}{\omega} \int zBU\, \mathrm{d}\ln z + \int F \frac{\delta c^2}{\overline{c^2}}\, \mathrm{d}\ln z + \int G \frac{\delta\gamma}{\bar{\gamma}}\, \mathrm{d}\ln z\,. \tag{2.7}$$

An explicit expression for the kernel F, in the case when γ is constant, is derived in the Appendix.

Notice that since f is independent of the sign of k, so are F and G. Thus the second and third terms on the right-hand side of Equation (2.7) are independent of the wave direction, whereas the first term has opposite signs for waves propagating in the positive and negative x directions. Hence, if local dispersion relations can be measured for waves propagating in opposite directions, the contributions to their variations induced by large-scale convection can be partially separated by computing sums and differences of those variations.

3. A Simple Illustration

We illustrate the result by considering an envelope of infinite depth whose horizontally averaged structure is that of a polytrope of index m. In addition we take γ to be constant.

In a polytropic model the square of the sound speed increases linearly with depth:

$$c^2 = \frac{\gamma g z}{m+1}, \tag{3.1}$$

where z is measured downwards from the top of the envelope. Adiabatic p modes of order n have a dispersion relation given by

$$\omega^2 = \frac{\gamma g |k|}{m+1} \left\{ n + \tfrac{1}{2}m + \left[(n + \tfrac{1}{2}m)^2 - \frac{m+1}{\gamma}\left(m - \frac{m+1}{\gamma} \right) \right]^{1/2} \right\} \tag{3.2}$$

(e.g. Gough, 1978). They are trapped in a region that extends roughly to the depth z_t at which $\omega^2 = k^2c^2$, namely where c^2 is such that pure plane acoustic waves of frequency ω and horizontal wave number k, if they could exist, would travel horizontally. They are also evanescent near the surface, very roughly at depths less than z_c at which ω is equal to Lamb's acoustical cutoff frequency $\gamma g/2c$ for an isothermal atmosphere. The depths z_c and z_t are essentially the turning points of a JWKB analysis, though they do not correspond precisely unless the variables are chosen judiciously.

It is evident from Equations (3.1) and (3.2) that z_t is insensitive to the structure of the envelope, and approaches nk^{-1} as n increases. On the other hand, z_c is roughly inversely proportional to the gradient of the sound speed, and also to n.

The velocity amplitude of the f mode is proportional to $e^{-|k|z}$, and the frequency is $(g\,|k|)^{1/2}$.

Kernels B and F are illustrated in Figures 1 and 2 for the f mode and the first five p modes. In this simple example, the kernels G are zero. As one might expect, the depths of the main contributions to the kernels increase with n, since z_t increases with n. The velocity kernels are, of course, positive everywhere. For p modes the kernels F are positive throughout most of the region, as one would expect because increasing c^2 tends to decrease the acoustical propagation time across the wave guide, thus increasing the resonant frequency. That is not so in the evanescent region at the top of the layer, however, because there an increase in c^2 results in a decrease in the acoustical cutoff frequency. This increases the extent of the region within which the waves can propagate, and so increases the sound travel time across the trapping region. No such reversal in the sign of F occurs at the bottom of the trapping region, since z_t is intensitive to c^2. The kernel F is identically zero for the f modes, because the frequencies of f modes in a plane parallel atmosphere are independent of stratification.

To model a possible giant cell, velocity and temperature fluctuations were obtained from a steady two-dimensional convective flow of a fluid with $\gamma = \frac{5}{3}$. The flow was computed by N. Hurlburt using the numerical programme described by Graham (1975). The Rayleigh number was 10^6 and the Prandtl number was unity, and the mean polytropic index characterizing the horizontally averaged density stratification was about 1.2. The convecting region extended over about 3.6 density scale heights. Amplitudes of the vertical component of velocity and the temperature fluctuation, W and δT, were estimated by taking at each value of z half the difference between the values of velocity and temperature at the two sides of the cell, where the velocity was constrained to be vertical. The horizontal velocity amplitude U was taken to be the horizontal component of velocity on the vertical line midway between the sides. These functions were used to represent the giant-cell flow in the bottom 3.6 scale heights of the convection zone. The relative temperature fluctuation $\delta T/T$ was used without modification and the velocity was scaled by a constant factor, chosen to make the maximum value of the rms vertical component equal to 50 m s^{-1}; this is roughly consistent with mixing-length estimates at the height at which the maximum occurs. Beneath the convection zone, which is assumed to be 200 Mm deep, convective fluctuations were presumed to vanish; above the region modelled by Hurlbert's solution the temperature fluctuation and the vertical velocity were set to zero, and the horizontal velocity was assumed to be constant and continuous with the flow beneath. The result is illustrated in Figure 3.

The contributions from U and $\delta T/T = \overline{\delta c^2/c^2}$ to the frequency deviation were computed from the appropriate terms in Equation (2.7), using the kernels depicted in

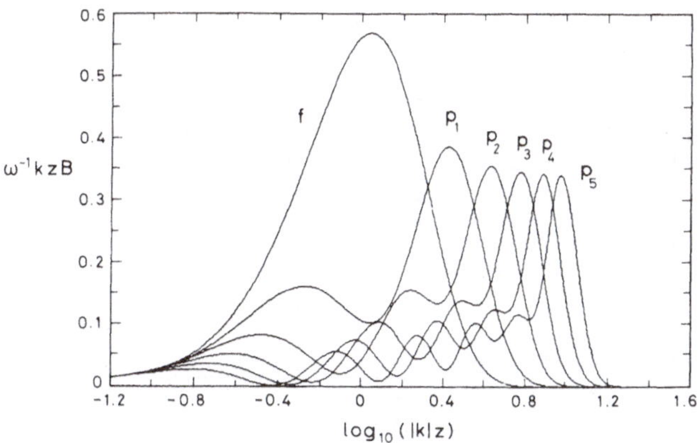

Fig. 1. Kernels $\omega^{-1}kzB$ for a polytrope of index $m = 1.2$ with $\gamma = \frac{5}{3}$, measured in units of $(k/g)^{1/2}$.

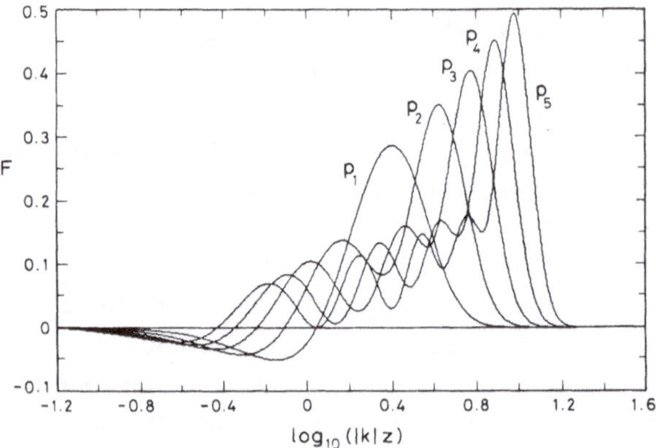

Fig. 2. Kernels F for a polytrope of index $m = 1.2$ with $\gamma = \frac{5}{3}$.

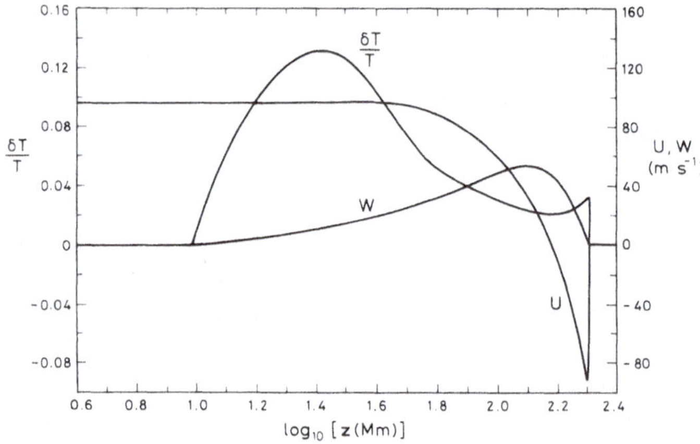

Fig. 3. Horizontal and vertical components (U, W) of velocity and relative temperature fluctuation $\delta T/T$ associated with the assumed large-scale convective flow.

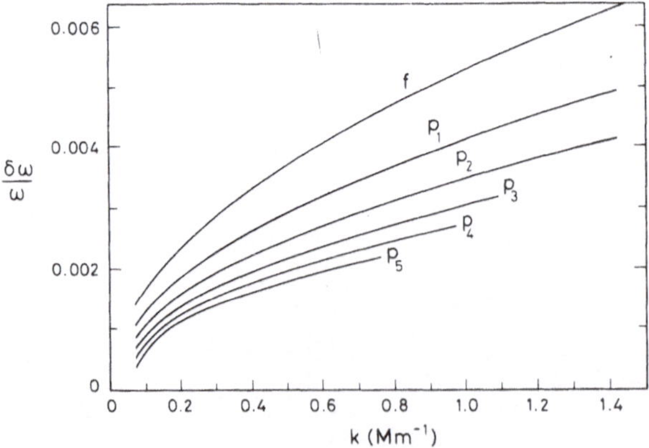

Fig. 4. Relative frequency perturbations for $k > 0$ produced by the horizontal component of velocity illustrated in Figure 3. The perturbations for $-k$ have the same magnitude, but opposite sign.

Figures 1 and 2. The results are shown in Figures 4 and 5. Notice that the contribution from the temperature fluctuation is such that it should be detectable by modern observations, for the half-widths of the bands in the power spectrum of Deubner *et al.* (1979) are only of order one per cent. The velocity contribution is also large enough to be interesting, especially since the modal computations of Latour *et al.* (1983) suggest that we may have underestimated the speed of the horizontal flow.

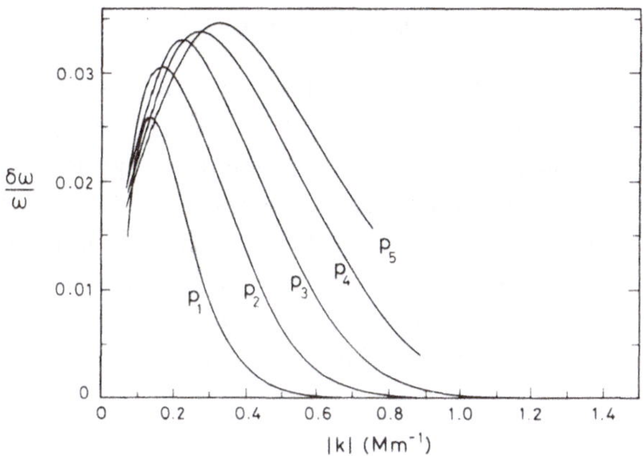

Fig. 5. Relative frequency perturbations produced by the temperature fluctuation illustrated in Figure 3. The perturbation is independent of the sign of k.

The forms of the frequency changes are also noteworthy. The modes are confined to a layer whose depth is about nk^{-1}, and only in that layer do they experience the mean structure of the envelope. Thus at high k the velocity contribution to $\Delta\omega/\omega$ is simply that of advection by the surface value of U. But as k decreases a more extensive average of U is taken. Since U decreases monotonically with depth, $\Delta\omega$ decreases with diminishing k, and at the lowest values of k (not shown in Figure 4) the frequency perturbations change sign. The relative temperature fluctuation, on the other hand, has a maximum at a depth of about 25 Mm, and its contribution to $\Delta\omega$ is greatest when k is such that the main weight of the kernel occurs also at this depth.

We have not attempted to estimate from these results how an actual power spectrum would be affected. It is evident that if the horizontal scale of giant cells is comparable with the region of the solar disk over which the observations are made, the convective fluctuations would both broaden and distort the bands of power. The distortions would vary with time as the cells are caused to drift across the field of view by the rotation of the Sun. Moreover, the different contributions to the distortion will be out of phase, the temperature contribution being a maximum roughly when the velocity contribution is zero. There is evidence for such behaviour in the observations of Hill *et al.* (1982), though the data are not sufficiently free from noise for us to be sure that they have been interpreted correctly.

4. Conclusion

Large-scale subphotospheric convective motion in the Sun is likely to be of a magnitude just sufficient to be detected by present observations of five-minute oscillations of high degree. The form of the distortion to the $k-\omega$ power spectra depends on the structure of the flow in the convective eddy. With longer more and precise observations it should be possible to measure some aspects of that flow.

Acknowledgements

We are grateful to Mr N. Hurlburt for making his model of convection available to us. The work reported in this paper was carried out whilst J.T. held a Science Research Council Senior Visiting Fellowship at the Department of Applied Mathematics and Theoretical Physics in Cambridge, and subsequently when D.O.G. was a visitor at the Joint Institute for Laboratory Astrophysics, which is operated jointly by the University of Colorado and the National Bureau of Standards. The research was also supported in part by the National Aeronautics and Space Administration by grants NSG–7511 and NAGW–91, and by the US Air Force Geophysics Laboratory by contract F196228–77C–0104.

Appendix: *F*: the c^2 Kernel

We restrict attention to constant γ, and start from Lamb's (1932) form of the equation describing adiabatic waves in a plane parallel atmosphere:

$$\left(\frac{\mathrm{d}^2}{\mathrm{d}z^2} - k^2\right)X + \left(\frac{\mathrm{d}\ln c^2}{\mathrm{d}z} + \frac{\gamma g}{c^2}\right)\frac{\mathrm{d}X}{\mathrm{d}z} + \\ + \left\{\frac{\omega^2}{c^2} - \frac{gk^2}{\omega^2}\left[\frac{\mathrm{d}\ln c^2}{\mathrm{d}z} - (\gamma-1)\frac{g}{c^2}\right]\right\}X = 0\,, \tag{A1}$$

where $X(z)$ is the (real) amplitude of $\chi = \operatorname{div}\mathbf{u}$, as in Equation (2.1), and z is measured downwards. We shall assume that $c^2 = 0$ at $z = 0$, and choose solutions X that are regular at $z = 0$ and that tend to zero as $z \to \infty$.

The equation may be cast into self-adjoint form by multiplying it by the factor $c^2\psi$, where

$$\psi = \exp\left[-\gamma g \int_z^Z c^{-2}(s)\,\mathrm{d}s\right] \tag{A2}$$

and Z is any positive constant. Multiplying by X and integrating then yields

$$\omega^4 \int_0^\infty X^2\psi\,\mathrm{d}z - \omega^2 \int_0^\infty c^2(X'^2 + k^2X^2)\psi\,\mathrm{d}z - \int_0^\infty \Omega^4 X^2\psi\,\mathrm{d}z = 0\,, \tag{A3}$$

where

$$\Omega^4 = gk^2\left[\frac{\mathrm{d}c^2}{\mathrm{d}z} - (\gamma - 1)g\right] \tag{A4}$$

and a prime denotes differentiation with respect to the argument. The equation provides a variational principle for the problem: the stationary values of $\omega^2(X)$ defined by Equation (A3) amongst all functions X that satisfy the boundary conditions are the eigenvalues of Equation (A1), and occur when X is an eigenfunction.

It is now a simple matter to compute from Equation (A3) the linearized change $\delta\omega^2$ in ω^2 that occurs when c^2 is changed to $c^2 + \delta c^2$. In view of the stationary property of (A3), variations in X that arise from perturbing the equilibrium state do not affect ω^2 to leading order, and therefore need not be calculated. The linearized perturbation to Equation (A3) is thus

$$\begin{aligned}&\gamma g \int_0^\infty [(\omega^4 - \Omega^4)X^2 - \omega^2 c^2(X'^2 + k^2X^2)]\psi \,\mathrm{d}z \int_z^Z c^{-4}(s)\delta c^2(s)\,\mathrm{d}s - \\ &-\omega^2 \int_0^\infty (X'^2 + k^2X^2)\psi\delta c^2 \,\mathrm{d}z + gk^2 \int_0^\infty (X^2\psi)'\,\delta c^2\,\mathrm{d}z - I\delta\omega^2 = 0\,,\end{aligned} \tag{A5}$$

where

$$I = \int_0^\infty [c^2(X'^2 + k^2X^2) - 2\omega^2X^2]\psi\,\mathrm{d}z\,. \tag{A6}$$

After interchanging the order of integration in the double integral, and taking the limit $Z \to \infty$, Equation (A5) may be rewritten

$$\frac{\delta\omega^2}{\omega^2} = 2\int_0^\infty F\,\frac{\delta c^2}{c^2}\,\mathrm{d}\ln z\,, \tag{A7}$$

where

$$\begin{aligned}F = \tfrac{1}{2}z\omega^{-2}I^{-1}\Big\{&\gamma g c^{-2}\int_0^z [(\omega^4 - \Omega^4)X^2 - \omega^2c^2(X'^2 + k^2X^2)]\psi\,\mathrm{d}s - \\ &- [\omega^2c^2(X'^2 + k^2X^2) - gk^2(\gamma g X + 2c^2X')X]\psi\Big\}\,.\end{aligned} \tag{A8}$$

In the case of the polytrope of index m, c^2 is given by Equation (3.1), ψ is proportional to z^{m+1}, $\Omega^4 = g^2k^2(m + 1 - m\gamma)/(m + 1)$, ω^2 is given by Equation (3.2) and $X = e^{-|k|z}L_n^{(m+2)}(2|k|z)$, where L is a Laguerre polynominal.

References

Berthomieu, G., Cooper, A. J., Gough, D. O., Osaki, Y., Provost, J., and Rocca, A.: 1980, in H. A. Hill and W. A. Dziembowski (eds.), *Nonradial and Nonlinear Stellar Pulsation*, Springer, Heidelberg, p. 307.

Deubner, F.-L., Ulrich, R. K., and Rhodes, Jr. E. J.: 1979, *Astron. Astrophys.* **72**, 177.

Gough, D. O.: 1978, in G. Belvedere and L. Paternò (eds.), *Proc. Workshop on Solar Rotation*, University of Catania Press, Catania, p. 255.

Graham, E.: 1975, *J. Fluid Mech.* **70**, 689.

Hill, F., Toomre, J., and November, L.: 1983, *Solar Phys.* **82**, 411 (this volume).

Lamb, H.: 1932, *Hydrodynamics*, Cambridge University Press, Cambridge.

Latour, J., Toomre, J., and Zahn, J.-P.: 1983, *Solar Phys.* **82**, 387 (this volume).

Lubow, S. H., Rhodes, Jr. E. J., and Ulrich, R. K.: 1980, in H. A. Hill and W. A. Dziembowski (eds.), *Nonradial and Nonlinear Stellar Pulsation*, Springer, Heidelberg, p. 300.

VARIABILITY IN THE POWER SPECTRUM OF SOLAR FIVE-MINUTE OSCILLATIONS*

(Invited Review)

FRANK HILL and JURI TOOMRE

*Department of Astrophysical, Planetary and Atmospheric Sciences**, and Joint Institute for Laboratory Astrophysics†, University of Colorado, Boulder, CO 80309, U.S.A.*

and

LAURENCE J. NOVEMBER

Air Force Geophysics Laboratory and AURA, Sacramento Peak Observatory, Sunspot, NM 88349, U.S.A.

Abstract. Two-dimensional power spectra of solar five-minute oscillations display prominent ridge structures in (k, ω) space, where k is the horizontal wavenumber and ω is the temporal frequency. The positions of these ridges in k and ω can be used to probe temperature and velocity structures in the subphotosphere. We have been carrying out a continuing program of observations of five-minute oscillations with the diode array instrument on the vacuum tower telescope at Sacramento Peak Observatory (SPO). We have sought to establish whether power spectra taken on separate days show shifts in ridge locations; these may arise from different velocity and temperature patterns having been brought into our sampling region by solar rotation. Power spectra have been obtained for six days of observations of Doppler velocities using the Mg I λ5173 and Fe I λ5434 spectral lines. Each data set covers 8 to 11 hr in time and samples a region $256'' \times 1024''$ in spatial extent, with a spatial resolution of $2''$ and temporal sampling of 65 s. We have detected shifts in ridge locations between certain data sets which are statistically significant. The character of these displacements when analyzed in terms of eastward and westward propagating waves implies that changes have occurred in both temperature and horizontal velocity fields underlying our observing window. We estimate the magnitude of the velocity changes to be on the order of 100 m s^{-1}; we may be detecting the effects of large-scale convection akin to giant cells.

1. Introduction

Ridge structures are evident in spatial and temporal (k, ω) power spectra of the observed five-minute oscillations of the Sun (e.g., Deubner *et al.*, 1979). Such ridges in power arise because acoustic waves trapped in the thermal structure below the solar surface possess a fairly specific relationship between their frequency ω and horizontal wavenumber k. Since the character of the oscillations is determined largely by the vertical stratification of temperature below the solar surface, the observed positions of the ridges could be compared with theory to determine at least the adiabat in the isentropic part of the convection zone (e.g., Berthomieu *et al.*, 1980; Lubow *et al.*, 1980). However, the dispersion relations linking k and ω will be perturbed by temperature and velocity structures associated with the convection, and so too by differential rotation. The purpose of this observational study is to attempt to detect shifts in ridge positions that

* Proceedings of the 66th IAU Colloquium: *Problems in Solar and Stellar Oscillations*, held at the Crimean Astrophysical Observatory, U.S.S.R., 1–5 September, 1981.

** Formerly Department of Astro-Geophysics.

† JILA is operated jointly by the University of Colorado and the National Bureau of Standards.

Solar Physics **82** (1983) 411–425. 0038–0938/83/0822–0411$02.25.

may be attributed to large-scale convective fluctuations of temperature and velocity. A preliminary theoretical assessment of the relative ridge displacements that may arise from such convection is presented in an accompanying paper by Gough and Toomre (1983).

The apparent frequency ω of an acoustic mode trapped below the surface can be modified in several ways. The simplest is that of a Galilean translation of the wave fronts: a uniform horizontal flow of amplitude U will advect the wave fronts, producing a change in ω with respect to a stationary observer. This change $\Delta\omega$ scales as kU, but its sign depends upon the direction of propagation of the wave relative to U. If this flow varies with depth z, the effective velocity that is sampled by each mode may be different for it depends on an average of $U(z)$ weighted with the kinetic energy density of the wave with depth (cf. Gough, 1978). In contrast, steady flows with a cellular pattern in the horizontal produce local dilations and contractions in the wavefronts, thereby modulating the local ω. However, if the horizontal scale of such a cellular flow is much larger than that of the trapped mode being sampled, then (k, ω) power spectra formed over different portions of the Sun should display reasonably simple ridge shifts in ω. Thus giant cells with proposed scales of 300 Mm or greater may be capable of being detected by using modes with horizontal wavelengths for instance between 5 to 25 Mm. On the other hand, supergranular flows with typical 50 Mm scales would primarily broaden such ridges in ω, owing to the large horizontal areas that must be sampled by the observations before the k resolution of the Fourier transforms becomes small enough to discern ridges. Finally, the presence of horizontal temperature variations across a cell also produces dilations and contractions in the wavefronts of the acoustic modes, most simply because of changes in the local sound speed. Once again the local change in ω is determined by an integral with depth of the perturbation in sound speed weighted by the kinetic energy density of the wave, but the sign of $\Delta\omega$ is independent of the propagation direction. Thus the oscillations could also serve as probes of temperature structures, though here too interpretations are simplest if the horizontal scale of the inhomogeneity is much larger than that of the modes.

Since the variations produced in ω by horizontal flows depend upon the wave propagation direction while those for temperature do not, it should be possible to separate the two effects of convection by measuring local dispersion relations for waves travelling in opposite directions. Theoretical details concerning the perturbations to the dispersion relations are given in the accompanying paper by Gough and Toomre (1983), along with illustrations of the kernels that enter into the integrals that determine $\Delta\omega$. An outline of the inversion procedure required to deduce the velocity and temperature variations with depth from observed ridge shifts is discussed in Gough (1978).

In this observational search for effects of large-scale convection on the spectrum of five-minute oscillations, we will restrict ourselves to modes of high degree. The oscillations are then trapped in a relatively shallow resonant cavity, or wave guide, just beneath the photosphere. The lower reflection occurs at a depth of about nR/l, where n and l are the order and degree of the mode and R is the radius of the Sun. The horizontal wavenumber of such modes is $k \simeq l/R$. The observations will be used to determine (k, ω) power diagrams for individual days, and the ridge structures in these will be compared from day to day to look for any significant displacements. We will concentrate on modes with horizontal wavenumbers ranging from about 0.25 to 1.25 Mm^{-1}, with the corresponding l values ranging from about 175 to 875, and for n values of 1 to 4.

2. Observations and Reduction

Our observations are made with the SPO diode-array, which operates at the exit of the echelle spectrograph attached to the vacuum tower telescope. Four strings of 128 diodes each are arranged parallel to the spectrograph slit. One pair of diode strings, with 0.052 Å masks, is located at ± 0.079 Å in the magnetically insensitive Fe I λ5434.5 line, which allows us to measure Doppler velocities in the lower photosphere (Altrock *et al.*, 1975). The second pair of diode strings, with 0.120 Å masks, is located at ± 0.060 Å in the Mg I λ5172.7 line, which is representative of the temperature minimum region (Altrock and Canfield, 1974). Two-dimensional intensity and velocity images are obtained by spatially scanning the solar image across the spectrograph slit. The scanning direction is parallel to the solar equator and covers 512 steps, producing nominally a $256'' \times 1024''$ image with a spatial resolution of $2''$. Such an observing window elongated in the east–west direction will permit detection of waves with a wide range of horizontal wavenumbers k which are propagating parallel to the equator; averaging of the data in the north–south direction will serve to largely filter out waves traveling in oblique directions. These scans are repeated every 65 s for periods as long as possible given the constraints of weather and the extent of the daylight hours; the observational sets typically span 8 to 11 hr. The scanning region remains centered on the disc throughout the observing day.

The determination of instantaneous Doppler velocities and intensities at each spatial position requires forming appropriate sums, differences and ratios of the diode measurements carried out on each side of the spectral line. Details of the procedures used are discussed in November *et al.* (1979, 1982), together with an analysis of the major instrumental errors affecting the velocity measurements. We find that sensitivity variations in the diodes and random seeing effects in the spectrograph are the dominant sources of instrumental error at spatial scales (in the scan direction) smaller than about 200 Mm, or spatial wavenumbers $k \gtrsim 3 \times 10^{-2}\ \text{Mm}^{-1}$; we estimate that the signal-to-noise ratio in our determination of instantaneous Doppler velocities is about 30 for such spatial scales. Noise is also introduced by spectrograph grating jitter, the latter driven by mechanical flexure of the spectrograph tube. As a consequence of our scanning procedure to build up to two-dimensional velocity image, such flexure at low temporal frequencies results in noise which may dominate the velocity data at spatial scales greater that about 200 Mm. Since the flexure is a stochastic process, noise is distributed over a broad range of temporal frequencies. Incremental stepping of the observing table to compensate for rotation of the solar image also excites some grating jitter. Because the stepping is carried out at precise intervals of 1 s in time and is synchronized with the start of each spatial scan, this noise appears as power at zero temporal frequency and at selected spatial frequencies.

The apparent position of a ridge in the (k, ω) domain will be affected by our ability to define accurately the temporal and spatial scales in our observations. We believe that our time scale is very stable, for the telescope control computer sequencing has access to a precision clock and we monitor the repetition timing for the scanning with care. The stability of the spatial reference scale is a more intricate issue. Basically this scale is set by the apparent diameter of the Sun. Changes in the magnification of the telescope or the spectrograph, variations in atmospheric refraction, and changes in the travel of the scanning platform for the spectrograph entrance slit can each modify the apparent spatial scale. Given the thermal and optical design of the vacuum telescope and the echelle spectrograph, we estimate that thermal expansion effects in the structures produce variations in the apparent solar diameter of less than 0.05%. Reproducibility of magnification in the diode-array apparatus also does not constitute a problem, since thermal stresses are negligible and all equipment has been undisturbed for the entire time span of the observations. Differential atmospheric refractive effects will change the image scale during the course of the observations in a fairly predictable manner (Simon, 1966; Saastamoinen, 1979), but this effect is only significant very close to the horizon. We have corrected for the shape of the Sun given its position in the sky, and estimate that residual errors in image scale due to uncorrected atmospheric refraction effects in our observations are less than 0.03%. In addition, we apply daily corrections for changes in the apparent radius of the Sun due to the Earth's orbital position. Further, the vacuum telescope is now equipped with a limbtracking guiding telescope that is locked to the main optical beam by an actively monitored laser system, and we find that the main telescope pointing is better than the seeing limit throughout the course of a day's observations. The precision with which we can return to the same starting point for the scans and the extent of travel of the scanning platform are now reproducible to the extent that the spatial reference scale may be modified at most by 0.02%.

The construction of (k, ω) power diagrams from our data proceeds in the following manner. The two-dimensional velocity raster formed every 65 s is averaged in the direction of the shorter dimension perpendicular to the equator. This has the effect of averaging out the signature of most waves propagating obliquely to the slit and preserves those waves traveling perpendicular to it, or parallel to the solar equator. We form a two-dimensional array with these collapsed rasters, with each row of the array containing the averaged data from one scan in time. We extend our data set with zeroes if necessary to a length of 644 points so that the resolution bin width in ω is constant for all runs to permit detailed comparison of results from different days. Thus the resolution bin width in ω is $1.5 \times 10^{-4}\ \mathrm{s}^{-1}$.

The length of our scans parallel to the equator is about one solar radius. We therefore have to correct our data for foreshortening effects, and do so by interpolating our data from its observed rectangular grid onto a projected spherical grid with equal longitude bins of width 0.133 degrees. The width of the corresponding horizontal wavenumber k bins is $7.6 \times 10^{-3}\ \mathrm{Mm}^{-1}$. The data are apodized or tapered to zero at the edges of the collapsed array by using either a cosine bell window or a variety of exponential functions extending typically 10% into the data. The apodization is also applied around any gaps that may be present due to clouds or other causes. The ridge locations are insensitive to the various apodizing procedures that we have examined, although power between the ridges is influenced by the functional form. After extending, interpolating and apodizing the data, we apply a two-dimensional fast Fourier transform to it and produce power and phase pictures from the complex transform.

We studied differences of ridge locations in (k, ω) power diagrams from two different days by at first blink comparing them on a color imaging system. This procedure suggested that there were relative changes in ridge positions, but these differed from ridge to ridge and quadrant to quadrant. However, this qualitative impression is easily influenced by changes in the width of the ridges and by the choice of scaling for the displays. Consequently, we developed the following quantitative procedure for determining differences in the position of a given ridge on two different days. First, a contour map of a (k, ω) diagram is produced on a graphics device. The cursor of the device is used to draw a polygonal box enclosing the ridge of interest, and the positions of the vertices of the box in the (k, ω) diagram are recorded. The region of the diagram enclosed in the box is extracted from the full diagram, and the power as a function of ω for each fixed k in this region is then examined. An integral moment method is used to define the centroid of power in a given ridge at each value of k: we adopted the ratio of the first moment to the zeroth moment to specify the centroid position. These points are then fitted with a least squares variable knot cubic spline. This fitting algorithm applies a least squares fit to the segments of data between knot locations. The cubic spline conditions are satisfied at the knots, and the positions of the knots are varied from some initial distribution until the least squares error is minimized. This error is the root mean square value of the deviation between the local ridge centroids and the calculated fit. We typically use 5, 10, and 28 knots in our cubic spline fits for data sets involving about 130 points on a given ridge

3. Results

Our data base currently consists of six days of observations carried out in February and March 1981. Details of these data sets are provided in Table I which lists the dates and

TABLE I

Summary of oscillation data sets

Date	Time extent	Time samples
* 22 Feb. 81 (Day 53)	9.4^h	522
23 Feb. 81 (Day 54)	6.9^h	384
* 24 Feb. 81 (Day 55)	9.9^h	547
* 23 Mar. 81 (Day 82)	11.4^h	632
25 Mar. 81 (Day 84)	10.0^h	512
26 Mar. 81 (Day 85)	10.7^h	550

corresponding day numbers in the year, the total extent in time of the observations, and the number of spatial raster scans (or time samples) available in each set. We have chosen to present here comparisons involving the three data sets marked with asterisks in Table I. These sets are the longest in time extent and show the least noise. Comparisons of ridge locations involving the other days yield results which are less certain because of increased observational noise, but these generally support the conclusions based on Days 53, 55, and 82. We presently have a further seven data sets that need to be analyzed from more recent observations in this continuing program.

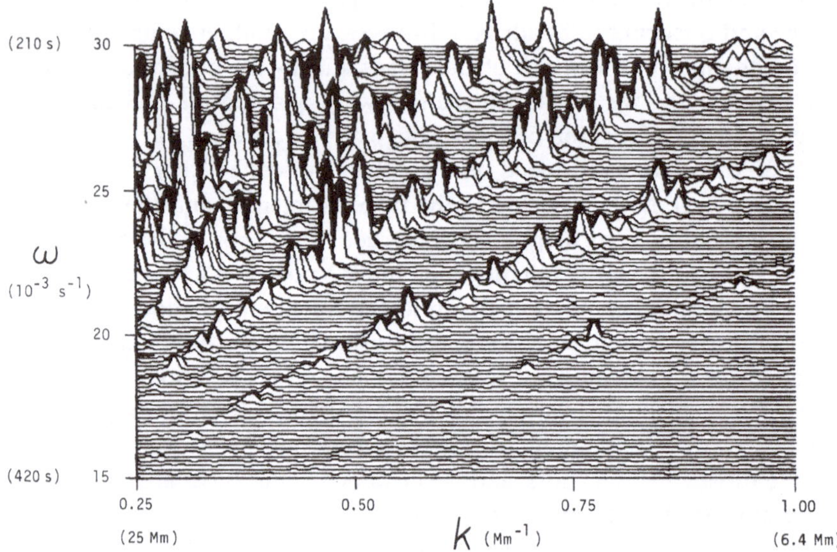

Fig. 1. Portion of positive ω quadrant of (k, ω) power diagram for high degree solar oscillations observed in Mg I Doppler velocities on Day 82 (23 March, 1981). The p_0 to p_4 ridges appear distinctly in the foreground, while other ridges are partially obscured in this perspective view. The power levels have been multiplied by the arbitrary factor k to enhance the ridges at high values of k. An uneven distribution of power is evident along the ridges. These variations are probably due to the effects of interference from mode beating within the resolution bins and bin coupling from seeing.

Figure 1 displays a portion of one of the quadrants of the (k, ω) power diagram for Doppler velocities observed in Mg I on Day 82. The section shown is an area with a high signal-to-noise ratio where instrumental error effects are negligible. We can easily discern four ridges in this perspective view, with another three ridges partially hidden in the background of the plot. A full contour plot of this quadrant clearly shows 11 ridges. The fundamental, or p_0, ridge possesses substantially less power than the other ridges, in keeping with previous observational results (e.g. Deubner *et al.*, 1979; Rhodes *et al.*, 1981). There is also considerable variation in power along any given ridge, with the power often appearing to be distributed in a succession of peaks and troughs. Such variations in power are partly due to beating between discrete modes whose separations in k and ω are too small to be resolved by our observations. The fact that we can only observe a finite region of the Sun for a finite length of time results in resolution elements (or bins) in k of $2\pi/X$ and in ω of $2\pi/T$, where X is the length of our observing raster and T is the time interval covered by one day of observations. Thus our sampling bins are larger than the separation in (k, ω) space of individual modes, each bin containing on the order of 30 modes, given the aspect ratio of our observing window and its efficiency in filtering out waves propagating obliquely to the equator. The presence of many modes with small separations in k and ω in each bin results in a complicated beating pattern in both space and time. In principle, mode beating by itself leads to net power that would probably show little correlation from bin to bin. Thus we would expect the power in a given bin to be different on different days, possibly even occasionally vanishing.

In addition to mode beating, seeing effects will distribute power unevenly along the ridges. Figure 1 shows regions along the ridges where power is concentrated into structures with scales of about 5 to 10 bins in k. It thus appears that power in adjacent bins along a ridge are not independently determined, in the manner that we might expect from mode beating effects. There is coupling or leakage between the bins on a scale that cannot be just attributed to our discrete sampling in k and ω, which may result in the coupling of about 3 bins together, rather than the 5 to 10 that we observe. The observed structures in power along the ridge are likely to be the results of seeing effects in the Earth's atmosphere. There are three stochastic effects of seeing that can couple bins in the observed (k, ω) power diagram. The first effect is that of blurring, which involves small relative displacements of the image and consequent blending of image elements. Such blurring removes power from small wavelengths or high k values; it also changes the observed phase of the wave, and thus smears the power in both k and ω. Further, blurring also alters abruptly the sensitivity of our velocity measurements, for differential image motion typically broadens the spectral line sampled by the diode detectors. This too serves to change the apparent phase in the velocity signal discontinuously, thereby coupling nearby bins in both k and ω. Finally, the most important effect is that of sudden displacements of larger portions of the solar image: lateral image displacements of 5″ or more are common, with the image motion being correlated on horizontal scales typically of about 100″ (e.g., Brandt, 1969; Tarbell and Smithson, 1981; Tarbell, 1981). Displacements of such large areas of the image will effectively transfer power from a bin to many of its neighbors. These effects of seeing contribute to the observed structure of power along the ridges, and must be recognized when trying to determine the centroid of a ridge.

Shifts in the locations of ridge centroids between Days 53 and 55 are shown in Figures 2 and 3. Cubic spline fitting procedures with 5 knots have been used to determine the ridge shifts displayed here, with the centroid positions at each k having been calculated using the integral moment method discussed earlier. Figure 2 presents frequency shifts for the p_1, p_2, p_3, and p_4 ridges in the positive ω quadrant, while Figure 3 shows the shifts for the corresponding ridges in the negative ω quadrant. The solid curves show shifts observed in Mg I velocity, the dotted curves are the shifts

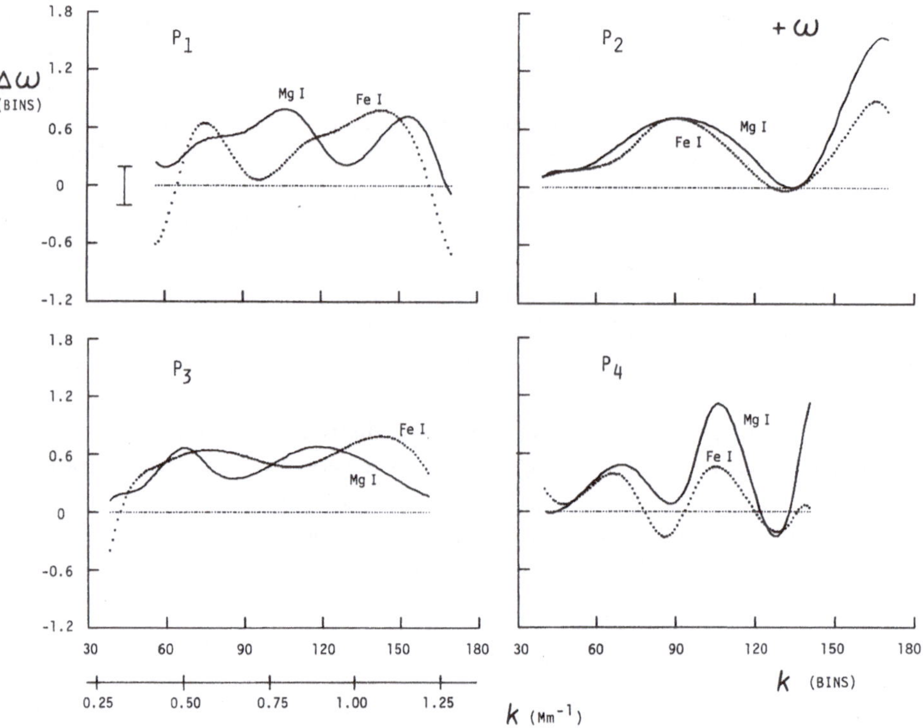

Fig. 2. Shifts in the location of the p_1 to p_4 ridges in the positive ω quadrant between Days 53 and 55 (22 February, 1981 and 24 February, 1981). These shifts result from changes in velocity and temperature perturbations experienced by waves propagating in the westward direction. The solid line is the shift observed in Mg I Doppler velocity data, the dotted line is the shift simultaneously observed in Fe I Doppler velocity. The abscissa is scaled by the value of k in spatial bin numbers, with one bin equal to 7.6×10^{-3} Mm^{-1}. In addition, a corresponding horizontal wavelength scale is superposed below. The ordinate display shifts in ω, or $\Delta\omega$, in terms of frequency resolution bins, with one bin corresponding to 1.5×10^{-4} s^{-1}. The thin dotted line demarks the reference level of a null shift. The overall observational errors are denoted by the vertical bars. The character of the shifts with k are different for the various ridges shown here, thus suggesting that the shifts are solar rather than instrumental in origin. The Fe I and Mg I data display similar trends in their ridge shifts.

reduced from simultaneous observations of Fe I velocity. The ordinate display shifts in ω in terms of frequency resolution bins, with one bin corresponding to 1.5×10^{-4} s^{-1}; the abscissa is scaled by the value of k measured in spatial bin numbers, with one bin corresponding to 7.6×10^{-3} Mm^{-1}. In addition, a horizontal wavelength scale is superposed on the abscissa.

When observed in the Mg I and Fe I spectral lines, the shifts for a given ridge in Figures 2 and 3 display the same general trends, with the possible exception of the p_1 ridge in the positive ω quadrant. This is comforting since the two spectral lines are formed at heights in the solar atmosphere differing by only about 400 km, and at both heights these acoustic wave modes are evanescent. Figures 2 and 3 indicate that the displacement of the different ridges is distinctive in character. The p_2 ridge shifts in both

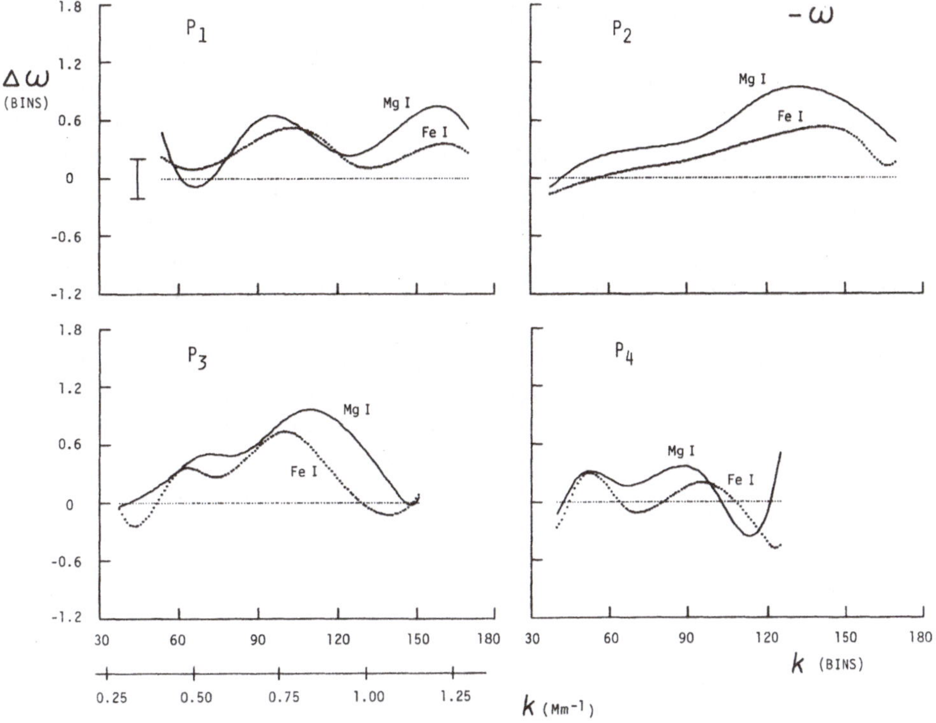

Fig. 3. Similar to Figure 2, but for ridge shifts observed in the negative ω quadrant corresponding to waves propagating in the eastward direction. Comparison of these shifts with those in Figure 2 reveal differences that are attributable to the combined effects of velocity and temperature perturbations.

quadrants show a small shift at low values of k and a large shift at high k. The p_4 ridge in the positive ω quadrant and the p_3 ridge in the negative ω quadrant have their maximum shifts near the middle of the ridge. The other ridges shifts display a number of bends along their length. The fact that the ridge shifts have different behavior suggests that the apparent shifts do not arise from instrumental effects which would tend to shift all ridges in a similar manner.

Our estimates of confidence level in the observed ridge displacements is indicated by the error bars shown in Figures 2 and 3. Shifts with magnitudes below that level are probably not significant. Estimating our confidence level in these shifts is somewhat intricate. One possible choice of such error estimates is to calculate the rms difference between the individual centroid determinations and the smoothed spline fits to these points, thereby defining a scatter about the fit. The error, σ_1, estimated from such scatter, stems from all the processes that control the signal-to-noise ratio in our velocity measurements, supplemented by effects of mode beating and bin coupling. We further must add to that our estimate of errors, σ_2, arising from the uncorrected instrumental changes of scale. An observed shift in ridge position would then be significant if the magnitude of the shift exceeds the combined error. However, the variations in power due to mode beating and the coupling of bins from effects of seeing suggest that a simple rms scatter between each centroid point and its corresponding fitted point will overestimate σ_1, for the centroid points are not statistically independent. Thus we believe that it is appropriate to locally smooth the centroid data, using a smoothing comparable to twice the bin coupling coherence length. We then apply one of the cubic spline fitting functions, recognizing that such functions are particularly responsive to the large-scale trends along the ridges, with these trends more likely to arise from solar effects. The estimates of σ_1 are then based on the rms difference between the cubic spline fits and the smoothed data points. Thus the total rms error in the ridge shifts is approximately ± 0.2 bins in ω compared to ± 0.7 bins when calculated on the basis of individual centroid points with no smoothing. The general trends with k for the ridge shifts are not overly sensitive to the number of knots used in the spline fits, nor to their initial distribution, provided we have reduced some of the effects of bin coupling.

The shifts in ridge location reflect physical structures below the solar surface. As shown in Gough and Toomre (1983), the proportional shifts $\Delta\omega/\omega$ in local wave properties can be related to subsurface velocity and temperature fields with spatial scales

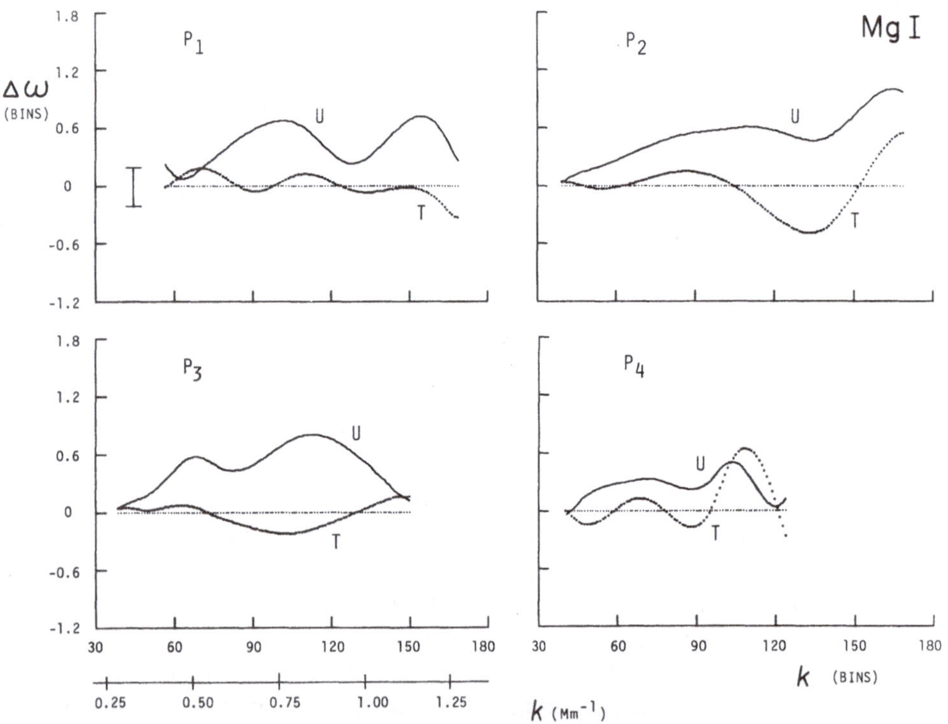

Fig. 4. Separation of ridge shifts with k into the displacements attributable to velocity and temperature perturbations. The combined shifts were observed in Mg I Doppler velocity and displayed in Figures 2 and 3. The solid line (labelled U) is the shift due to velocity, the dotted line (T) is that due to temperature. All scales are as in Figure 2. The velocity perturbation effects are generally larger than those due to temperature. The displacements related to velocity differences can be associated with a depth averaged velocity $\tilde{U}$ weighted by the kinetic energy density of each wave mode. Table II provides some interpretation of such velocity differences.

much larger than the wavelengths of the observed modes. The inhomogeneities arising from convection are linked to the ridge shifts by a set of integrals possessing kernels that are fairly complicated functions of depth, and quite unlike delta functions. The proportional frequency shifts arising from advection of wave fronts by horizontal flows depend in sign upon the wave propagation direction, whereas the shifts due to temperature modifying the local sound speed do not. Thus we can separate the velocity and temperature contributions to the ridge shifts by adding and subtracting the shifts for the positive ω and negative ω ridges at a given value of k. The results of this separation are shown in Figures 4 and 5. These figures present the velocity perturbation of $\Delta\omega$ as a solid line and the temperature perturbation as a dotted line. Figure 4 displays these perturbations for the Mg I data, and Figure 5 shows the corresponding curves for those of Fe I; again we are here comparing Days 53 and 55. The general trends with k are quite similar for Mg I and Fe I. The frequency shifts due to changes in velocity are typically greater than those for temperature. These convective velocity and temperature perturbations are superimposed on top of differential rotation effects, and hence cannot be isolated without comparing two or more days of observations. It is in this regard that the estimates of differential rotation reported by Deubner *et al.* (1979) may be brought somewhat into question, for they used data from individual days without seeking to isolate effects that may arise from convection.

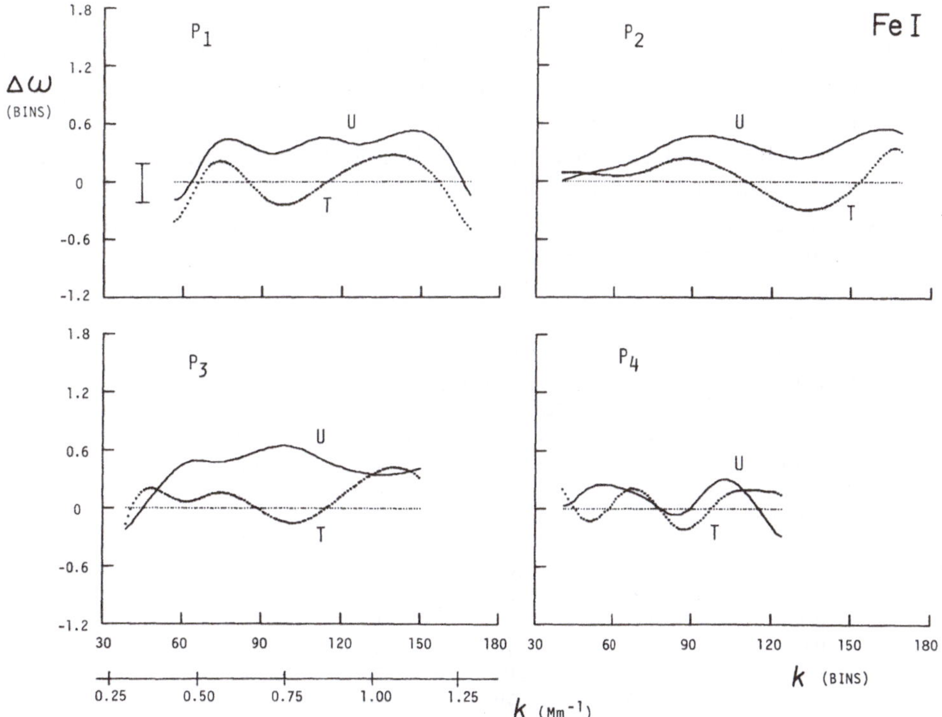

Fig. 5. Similar to Figure 4, but the separation of ridge shifts into velocity (U) and temperature (T) effects has been carried out for the Fe I data.

We recognize that deducing specific temperature and velocity estimates as functions of depth from the observed ridge shifts requires careful application of inverse theory. Although we have not yet carried out such inversion procedures, we can at least provide some estimates of the velocities that may produce the observed shifts. Namely, for a mode with wavenumber k, the frequency shift $\Delta\omega$ arising from simple advection of wave fronts by a horizontal velocity $U(z)$ is just $\Delta\omega = k\tilde{U}$, where $\tilde{U}$ is the average of U weighted with the kinetic energy density of the wave. As shown in Gough and Toomre (1983), a given mode clearly samples a range of depths, and thus its associated $\tilde{U}$ does not correspond to a single velocity at a specific depth. However, the high degree modes that we consider possess lower reflection points at depths of about n/k, and thus the depths sampled range from about 1 Mm to 16 Mm. We present in Table II the differences in the average velocities $\tilde{U}$ inferred from the observed shifts between Days 53 and 55 in the positions of the 4 ridges at selected values of k for the Mg I velocity data. The absence of entries in the table represents locations where ridge centroids could not be reliably determined. Errors in determining $\tilde{U}$ are directly related to the rms error in locating the ridge centroids; we estimate that such errors in the $\tilde{U}$ entries in Table II increase linearly with k, from a value of ± 15 m s^{-1} at $k = 0.3$ Mm^{-1} to ± 50 m s^{-1} at $k = 1.2$ Mm^{-1}.

We note that all the differences in velocity $\tilde{U}$ in Table II are of the same sign. Our observational setup and subsequent use of the Fourier transform assigns the positive ω quadrant to those wave modes propagating in the westward direction. This also means that a positive velocity difference in Table II indicates an increase in the westward velocity being experienced by a mode between the two days being compared. Since we are sampling somewhat different regions on the Sun on the two days, we must emphasize that such differences in velocity probably do not represent temporal changes, but rather arise from the sampling of different convective cells. Table II reveals differences in various depth averaged horizontal velocities that range from 30 m s^{-1} to 170 m s^{-1} for the ridge intervals that were analyzed.

TABLE II

Differences in averaged horizontal velocities $\tilde{U}$ (in m s^{-1}) inferred from ridge shifts in Mg I

k (Mm^{-1})	p_1	p_2	p_3	p_4
0.3	–	–	30	–
0.4	–	70	80	85
0.5	30	100	170	90
0.6	90	120	110	70
0.7	130	120	110	50
0.8	125	110	140	95
0.9	70	100	130	–
1.0	40	70	80	–
1.1	80	80	30	–
1.2	90	115	–	–

Comparison of the ridge positions between Days 53 and 82 again show significant shifts, while those between Days 55 and 82 do not. This suggests that there may be something distinctive about the region of the Sun spanned by our observing window on Day 53. Having examined synoptic magnetic maps of the solar surface for the months of February and March 1981, we find that the observing area for Day 53 spans a region on the quiet Sun showing fairly rapid evolution of the large-scale magnetic field pattern. It has been suggested that giant cells may be revealed by the evolution of magnetic structures, though more direct measurements of such cells have been inconclusive (e.g., Bumba and Howard, 1969; Bumba, 1970; McIntosh, 1980). We believe it is possible that a structure akin to a giant cell was fortuitously placed beneath our observing area on Day 53, and was then moved to an unobservable location by solar rotation on Days 55 and 82. In support of this conjecture, we note that our method of probing responds best to subphotospheric velocity and temperature structures which have a spatial scale that is a significant fraction of the ~ 700 Mm length of our observing window. There are ample theoretical reasons for expecting large-scale convection cells to be present that would possess such dimensions (e.g., Gilman, 1980; Toomre, 1980; Latour *et al.* 1981, 1983), and a number of such cells should be active at any given time. However, our method of probing these cells by observing local changes in wave properties has a sensitivity that varies with the orientation and scale of these features within the observing window. This may lead to only occasional days showing distinct shifts in (k, ω) ridges, and thus an extended program of observations is required to establish the statistics of such events.

4. Conclusions

Our observations have revealed that significant changes in the positions of ridges for five-minute oscillations of high degree can be detected by comparing two or more data sets. The only previous study to have compared (k, ω) diagrams from two different days reported no changes in ridge locations (Rhodes *et al.*, 1981). However, the method used there was one of simple subtraction of one power diagram from the other, rather than seeking to fit the ridge positions in some quantitative manner. Such a subtraction method suffers from the effects of mode beating and from fluctuations in overall power levels caused by seeing variations, differences in data set lengths and varied instrumental effects. In addition, the shifts we have reported here are small and can be easily masked by the choice of scaling used to examine the differences. Any one of these factors could serve to hide a small displacement in ridge positions,and thus it is not surprising that Rhodes *et al.* (1981) did not detect shifts.

We believe that our detection of ridge shifts on one day out of three suggests that such observational procedures to probe for subphotospheric velocity and temperature structures holds out considerable promise. We recognize that major improvements are required in the instrumentation to reduce the noise component in the observed power diagrams, for the successful implementation of inverse theory is likely to hinge on that factor. Our suggestion that the observed ridge shifts may be related to the elusive giant cells is speculative at best, though it is clear that only velocity and temperature features with a horizontal scale that is a substantial fraction of our observing window could produce such ridge shifts. We think it also reasonable that only the occasional data set may show discernible shifts in ridges, for our ability to detect shifts is influenced by the orientation and scale of the convective structures. For instance, the presence of several flow patterns within our observing window may lead to dilation and contraction of wave fronts that appear to just broaden the ridges, with little perceptible shift in ridge centroids. Clearly the data base of such observations must be considerably extended, and we are so engaged.

Acknowledgements

We thank Dr Douglas Gough for extensive scientific advice and discussions concerning the interpretation of these observations, and Drs Dick Dunn, Jack Harvey, and George Simon for suggestions about potential sources of error inherent in carrying out such observations. Patrick McIntosh kindly helped us to interpret synoptic maps of magnetic field evolution. Richard Mann, Horst Mauter, and Fritz Stauffer contributed vital observing skills and patience in calibrating and stabilizing the diode array at SPO. F.H. and J.T. were Guest Observers during the course of the observations at SPO, the latter being operated by the Association of Universities for Research in Astronomy, Inc., (AURA), under contract with the National Science Foundation. L.J.N. was supported as a NAS/NRC Research Associate at Air Force Geophysics Laboratory at SPO. This research was supported in part by the National Aeronautics and Space Administration through grants NSG–7511 and NAGW–91 by the Air Force Geophysics Laboratory through contract F19628–77–C–0104 to the University of Colorado.

References

Altrock, R. C. and Canfield, R. C.: 1974, *Astrophys. J.* **194**, 733.

Altrock, R. C., November, L. J., Simon, G. W., Milkey, R. W., and Worden, S. P.: 1975, *Solar Phys.* **43**, 33.

Berthomieu, G., Cooper, A. J., Gough, D. O., Osaki, Y., Provost, J., and Rocca, A.: 1980, in H. A. Hill and W. A. Dziembowski (eds.), *Nonradial and Nonlinear Stellar Pulsation*, Lecture Notes in Physics, Vol. 125, Springer, Berlin, p. 307.

Brandt, P. N.: 1969, *Solar Phys.* **7**, 187.

Bumba, V.: 1970, *Solar Phys.* **14**, 80.

Bumba, V. and Howard, R.: 1969, *Solar Phys.* **7**, 28.

Deubner, F. L., Ulrich, R. K., and Rhodes, E. J., Jr.: 1979, *Astron. Astrophys.* **72**, 177.

Gilman, P. A.: 1980, *Highlights Astron.* **5**, 91.

Gough, D. O.: 1978, in G. Belvedere and L. Paterno (eds.), *Workshop on Solar Rotation*, Astrophys. Obsv. Catania, Palermo, p. 255.

Gough, D. O. and Toomre, J.: 1983, *Solar Phys.* **82**, 401 (this volume).

Latour, J., Toomre, J., and Zahn, J.-P.: 1981, *Astrophys. J.* **248**, 1081.

Latour, J., Toomre, J., and Zahn, J.-P.: 1983, *Solar Phys.* **82**, 387 (this volume).

Lubow, S. H., Rhodes, E. J., Jr., and Ulrich, R. K.: 1980, in H. A. Hill and W. A. Dziembowski (eds.), *Nonradial and Nonlinear Stellar Pulsation*, Lecture Notes in Physics, Vol. 125, Springer, Berlin, p. 300.

McIntosh, P. S.: 1980, in M. Dryer and E. Tandberg-Hanssen (eds.), *Solar and Interplanetary Dynamics*, D. Reidel Publ. Co., Dordrecht, Holland, p. 25.

November, L. J., Toomre, J., Gebbie, K. B., and Simon, G. W.: 1979, *Astrophys. J.* **227**, 600.

November, L. J., Toomre, J., Gebbie, K. B., and Simon, G. W.: 1982, *Astrophys. J.* **258**, 846.

Rhodes, E. J., Jr., Ulrich, R. K., Harvey, J. W., and Duvall, T. L.: 1981, in R. B. Dunn (ed.), *Solar Instrumentation: What's Next?*, Sacramento Peak Obs., Sunspot, p. 37.

Saastamoinen, J.: 1979, in E. Tengstrom and G. Teleki (eds.), 'Refractional Influences in Astrometry and Geodesy', *IAU Symp.* **89**, 73.

Simon, G. W.: 1966, *Astron. J.* **71**, 190.

Tarbell, T. D.: 1981, private communication.

Tarbell, T. D. and Smithson, R. C.: 1981, in R. B. Dunn (ed.), *Solar Instrumentation: What's Next?*, Sacramento Peak Obs., Sunspot, p. 491.

Toomre, J.: 1980, *Highlights Astron.* **5**, 571.

MAUNDER CONVECTION MODE ON THE SUN AND LONG SOLAR ACTIVITY MINIMA*

V. A. DOGIEL

P. N. Lebedev Physical Institute, Academy of Sciences, Moscow, U.S.S.R.

Abstract. A model of velocity field oscillations in the solar convective zone is suggested. The system of convective equations is investigated for a thin rotating spherical envelope when the rotation velocity is depended on the coordinates. It is shown that two different structures of convective cells (longitudinal, or latitudinal) can exist in the envelope depending on gradients values of the rotation velocity and Prandtl number.

It is supposed that two different regimes of convection (stationary and autofluctuating) are possible in the envelope when the angular velocity gradients are determined by the convection itself. In the case of autofluctuating regime the alternation of longitudinal and latitudinal structure of convection is realized.

If one assumes that on the Sun there exists an autooscillating convection regime, then the periods of the existence of latitudinal convection structure may be associated with long periods of activity minima since according to Cowling's theorem, the action of the axisymmetric magnetic field generation mechanism is impossible under conditions of axisymmetric velocity structures.

In recent years attention has been payed to rather unusual phenomena proceeding on the Sun. They manifest themselves in the fact that long periods of anomalously low solar activity take place. The most investigated of such minima is the last one, the so-called Maunder minimum, which was accompanied by various unusual phenomena in the terrestrial and solar atmospheres and in the interplanetary space. But the Maunder minimum manifested itself first of all in almost a complete disappearance of spots from the surface of the Sun. For seventy years of the minimum (1645–1715) there were less spots than for one 'normal' 11-year solar cycle (Eddy, 1976).

The most convincing proof of the reality of this phenomenon was obtained from the study of the amount of isotope C^{14} in tree rings, that is produced in interaction between a flux of highenergy cosmic rays and the matter of the Earth's atmosphere. From these data it is seen that during the whole period of the Maunder minimum the cosmic ray intensity was anomalously large (see, e.g., Kocharov *et al.*, 1979). This points to some interplanetary magnetic field variations which make it possible for cosmic rays to penetrate more freely into interplanetary space. This was observed not only in the Maunder but also in some other earlier solar activity minima. As can be seen from radio-carbon data, the time distribution of these minima is extremely inhomogeneous. In particular, the last two solar activity minima (Maunder and Spörer) took place with about a two hundred-year interval, whereas after the Maunder minimum it is already alsmost three hundred years that normal 11-year cycles without any essential lowering of solar activity are observed.

* Proceedings of the 66th IAU Colloquium: *Problems in Solar and Stellar Oscillations*, held at the Crimean Astrophysical Observatory, U.S.S.R., 1–5 September, 1981.

Solar Physics **82** (1983) 427–436. 0038–0938/83/0822–0427$01.50.

The question of the causes of this phenomenon remains open. In a number of papers it is supposed that the activity minima are due to a sharp decrease in the peaks of 11-year activity which is caused by the position of the minima of hypothetic 80-year and 170-year periodicities of solar activity (see, for example, Link, 1978). However, the analysis of statistical properties of solar activity does not confirm such an assumption (Willis and Tulunay, 1979).

Leighton (1969) suggested that during Maunder minimum the toroidal magnetic fields were too weak in the solar subphotosphere zone so they could not emerge. The active regions were absent on the photosphere and for this reason the dynamo mechanism did not operate.

Rusmaikin and Zel'dovich (1980) assumed that in the period of long activity minima the magnetic fields in the convective zone were able to suppres the-α-effect, and as a consequence the action of the dynamo mechanism on the Sun ceased.

Another explanation was given by Yoshimura (1978). He assumed that for the periods of long activity minima the convection mechanism was switched off, for which reason the magnetic fields were not generated any longer. According to this model the solar rotation differentiality must increase in the period of activity minimum, since the restraining effect of magnetic fields on differential rotation disappeared. It should be noted however that differentiality of solar rotation is most likely to be due to the presence of convection (Gilman, 1975), therefore the convection having being switched off, the differential rotation would rather damp than be amplified. Besides, according to Dearborn and Neuman (1978), not only switching off but even a small decrease in efficiency of convective heat transfer would have led to catastrophic changes in the climate on the Earth. Proceeding from what has been said above, one should assume that convective heat transfer does exist also in the periods of long activity minima and its efficiency changes quite little, if at all. In our opinion long solar activity minima are caused by changes in the structure of hydrodynamic flows in the convective zone (Dogiel and Syrovatskii, 1977, 1979a). In the periods of minima the convection structure is such that it proves to be inefficient for magnetic field generation. Parker (1976) suggested that such a convection mode should be called Maunder mode.

The convection structure in the rotating spherical shell has been investigated in many papers. First of all the action of rotation on convective motions was shown to lead to the fact that convective cells try to stretch along the rotation axis (Cowling, 1953; Chandrasekhar, 1961). Under the conditions of a spherical shell rotating at a constant angular velocity in the initial state, such a longitudinal structure of motion convection leads to rotation momentum redistribution as a result of which angular velocity gradients appear (Busse, 1972). The results of numerical calculations show that such a stationary state can be achieved in which there remains only one convective mode characterized by the spherical function $Y_l^l(\vartheta, \varphi) = P_l^l(\cos\vartheta)e^{il\varphi}$ (Durney, 1970). There forms a longitudinal structure of convective cells stretching from pole to pole.

The longitudinal structure of motions turns out to be effective for the dynamo mechanism to be realized in a rotating shell. The time development of magnetic fields in such a shell that models the behaviour of magnetic fields on the Sun was investigated in numerical calculations (Yoshimura, 1975).

Usually the convective structure is investigated under the condition of rigid rotation. The only nonlinear term taken into account in the problem is an averaged term of convection modes interaction. It is responsible for the appearance of angular velocity gradients in the shell. Thus, the development of convection proceeds under condition of differential and not rigid rotation. Interaction of convective motions with shear flow (differential rotation) can lead to changes in the convective structure. To take into account this effect, we have solved the problem in which we have sought the most unstable convection mode (i.e. the mode excited at the lowest difference of temperatures between the internal and external spheres) under condition of differential rotation. The angular velocity was set up in the form

$$\Omega(\vartheta) = \Omega_0(1 + p\cos^2\vartheta)\,, \tag{1}$$

where Ω_0 = const., p = const. and ϑ is the latitude.

We have solved a set of equations for dimensionless fluctuations of velocity $\mathbf{v}$, temperature θ, and pressure p' at a given difference of temperatures ΔT = const. between the internal and external spheres and against the background of the established temperature distribution

$$\nabla^2 T_0 = 0\,. \tag{2}$$

From the noncompressibility condition, $\mathbf{v}$ is presented in the form (Chandrasekhar, 1961)

$$\mathbf{v} = \operatorname{rot}\operatorname{rot}\mathbf{r}\,v + \operatorname{rot}\mathbf{r}\,w\,. \tag{3}$$

Then for v, w, and θ we have the equations (Busse, 1971)

$$\left[L^2\left(\nabla^2 - \frac{\partial}{\partial t}\right) + \lambda\mathbf{k}\times\mathbf{r}\cdot\nabla\right]\nabla^2 v + \lambda Q w - \frac{1}{r_0}RL^2\theta = $$

$$= \bar{\mathbf{r}}\operatorname{rot}\operatorname{rot}[\mathbf{u}\times\nabla\times\mathbf{v} + \mathbf{v}\times\nabla\times\mathbf{u}]\,, \tag{4}$$

$$\left[L^2\left(\nabla^2 - \frac{\partial}{\partial t}\right) + \lambda\mathbf{k}\times\mathbf{r}\cdot\nabla\right]w - \lambda Q v = $$

$$= \mathbf{r}\operatorname{rot}[\mathbf{u}\times\nabla\times\mathbf{v} + \mathbf{v}\times\nabla\times\mathbf{u}]\,, \tag{5}$$

$$\nabla^2\theta + Pr\frac{\partial}{\partial t}\theta + \frac{L^2 v}{r_0} = Pr(\mathbf{u}\nabla)\theta\,, \tag{6}$$

with the boundary conditions

$$v = \frac{\partial^2 v}{\partial r^2} = \frac{\partial w}{\partial r} = \theta = 0\,, \qquad r = r_0\,, \quad r_0 + h\,. \tag{7}$$

Here h is the shell thickness, r_0 is its internal radius, ν and k are coefficients of viscosity and heat conductivity, α is a volume expansion coefficient, g_0 is gravity acceleration, $\lambda = (2\Omega_0 h^2)/\nu$ is Taylor number, $R = (\alpha g_0 \Delta T h^3)/k\nu$ is Rayleigh number, $Pr = \nu/k$ is Prandtl number, $\mathbf{u} = [\boldsymbol{\omega}\mathbf{r}]$ is shear velocity, where $\omega = (p\lambda \cos^2\vartheta)/2$, $\mathbf{k}$ is the direction of rotation,

$$L^2 = -\frac{1}{\sin\vartheta}\frac{\partial}{\partial\vartheta}\sin\vartheta\frac{\partial}{\partial\vartheta} - \frac{1}{\sin^2\vartheta}\frac{\partial^2}{\partial\varphi^2}, \tag{8}$$

$$Q = \mathbf{k}\nabla - \tfrac{1}{2}(L^2\mathbf{k}\nabla + \mathbf{k}\nabla L^2). \tag{9}$$

In the thin-shell approximation, $h/r_0 \ll 1$,

$$\mathbf{r}\,\mathrm{rot}\,\mathrm{rot}(\mathbf{u}\times\nabla\times\mathbf{v} + \mathbf{v}\times\nabla\times\mathbf{u}) \approx \frac{p\lambda}{2}\frac{\partial}{\partial\varphi}(L^2\cos^2\vartheta\,\nabla^2 v), \tag{10}$$

$$\mathbf{r}\,\mathrm{rot}(\mathbf{u}\times\nabla\times\mathbf{v} + \mathbf{v}\times\nabla\times\mathbf{u}) \approx -\frac{p\lambda}{2}\frac{1}{\sin\vartheta}\frac{\partial}{\partial\vartheta}\left(\sin\vartheta\cos^2\vartheta\frac{\partial}{\partial r}L^2 v\right). \tag{11}$$

The unknown quantities v, w, and θ are presented in the form of a series of the spherical functions $Y_l^m(\vartheta, \varphi) = P_l^m(\cos\vartheta)e^{im\varphi}$. The time dependence t is sought in the form $e^{i\sigma t}$. The problem is investigated within the stability limit, i.e. σ has only real values. The problem is reduced to obtaining the mode excited at the lowest temperature difference. In other words, from the conditions of the minimal Rayleigh number it is necessary to find the values of l and m.

The solution of the problems (3)–(7) is sought for in the approximation of slow rotation at $\lambda \ll 1$ and small perturbation amplitudes. All the unknown quantities are presented in power series of λ:

$$R = \sum_n R_n\lambda^n, \qquad \sigma = \sum_n \sigma_n\lambda^n, \qquad v = \sum_n v_n\lambda^n, \quad \text{etc.} \tag{12}$$

We start by considering the order λ^0 (Busse, 1972):

$$v_0 = \sin(\pi\Delta r)P_l^m(\cos\vartheta)e^{im\varphi}, \tag{13}$$

$$\Delta r = \frac{r - r_0}{h},$$

$$\theta_0 = r_0\frac{\nabla^4 v_0}{R_0}, \tag{14}$$

$$R_0 = \frac{\left[\pi^2 + \dfrac{l(l+1)}{r_0^2}\right]^3}{\dfrac{l(l+1)}{r_0^2}}, \tag{15}$$

$$\sigma_0 = w_0 = 0. \tag{16}$$

From the condition of the minimum of R_0 one can find the value of l,

$$\frac{l(l+1)}{r_0^2} = \frac{\pi^2}{2} . \tag{17}$$

In the first approximation in λ

$$v_1 = \sin(\pi \Delta r) e^{im\varphi}(K_1 P_{l+2}^m + K_2 P_l^m + K_3 P_{l-2}^m) , \tag{18}$$

$$\Theta_1 = \sin(\pi \Delta r) e^{im\varphi}(G_1 P_{l+2}^m + G_2 P_l^m + G_3 P_{l-2}^m) , \tag{19}$$

$$\begin{aligned} w_1 = e^{im\varphi} \Bigg\{ & \frac{C_1}{2l+5} 4p\pi(l-m+3)\cos(\pi\Delta r) P_{l+3}^m + \\ & + \frac{l-m+1}{(2l+1)\beta(l+2)} \left[\left[\pi(1-p) + 4\pi p \left(C_1 \frac{(2l+1)(l+m+2)}{(2l+5)(l-m+1)} + C_2 \right) \right] \times \right. \\ & \left. \times \cos(\pi\Delta r) - \frac{l}{r_0} \sin(\pi\Delta r) - \frac{l}{r_0} \frac{\pi \operatorname{ch}\left[\frac{\alpha(l+1)}{r_0} (\Delta r - \frac{1}{2}) \right]}{\alpha(l+1) \operatorname{sh} \frac{\alpha(l+1)}{2r_0}} \right] P_{l+1}^m + \\ & + \frac{l+m}{(2l-1)\beta(l-1)} \left[\left[\pi(1-p) + 4\pi p \left(C_2 + C_3 \frac{(l-m+1)(2l+1)}{(l+m)(2l+3)} \right) \right] \times \right. \\ & \times \cos(\pi\Delta r) + \frac{l+1}{r_0} \frac{\pi \operatorname{ch}\left[\frac{\alpha(l-1)}{r_0} (\Delta r - \frac{1}{2}) \right]}{\alpha(l+1) \operatorname{sh} \frac{\alpha(l+1)}{2r_0}} + \\ & \left. + \frac{l+1}{r_0} \sin(\pi\Delta r) \right] P_{l-1}^m + \\ & + C_3 \frac{4\pi p}{2l-3} (l+m-2)\cos(\pi\Delta r) P_{l-3}^m \Bigg\} , \end{aligned} \tag{20}$$

$$\alpha^2(n) = n(n+1) , \qquad \beta(n) = \frac{n(n+1)}{r_0^2} + \pi^2 , \tag{20a}$$

$$\sigma_1 = \frac{m}{\alpha^2(l)(1+Pr)}\left[1 - \frac{p}{r_0}\,\frac{\alpha^2(l)}{2l+3}\left(1 + \frac{2(l+1)(l^2-m^2)}{4l^2-1}\right)(1+Pr)\right], \tag{21}$$

$$C_1 = \frac{p}{2}\,\frac{(l-m+1)(l-m+2)}{(2l+1)(2l+3)}, \tag{22}$$

$$C_2 = \frac{p}{2(2l+3)}\left[1 + \frac{2(l^2-m^2)}{2l-1}\right], \tag{23}$$

$$C_3 = \frac{p}{2}\,\frac{(l+m)(l+m+1)}{(2l+1)(2l+3)}, \tag{24}$$

$$K_1 = \frac{im\,C_3\,\beta(l+2)\beta(l)}{\left[\dfrac{\alpha^2(l+2)}{\alpha^2(l)}\beta^3(l) - \beta^3(l+2)\right]}, \tag{25}$$

$$K_2 = \frac{im}{2\alpha^2(l)}\,\frac{Pr}{\beta(l)(1+Pr)}, \tag{26}$$

$$K_3 = im\,C_3\,\frac{\beta(l-2)\beta(l)}{\left[\dfrac{\alpha^2(l-2)}{\alpha^2(l)}\beta^3(l) - \beta^3(l-2)\right]}, \tag{27}$$

$$G_1 = \frac{1}{r_0\beta(l+2)}\left[\alpha^2(l+2)K_1 - im\,Pr\,C_1\,\frac{\alpha^2(l)}{\beta(l)}\right], \tag{28}$$

$$G_2 = -\frac{\beta(l)}{R_0}K_2, \tag{28a}$$

$$G_3 = \frac{1}{r_0\beta(l-2)}\left[\alpha^2(l-2)K_3 - im\,Pr\,C_3\,\frac{\alpha^2(l)}{\beta(l)}\right]. \tag{29}$$

In the second approximation

$$R_2 = \frac{m^2}{2\alpha^2(l)}\,\frac{Pr^2}{(1+Pr)^2} + \frac{\alpha^2(l)\zeta(l+1)}{2\beta(l+1)}\left\{\pi^2\left[1 + 2p(\zeta(l+2) + \frac{1}{2l+3}\times\right.\right.$$

$$\left.\times\left[1 - \frac{2(l^2-m^2)}{2l-1}\right]\right) - p\Bigg] + \frac{l(l+2)}{r_0^2}\times$$

$$\left.\times\left[1 + \frac{2\pi^2 r_0}{\alpha(l+1)\beta(l+1)}\,\mathrm{cth}\,\frac{\alpha(l+1)}{2r_0}\right]\right\} + \frac{\alpha^2(l)\zeta(l)}{2\beta(l-1)}\times$$

$$\times \left\{ \pi^2 \left[1 + 2p(\zeta(l-1) + \frac{1}{2l+3} \left[1 + 2p(\zeta(l-1) + \frac{1}{2l+3} \times \right.\right.\right.$$

$$\left.\left.\left. \times \left[1 + \frac{2(l^2 - m^2)}{2l-1} \right] - p \right] + \frac{(l^2-1)}{r_0} \quad \frac{1 + 2\pi^2 r_0 \operatorname{cth} \dfrac{\alpha(l-1)}{2r_0}}{\alpha(l-1)\beta(l-1)} \right] \right\} +$$

$$+ p^2 \frac{\pi^2}{4} \alpha^2(l)[(l+4)\zeta(l+3)\zeta(l+2)\zeta(l+1) -$$

$$- (l-3)\zeta(l-2)\zeta(l)\zeta(l-1)] -$$

$$- p \frac{\pi^2}{2} \left[1 + 2p \left[\zeta(l+2) + \frac{1}{2l+3} \left(1 + \frac{2(l^2-m^2)}{2l-1} \right) \right] - p \right] \times$$

$$\times \frac{(l+1)^2}{\beta(l+1)} \zeta(l+1) \left[(l+1)\zeta(l+2) - \frac{(l+2)}{2l+3} \left(1 + \frac{2(l^2-m^2)}{2l-1} \right) \right] +$$

$$+ p \frac{\pi^2}{2} \left[1 + 2p \left[\zeta(l-1) + \frac{1}{2l+3} \left(1 + \frac{2(l^2-m^2)}{2l-1} \right) \right] - p \right] \times$$

$$\times \frac{l^2}{\beta(l-1)} \zeta(l) \left[l\zeta(l-1) - \frac{(l-1)}{2l-3} \left(1 + \frac{2(l^2-m^2)}{2l-1} \right) \right] -$$

$$- \frac{m^2 p^2}{8} \alpha^2(l)\beta(l) \left\{ \zeta(l+2)\zeta(l+1) \left[\frac{\beta^2(l+2)}{\xi(l+2)} + Pr \frac{\beta(l)}{\beta(l+2)} \times \right.\right.$$

$$\left. \times \left[\frac{\alpha^2(l+2)}{\alpha^2(l)} \frac{\beta(l+2)(\beta(l)}{\xi(l+2)} - \frac{Pr}{\beta(l)} \right] \right] - \zeta(l)\zeta(l-1) \left[\frac{\beta^2(l-2)}{\zeta(l-2)} + \right.$$

$$\left.\left. + Pr \frac{\beta(l)}{\beta(l-2)} \left[\frac{\alpha^2(l-2)}{\alpha^2(l)} \frac{\beta(l-2)\beta(l)}{\xi(l-2)} - Pr \frac{1}{\beta(l)} \right] \right] \right\}, \tag{30}$$

$$\zeta(n) = \frac{n^2 - m^2}{(2n-1)(2n+1)}, \tag{31}$$

$$\xi(n) = \frac{\alpha^2(n)}{\alpha^2(l)} \beta^3(l) - \beta^3(n). \tag{32}$$

From expression (30) one can obtain the value of m by determining the minimum possible Rayleigh number R. Investigation of expression (30) show that at sufficiently low values of the number p, $|p| < p_{cr}$ (p_{cr} is some critical value) the most dangerous is the mode with $m = l$. It has the form of longitudinal cells extended along meridians. It should be noted that as has been said above, angular velocity gradients are produced just at such a convection structure. At a realization of this cell structure the decisive factor is the presence of Coriolis forces.

If $|p| > p_{cr}$, there exists an axisymmetric latitudinal structure of convective cells which are toroids ($m = 0$), whose axis coincide with the rotation axis. The decisive factor in the formation of such a structure of motions is the presence of shear flow (differential rotation). An analogous result was obtained in the problem of convection in a plane sheet in the presence of shear flow. In this case convective cells stretch along the direction of shear velocity (Kuo, 1963; Deardorff, 1965).

Let us try to imagine how the time development of convection structure may proceed when angular velocity gradients are created by the convection itself.

As long as latitudinal gradients of angular velocity are small or absent, the convection structure is longitudinal. Redistributing the momentum in the shell, convective motions create angular velocity gradients. These gradients can grow until the momentum transfer by convection is compensated by opposite momentum transfer due to viscosity which tries to level the angular velocity.

As a result the state is achieved when the longitudinal convection structure exists against the background of stationary angular velocity gradients characterised by the value p_{st}. Such a state is possible if $|p_{st}| < p_{cr}$.

In case $|p_{st}| > p_{cr}$ the increase of angular velocity gradients created by the longitudinal convection structure proceeds only up to the value p_{cr} after which the convection structure changes from longitudinal to latitudinal. The latitudinal convection structure is not able to maintain the angular velocity gradients that caused its appearance, and therefore the gradients are decreased due to viscosity, and the system goes back to the state with the longitudinal structure of convection. So, an autooscillating convection regime with alternating latitudinal and longitudinal structure of convective cells is realized. The lifetime of longitudinal convection structure is the time of angular velocity gradient increase, the lifetime of latitudinal convection structure is the time of dissipation of these gradients.

If one assumes that on the Sun there exists an autooscillating convective regime, then the periods of the existence of latitudinal convection structure may be associated with long periods of activity minima since according to Cowling's theorem the action of axis-symmetric magnetic field generation mechanism is impossible under the conditions of axis-symmetric structure of motions. A high solar activity corresponds to the periods of the existence of a longitudinal structure of motion.

It should be noted that the alternation of longitudinal and latitudinal structure of convection cells could be happened irregularly due to the nonlinear interaction of different modes of motion (Dogiel and Syrovatskii, 1979b; Dogiel, 1980).

The behaviour of convection structure in time could be investigated with the help of numerical methods. Up to now there have been several investigations of this kind. Durney (1970) has investigated the system of convective equations for a rotating envelope when nonlinear interactions of motion modes was presented only in equations for a large scale modes. It was shown that the stationary case was realised in this case and the structure of convection was pure longitudinal.

Recently Gilman (1978, 1980) has carried out more complex calculations for convection in a rotating envelope. He included in the equations from 16 to 24 longitudinal wave numbers and nonlinear interaction was present for all of these modes. As a result the stationary structure of convection was also realised for investigated range of parameters of the envelope but the structure of the motion was more tangled. It was discovered that mainly longitudinal convection was realised in the equatorial zone and mainly latitudinal convection was realised near the poles of envelope. We think that the question of the existence of a range of parameters where the latitudinal structure of the convection could sometimes be realised in the whole envelope that is necessary for an autofluctuating regime, is open. As it is possible to see from the paper of Gilman (1980) the inclusion of different new effects in the problem of convection in a rotating envelope can drastically change the convection picture. In this situation we think that the realisation of an autofluctuating regime is possible and further special investigations are necessary. It seems that in more real situations we have mixture of longitudinal and latitudinal convective modes in the rotating envelope. The autofluctuating regime in this case one can imagine himself as a alternating of an amplitudes increase of longitudinal and latitudinal convective modes and accordingly the increase and decrease of the efficiency of a magnetic field generation.

References

Busse, F. H.: 1972, *Astron. Astrophys.* **28**, 27.
Cowling, T. G.: 1951, *Astrophys. J.* **114**, 272.
Chandrasekhar, S.: 1961, *Hydrodynamic and Hydromagnetic Stability*, Oxford, Clarendon Press.
Deardorff, J. W.: 1965, *Phys. Fluids* **8**, 1027.
Dearborn, D. S. P. and Newman, M. J.: 1978, *Science* **201**, 150.
Dogiel, V. A.: 1980, *Usp. Fiz. Nauk* **132**, 691.
Dogiel, V. A. and Syrovatskii, S. I.: 1977, *Proc. IX Leningrad's Seminar of Cosmophysics*, p. 14.
Dogiel, V. A. and Syrovatskii, S. I.: 1979a, *Izv. Acad. Nauk, ser. fizich.* **43**, 716.
Dogiel, V. A. and Syrovatskii, S. I.: 1979b, *Proc. IX Leningrad's Seminar of Cosmophysics*, p. 15.
Durney, B. R.: 1970, *Astrophys. J.* **161**, 1115.
Eddy, J. A.: 1976, *Science* **192**, 1189.
Gilman, P. A.: 1975, in V. Bumba and J. Kleczek (eds.), 'Basic Mechanisms of Solar Activity', *IAU Symp.* **71**, 207.
Gilman, P. A.: 1978, *GAFD* **11**, 157.
Gilman, P. A.: 1980, *Highlights Astronomy* **5**, 91.
Kocharov, G. E., Vasiliev, V. A., Dergachev, V. A., and Mikhalchenko, N. G.: 1979, *Proc. XVI Intern. Cosmic Ray Conf.* (Japan) **2**, 256.
Kuo, H. L.: 1963, *Phys. Fluids* **6**, 195.
Leighton, R. B.: 1969, *Astrophys. J.* **156**, 1.
Link, F.: 1978, *Solar Phys.* **59**, 175.

Parker, E. N.: 1975, in V. Bumba and J. Kleczek (eds.), 'Basic Mechanisms of Solar Activity', *IAU Symp.* **71**, 3.

Willis, D. M. and Tulunay, Y. K.: 1979, *Solar Phys.* **64**, 237.

Yoshimura, H.: 1975, *Astrophys. J. Suppl.* **29**, 467.

Yoshimura, H.: 1978, *Astrophys. J.* **220**, 692.

Zeldovich, Ya. N. and Ruzmaikin, A. A.: 1980, *Dynamo Problems in Astrophysics*, Preprint, Inst. Appl. Mathem., the USSR Academy of Sciences, No. 52.

TORSIONAL OSCILLATIONS OF THE SUN*

(Invited Review, Abstract)

ROBERT HOWARD

Mount Wilson and Las Campanas Observatories of the Carnegie Institution of Washington, U.S.A.

Abstract. A series of digitized synoptic observations of solar magnetic and velocity fields has been carried out at the Mount Wilson Observatory since 1967. In recent studies (Howard and LaBonte, 1980; LaBonte and Howard, 1981), the existence of slow, large-scale torsional (toroidal) oscillations of the Sun has been demonstrated. Two modes have been identified. The first is a travelling wave, symmetric about the equator, with wave number 2 per hemisphere. The pattern–alternately slower and faster than the average rotation–starts at the poles and drifts to the equator in an interval of 22 years. At any one latitude on the Sun, the period of the oscillation is 11 years, and the amplitude is 3 m s^{-1}. The magnetic flux emergence that is seen as the solar cycle occurs on average at the latitude of one shear zone of this oscillation. The amplitude of the shear is quite constant from the polar latitudes to the equator. The other mode of torsional oscillation, superposed on the first mode, is a wave number 1 per hemisphere pattern consisting of faster than average rotation at high latitudes around solar maximum and faster than average rotation at low latitudes near solar minimum. The amplitude of the effect is about 5 m s^{-1}. For the first mode, the close relationship in latitude between the activity-related magnetic flux eruption and the torsional shear zone suggests strongly that there is a close connection between these motions and the cycle mechanism. It has been suggested (Yoshimura, 1981; Schüssler, 1981) that the effect is caused by a subsurface Lorentz force wave resulting from the dynamo action of magnetic flux ropes. But, this seems unlikely because of the high latitudes at which the shear wave is seen to originate and the constancy of the magnitude of the shear throughout the life time of the wave.

References

Howard, R. and LaBonte, B. J.: 1980, *Astrophys. J. Letters* **239**, L33.
LaBonte, B. J. and Howard, R.: 1982, *Solar Phys.* **75**, 161.
Schüssler, M.: 1981, *Astron. Astrophys.* **94**, L17.
Yoshimura, H.: 1981, *Astrophys. J.* **247**, 1102.

* Proceedings of the 66th IAU Colloquium: *Problems in Solar and Stellar Oscillations*, held at the Crimean Astrophysical Observatory, U.S.S.R., 1–5 September, 1981.

Solar Physics **82** (1983) 437. 0038–0938/83/0822–0437$00.15.

EVIDENCE FOR THE PHI-DEPENDENT ROTATION-OSCILLATION OF THE SUN (AND FOR THE DRIVING MECHANISM OF THE ASYMMETRIC DYNAMO)*

I. K. CSADA

Konkoly Observatory, Budapest, Hungary

Abstract. Longitude-dependent oscillations of the solar rotation are derived from the 27-day averages of the photospheric velocity data. Two pairs of prominent periods are obtained. Their harmonic means correspond to a semiannual variation and to the first harmonic of the latter. To explain the origin of the oscillation the corona and the interplanetary material are supposed to rotate parallel to the planetary plane with an inclination to the solar equator. The non-uniform shearing around the equator is assumed to result in oscillation with a period of half of a year.

A preliminary analysis expressed in terms of an asymmetric solar dynamo suggests the existence of φ-dependent oscillations of the angular velocity (Csada, 1980). In the present report the observational evidence is summarized and a physical explanation of the variation in solar rotation is suggested.

Let us take a model example to understand the basic feature of a φ-dependent oscillation. Consider a fictive external body with a supposed tidal effect acting on the Sun. Alternate increasing and decreasing rotational velocities manifest themselves as φ-dependent oscillations corresponding to the rising and falling of the photosphere. If the external body is supposed to be fixed in space, the oscillations are stationary with respect to an inert frame, and two maxima and two minima will be observed in the rotational velocity over a terrestrial year.

This hypothetical physical mechanism will be used to construct the mathematical form of the variation. It is proposed that the corona and the interplanetary matter rotate parallel to the planetary plane with an inclination to the solar equator. Thus φ-dependent oscillations are induced, but the formulation is much more difficult than that in terms of a solid fictive body. In particular, the driving of the oscillation must be sustained by some energy source which is able to compensate for the decay of the magnetic field.

The oscillations are superimposed on the mean rotation and on random oscillations due to the granular pattern. To eliminate the random term, the daily velocity data were averaged over 27 days. The φ-dependent terms were separated from the mean rotation by taking averages over the solar disc according to scheme

$$\bar{v} = -\frac{1}{2\Delta\varphi}\int_{-A-\Delta\varphi}^{-A+\Delta\varphi} \omega_1 \sin\varphi \, \mathrm{d}\varphi + \frac{1}{2\Delta\varphi}\int_{A-\Delta\varphi}^{A+\Delta\varphi} \omega_2 \sin\varphi \, \mathrm{d}\varphi .$$

* Proceedings of the 66th IAU Colloquium: *Problems in Solar and Stellar Oscillations*, held at the Crimean Astrophysical Observatory, U.S.S.R., 1–5 September, 1981.

Solar Physics **82** (1983) 439–442. 0038–0938/83/0822–0439$00.60.

On the assumption that the variation of the angular velocity is caused by an elliptical distortion of the photospheric level, we can write that

$$\omega_1 = \omega_0 + B \sin(\lambda + \varphi),$$

$$\omega_2 = \omega_0 - B \sin(\lambda - \varphi),$$

where λ is the (heliographic) longitude of the central meridian relative to the fictive body, and φ is taken increase eastwards from the central meridian.

The photospheric velocity data observed in the Mount Wilson Observatory from 1 January, 1975 to 27 December, 1978 were collected by Dr Howard for the study of long period oscillations. The list constains velocity residuals (obtained after the subtraction of the Earth's motion, the solar rotation, the limb shift and the 'ears') in the azimuth zones $30° \pm 2°$ on either side of the central meridian. The west zone values are subtracted from the east zone values. The residuals were averaged over 27 day intervals. The material was analysed in the Konkoly Observatory to find significant periods from 27 to 250 days. For data fit the formula

$$\langle v \rangle = 0.433 + B \sin 2\omega t$$

was used and ω was tested by steps of $2\pi/100$.

The result of the analysis shows prominent periods of 199.2, 165.1, 93.1, and 86.9 days with a less than 10% error of the printed values in Table I. The method of computation consists in comparing three successive values of the mean deviation of the observed from the calculated values. If the significant periods show normal distribution the interpolation of the logarithm of the mean deviations by a parabola yields the periods.

The harmonic mean of the values printed in the first two columns of Table I is 180.4 days indicating an oscillation of a period of half of a year while the mean value of the last two columns, being 90.2 days, corresponds approximately to periods of a quarter of a year.

Since the velocity residuals are corrected for the Earth's motion, the apparent half year variation and its harmonics are eliminated from the data. However, the double periods suggest that the normal distribution expected in the analysis for the apparent variation is smaller than manifested by the observations. This suggests the existence of yet another mechanism which generates oscillations. This mechanism could be as follows: the corona rotates with the interplanetary material in the planetary plane and thus it is connected to the Sun over a great circle where the rotational displacement on the surface is larger than that of the corona at the nodes and smaller perpendicular to them. A shearing between the photosphere and the corona has a tendency to equalize the motion and results in an oscillation of the angular velocity which is stationnary relative to an inertial frame. The oscillations are fed by the angular momentum of the Sun which is transported via the friction of the interplanetary matter in the shearing area.

To verify the reality of the rest of the printed periods (Table I) other material, published at the Stanford Observatory (Scherrer *et al.*, 1980), was chosen for analysis

TABLE I

Significant periods printed for the φ-dependent rotation-oscillation of the Sun as function of the heliographic latitudes. All the periods and their errors are given in days.

40	198.3 ± 21.6	169.4 ± 25.4	151.1 ± 30.6	134.7 ± 24.4		114.8 ± 96.0		92.7 ± 7.5	85.9 ± 7.7
35	199.7 ± 42.6	165.8 ± 20.3	144.7 ± 33.6	133.0 ± 36.2	120.5 ± 16.4		108.8 ± 17.8	96.8 ± 16.3	89.5 ± 12.7
30	206.0 ± 29.4	162.6 ± 10.1	144.3 ± 7.3		124.3 ± 10.1		107.5 ± 7.6	98.8 ± 6.0	87.0 ± 9.3
25	192.1 ± 26.1	160.2 ± 65.8		131.4 ± 17.9		113.1 ± 22.1	103.0 ± 5.4	93.2 ± 5.5	86.0 ± 6.1
20	193.1 ± 71.7	165.7 ± 20.2		132.3 ± 15.0	120.7 ± 39.3			99.4 ± 11.4	83.5 ± 7.8
15	210.2 ± 50.3	163.9 ± 22.4		137.5 ± 24.0	124.6 ± 13.2	114.2 ± 16.7	104.8 ± 8.4	95.7 ± 9.7	85.3 ± 4.4
10	(206.8)	158.3 ± 32.8		136.2 ± 11.7	123.4 ± 11.9	110.5 ± 10.4	100.8 ± 10.0		89.1 ± 8.2
5	202.4 ± 23.8	173.9 ± 16.7		129.3 ± 6.4		112.9 ± 5.6	98.3 ± 5.5	90.7 ± 6.6	83.6 ± 3.9
0	219.7 ± 94.6	159.4 ± 45.4	145.4 ± 87.8	137.0 ± 47.9	122.5 ± 34.6	110.8 ± 20.9	100.6 ± 29.9	95.9 ± 27.5	85.3 ± 15.5
−5	205.3 ± 25.2	181.3 ± 32.0	157.7 ± 11.7	136.1 ± 11.6			104.4 ± 6.6	92.3 ± 14.3	86.7 ± 5.2
−10	212.4 ± 55.3	163.9 ± 17.1		138.7 ± 15.1	119.6 ± 28.4	113.6 ± 23.1	104.1 ± 7.9	95.5 ± 5.0	88.9 ± 6.1
−15	197.6 ± 40.9		149.9 ± 20.7	127.8 ± 28.5		110.8 ± 19.2		95.6 ± 5.9	87.6 ± 10.9
−20	185.0 ± 13.5	161.5 ± 13.3	144.6 ± 14.4	131.6 ± 11.2	119.5 ± 8.6		104.1 ± 12.2	90.4 ± 4.6	84.4 ± 3.8
−25	196.8 ± 78.9		152.3 ± 52.0		123.7 ± 12.7	109.6 ± 10.4		95.9 ± 6.1	88.0 ± 7.0
−30	198.3 ± 38.1		145.4 ± 19.1		124.6 ± 8.3		103.1 ± 5.7	93.7 ± 9.7	86.5 ± 11.1
−35	193.9 ± 22.8	159.5 ± 32.2	141.2 ± 21.0	127.9 ± 98.1	117.6 ± 12.1	107.7 ± 8.1	101.6 ± 15.2	93.0 ± 9.8	83.6 ± 9.1
−40		160.7 ± 18.4		136.4 ± 47.2		112.4 ± 8.1	103.8 ± 9.2	95.9 ± 9.7	86.7 ± 11.1

in the Konkoly Observatory. The coordinates of the whole disc velocity data were read directly from the diagram and the analysis leads to periods of 90.6, 146.8, and 241.1 days. The oscillation of a quarter of a year, which could be the first harmonic of the model, is confirmed and also the 146.8 day period is printed in the table as 147.8. Since the error of the last value is about 14 days and that of the first data is probably larger, we think that this period also exists but its origin in unknown.

Finally, we note that the 199.3 day period (Table I) is as long as half the Jupiter's synodic revolution time. This permits to see the planetary influence on the solar dynamo. Direct gravitational effect is an absurdity but its trace on the printed periods could be possible. The small deviation of the 180.5 day period from the semiannual value (182.6 days) could be explained in this way.

References

Csada, I. K.: 1981, *Solar Phys.* **74**, 103.
Scherrer, P. H., Wilcox, L. M., and Svalgaard, L.: 1981, *Astrophys. J.* **241**, 811.

THE OBSERVATIONS OF 80-MIN OSCILLATIONS IN THE QUIESCENT PROMINENCES*

V. S. BASHKIRTSEV, N. I. KOBANOV, and G. P. MASHNICH

Siberian Institute of Terrestrial Magnetism, Ionosphere and Radio Wave Propagation, (SibIZMIR), Irkutsk 33, P.O. Box 4, U.S.S.R.

Abstract. Oscillations of the line-of-sight velocities with periods $82^{m}_{\cdot}2$ and $76^{m}_{\cdot}7$ were detected in quiescent prominences, with coordinates $\varphi = -75°$ W and $\varphi = -18°$ W, respectively.

At Sayan Observatory, Bashkirtsev and Mashnich have been successful in taking the first observations of line-of-sight velocity oscillations in two limb prominences with coordinates: (1) $\varphi = -75°$ W and (2) $\varphi = -18°$ W on 30 and 31 March, 1981.

Within the least years, the highly sensitive differential methods (Kalinyak and Vasilyeva, 1971; Kotov *et al.*, 1978, 1982) have been extensively used for line-of-sight velocity measurements. We employed a differential technique designed for the study of local quasi-periodical motions of solar matter (Kobanov, 1983).

The principle underlying our method is that we measure the difference of Doppler shift of two line profiles from two areas on the Sun's surface using the same spectral line, in our case the Hβ-line. The separation of the two measured areas, each having the size $5''.0 \times 4''.4$ in the prominences was 4" along the radius. The apparent height above limb at the site of measurement was respectively $\sim 1.5 \times 10^4$ km and $\sim 10^4$ km for prominences (1) and (2). As revealed by an inspection of the spectrum, the two prominences displayed no significant line-of-sight velocities or apparent changes in their profiles for the time of recording the oscillations. Therefore one can consider them as quiescent objects.

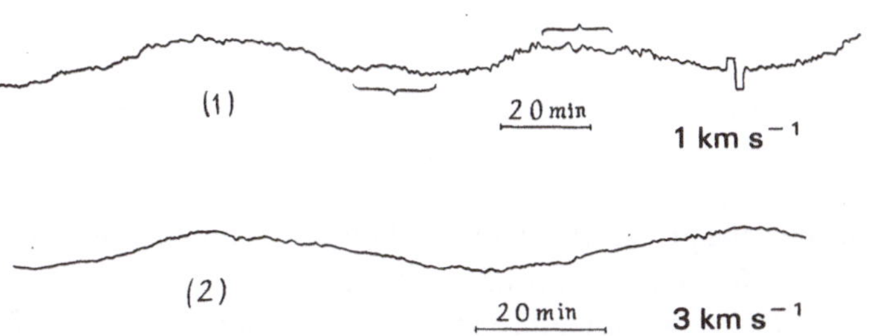

Fig. 1. Copies of records of prominence line-of-sight velocity oscillations: (1) $\varphi = -75°$ W, 30 March, 1981, $02^h54^m \leq UT \leq 06^h00^m$. Parts of the record marked by the braces have been obtained through cirrus clouds. The calibration procedure appears at the end of the record: (2) $\varphi = -18°$ W, 31 March, 1981, $02^h16^m \leq UT \leq 04^h10^m$.

* Proceedings of the 66th IAU Colloquium: *Problems in Solar and Stellar Oscillations*, held at the Crimean Astrophysical Observatory, U.S.S.R., 1–5 September, 1981.

Solar Physics **82** (1983) 443–445. 0038–0938/83/0822–0443$00.45.

We made measurements with a magnetograph, with one photomultiplier tube and automatic compensation of brightness fluctuations and information output on a strip-chart recorder (Lebedev *et al.*, 1972). Prominence image on the spectrograph entrance slit was kept by a photoelectric guide to an accuracy of 1 arc sec.

The record copies of the line-of-sight prominence velocities are shown in Figure 1. The straight lines indicate time and amplitude scales. The registration length for prominence (1) is longer than two oscillation periods (3^h06^m), and for prominence (2), one period and a half (1^h54^m).

The reduction of the oscillation records was carried out using correloperiodogram analysis (Kopecký and Kuklin, 1971). The reduction used data series consisting of 840 and 520 points for prominences (1) and (2), respectively. One digitizing step is 12.5 s.

The reduction results are shown in Figure 2, where T is the period in seconds and ρ the correlation coefficient that had been inferred from a comparison of our results with a harmonic oscillation. For prominence (1) the period was 82.2 min while that for prominence (2) was 76.7 min.

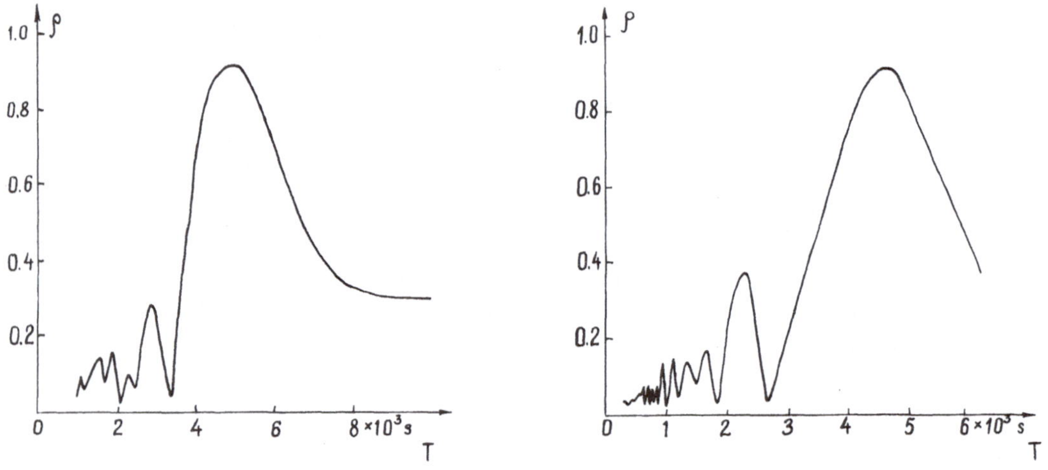

Fig. 2. Prominence correloperiodograms (1) and (2).

It is interesting that the prominence did not show any 5-min periods (or close to them) reported in literature for different prominence types.

In our opinion, it is now premature to given some interpretation of the observed oscillation periods in prominences. However detection of long-periodic oscillations in quiescent prominences is of great interest.

At the time of reduction and preparation of our observational data for publication we happened to know of a just published paper co-authored by Malville and Schindler (1981) who observed radial and torsional oscillations with a period about 75^m in a loop prominence approximately 90 min before the onset of a limb flare. These oscillations have been interpreted in terms of a kink instability of a current carrying flux tube before the flare. It will be noted here that the prominences we observed were stable and not connected with flare phenomena. Therefore the above mentioned interpretation of the oscillations is not the only one possible.

We feel that further observations of the prominence velocity oscillations are needed at present.

Acknowledgements

We would like to thank Prof. V. E. Stepanov, Drs N. V. Klochek, G. V. Kuklin, M. N. Nikonova, Yu. M. Palachev, and V. G. Mikhalkovsky for their assistance with this work and helpful comments. We are also grateful to the anonymous referee for critical and helpful comments on our paper.

References

Kalinyak, A. A. and Vasilyeva, G. J.: 1971, *Solar Phys.* **16**, 37.

Kobanov, N. I.: 1983, *Solar Phys.* **82**, 237 (this volume).

Kopecký, M. and Kuklin, G. V.: 1971, *Issledovaniya po geomagnetizmu, aeronomii i fizike Solntsa* **2**, 167.

Kotov, V. A., Severny, A. B., and Tsap, T. T.: 1978, *Monthy Notices Roy. Astron. Soc.* **183**, 61.

Kotov, V. A., Severny, A. B., and Tsap, T. T.: 1982, *Izv. Krymsk. Astrofiz. Obs.* **65** (in press).

Lebedev, N. N., Grigoryev, V. M., Klochek, N. V., and Kobanov, N. I.: 1972, Certificate of authorship No. 335652 (U.S.S.R.). A technique for magnetic field strength measurements.

Malville, J. M. and Schindler, M.: 1981, *Solar Phys.* **70**, 115.

ATMOSPHERIC INTERNAL GRAVITY WAVES AS A SOURCE OF QUASIPERIODIC VARIATIONS OF THE COSMIC RAY SECONDARY COMPONENT AND THEIR LIKELY SOLAR ORIGIN*

(Invited Review)

A. M. GALPER, V. G. KIRILLOV-UGRYUMOV, N. G. LEIKOV, and B. I. LUCHKOV

The Moscow Physical Engineering Institute, 115409, Moscow

Abstract. Hard gamma-radiation fluctuations with the periods from 4 to 60 min were investigated in the course of balloon flights at altitudes of 30–40 km. Quasiperiodic intensity variations (QPV) were observed with periods of 5 min, 12–15 min, and 23–26 min, those of 5 min predominating. QPV last no longer than several hours, their associated amplitudes ranging from 5 to 20%. QPV were observed both in mid-latitudes and in the tropics, their detection probability for 3^h exposure being 0.3. In the total charged component QPV with comparable amplitudes were not registered. Synchronous atmospheric pressure variations were recorded practically with an amplitude 20 times less than that of gamma-radiation. This suggest short internal gravity waves (IGW) in the stratosphere in the range from 10 to 100 km as the most likable source of QPV. Since the temperature profile of the Earth atmosphere provides conditions for superdistant waveguiding propagation of short IGW with a period of ~ 5 min at altitudes of 110 and 30 km, the source of waves can be well away from the point of their registration. The IGW generation in the stratosphere can be attributed to the resonance caused by global solar oscillations with low l modes. The resonance probability is likely to be due to the hard solar radiation variations which are absorbed in the ozone layer. The coincidence of the frequency oscillation range in the chromosphere and that of IGW in the stratosphere suggests an IGW resonant excitation mechanism in the Sun–Earth system.

The observations at 46° N geomagnetic latitude (cutoff rigidity $R = 3.5$ GV) and 8° N geomagnetic latitude ($R = 16.9$ GV) by high altitude balloons in 1972–79 recorded quasiperiodic gamma-ray intensity variations (QPV) with an energy of more than 40 MeV (Galper *et al.*) Measurements made with a spark chamber telescope, with a geometrical factor of 115 cm^2 ster and an effective aperture of 40° (Galper *et al.*, 1974) were carried out at altitudes from 4 to 10 g cm^{-2} of the residual atmosphere, the measuring time lasting from 5 to 10 hr.

Simultaneously, the intensity of charged particles was recorded by means of a single counter (the threshold energy of ~ 1 MeV) and a directed counter telescope detecting electrons with energies of more than 60 MeV and protons with energies of more than 300 MeV. The atmospheric pressure was also recorded, a sensor with a precision of 10^{-2} mbar being used in one of the balloon flights.

Time-variation analysis made in the frequency range from 3×10^{-4} to 4×10^{-3} Hz (the associated periods from 4 to 60 min) revealed the following regularities of QPV that might suggest their origin:

* Proceedings of the 66th IAU Colloquium: *Problems in Solar and Stellar Oscillations*, held at the Crimean Astrophysical Observatory, U.S.S.R., 1–5 September, 1981.

Solar Physics **82** (1983) 447–449. 0038–0938/83/0822–0447$00.45.

(1) QPV with the periods of 5 min, 12–15 min, and 23–26 min are observed, those of 5 min predominating;

(2) QPV last no longer than several hours, their associated amplitudes ranging from 5 to 20%;

(3) the QPV detection probability for 3 hr exposure is about 0.3;

(4) there are no variations in the charged component, at least with the amplitudes more than 5%;

(5) atmospheric pressure variations were recorded which are prectically synchronous with QPV of gamma-radiation, but with amplitudes smaller by a factor of 20.

It is possible to explain the observed QPV data. Hard gamma-radiation in the stratosphere is likely to be due to the interaction of primary cosmic rays with the matter of the atmosphere. Gamma-ray intensity can be expressed as: $I_\gamma = AI_{CR}(>R)t$, where $I_{CR}(>R)$ is the primary cosmic ray intensity with an energy higher than the local cutoff rigidity R, $t = \int_{z_0}^{\infty} \rho(z)\, dz$ is the thickness of the residual atmosphere, $\rho(z)$ is air density, and z_0 is the balloon altitude. Since the air density changes exponentionally as $\rho(z) \sim e^{-z/H}$, only the thickness of the atmosphere comparable to the standard height $H = 7$ km is essential. The interpretation of QPV in terms of I_{CR} or R variations seems unlikely since no marked variations in the charged component are recorded. In view of modern ideas of internal gravity waves (IGW) it is, however, possible to interpret the obtained data in terms of thickness t variations.

IGW are exhibited in periodic variations of atmospheric parameters, air density in particular (Gossard and Hook, 1978; Hines, 1960). These variations would, in fact, amount to t variations. The representation of an isothermal atmosphere where IGW behave as flat waves will hold as the first approximation. Short IGW with wavelengths of about 100 km are essentially transversal (Golitzin, 1965) and their relative pressure fluctuations are much smaller than those of density and temperature. The approximated dispersion ratio is $\omega = \omega_B \cos\alpha$, where α is the phase velocity horizontal angle, and ω_B is the Brunt frequency (Grossard and Hook, 1978). The corresponding B-period at balloon altitudes is 4.8 min in mid-latitudes and 4.5 min in the tropics.

Variability of t, and hence of I, will be maximum in the direction along the wave phase plane. Therefore, if the gamma-telescope axis orients to the zenith, as it did in our case, the observation would give predominently QPV due to horizontal IGW with a period of about 5 min.

The ratio $\Delta\rho/\rho \sim \sqrt{\rho}$ holds for the isothermal atmosphere, from where gamma-ray intensity has to change as $\Delta I_\gamma/I_\gamma = \Delta t/t = \beta(\Delta\rho/\rho)_{z_0}$, where $\beta = 2$, which implies that QPV relative amplitude is twice as high as that of density variations at altitude z_0. For the real atmosphere $I < \beta < 2$ since short IGW undergo scattering at the altitudes between 30 and 50 km (and at altitude of 110 km) due to a high temperature gradient involved. This effect reduces QPV amplitude and at the same time provides conditions for superdistant waveguiding propagation of short IGW (Dikii, 1969) and, hence, increases the chance of their observation.

The fact that no marked variations of the charged component intensity (I_c) have been recorded seems natural in the frame of IGW explanation because of a weak altitude dependence of $I_c(z)$ which makes that $\Delta I_c/I_c \ll \Delta t/t$. Some modulation effects due to IGW must, however, be observable in the secondary charged component which is really the case (Komoda *et al.*, 1975).

The QPV–IGW association may suggest an explanation for the origin of IGW at altitudes of 30–40 km in the Earth's atmosphere. The observations provide that IGW are generated with periods of 5 min, 12–15 min, and 23–26 min while the theory of IGW propagation in the real atmosphere predicts only a *B*-period of about 5 min. On the other hand, the coincidence of QPV periods with those of solar oscillations (Hill *et al.*, 1978) has already been discussed and their common genetic origin has been postulated (Galper *et al.*, 1977) which strongly suggest the association of IGW of the Earth's atmosphere with short periodic solar oscillations. It is not yet known how the energy of solar oscillations reaches the Earth's atmosphere but it can be qualitatively postulated that the atmosphere of the Earth may experience resonant vibrations provoked by solar oscillations. These resonances can be most easily realized at radial global solar oscillations of low l-modes ($l = 0, 1, 2$). These modes of global solar oscillations with the period of about 5 min have been recently recorded (Grec *et al.*, 1980; Claverie *et al.*, 1979). Ultraviolet radiation with a wavelength ranging from 2400 to 2900 Å, which is known to undergo strong fluctuations (as compared with extremely weak ones in the optic rang) and is absorbed in the ozone layer, that is, practically at the same altitudes 30–40 km, may provide the mechanism of the solar oscillation propagation. The fact that during a solar eclipse atmospheric IGW has been detected with a period of 20 min speaks in favour of this hypothesis (Goodwin and Hobson, 1978). The assumptions made need experimental verification.

Acknowledgement

We thank Dr L. P. Gorbatchev for very useful discussions.

References

Claverie, A., Isaak, G. R., McLeod, C. P., van der Raay, H. B., and Roca Cortes, T.: 1979, *Nature* **282**, 591.

Dikii, L. A.: 1969, *The Theory of Vibrations of the Earth Atmosphere*, Leningrad.

Galper, A. M., Kurochkin, A. V., Leikov, N. G., Luchkov, B. I., Yurkin, Yu. T.: 1974, *Pribori i technika experimenta* **1**, 50.

Galper, A. M., Kirillov-Ugryumov, V. G., *et al.*: 1977, *15th Intern. Cosmic Ray Conference*, Plovdiv, Vol. 4, p. 341.

Galper, A. M., Kirillov-Ugryumov, V. G., Kurochkin, A. V., Leikov, N. G., and Luchkov, B. I.: 1979, *Pis'ma v JETF* **30**, 631.

Galper, A. M., Kirillov-Ugryumov, V. G., Leikov, N. G., and Luchkov, B. I.: 1980, *Pis'ma v JETF* **31**, 693.

Golitzin, G. S.: 1965, *Izvestiya Akademii Nauk USSR* **1**, 136.

Goodwin, G. L. and Hobson, G. J.: 1978, *Nature* **275**, 109.

Gossard, E. and Hook, U.: 1978, *Waves in the Atmosphere*, Moscow.

Grec, G., Fossat, E., Pomerantz, M.: 1980, *Nature* **288**, 541.

Hill, H. A., Stebbins, R. T., and Brown, T. M.: 1978, *Astrophys. J.* **223**, 324.

Hines, C. O.: 1960, *Canad. J. Phys.* **38**, 1441.

Kodoma, M., Sakai, T., *et al.*: 1975, *14th Intern. Cosmic Ray Conference*, München, Vol. 3, p. 1120.

160 m PULSATIONS IN THE MAGNETOSPHERE OF THE EARTH POSSIBLY CAUSED BY OSCILLATIONS OF THE SUN*

B. M. VLADIMIRSKY, V. P. BOBOVA, N. M. BONDARENKO, and V. K. VERETENNIKOVA

Crimean Astrophysical Observatory, Nauchny, Crimea 334413, U.S.S.R.

Abstract. The measurements of the amplitudes envelope of Pc 3–4 geomagnetic micropulsations obtained at the Borok Geophysical Observatory were analysed by the cosinor method to search for magnetospheric pulsations with a period of about 160 m. 216 days of observations in 1974–1978 were used. It was found that Pc 3–4 amplitudes are modulated by the period 160.010 m with a stable phase. The maximum of the Pc 3–4 amplitudes follows approximately 20 m after the maximum of the solar expansion velocity (for the center of the disk) in the optical observations of Severny *et al.* This modulation of the Pc 3–4 amplitudes could be caused by the presence of an oscillating component in solar UV radiation over the wavelength range 100–900 Å. The amplitude of the UV flux variation may be as large as 2–4%.

1. Introduction

After the discovery of oscillations of the Sun with a period of 160 m (Severny *et al.*, 1976; Brookes *et al.*, 1976) it was suggested that it might be possible to find some effect of these oscillations in the magnetosphere of the Earth. At least two mechanisms which may transfer the effect of the oscillations to the Earth may be considered (see Gul'elmi *et al.*, 1977; Vladimirsky *et al.*, 1981): (1) The oscillating Sun can generate long-period MHD waves. These waves could be carried by the solar wind to influence the magnetosphere. (2) Solar oscillations are probably accompanied by small temperature variations. A corresponding modulation of UV radiation might be expected in this case. Thus, the presence of a 160 m periodic component in the ionospheric current systems is possible because of the electron density modulation.

Some authors have detected a presence of 160 m periods in magnetospheric phenomena such a s the H component of the geomagnetic field (Winch *et al.*, 1963; Toth, 1977), the occurrence of substorms (Tverskaia and Chorosheva, 1975), and *AE*-index (Gul'elmi *et al.*, 1977; Gaivoronskaia and Ljachova, 1979).

2. Observational Data and Processing

To search for possible magnetospheric effects of the oscillations of the Sun, an amplitude envelope of Pc 3–4 geomagnetic micropulsations was used. These micropulsations are quasi-sinusoidal oscillations of the geomagnetic field in approximately the

* Proceedings of the 66th IAU Colloquium: *Problems in Solar and Stellar Oscillations*, held at the Crimean Astrophysical Observatory, U.S.S.R., 1–5 September, 1981.

Solar Physics **82** (1983) 451–455. 0038–0938/83/0822–0451$00.75.

0.1–0.01 Hz frequency range. They are dayside phenomena. Pc 3 can be generated at the boundary of the magnetosphere (Gul'elmi, 1976), so their amplitudes would depend on the parameters of the solar wind and IMF. It is known that the ionosphere also influences micropulsations. Pc 3–4 micropulsation amplitudes were measured at Borok Geophysical Observatory using standard flux-meter instrumentation. Envelope data were obtained from the usual records in 3 m steps from 02^h to 10^h UT. This time is optimal for Pc 3–4 measurements at this observatory. The data were taken for days which included those when optical observations of oscillations of the Sun were made in the Crimea by Severny *et al.* (1976). 216 days of observation in 1974–1978 were analysed. The linear trend was removed from the data to reduce the diurnal variation in solar time. These corrected envelope data were then reduced by cosinor· analysis (Emel'janov, 1976). Each daily segment of the data is approximated in this method by relation $x(t) = A \cos(2\pi/T)(t - \varphi) + h$, using a least-squares best fit with a reference time of 00^h00^m January 1, 1974. Thus, an estimate of amplitude A and phase φ may be obtained for each day, with its error, for any period T to be studied. The distribution of the phases is a very good way of investigating any oscillation features. The mean values of $\overline{A}$ and $\overline{\varphi}$ were also calculated for some groups of days (month, year).

The same computer program was also used for processing the data from the optical observations of oscillations of the Sun (the measurements were presented by A. Severny, V. Kotov, and T. Tsap for the interval 1974–1980). Very good detailed agreement was found between the cosinor analysis results and those obtained by the superposed epoch method (Scherrer *et al.*, 1980).

3. Results

The modulation of Pc 3–4 amplitudes with period 160 m is seen in the power spectra for nearly every day (examples in Figure 1) but significant values of A and φ were obtained for only individual days. To find the precise value of the period, the relationship was obtained between a trial period and the deviation of its distribution of phase estimates from uniform, as measured by χ^2 (5 degrees of freedom) calculated for all the data. This relationship is presented in Figure 2a. The distribution of the phases for some periods is shown in Figure 2b. As one can see in Figure 2a, there are two maxima for the periods $T = 160$ m and $T = 160.010$ m. The first, large maximum is caused by the ninth harmonic of the day and probably by pulsations with a period ≈ 180 m which occurred during local time (see Nikol'ski, 1980; Galkin, 1980). The second maximum (with a lower significance) corresponds precisely to the period which was found in the optical observations. If this maximum is real, a systematic drift of the mean phase $\overline{\varphi}$ from year to year might be revealed for the period $T = 160$ m. Evidence for such a drift is shown in Figure 3. All the data are used here. The line is a least-squares best fit. The shift of mean phase $\overline{\varphi}$ is equal to 1.69 ± 0.36 radians per year. The real error of the slope of the line is probably greater if one takes into account the small dependence of the phase on the geomagnetic activity level. The measured value of the phase shift is not far from the 1.24 radians per year, which would correspond to a period of 160.010 m.

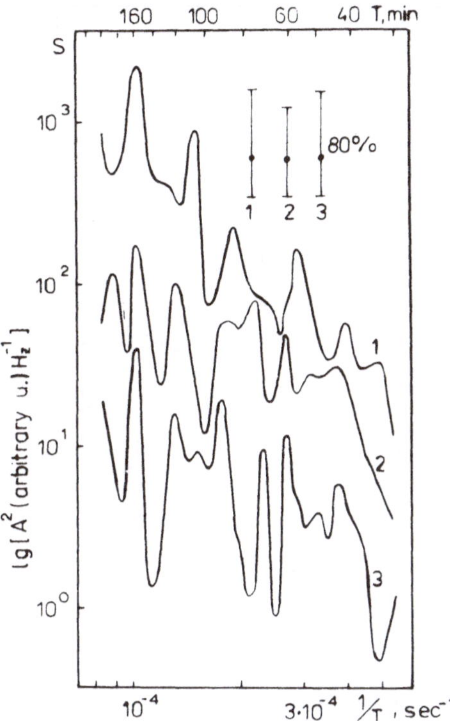

Fig. 1. The modulation power spectrum of the amplitudes of micropulsations Pc 3–4 (Borok Observatory; 32 s data from Gul'elmi *et al.* (1977)). *S* is spectral density, and errors correspond to a 80% confidence level. (1) 24 September, 1973; 06^h05^m–12^h00^m UT. (2) 12 November, 1973; 05^h00^m–08^h56^m UT. (3) 22 April, 1972; 07^h00^m–11^h00^m UT.

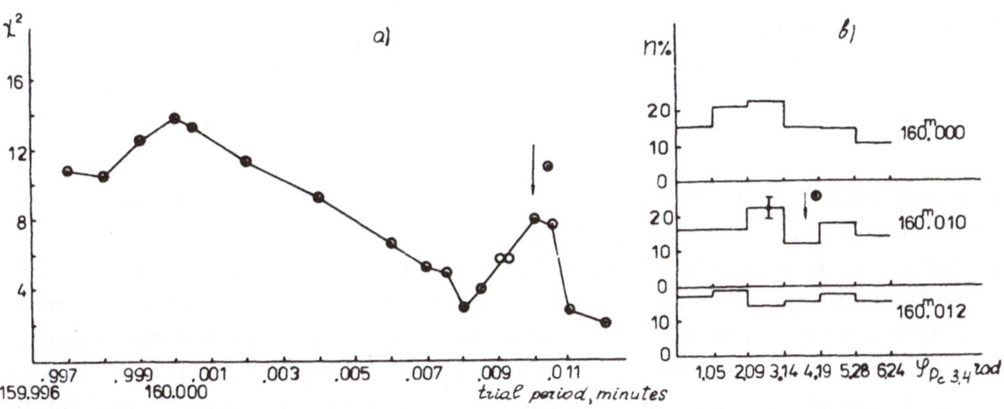

Fig. 2. (a) The relationship between the trial period and the phases distribution deviation from uniform (as measured by χ^2 for 5 degrees of freedom). The value of the period for solar oscillations derived from optical observations is indicated by an arrow. (b) The distribution of the phases for the periods 160 m; 160.010 m and 160.012 m. The stable phase for the period of 160.010 m is about 4.5 radians. The phase of the maximum of solar expansion velocity (for the center of the solar disk) according to Severny *et al.* (1976) is indicated by an arrow.

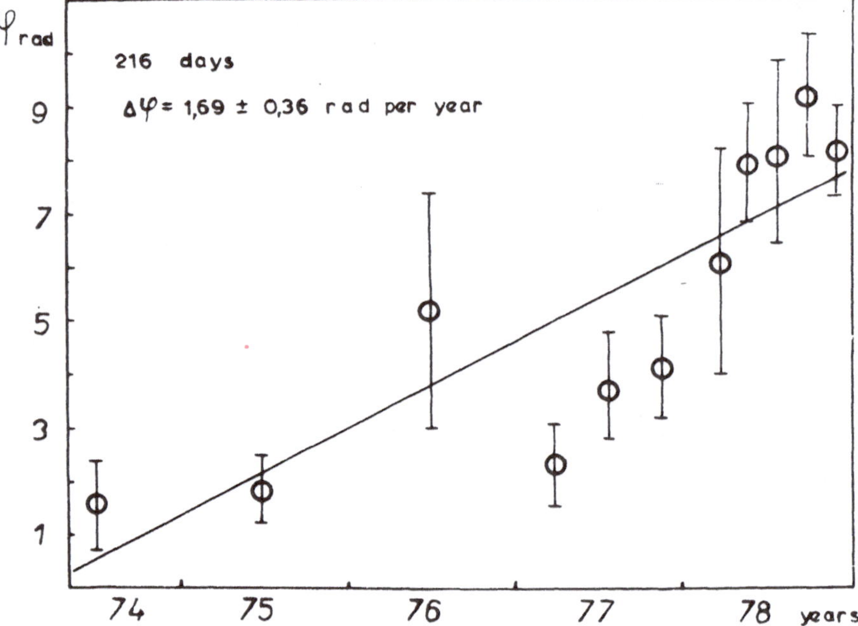

Fig. 3. The drift of the mean phase φ for the period 160 m calculated for the whole set of data. The line is a least-squares best fit and the slope is equal to 1.69 ± 0.36 radians per year.

It is hard to estimate with confidence the value of the stable (non-drifting) phase for the period 160.010 m. This phase is probably equal to ≈ 4.5 radians. If this estimate is correct, the maximum of the Pc 3–4 amplitudes follows approximately 20 m after the maximum of the solar expansion velocity (for the center of the solar disk) in the optical observations of Severny *et al.* (3.70 ± 0.24 radians in our notation).

4. Discussion

The most simple explanation for the existence of Pc 3–4 modulations with a stable phase for $T = 160.010$ m is to assume the presence of a variation in the ionospheric electron density caused by a periodic change in the solar UV radiation. If Pc 3–4 amplitude is dependent on the parameters of the F1 ionosphere, UV flux variations would take place over the wavelength range 100–900 Å. The relative amplitude of Pc 3–4 modulation is equal (in some approximation) to the relative change of electron density and, consequently, to the relative variation of UV emission. The average amplitude of modulation is about 0.04. According to the measurements of Titheridge (1971), the variation of the integral concentration of electrons over the periods from 02^h to 04^h is about 0.02 – this is in good agreement with our estimate. The delay time between the increase of UV flux and the increase of Pc 3–4 amplitude is small, so the maximum of UV radiation follows ≈ 23 m after the maximum of the expansion velocity in the optical observations.

5. Conclusions

Strong evidence has been obtained that Pc 3–4 micropulsation are modulated by a period nearly coinciding with the period of solar oscillations (160.010 m). The modulation might be caused by the presence of an oscillating component in the solar UV radiation over the wavelength range 100–900 Å. The amplitude of such variations may be as large as 2 to 4%.

All these results are based on the limited data from one observatory; they are therefore in need of confirmation.

References

Brookes, J. R., Isaak, G. R., and van der Raay, H. B.: 1976, *Nature* **259**, 92.

Emel'janov, I. P.: 1976, *Formy kolebanii v bioritmologii*, Nauka, Novosibirsk.

Gaivoronskaia, T. V. and Ljachova, L. N.: 1979, *Geomagnetizm i aeronomia* **19**, 934.

Galkin, R. M.: 1980, *Proc. (Trudi) of Arctic – Antarctic Res. Institute* **366**, 111.

Gul'elmi, A. V.: 1976, *Geomagnetizm i aeronomia* **16**, 744.

Gul'elmi, A. V., Vladimirsky, B. M., and Repin, V. N.: 1977, *Geomagnetizm i aeronomia* **17**, 930.

Nikol'ski, A. P.: 1980, *Geomagnetizm i aeronomia* **20**, 889.

Scherrer, P. H., Wilcox, J. M., Severny, A. B., Kotov, V. A., and Tsap, T. T.: 1980, *Astrophys. J.* **237**, L97.

Severny, A. B., Kotov, V. A., and Tsap, T. T.: 1976, *Nature* **259**, 87.

Titheridge, J. E.: 1971, *Planet. Space Sci.* **19**, 1593.

Toth, P.: 1977, *Nature* **270**, 159.

Tverskaia, L. V. and Chorosheva, O. V.: 1975, *Geomagnetizm i aeronomia* **15**, 573.

Vladimirsky, B. M., Bobova, V. P., Repin, V. N., and Veretennikova, V. K.: 1981, *Izv. Krymsk. astrofiz. obs.* **64**, 211.

Winch, D. E., Bolt, B. A., and Slaucitajs, L.: 1963, *J. Geophys. Res.* **68**, 2685.

AN ESSAY ON STELLAR OSCILLATIONS AND EVOLUTION* **

(Invited Review)

ICKO IBEN, Jr.

Dept. of Astronomy, University of Illinois, Urbana, Illinois, U.S.A.

Abstract. It is cautioned that solar models adjusted in such a way as to achieve a match between theoretical solar oscillation characteristics and observed ones may produce neutrino fluxes inconsistent with the observations and that this is likely to be explicable as a deficiency in modeling that portion of the envelope which is most strongly affected by uncertainties in the treatment of convection. Then follows a summary of how the results of pulsation theory and of stellar evolution theory have been used together to learn about the structure and evolution of RR Lyrae stars, classical Cepheids, and high luminosity AGB stars.

1. Introduction

I have been assigned the task of showing how observations of stellar oscillations may be used to help solve problems in stellar evolution. The store of experience on stellar 'seismology' from which I will draw is restricted to large amplitude pulsators which lie in the classical instability strip (outside of which the Sun finds itself). For these pulsators, the theoretical apparatus necessary to use the observations to learn something about global properties is in most instances far less sophisticated than is required for interpreting solar oscillations. Usually only one mode is excited at large amplitude, and this one mode is radial. A beautiful review of the whole subject is given by Cox (1974).

Before beginning the main story, I will make several observations concerning attempts to deduce information about the solar interior from comparison between the mode structure of solar models and observed solar oscillatory characteristics.

2. A Cautionary Note Concerning Interpretations of Solar Structure Based on Observed Oscillatory Properies of the Sun

The choice of the Crimea for this conference is most fitting since the evidence for a coherent solar oscillation at a period of about 160 min was first discovered here (Severny *et al.*, 1976). I recall reading the discovery paper and being very skeptical about the alleged demonstration of periodic motion. I was even more skeptical about the interpretation of the reported period as an indication that the Sun's effective polytropic index is quite different from 3. As reported at this meeting, the quality of the data both

* Proceedings of the 66th IAU Colloquium: *Problems in Solar and Stellar Oscillations*, held at the Crimean Astrophysical Observatory, U.S.S.R., 1–5 September, 1981.

** Supported in part by the National Science Foundation Grant AST 81–15325.

Solar Physics **82** (1983) 457–467. 0038–0938/83/0822–0457$01.65.

from the Crimea and from other stations around the world has improved to such an extent that it is difficult to remain skeptical about the existence of the 160 min oscillation. However, the interpretations of the period in terms of oscillatory characteristics of models of solar structure have not yet reached the same level of believability.

To dramatize the danger of taking too seriously inferences about solar structure based on comparison between observed oscillatory characteristics and results of a pulsational analysis of solar models, adopt the absurd assumption that the period, radius, and surface gravity of the Sun are related in the same way as in a pendulum for which $P \simeq 2\pi\sqrt{l/g}$. Here P is the period, l is the length, and g is the gravitational acceleration. Setting $l = R_\odot$ and $g = GM_\odot/R_\odot^2$, where $R_\odot$ and $M_\odot$ are the solar radius and mass, respectively, one has

$$P \cong 10\,000 \text{ s} = 167 \text{ min}.$$

On comparison with the observed 160 min period of the Sun, one might infer that *the Sun is a pendulum*!

As another (slightly more serious) illustration of the dangers of interpretation, consider the oscillations of periods between about 5 and about 60 min that have been variously reported in the literature. Evidence for an oscillation at a period of about 40 min has in the past been published by at least four groups. Standard solar model calculations demonstrate that the period of the first solar radial overtone is about 40 min and is essentially independent of model details. One might therefore be tempted to infer that theory and observation have together revealed that the first radial overtone is excited in the Sun. However, at this conference, several of the authors have stated in private that their published 'observational' identifications are spurious. Several other period identifications have undergone evolutionary changes in the literature. Thus, the confidence one might have felt at relating a model-independent result to an author-independent, published observational result has in this case shown itself to be misplaced; confidence gives way to severe skepticism. The danger illustrated by this parable is that agreement between 'observation' and a theoretical interpretation does not always mean that either the theoretical interpretation or the observation is correct.

To pursue this theme, suppose that someone were to report the detection of a long period oscillation with a period in the neighborhood of 60 min. A stellar astronomer acquainted with classical Cepheids and RR Lyrae stars might well jump to the conclusion that the period of the fundamental radial mode of the Sun had been detected. He might well construct a series of solar models as a function of uncertain parameters in an effort to use the 'observed' period as a constraint on these parameters. In Table I (Iben and Mahaffy, 1976) is the possible outcome of such an exercise. In this table, Z and X are the abundances by mass of the heavy elements and of hydrogen, respectively, in the initial Sun; T_c and ρ_c are the current central temperature and density in units of 10^6 K and g cm^{-3}, respectively, and N_ν is the solar neutrino flux at the Earth in SNU's. The mixing length to pressure scale height ratio is given by l/H; ΔR and ΔM are the radial thickness and mass of the convective envelope, both in solar units; and T_{BASE} is the temperature at the base of the convective envelope in units of 10^6 K. Finally, P_0

is the period in minutes of the fundamental radial mode. In all cases the solar model is of age 4.6×10^9 yr and its luminosity and radius are the current solar values.

It is simple to understand qualitatively the trends in Table I. The larger Z, the larger is the interior opacity, and therefore the larger must be central temperature be in order to permit the outward diffusion of a given solar flux. Larger central temperatures and densities are accomplished by reducing the number of particles per gram – this means a smaller abundance of hydrogen. Higher interior temperatures mean a larger neutrino flux.

TABLE I

Dependence of several solar model characteristics on composition choice

Z	X	P_0	T_c	ρ_c	N_ν
0.01	0.82	56	14.5	146	4.4
0.02	0.74	62	15.2	165	10.3
0.03	0.67	66	15.9	192	24.7

Z	X	P_0	l/H	ΔR	ΔM	T_{BASE}
0.01	0.82	56	0.69	0.16	1.3×10^{-3}	0.91
0.02	0.74	62	1.01	0.23	1.0×10^{-2}	1.6
0.03	0.67	66	1.26	0.32	4.1×10^{-2}	2.7

For fixed l/H, increasing Z causes the model solar radius to increase (thanks to an increase in opacity). In order to maintain the model radius fixed at the solar value, an increase in Z must be compensated for by an increase in l/H, thereby increasing the efficiency of convection. More efficient convection in the solar envelope means a shallower temperature gradient there and a deeper and more massive convective envelope. Finally, since it is proportional to $T^{1/2}$, sound velocity increases inward through the envelope more slowly for larger Z. Since most of the time required by a sound wave to travel from center to surface is spent in the outer envelope where velocities are considerably smaller than in the deep interior, the larger Z, the longer is the period of the fundamental mode.

Conventional wisdom has it that $Z_\odot \simeq 0.015$, implying that $N_\nu \simeq 6$ SNU, $T_{\mathrm{BASE}} \simeq 1.25 \times 10^6$ K, and $P_0 \simeq 59$ min. If neutrinos have mass, then theory has it that the electron neutrino is just one of two or three states of a more universal neutrino and that, therefore, the actual counting rate should be $N_\nu/(2\text{–}3) \simeq 3\text{–}2$ SNU. This is reasonably close to the most recent experimental estimate of about 2 SNU (Davis, 1979).

Note, however, that T_{BASE} is too small for the element Li to burn within the solar lifetime, making it difficult to understand why the abundance of Li is so small at the stellar surface (compared to abundances 100 times larger in young stars). Within the framework of the simple mixing length theory, the only way to solve the Li problem is to assume a larger than conventional value of Z (by about a factor of 2) and a similarly larger value of l/H. However, the associated value of X is so much larger than found in the interstellar gas (or even in conventional theoretical models of the Big Bang) that it is unacceptable, and the derived value of N_ν can be reconciled with experiment only by, say, adopting the assumption of a 10 or 12 component neutrino; but this even the most adventuresome of the theoretical physicists would be reluctant to do.

Suppose, however, that a reliable solar oscillation experiment actually gave $P_0 \simeq 66$ min? One should actually not find this result either surprising or alarming. It would certainly be a mistake to infer that the 'observed' period demonstrates an error in at least one of: (1) estimates of Z; (2) the opacity associated with Z; (3) the solar neutrino experiment; (4) elementary particle theory. Although none of these possibilities for error is excluded, it is much more likely that the efficiency of envelope convection given by the mixing length recipe is totally in error. The correct inference would be that both P_0 and T_{BASE} reflect properties of the outer convective zone and that, as we have all known from the beginning, the mixing length recipe for convection is an utter failure when it comes to describing locally the actual physics of energy flow in an outer convective region. Appropriate action woul be to see if models using more sophisticated treatments of convection could produce the observed P_0 and N_ν simultaneously.

It is sobering to realize that it is the uncertainty in our treatment of convection that prevents us from estimating via pulsation theory and an observed period and radius the mass of a Main Sequence star with a deep convective envelope. If it were not for this uncertainty, one might expect a relationship of the form

$$M = P^2 R^{-3} f(R),$$

where f is weakly dependent on the radius R, to hold. From Table I, one sees that an uncertainty of at least 20% in P_0 is associated with the uncertainty in the treatment of convection. What this means is that, even if P_0 were known, there is a 40% uncertainty in the inferred stellar mass!

3. Classical Large-Amplitude Pulsators (Now for the Main Story)

In the Hertzsprung–Russell diagram one may draw a line along which hydrogen at the stellar photosphere is $\frac{1}{2}$ ionized and $\frac{1}{2}$ neutral; due to the high opacity in the zone of partially ionized hydrogen, convection extends inward from somewhat above the photosphere. With increasing luminosity, this line slopes gently toward lower surface temperatures (from $\log T_e \simeq 3.9$ at $\log L \simeq 1$ to $\log T_e \simeq 3.7$ at $\log L \simeq 5$). As one moves to the right of the line, the base of the convective envelope extends ever deeper, reaching and extending beyond the region of partial helium ionization. As one moves to the left of the line, the region of partial hydrogen ionization moves deeper and deeper below the photosphere; and the base of the surface convective zone recedes outward. Eventually, extensive envelope convection disappears and convection is restricted to small portions of the deep hydrogen and helium ionization layers. A star within a narrow strip parallel to and slightly to the left of the line will oscillate at large amplitude in either the fundamental or first overtone radial mode. That it will oscillate is known from the observations. That it oscillates in one of two radial modes (in most cases) is known from a comparison between results of radial pulsation theory and observation. How a star gets into the instability strip is known from stellar evolution theory. What is responsible for the oscillation is known from pulsation theory (driving in the helium and hydrogen ionization zones). The right hand limit to the instability strip is presumably due to the development of a deep convective envelope which damps out the instability mechanism in the ionization zones. The left hand limit is due to the positioning of the driving zones so far below the stellar surface that they cannot 'lift' the layers above them.

In practice, the interplay between the observations, pulsation theory, and stellar evolution theory has led both to changes in *interpretations* of observational results (which have been mistaken for observational results themselves) and to changes in the assumptions that must be made in order to produce concrete theoretical calculations. Thus, analysis of the pulsational characteristics of classical pulsators has led to improved interpretations of the observations and to a better understanding of the internal structure of evolving stars.

As a first example, consider those stars in galactic globular clusters that are burning helium in their cores while on the horizontal branch. Those stars that are at the same time in the instability strip which intersects this branch are known as RR Lyrae stars. They fall into two groups: those with short period and symmetric light curves known as *c*-type variables and those with longer periods and asymmetric light curves known as *ab*-type variables. The *c*-type variables occupy a region in the left hand or hotter portion of the instability strip and the *ab*-type variables occupy the right hand portion.

How does one know that variables in the shorter period group are oscillating in the first overtone while those in the long period group are oscillating in the fundamental mode? Using pulsation theory and model stellar envelopes one can construct a relationship between period P_F in the fundamental radial mode, the radius of the envelope, and the assumed mass of the underlying stellar interior. This relationship is roughly (see, e.g., Böhm-Vitense *et al.*, 1975)

$$P_F^2 \sim 0.00716\,(R^3/M)(R/M)^{0.445}\,, \tag{1}$$

and it is to a very good approximation independent of the composition. The period P_H of the first overtone or first 'harmonic' is related to P_F by $\log P_H \simeq \log P_F - 0.13$. From the photometric data, as filtered through a stellar atmosphere analysis, one may determine the relative radii of stars in the instability strip and determine $\mathrm{d}P/\mathrm{d}R$ for the observed stars. One finds that $(\mathrm{d}P/\mathrm{d}R)_{\mathrm{obs}}$ agrees with the theoretical result with $M \sim$ constant and, if 0.13 is added to the log of the period of all of the *c*-type variables, they fall on a continuum in the $\log P - \log R$ plane with the *ab*-type variables. It follows, of course, that we have evidence that the RR Lyrae variables are radial pulsators with the *c*-type variables oscillating in the first overtone and the *ab* variables oscillating in the fundamental mode. The clincher comes when one constructs theoretical full amplitude nonlinear light curves showing the symmetry properties consistent with the observations and with the classification into fundamental and first overtone oscillators obtained simply by using results of linear pulsation analysis (see, e.g., Christy, 1966).

So far, we have learned nothing about stellar structure and evolution. This is achieved by estimating RR Lyrae masses with the help of Equation (1) and an assumption about the absolute luminosity of RR Lyrae stars (needed to estimate R absolutely). This may be done in one of two ways. Either one resorts to main sequence fitting (dangerous because the location of theoretical Main Sequences are composition dependent and because the globular cluster heavy element abundances Z are an order of magnitude smaller than the Z appropriate to comparison clusters whose distances are known by parallax) or one uses the predictions of stellar evolution theory (dangerous because these predictions are functions of assumed Z and Y).

In any case, making use of either scheme, one finds that $\log L_{HB} \simeq 1.6 \pm 0.1$ (e.g. Sweigart and Gross, 1976; Sandage *et al.*, 1981). Using this luminosity in conjunction with an observed surface temperature one may find R for any given RR Lyrae star. Inserting this R and an observed period into Equation (1) then gives the mass $M = M_{RR} \simeq 0.6\, M_\odot$. This is substantially less than the mass $M \simeq 0.8\, M_\odot$ of a star near the Main Sequence turnoff in a globular cluster. Thus one discovers that, somewhere in the interim between the core hydrogen burning phase (the Main Sequence) and the core helium burning phase (the horizontal branch), a low mass star loses a substantial fraction of its original mass. For a popular discussion of this result, see Iben (1970, 1971).

A comparison between linear non adiabatic pulsation theory and the observations can give further information about the bulk properties of RR Lyrae stars and their progenitors. Non-adiabatic pulsation theory provides a composition-dependent estimate of the location of the blue edge of the instability strip for pulsation in any given radial mode. For horizontal branch stars the relevant mode is the first harmonic. One finds (Tuggle and Iben, 1973)

$$(0.2 - Y) + 4.9\,(\log T_{HBE} - 3.859) + 0.35\,(\log P_{HBE} + 0.55) \simeq 0\,, \qquad (2)$$

where Y is the abundance by mass of helium, T_{HBE} is the surface temperature at the blue edge of the distribution of c-type variables, and P_{HBE} is the mean period of stars near this edge. Expression (2) is, to first order, independent of Z. Adopting $\log T_{HBE} \simeq 3.863 \pm 0.01$ (probably too large an error estimate by a factor of 2) and $\log P_{HBE} \simeq -0.55 \pm 0.05$, one obtains $Y \simeq 0.22 \pm 0.07$ (± 0.05 is perhaps more reasonable).

Thus, with the help of the observational and theoretical properties of RR Lyrae stars, we have been able not only to establish the existence of a mass loss process (which theory cannot as yet produce from first principles) but have been able to show that the abundance of helium in the odest stars is substantially greater than zero. This latter result provides support for the inference from the characteristics of the 3 K background radiation that nucleosynthesis in the early Universe produced a helium abundance of about 23% by mass.

As a next example, consider the classical Cepheids which have been used in establishing the second step in the extragalactic distance scale. These are Cepheids in galactic clusters whose absolute distances and luminosities can be estimated by fitting the cluster Main Sequence to an extrapolation of the Hyades main sequence (the traditional first step in the distance scale).

Because they are pulsating, one may obtain an estimate of the mass of each Cepheid with the help of the results of pulsation theory. For stars with characteristics in the ranges ($2 < M/M_{\odot} < 10$, $2.75 < \log L < 4.75$, $3.65 < \log T_e < 3.85$), linear non-adiabatic pulsation theory gives (Iben and Tuggle, 1975)

$$\log P_F \simeq 0.71 - 3.34\,(\log T_e - 3.75) + 0.85\,(\log L - 3.25) - 0.62\,(\log M - 0.7) \tag{3}$$

or

$$P_F \simeq \text{constant}\,(R^3/M)^{1/2} R^{0.2}/M^{0.12}\,, \tag{3'}$$

again roughly independent of composition. Here P_F is the pulsation period in days, T_e is the surface temperature in Kelvins, and M, L, and R are stellar mass, luminosity, and radius, all in solar units. Clearly, if P, T_e, and L are known, M may be determined from Equation (3). The mass $M = M_{\text{pulse}}$ estimated in this way may then be compared with the mass $M = M_{\text{evo}}$ of a model intermediatemass star which evolves through the classical instability strip on a long-time scale during its core helium burning phase. By comparing theoretical evolutionary tracks with the location of the theoretical instability strip, one finds the following relationship between M_{evo}, L, and composition of a model crossing the strip (Becker *et al.*, 1977 = BIT):

$$\log L \simeq (0.46 - 41\Delta Z + 6.6\Delta Y) + (3.68 + 21\Delta Z - 4.5\Delta Y)\log M_{\text{evo}}\,, \tag{4}$$

where $\Delta Z = Z - 0.02$, and $\Delta Y = Y - 0.28$. For the 13 Cepheids in question, $\Delta Z = \Delta Y = 0$ are reasonable choices, and insertion into Equation (4) of L obtained by Main Sequence fitting gives a first estimate of M_{evo}. When luminosities estimated by adopting a distance modulus of ~ 3.05 mag. for the Hyades are used, one finds that M_{pulse} is typically 20–40% smaller than M_{evo} (Cogan, 1970; Rogers, 1970; Iben and Tuggle, 1972) and one might infer once again that evidence for considerable mass loss during the interim between the Main Sequence and the core helium burning phases has been uncovered. However, in this case, the lifetime of the red giant phase which precedes the Cepheid phase (and which is the most likely place for extensive mass loss to occur) is so short that one would require mass loss rates on the order of several times $10^{-5}\,M_{\odot}\,\text{yr}^{-1}$ to explain a 30% decrease in mass (a typical Cepheid mass is 3–6 $M_{\odot}$) and such high rates for normal intermediate-mass red giants (of luminosity $\log L \sim 3$–4) are not found observationally.

Inspection of Equations (3) and (4) shows that a modest increase of about 0.25 mag. in the assumed distance modulus to the Hyades will remove the discrepancy between M_{pulse} and M_{evo} (Iben and Tuggle, 1972). In the last decade, a number of investigations have shown that such an increase is not inconsistent with the data used to derive the Hyades distance (parallax measurements, convergent point assumption, etc.). References to the relevant literature are given by Cox (1980).

Thus, pulsation theory and evolution theory have participated in the establishment of a reliable measure of the traditional first step in the distance scale. They may participate also in the establishment of the second step, namely the distance to Andromeda. In addition to producing Equation (3), linear non-adiabatic pulsation theory can provide a relationship between T_e, L, M, P, and composition along the blue edge for pulsation in the fundamental mode (relevant for the brightest Cepheids). It is clear that from this relationship and Equations (3) and (4) one may eliminate three unknowns to obtain a relationship between L, P, and composition. In practice it is more reliable (and easier) to draw theoretical blue edges in the H–R diagram for each choice of M and composition, adopt a width of, say, $\Delta \log T_e \simeq 0.04$ for the instability strip, and find where the evolutionary track for a model of the chosen mass and composition crosses the semiempirical instability strip. In this way one finds that (BIT)

$$\langle \log L \rangle \simeq 2.51 + 1.25 \langle \log P \rangle , \tag{5}$$

where $\langle \log L \rangle$ is the average luminosity in the strip and $\langle \log P \rangle$ is the average period.

A remarkable feature of relationship (5) is that it holds reasonably well *for all compositions*, thus providing at long last theoretical justification for the observers *assumption* that the period-luminosity relationship is independent of composition. Note further that the *theoretical* relationship is indeed that: only through the adoption of a width for the instability strip have results of observations entered (weakly) into establishing the relation. I believe that this is a triumph of both stellar evolution and pulsation theory of a magnitude that has yet to be appreciated.

In any case, using Equation (5) to establish the distance to Andromeda requires observational estimates of P and apparent bolometric magnitude for a number of Cepheids, and the equivalent of (5) that has been derived by normalizing to the 13 galactic Cepheids whose distances can be estimated by Main Sequence fitting is (when corrected by 0.25 mag.!) just as good a relationship as (5). Given this, perhaps the best conclusion to draw is that the remarkable similarity between the theoretical P–L relationship and the observationally based relationship tells us that our interior models of core helium burning stars of intermediate mass are perhaps believable.

As a final example, consider stars in the asymptotic giant branch (AGB) phase of evolution. These are stars with dense carbon-oxygen cores in which electrons are highly degenerate. Nuclear burning occurs alternatively in two shells outside of this core. Most of the time hydrogen-burning in the outermost shell provides the luminosity escaping from the surface. However, at regular intervals, helium-burning in the innermost shell becomes unstable and the star experiences a *thermal pulse* (Schwarzschild and Härm, 1965; Weigert, 1965). The time between pulses varies from $\sim 10^5$ yr when core mass $M_c = 0.6\, M_\odot$ to $\sim 10^3$ yr when $M_c = 1\, M_\odot$ and the quiescent helium burning phase which follows each pulse last approximately one fourth as long as the quiescent helium burning phase. The pulse itself lasts from 300 yr ($M_c = 0.6\, M_\odot$) to 30 yr ($M_c = 1\, M_\odot$).

There exists a simple relationship between the maximum luminosity of an AGB star during the quiescent hydrogen-burning phase and the core mass. In good approximation

(Paczynski, 1971; Uus, 1971)

$$L = 6 \times 10^4 L_\odot (M_c/M_\odot - 0.5)\,. \tag{6}$$

From this relationship (with the knowledge that $M_c^{\min} \simeq 0.53\, M_\odot$ and $M_c^{\max} = 1.4\, M_\odot$) one sees that AGB stars occur typically at magnitudes in the range

$$-7.3 < M_{\mathrm{BOL}} < -3.3.$$

AGB stars spend most of their lives far to the red of the classical instability strip and they possess extremely deep convective envelopes. However, because of the low density in the envelopes, convection is not efficient and the temperature gradient is highly superadiabatic. Driving occurs in the hydrogen ionization zone and the stars oscillate as Miras (low luminosity) or as long period variables (high luminosity). Unfortunately, convection once again interferes with understanding. Because the degree of superadiabaticity is large over much of the envelope and is quite sensitive to the treatment of convection (this time, the time dependence is also important), a unique PLT$_e$M relationship does not exist for any mode of oscillation. Furthermore, results depend also on uncertain molecular opacities and on surface boundary conditions (Fox and Wood, 1981). Hence, not only can one not use pulsation theory to estimate mass, as in the case of RR Lyrae stars and Cepheids, there is even controversy over the mode in which an AGB star oscillates (see, e.g., Willson, 1981; Wood, 1981).

Given the difficulty of ascertaining directly the global chacteristics of AGB stars by using observed pulsation properties in conjunction with an adequate pulsation theory, it is encouraging that there exists an indirect way of using pulsation theory to learn something interesting about AGB stars. This way relies on the fact that every Cepheid in the period range 2–10 days should ultimately turn into an AGB star of high luminosity. That this is true is illustrated in Table II, which gives various properties of 12 Cepheids used in the classic derivation of the distance scale. Relevant observational data used

TABLE II

Properties of 12 classical Cepheids in galactic clusters

Star	M_{BOL}	$\log M$	$\log P$	Age	M	$M_{\mathrm{CO}}^{\min}$	$M_{\mathrm{BOL}}^{\mathrm{AGB}}$
EV Sct	−2.90	0.615	0.489	11.7	4.12	0.74	−5.6
CE Casb	−3.49	0.676	0.651	8.8	4.74	0.82	−5.9
CE Cas	−3.39	0.658	0.687	8.2	4.55	0.79	−5.8
CE Casa	−3.58	0.710	0.710	7.9	5.13	0.84	−6.0
UY Per	−3.83	0.738	0.729	7.6	5.47	0.86	−6.2
VY Per	−4.19	(∼0.70)	(0.742)	(<7.4)	∼5	0.83	−6.0
U Sgr	−4.21	0.758	0.828	6.4	5.73	0.88	−6.1
DL Cas	−4.17	0.754	0.903	5.6	5.68	0.88	−6.1
S Nor	−4.37	0.757	0.989	4.8	5.71	0.88	−6.1
VX Per	−4.63	0.748	1.037	4.4	5.62	0.87	−6.1
SZ Cas	−5.01	0.781	1.134	3.7	6.04	0.90	−6.2
RS Pup	−6.31	1.019	1.616	1.6	10.4	supergiant	

to prepare this table are from Sandage and Tammann (1969), Sandage and Gratton (1963), and Rogers (1970). The bolometric magnitudes are 0.25 mag. brighter and the logarithms of the masses are 0.1 larger than given in Table II of Iben and Tuggle (1972). Cepheid ages (in 10^7 yr) are estimated using Equation (20) in BIT. Of primary interest here are the core mass M_{CO}^{min} and the bolometric magnitude M_{BOL}^{AGB} which each Cepheid should ultimately attain when it first transforms into an AGB star and begins to pulse thermally. The relationship between initial mass M and M_{CO}^{min} is taken from Iben (1981) and the bolometric magnitude follows from Equation (6).

Since the theoretical lifetime of a thermally pulsing (TP) AGB star and the lifetime of a typical Cepheid are roughly the same ($\sim 10^6$ yr), one expects to find approximately a dozen TP–AGB stars in the galactic clusters containing the studied Cepheids. As Table II shows, many of these AGB stars should be brighter than $M_{BOL} = -6$.

Theory also predicts that TP–AGB stars with cores as large as those in Table II should be dredging up carbon to their surfaces (see, e.g., Iben, 1981, for a recent discussion) and that roughly half of their AGB lifetime should be spent as carbon stars (abundance of carbon greater that the abundance of oxygen at the surface). Significantly, not one of the galactic clusters chosen for study contains a carbon star, whereas there should be altogether about 6.

This point is made even more graphically by the bright stars in the cluster NGC 1866 in the large Magellanic Cloud. This cluster contains 7 Cepheids and at least four AGB stars ranging in brightness from $M_{BOL} = -5$ to $M_{BOL} \simeq -6.2$ (Frogel and Blanco, 1981). Using pulsation theory and the relevant observed properties of the Cepheids one finds that the average Cepheid mass is $M_{Ceph} \sim 4$–$5\ M_\odot$ and $M_{BOL}^{AGB} \cong -5.6$ to -6.1 mag. Not one of the AGB stars is a carbon star.

There are at least two possible interpretations of these results. Either evolution theory (which says that about half of the bright TP–AGB stars should be carbon stars) is wrong; or, when it becomes a carbon star, the lifetime of a bright AGB star is short compared to its lifetime as a non-carbon star. The existence of a class of carbon stars in our own Galaxy with estimated mass loss rates greater than $10^{-5}\ M_\odot\ \text{yr}^{-1}$ (Knapp *et al.*, 1982) suggests that the second possibility is the more likely one. The inference is that, when C exceeds O at the stellar surface, a 'superwind' is engendered (perhaps due to radiation pressure on carbon grains).

To summarize, the fact that, in those clusters containing Cepheids, the brightest AGB stars are not carbon stars is support for the theoretical prediction that bright AGB stars will become carbon stars! Since theoretically C exceeds O at the surface well before the core mass reaches the Chandrasekhar limit of about 1.4 $M_\odot$, we have also learned that single stars of intermediate mass do not become supernovae! These are rather sweeping conclusions, and they have been made possible with the help of a seismological analysis of classical variables.

May the helioseismologists ultimately be as successful!

References

Becker, S. A., Iben, I. Jr., and Tuggle, R. S.: 1977, *Astrophys. J.* **218**, 633.
Bohm-Vitense, E., Szkody, P., Wallerstein, G., and Iben, I. Jr.: 1974, *Astrophys. J.* **194**, 125.
Christy, R.: 1966, *Astrophys. J.* **144**, 108; **145**, 337.
Cogan, B. C.: 1970, *Astrophys. J.* **162**, 129.
Cox, J. P.: 1974, *Rep. Prog. Phys.* **37**, 563.
Cox, A. N.: 1980, *Ann. Rev. Astron. Astrophys.* **18**, 15.
Davis, R.: 1981, private communication.
Fox, M. W. and Wood, P. R.: 1981, preprint.
Frogel, J. A. and Blanco, V. M.: 1982, private communication.
Iben, I. Jr.: 1970, *Scientific American*, July issue.
Iben, I. Jr.: 1971, *Publ. Astron. Soc. Pacific* **83**, 697.
Iben, I. Jr.: 1975, *Astrophys. J.* **197**, 39.
Iben, I. Jr.: 1981, *Astrophys. J.* **246**, 278.
Iben, I. Jr. and Mahaffy, J.: 1976, *Astrophys. J.* **209**, L39.
Iben, I. Jr. and Tuggle, R. S.: 1972, *Astrophys. J.* **178**, 433.
Knapp, G. R., Phillips, T. G., Leighton, R. B., Lo, K. Y., Wannier, P. G., Wooten, H. A., and Huggins, P. J.: 1982, *Astrophys. J.* **252**, 616.
Paczynski, B.: 1970, *Acta. Astron.* **20**, 47, 287.
Rogers, A. W.: 1970, *Monthy Notices Roy. Astron. Soc.* **151**, 133.
Sandage, A. R. and Gratton, L.: 1963, *Star Evolution*, Academic Press, New York, p. 11.
Sandage, A. R. and Tammann, G. A.: 1969, *Astrophys. J.* **157**, 683.
Sandage, A., Katem, B., and Sandage, M.: 1981, *Astrophys. J.* **46**, 41.
Schwarzschild, M. and Härm, R.: 1965, *Astrophys. J.* **142**, 855.
Severny, A. B., Kotov, V. A., and Tsap, T. T.: 1976, *Nature* **259**, 87.
Sweigart, A. V. and Gross, P. G.: 1976, *Astrophys. J.* **32**, 367.
Sweigart, A. V. and Gross, P. G.: 1978, *Astrophys. J.* **36**, 405.
Tuggle, R. S. and Iben, I. Jr.: 1973, *Astrophys. J.* **186**, 593.
Uus, U.: 1970, *Nauch. Inform. Acad. Nauk USSR* **17**, 3, 25, 35, 48.
Weigert, A.: 1966, *Z. Astrophys. J.* **64**, 395.
Willson, L. A.: 1981, in I. Iben Jr. and A. Renzini (eds.), *Physical Processes in Red Giants*, D. Reidel Publ. Co., Dordrecht, Holland, p. 225.
Wood, P. R.: 1981, in I. Iben Jr. and A. Renzini (eds.), *Physical Processes in Red Giants*, D. Reidel Publ. Co., Dordrecht, Holland, p. 205.

STELLAR 5 MIN OSCILLATIONS*

JØRGEN CHRISTENSEN-DALSGAARD

*Advanced Study Program, National Center for Atmospheric Research***

and

SØREN FRANDSEN

*High Altitude Observatory, National Center for Atmospheric Research**, Boulder, Colorado*

and

Astronomisk Institut[†], *Aarhus Universitet, DK-8000 Aarhus C, Denmark*

Abstract. Estimates are given for the amplitudes of stochastically excited oscillations in Main Sequence stars and cool giants; these were obtained using the equipartition between convective and pulsational energy which was originally proposed by Goldreich and Keeley. The amplitudes of both velocity and luminosity perturbation generally increase with increasing mass along the Main Sequence as long as convection transports a major fraction of the total flux, and the amplitudes also increase with the age of the model. The 1.5 $M_\odot$ ZAMS model, of spectral type F0, has velocity amplitudes ten times larger than those found in the Sun. For very luminous red supergiants luminosity amplitudes of up to about $0\overset{m}{.}1$ are predicted, in rough agreement with observations presented by Maeder.

1. Introduction

There would be an obvious intrinsic interest in the discovery of stellar analogues to the solar 5 min oscillations of low degree. More important, however, is the fact that detection and detailed observation of such oscillations might enable the extension to other stars of the seismological investigations (e.g. Christensen-Dalsgaard and Gough, 1976) which are now beginning to yield information about the structure of the Sun (Scuflaire *et al.*, 1981; Christensen-Dalsgaard and Gough, 1980b, 1981; Shibahashi and Osaki, 1982). In addition, the variation of oscillation amplitude with stellar parameters would be of great interest in connection with the determination of the excitation mechanism for these oscillations.

An immediate problem facing any attempt to detect such oscillations, at least in solar-type stars, is their very small amplitude. Thus the velocity amplitudes of at most 15–40 cm s^{-1} for each mode of oscillation which is observed in the Sun (Grec *et al.*, 1980; Claverie *et al.*, 1981) are probably below the limit of present spectroscopic techniques; the relative luminosity amplitudes of $2–4 \times 10^{-6}$ (Deubner, 1981; Woodard and Hudson, 1983) would be observable in a star (Deubner found evidence

* Proceedings of the 66th IAU Colloquium: *Problems in Solar and Stellar Oscillations*, held at the Crimean Astrophysical Observatory, U.S.S.R., 1–5 September, 1981.

** NCAR is sponsored by the National Science Foundation.

[†] Permanent address.

Solar Physics **82** (1983) 469–486. 0038–0938/83/0822–0469$02.70.

for the oscillations in sunlight reflected from Neptune), but only in observations dedicated to a single star for several nights.

Traub *et al.* (1977) attempted to detect stellar 5 min oscillations using the PEPSIOS spectrometer. The criterion used to select the stars observed was the presence of Ca emission, which was taken as evidence for waves heating a chromosphere. These observations clearly showed the solar 5 min oscillations, but failed to find oscillations in the other stars, giving an upper limit of about 5 m s^{-1} on the total velocity amplitude. In addition M. A. Smith (private communication) has found evidence for velocity fluctuations in two giants (Arcturus and Aldebaran) which he tentatively interprets as oscillations that may be related to the solar 5 min oscillations.

The obvious difficulty in observing these oscillations, and the long observing sequences needed for each star, make desirable some guidance for the choice of stars to observe. Such stars should clearly have as large amplitudes as possible. In addition the periods of oscillation should be sufficiently short to allow removal of drift in the observations (e.g. caused by variations in the electronics, or by atmospheric extinction). Finally, if frequency resolved observations are sought, the frequency separation between the dominant peaks in the spectrum of oscillation should ideally be large enough to be resolved with a single night's observations.

The purpose of the present paper is to give theoretical estimates of amplitudes, periods and frequency separations for a selection of stellar models. The periods, and hence the frequency separations, for a given model are easily calculated from linear theory. To obtain estimates of oscillation amplitudes a model for the excitation of the oscillations is needed, and this is at present far less certain. We shall adopt as premise that the solar 5 min oscillations are not self-excited (i.e. that they are stable according to linear theory), but that they are excited stochastically by convection. It is true that the linear non-adiabatic calculations of Ando and Osaki (1975, 1977) showed instability of the solar 5 min oscillations. However these calculations assumed that the equilibrium value of the mean intensity of radiation was equal to the Planck function everywhere (cf. Christensen-Dalsgaard and Frandsen, 1983), and, probably more importantly, they neglected the Lagrangian perturbation in the divergence of the convective flux. Later calculations by Berthomieu *et al.* (1980) and Baker and Gough (cf. Gough, 1980) that included the perturbation in the convective flux found the modes in the 5 min range to be stable. Under these circumstances the hypothesis of stochastic excitation appears to be the most likely alternative.

Goldreich and Keeley (1977b) made a simplified analysis of this excitation mechanism. They found that the resulting amplitude of oscillation is such that there is approximate equality between the energy, integrated over the star, in one mode of oscillation and the kinetic energy in one convective eddy whose time scale is the same as the period of the oscillation. Keeley (1977, 1980) and Gough (1980) showed that this equipartition of energy predicts roughly the observed amplitudes of the solar 5 min oscillations (cf. also Christensen-Dalsgaard and Gough, 1982). We shall use it, in a form made precise below, to estimate oscillation amplitudes for other stars.

2. The Calculation

Complete stellar models were calculated using the code described by Christensen-Dalsgaard (1982; in the following C-D82) with a small modification in the way the variation in the super-adiabatic gradient was taken into account in the determination of the mesh. The Cox and Tabor (1976) opacity tables were used, and the parameters were the same as for Model sequence 1 of C-D82. The models comprised a set of Zero-Age Main Sequence models with masses between 0.8 and 1.8 $M_{\odot}$ ($M_{\odot}$ being the mass of the Sun), as well as a continuation of sequence 1 of C-D82 to well into the hydrogen shell-burning phase.

In addition to the complete models envelope models were calculated, to explore the properties of stars further from the Main Sequence. These envelopes were assumed to be chemically homogeneous and with constant luminosity; the physics was the same as in the calculation of the complete models. To approximate stars in the hydrogen shell-burning phase sets of envelopes with varying effective temperature, but constant mass M and luminosity L were computed, the relation between M and L being approximately as found in Iben's (1964) evolution calculations.

The complete models were transferred to a mesh more suitable for pulsation calculations using four point Lagrangian interpolation. This mesh was based partly on the variation in pressure and temperature, partly on the asymptotic properties of high-order acoustic oscillations, in the manner of Christensen-Dalsgaard (1977); the same mesh was used directly in the calculation of the envelope models.

Observations of stellar oscillation, by necessity made in integrated light, are dominated by modes with values of the degree l less than about 4 (Dziembowski, 1977; Christensen-Dalsgaard and Gough, 1980a), and the relevant properties of such modes depend little on l (Christensen-Dalsgaard and Gough, 1982). Hence we have only calculated radial modes of oscillation, by solving the equations of linear non-adiabatic oscillation. Radiation was treated in the Eddington approximation (e.g. Unno and Spiegel, 1966), and we neglected $\delta(\mathrm{div}\,\mathbf{F}_c)$, where δ denotes Lagrangian perturbation and $\mathbf{F}_c$ is the convective flux. The mechanical surface boundary condition was derived from matching the solution onto the outward decaying solution of the adiabatic wave-equation in an isothermal atmosphere and was applied at optical depth $\tau = 0.01$. At the inner boundary the oscillations were assumed to be adiabatic; in the complete models a second inner boundary condition was obtained by expansion around the centre, whereas the displacement was assumed to vanish at the bottom of the envelope models. With these boundary conditions the frequencies of oscillation are determined as eigenvalues of the pulsation equations.

For every model considered we have calculated all modes of oscillation in a range sufficiently large to determine the maximum amplitudes in velocity and luminosity perturbation, and to study in some detail the variation in the amplitudes around the maximum. As an upper bound on the frequencies we have used Lamb's (1909) acoustical cut-off frequency

$$\nu_c = \frac{\Gamma_1 g_s}{4\pi c_s}, \tag{2.1}$$

where $\Gamma_1 = (\partial \ln p/\partial \ln \rho)_s$ is an adiabatic exponent, p, ρ, and s being pressure, density and specific entropy, respectively, g_s is the surface gravity and c_s is the sound speed in the (assumed) isothermal atmosphere bounding the model. The energy equipartition between convection and pulsation derives from the fact that turbulent viscosity appears to dominate the linear damping of the oscillations (Goldreich and Keeley, 1977a); thus the dynamics of convection controls both the excitation and the damping of the oscillations. However there is a tendency for modes with frequency ν close to or above ν_c to behave like running waves in the atmosphere; this leads to a relative increase in the atmospheric amplitude of such modes and so for these the radiative atmospheric damping may become dominant (see also Christensen-Dalsgaard and Frandsen, 1983). This would cause their amplitudes to be smaller than predicted by energy equipartition. (It might be noted that this argument suggests that claims of detection of stellar oscillations with frequencies significantly exceeding ν_s, which can of course easily be estimated, should be viewed with some suspicion. Such oscillations probably do not represent large-scale pulsation of the star).

The oscillation amplitude calculated on the basis of energy equipartition depends on the details of the description of convection and the precise definition of the convective time scale. We have adopted the mixing length formulation for a static model given by Gough (1977), with parameters chosen to make the convective flux agree with that of Böhm-Vitense (1958). The time scale was taken to be the mean lifetime τ_c of a convective eddy (Gough, Equations (4.23) and (4.26)); the mean kinetic energy in an eddy was calculated as

$$E_c = \tfrac{1}{2} p_t l_c^3, \tag{2.2}$$

where p_t is the mean turbulent pressure (Gough, Equation (3.16)) and l_c is the mixing length, as usual taken to be a constant (in this case 1.6364) multiple of the pressure scale height. This choice of τ_c and E_c is clearly not unique; however as shown in Section 3 it does appear to give roughly correct amplitudes for the Sun, without the introduction of additional scaling factors. Furthermore the variation with stellar parameters in the predicted amplitude is probably not very sensitive to the precise formulation.

In a given model E_c and τ_c can thus be calculated as functions of the distance r from the centre of the model. At the edges of the convection zone the velocity tends to zero and τ_c tends to infinity; τ_c has a minimum $\tau_{c,\min}$ which is generally close to where the superadiabatic gradient is largest. For a mode with period Π greater than $\tau_{c,\min}$ there are at least two points, with $r = r_i$, say, in the convection zone where $\tau_c = \Pi$; we have calculated the oscillation amplitude by demanding that the amplitude of the kinetic energy of oscillation be equal to the sum of E_c over these points, that is

$$E_{\rm osc} \equiv \tfrac{1}{2} \int_V \rho |\mathbf{v}_{\rm osc}|^2 \, dV = \sum_i E_c(r_i) \, . \tag{2.3}$$

Here $\mathbf{v}_{\mathrm{osc}}$ is the amplitude of the oscillation velocity and the integral is over the volume V of the equilibrium model. The results would change little if the maximum over the energies of the resonant eddies, rather than their sum, had been used.

The present, grossly simplified, model of stochastic excitation predicts that modes with periods shorter than $\tau_{c,\mathrm{min}}$ are not excited. In reality there would be a contribution to the excitation of these modes from convective eddies with longer time scales, and from smaller eddies resulting from the turbulent breakdown of the dominant eddies. Thus one would expect a gradual decrease in amplitude at periods shorter than $\tau_{c,\mathrm{min}}$, rather than the sharp cut-off predicted here.

It is convenient to express the pulsational energy as

$$E_{\mathrm{osc}} = \tfrac{1}{2} M v_s^2 \mathscr{E}_{\mathrm{osc}} , \tag{2.4}$$

where v_s is the radial component of the surface velocity amplitude and

$$\mathscr{E}_{\mathrm{osc}} = \int_V \rho |\boldsymbol{\xi}|^2 \, \mathrm{d}V / M \xi_r(r_s)^2 , \tag{2.5}$$

$\boldsymbol{\xi}$ being the eigenfunction of linear oscillation, ξ_r its radial component and r_s the surface radius of the star (notice that Equations (2.3)–(2.5) can be applied to non-radial as well as to radial oscillations). Thus $\mathscr{E}_{\mathrm{osc}}$ can be found from linear theory, and then

$$v_s = \left[\frac{2 \sum_i E_c(r_i)}{M \mathscr{E}_{\mathrm{osc}}} \right]^{1/2} . \tag{2.6}$$

Finally the relative surface luminosity perturbation can be calculated from v_s as

$$\delta L_s / L_s = \lambda_s v_s , \tag{2.7}$$

where $\lambda_s = (\delta L_s / L_s)/v_s$ may be found from a linear calculation.

3. Results

We first consider the calibration against observations of solar oscillations. It might be argued that the criterion for 'resonance', viz. $\Pi = \tau_c$, is arbitrary and should be replaced by $\Pi = \gamma \tau_c$, where γ is a factor to be calibrated against the solar data. On Figure 1 are shown the predicted velocity amplitudes, as functions of the frequency, for different values of γ. In each case the amplitude increases monotonically with ν until the cut-off frequency $\nu_{\mathrm{max}} = 1/(\gamma \tau_{c,\mathrm{min}})$, and the dominant effect of changing γ is clearly to shift ν_{max}. For $\gamma = 1$, ν_{max} almost coincides with the observed position of maximum power, and the maximum velocity, about 15 cm s^{-1}, is consistent with the observed value. Of course the observed spectra show a gradual decrease in power towerds higher frequency, rather than the sharp cut-off found here; but as argued in the preceeding section this would be smoothed in a more detailed description of the excitation. Thus we have used $\gamma = 1$ in the following. The maximum amplitude of the relative luminosity perturbation

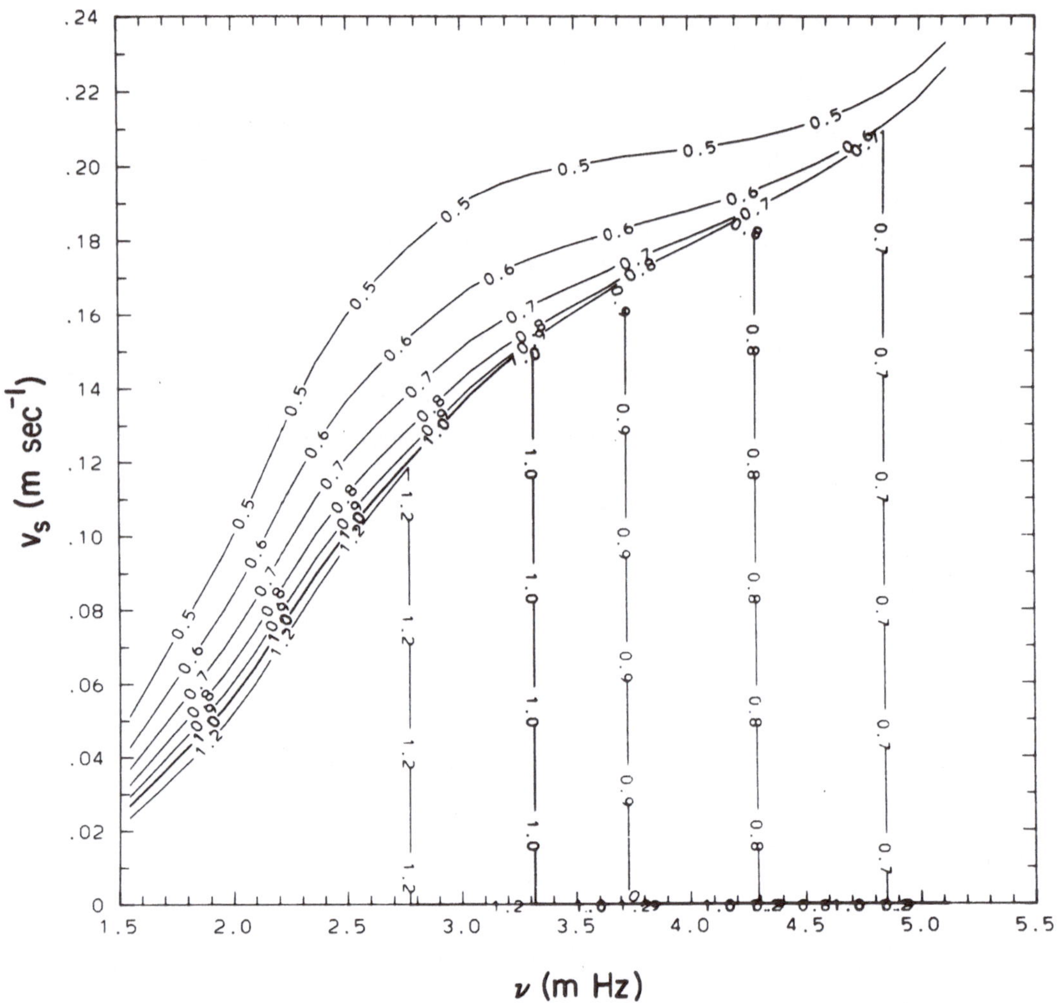

Fig. 1. Predicted surface velocity amplitudes v_s in a model of the present Sun, as functions of the cyclic frequency ν of oscillation, for various values of the ratio γ between the pulsation period and the time scale of the 'resonating' convective eddy. For clarity the values for the discrete modes of oscillation have been connected with continuous lines. The curves are labelled with the value of γ.

is then 3.5×10^{-6}, fairly close to the value observed by Woodard and Hudson (1983). It should also be noticed that the ratio found here between the luminosity perturbation and the velocity is in reasonable agreement with the value obtained by Gough (1980), who treated the radiation in the diffusion approximation but included the perturbation in the convective flux.

The main results concerning the complete stellar models are presented in Table I for the ZAMS models and in Table II for the 1 $M_\odot$ evolution sequence. The value given for $\Delta\nu$ is the frequency difference between two consecutive modes at the velocity maximum; as the frequency spacing is nearly uniform for high-order acoustic modes

TABLE I

Properties of oscillations of ZAMS models. M, T_{eff} and L are the mass, effective temperature and luminosity of the model, respectively, $M_\odot$ and $L_\odot$ being the solar values; $v_{s,max}$ and $(\delta L_s/L_s)_{max}$ are the maximum values of the surface velocity and relative luminosity perturbation, and Π_{max} is the period corresponding to, and $\Delta\nu$ the frequency difference between two adjacent modes at, the maximum velocity.

$M/M_\odot$	T_{eff}	$L/L_\odot$	$v_{s,max}$ (cm s^{-1})	$(\delta L_s/L_s)_{max}$	Π_{max} (min)	$\Delta\nu$ (μHz)
0.8	4880	0.25	3	6×10^{-7}	4	204
0.9	5285	0.44	6	1.5×10^{-6}	4	184
1.0	5646	0.71	10	2.5×10^{-6}	4	165
1.1	5916	1.13	15	3.4×10^{-6}	5	142
1.2	6178	1.70	20	4.2×10^{-6}	5	124
1.3	6457	2.49	25	5.1×10^{-6}	5	110
1.4	6778	3.49	37	6.2×10^{-6}	10	96
1.5	7184	4.74	144	1.3×10^{-5}	18	89
1.6	7640	6.26	72	1.0×10^{-5}	9	94
1.7	8072	8.06	71	2.7×10^{-6}	10	93
1.8	8475	10.2	72	2.7×10^{-5}	7	90

TABLE II

Properties of oscillations of the models in a 1 $M_\odot$ evolution sequence. The notation is as in Table I.

Age (10^9 yr)	T_{eff}	$L/L_\odot$	$v_{s,max}$ (cm s^{-1})	$(\delta L_s/L_s)_{max}$	Π_{max} (min)	$\Delta\nu$ (μHz)
0	5646	0.71	10	2.5×10^{-6}	4	165
2.65	5713	0.85	13	3.0×10^{-6}	5	150
4.75	5770	1.00	15	3.5×10^{-6}	5	137
9.68	5784	1.68	26	6.2×10^{-6}	8	96
11.71	5359	2.35	34	9.3×10^{-6}	15	59

(e.g. Vandakurov, 1967) this value is representative for the range of frequencies where the amplitudes are large.

The ZAMS models extend well into the region of the δ Scuti stars; as in other linear calculations (see e.g. the review by Dziembowski, 1980) a number of modes were found to be unstable in the models with $M \geq 1.4\ M_\odot$. For such modes the amplitude estimates based on energy equipartition are presumably invalid. However, except at 1.5 $M_\odot$, the predicted maximum amplitude occurs for modes that were found to be linearly stable.

There is clearly a marked increase in the predicted velocity amplitudes with increasing stellar mass until 1.5 $M_\odot$, and a similar, but more erratic, increase in the luminosity amplitude. Furthermore there is also a significant increase in the amplitudes as the 1 $M_\odot$ model evolves. To understand this behaviour we must study the dependence of the pulsational and convective energy on stellar parameters. The details are clearly quite complicated, but it is possible to get a qualitative understanding of the dominant features. The relation between the energy and the surface amplitude of the oscillations may be estimated from asymptotic theory. The modes are evanescent in a region close to the surface whose depth decreases with increasing oscillation frequency. Thus the energy density in the mode increases with increasing depth until the oscillatory region, where the amplitude of the energy density is roughly constant, is reached; the increase in the energy density decreases with increasing ν, and so does $\mathscr{E}_{\rm osc}$. From the asymptotic theory for high-order acoustic modes (e.g. Vandakurov, 1967), modified to take into account the nonvanishing surface temperature of a realistic model (Christensen-Dalsgaard and Gough, 1980b) one may show that

$$\mathscr{E}_{\rm osc} \approx \frac{3}{4\pi^2} \frac{\rho_s}{\bar{\rho}} \frac{c_s}{r_s} \mathscr{F}(\nu/\nu_c) \int_0^{r_s} \frac{\mathrm{d}r}{c} , \tag{3.1}$$

where ρ_s is the photosperic, and $\bar{\rho}$ the mean, density of the model, and c_s is the value in the atmosphere of the sound speed c; the dependence of $\mathscr{E}_{\rm osc}$ on ν is determined by $\mathscr{F}(\nu/\nu_c)$, where the acoustical cut-off frequency ν_c is given in Equation (2.1), and

$$\mathscr{F}(z) = \frac{1 + \Phi(z)^2}{\zeta[J_m(\zeta) - \Phi(z) Y_m(\zeta)]^2} , \tag{3.2}$$

$$\Phi(z) = \frac{J_{m+1}(\zeta) - \alpha(z) J_m(\zeta)}{Y_{m+1}(\zeta) - \alpha(z) Y_m(\zeta)} , \tag{3.3}$$

and

$$\alpha(z) = z^{-1} - (z^{-2} - 1)^{1/2} ; \qquad \zeta = (m + 1)z . \tag{3.4}$$

Here m is the effective polytropic index of the region close to the surface and J_m and Y_m are Bessel functions. When $\nu/\nu_c \ll 1$,

$$\mathscr{F}(\nu/\nu_c) \approx \tfrac{1}{2}\Gamma(m + 1)^2 [\tfrac{1}{2}(m + 1)(\nu/\nu_c)]^{-(2m+1)} ,$$

where Γ is the Gamma function; this rapid increase of $\mathscr{E}_{\rm osc}$ with decreasing ν reflects the increasing depth of the outer evanescent region.

The behaviour of the convective energy is more difficult to analyze, but the general trend may be understood in simple terms. Clearly E_c may be expected to increase roughly in proportion to the convective flux and the volume of the dominant convective eddy. A more detailed analysis shows that when the heat loss from a convective eddy during its lifetime is small,

$$\begin{aligned} E_c &\sim T^{5/2}(\nabla - \nabla_{\rm ad})^{-1/2} g_s^{-3} F_c \\ &\sim \rho^{1/3} F_c^{2/3} (T/g_s)^3 , \end{aligned} \tag{3.5}$$

where $\nabla = \mathrm{d}\ln T/\mathrm{d}\ln p$ and $\nabla_{\rm ad} = (\partial \ln T/\partial \ln p)_s$; when evaluated at the depth corresponding to $\tau_{c,\min}$, $\rho^{1/3}T^3$ roughly scales as its photospheric value. The dependence of E_c on g_s reflects the variation in the volume of the convective eddy.

We can now qualitatively account for the variation in the oscillation velocity found in Tables I and II. Along the ZAMS g_s decreases with increasing mass for $M \lesssim 1.2\, M_\odot$

and is roughly constant for larger masses, whereas T_{eff} increases. Furthermore ρ_s decreases with increasing T_{eff}; in fact we have approximately

$$\rho_s \approx \beta \frac{\mu_s g_s}{\mathscr{R} \varkappa_s T_{eff}} \tag{3.6}$$

(e.g. Schwarzschild, 1958), where μ_s and $\varkappa_s$ are the mean molecular weight and the opacity in the photosphere, $\mathscr{R}$ is the gas constant and β is of order unity, and in the relevant temperature range $\varkappa$ is a rapidly increasing function of T. The net effect is an increase in E_c with increasing mass, until the point where convection ceases to transport

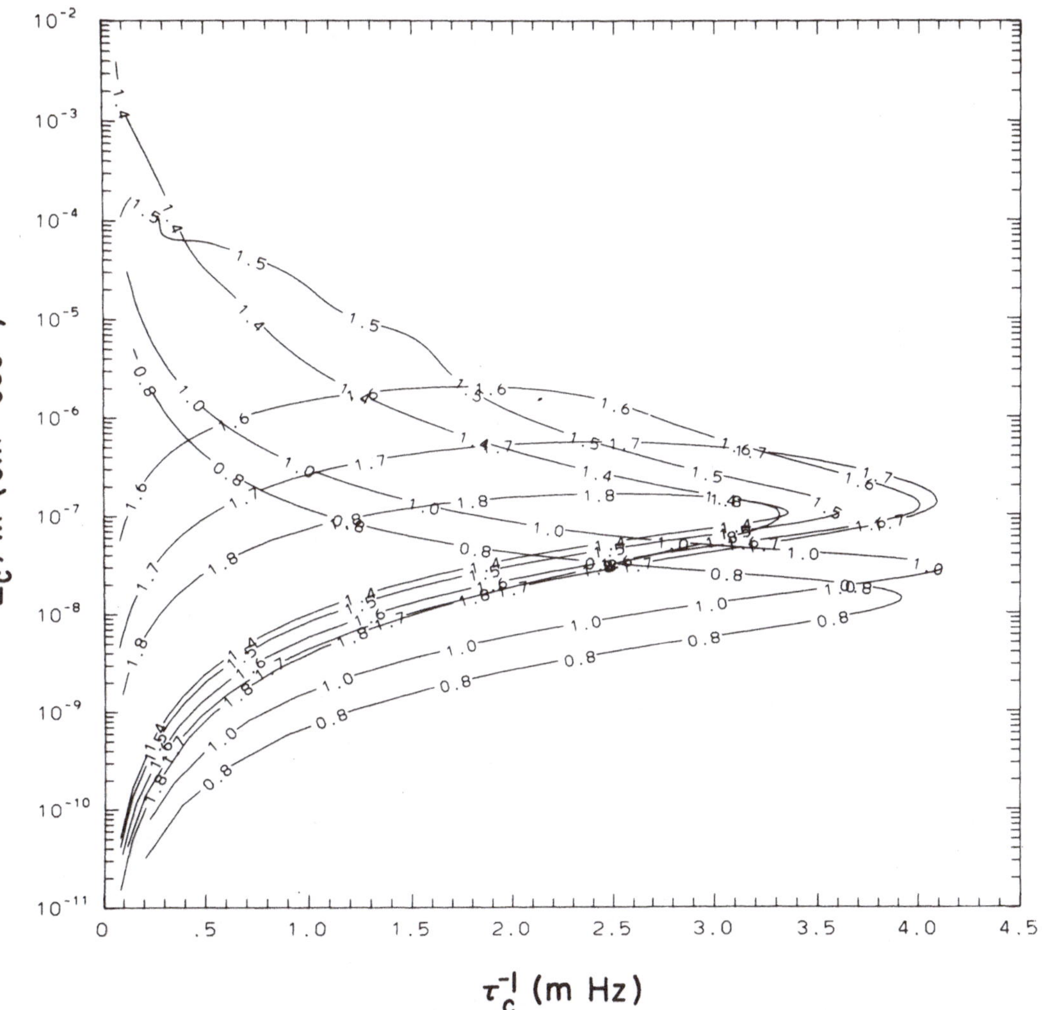

Fig. 2. The variation of E_c/M and $1/\tau_c$, E_c being the energy in a convective eddy and τ_c its time scale, through the upper convection zones in a selection of ZAMS models. The curves are labelled with $M/M_\odot$, where M is the mass of the model and $M_\odot$ the mass of the Sun.

the major part of the total energy flux. The variation in $\mathscr{E}_{\rm osc}$ is dominated by the decrease in ρ_s with increasing mass (cf. Equation (3.1)). Thus the variations in E_c and $\mathscr{E}_{\rm osc}$ both contribute to the general increase in v_s with M shown in Table I. The beginning decrease in the most massive models is caused by a decrease in the efficiency of convection, which in the 1.8 $M_\odot$ model carries at most about 40% of the flux.

It is of some interest to study in more detail the variation of the computed quantities. Figure 2 shows the run of τ_c^{-1} and E_c/M in a number of ZAMS models. In each case the upper edge of the convection zone corresponds to large τ_c and small E_c. With increasing depth in the convection zone τ_c decreases and E_c increases, until τ_c reaches a minimum whose position is generally close to that of the maximum in the superadiabatic

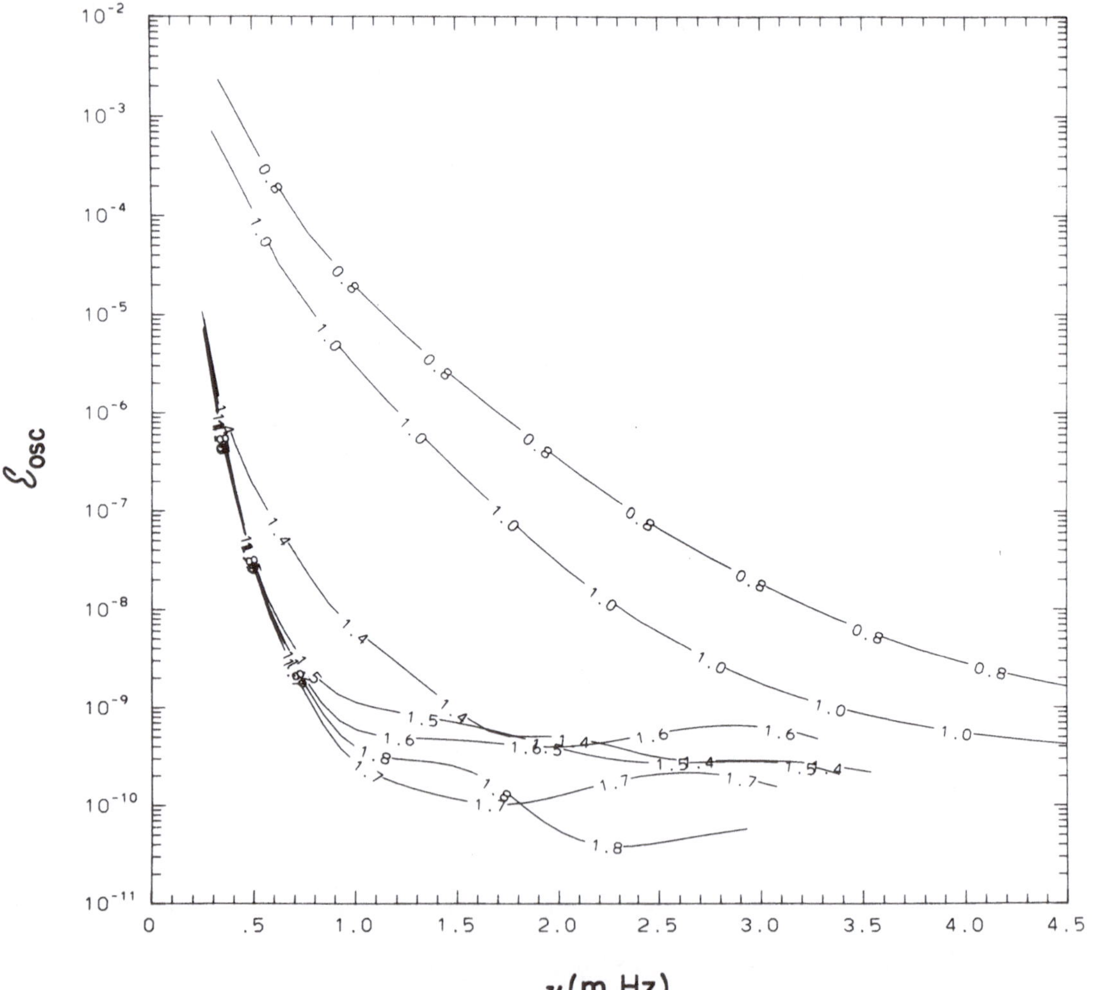

Fig. 3. The dimensionless pulsation energy $\mathscr{E}_{\rm osc}$ (defined in Equation (2.5)) for radial modes in the models on Figure 2, as functions of the cyclic oscillation frequency ν. For clarity the values for the discrete modes have been connected with continuous lines. The curves are labelled with $M/M_\odot$.

gradient. In the low-mass models, with deep convection zones, E_c continues to increase until close to the lower boundary of the convection zone where it tends to zero with the convective velocity. In high-mass models, on the other hand, convection only carries a significant fraction of the total flux near the middle of the convection zone, and E_c begins to decrease at a depth only slightly larger than the depth corresponding to the minimum in τ_c. In the models of highest mass in Table I mixing length theory predicts the existence of two separate convection zones, one corresponding to the ionization of H and the first ionization of He, and the other to the ionization of He^+. For these models only the H convection zone is included in Figure 1; the He^+ convection zone has time scales longer than the periods of the relevant oscillations.

The variation in $\mathscr{E}_{\text{osc}}$ with oscillation frequency, for the same models, is presented on Figure 3. All radial modes up to the acoustical cut-off frequency, or within the range of the convective time scales, have been included; for clarity the discrete values have been connected by continuous lines. The figure clearly shows the rapid increase in $\mathscr{E}_{\text{osc}}$ at small frequencies, predicted by Equation (3.1), as well as the decrease in $\mathscr{E}_{\text{osc}}$ with increasing stellar mass.

Finally Figure 4 shows the predicted velocity amplitudes as functions of the oscillation frequency. Both the total convective energy and the pulsational energy generally increases with increasing time scale, and hence the variation of the velocity with frequency depends on the balance between these two effects. For the lower-mass stars the variation in $\mathscr{E}_{\text{osc}}$ dominates, and the velocity is largest at the cut-off frequency $\tau_{c,\text{min}}^{-1}$. For higher mass, however, v_c is reduced (partly because of the reduction in the mean molecular weight and in Γ_1 caused by ionization in the atmosphere) so much that $\mathscr{E}_{\text{osc}}$ varies relatively little for frequencies close to $\tau_{c,\text{min}}^{-1}$. Hence here the increase in E_c with increasing τ_c dominates, leading to an amplitude maximum at intermediate frequencies; this is especially pronounced in the 1.5 $M_\odot$ model. For models of even higher mass, with two separate convection zones, the decrease in E_c at the lower boundary of the upper zone contributes to the decrease in the velocity at low frequencies and shifts the velocity maximum to somewhat higher frequencies.

In models with masses up to 1.6 $M_\odot$ the variation in λ_s is small, and the luminosity amplitudes follow the velocities fairly closely. At higher masses the behaviour of λ_s is less regular. Thus the 1.7 $M_\odot$ model has an exceptionally low value of λ_s, leading to the small $(\delta L/L)_s$ shown in Table I, whereas λ_s in the 1.8 $M_\odot$ model is large, causing a relatively high luminosity amplitude for this model. The reason for this behaviour of λ_s is not clear, but it may be related to differences between the two models in the structure of the convection zone. In particular the maximum efficiency of convection in the upper convection zone is reduced markedly from the 1.7 $M_\odot$ to the 1.8 $M_\odot$ model. Furthermore the upper edge of the convection zone is at a relatively small optical depth in these models, so that the surface luminosity perturbation is probably strongly affected by the behaviour of the luminosity perturbation in the convection zone. This also implies that the predicted luminosity perturbation might be quite sensitive to the neglect of the convective flux perturbation, and it should therefore be regarded as preliminary for the higher-mass models. In the lower-mass models, including the Sun, the upper boundary

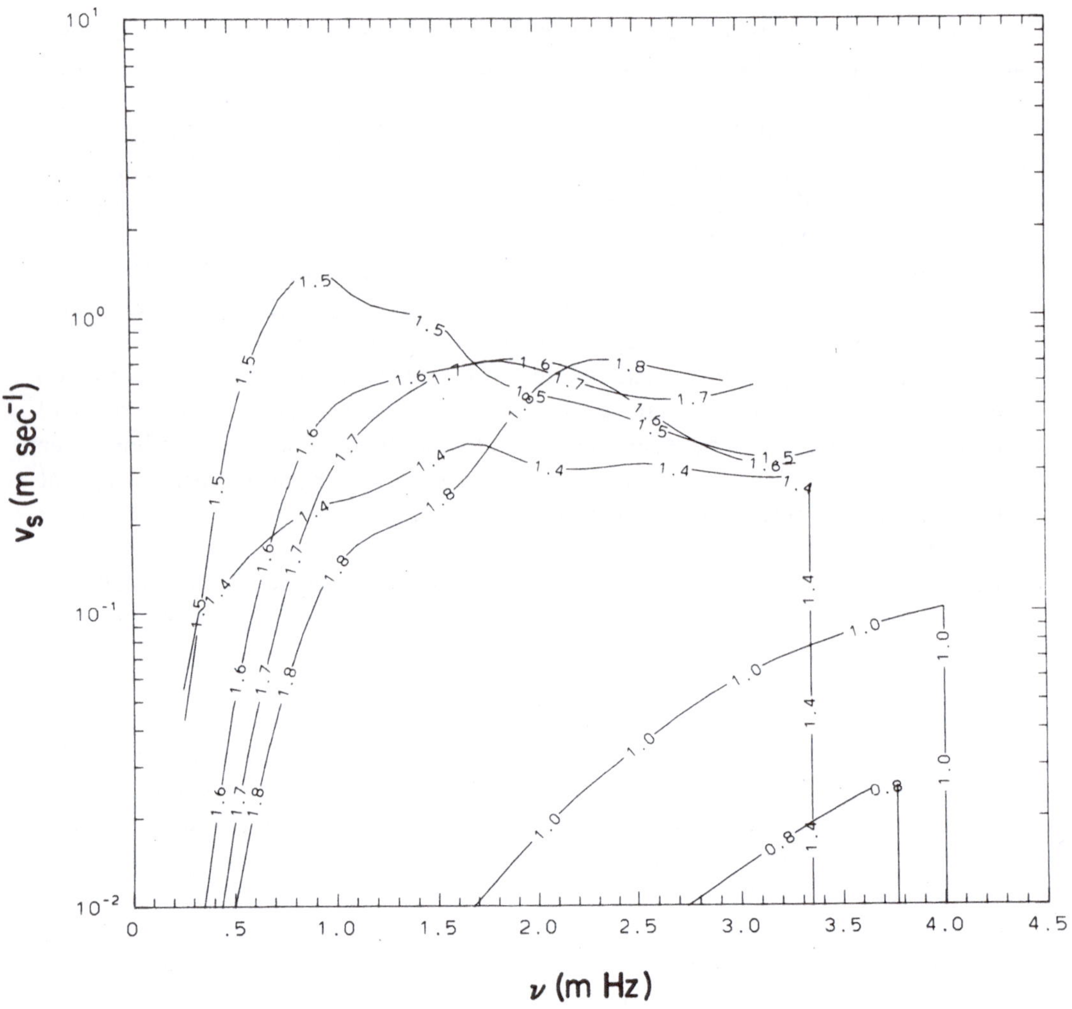

Fig. 4. Predicted surface velocity amplitudes v_s for the models on Figure 2, as functions of ν (cf. the caption to Figure 3).

of the convection zone is deeper and the effects of the neglect of the convective flux perturbation are probably less important. However a computation including the convective flux is clearly needed.

The mean frequency separation $\Delta\nu$ decreases with increasing mass of the model. This is largely an effect of the increasing radius. In fact $\Delta\nu$ can be estimated from asymptotic theory (e.g. Vandakurov, 1967) as

$$\Delta\nu \approx \left(2\int_0^{r_s} \mathrm{d}r/c\right)^{-1}, \tag{3.7}$$

and the variation in the sound speed with mass is relatively small.

Along the 1 $M_\odot$ evolution sequence (cf. Table II) the increase in velocity amplitude comes predominantly from an increase in E_c, caused by the decrease in surface gravity (cf. Equation (3.5)); in all cases the maximum amplitude occurs at the cut-off frequency $\tau_{c,\min}^{-1}$. The change in the behaviour of λ_s is small, so that the variation in the luminosity perturbation follows the surface velocity closely. The decrease in $\Delta\nu$ with age largely reflects the increase in the radius.

From Equation (3.5) the amplitudes might be expected to generally increase as one moves away from the Main Sequence. In fact, as the luminosity varies much more rapidly than the mass, the amplitude is predicted to increase rapidly with increasing luminosity at fixed effective temperature. For variations in T_{eff} at fixed mass and luminosity the change in surface gravity dominates over the change in flux, and so the amplitudes are predicted to increase with decreasing T_{eff}. These predictions are largely confirmed by our results on envelope models of high luminosity, presented in Table III. The very long time scales and small frequency separations for these models are clearly caused by their low surface gravities and large radii. Furthermore the acoustical cut-off frequency decreases more rapidly with increasing radius than does the frequency ν_1 of the fundamental radial oscillation; hence the number of modes with $\nu < \nu_c$ is decreased, and in fact for several of the models in Table III the maximum amplitude is found for the fundamental radial mode.

TABLE III

Properties of oscillations of envelope models. The relation between mass and luminosity is derived from Iben's (1964) evolution calculations. The notation is as in Table I.

$M/M_\odot$	$L/L_\odot$	T_{eff}	$v_{s,\max}$ (km s^{-1})	$(\delta L_s/L_s)_{\max}$	$\Pi_{\max}$ (days)	$\Delta\nu$ (μHz)
5	10^3	6800	0.014	1.1×10^{-4}	0.32	2.65
–	–	5770	0.015	1.7×10^{-4}	0.56	1.82
–	–	5000	0.021	3.1×10^{-4}	3.5	1.13
–	–	4500	0.015	3.6×10^{-4}	1.2	0.75
9	10^4	6800	0.021	2.2×10^{-4}	2.0	0.64
–	–	5770	0.046	5.2×10^{-4}	2.9	0.43
–	–	5000	0.095	1.9×10^{-3}	5.2	0.30
–	–	4500	0.12	2.5×10^{-3}	37	0.21
–	–	4000	0.054	1.4×10^{-3}	60	0.15
12	3×10^4	5000	0.093	1.8×10^{-3}	12	0.16
–	–	4500	0.20	5.2×10^{-3}	18	0.12
–	–	4000	0.25	7.3×10^{-3}	130	0.085
–	–	3750	0.16	5.0×10^{-3}	170	0.064
15	10^5	6800	0.011	6.4×10^{-5}	45	0.11
–	–	5770	0.085	7.9×10^{-4}	20	0.086
–	–	5000	0.13	2.4×10^{-3}	31	0.074
–	–	4500	0.19	5.0×10^{-3}	230	0.064
–	–	4000	0.44	0.018	370	0.052
–	–	3750	1.0	0.046	480	0.040
–	–	3500	1.8	0.074	110	0.027

4. Observational Implications

Given the results of the preceeding section, we may consider the prospects for observing stellar analogues to the solar 5 min oscillations. On the basis of the amplitudes alone stars somewhat hotter or more evolved than the present Sun appear to offer the best chances. For main sequence stars with an effective temperature of 7000–7500 K velocity amplitudes of up to 10 times higher, and luminosity amplitudes up to 5 times higher, than for the present Sun were obtained. The amplitudes are still too small to be easily detectable; however the fact that evidence for the solar luminosity oscillations, reflected in the light of Neptune, was obtained by Deubner (1981) suggests that one should be able to observe luminosity oscillations with reasonable certainty in stars of spectral types somewhat earlier than the Sun. Velocity measurements are probably beyond present-day techniques, but may eventually become feasible.

For the detection of the oscillations high frequency resolution is not needed. However to be of use for seismology the observations must provide values of individual frequencies, and this puts constraints on the frequency separation between the individual modes. As shown by e.g. Loumos and Deeming (1978) the frequency resolution is $1.5/T$ for a single observation of length T; if spectra resulting from many such observations are averaged the resolution is improved to $1/T$ (Christensen-Dalsgaard and Gough, 1982). For observations with $T = 8$ hr, which is probably about the maximum duration of stellar observations from moderate latitudes, the corresponding resolutions are 52 and 35 μHz, respectively. When comparing these values with the theoretical estimates of $\Delta\nu$ given in Tables I and II should be kept in mind that the latter refer only to the radial modes. For measurements of luminosity or velocity oscillations in integrated starlight modes of degree 0, 1, and 2 are expected to dominate (Dziembowski, 1977). Unless very high resolution is achieved one would therefore (cf. Christensen-Dalsgaard and Gough, 1982) expect the observed spectrum to consist of almost uniformly spaced peaks corresponding alternately to modes of degree 0 and 2, and to a mode of degree 1, the spacing between adjacent peaks being half the value of $\Delta\nu$ given in Tables I or II. Thus to be resolved with a single night of observation, or an average of several such nights, the spectrum of oscillations of a star should have a $\Delta\nu$ of at least about 100 or 70 μHz, respectively. Evidently the higher-mass ZAMS stars or the more evolved 1 $M_\odot$ stars are close to these limits. For evolved stars of mass significantly greater than 1 $M_\odot$ frequency resolution is probably impossible if only single-night observations, analyzed separately, are used. Thus one clearly has to find a balance between the demands of high amplitude and large frequency separation, the optimum being probably slightly evolved stars somewhat, but not too much, hotter than the Sun.

With more extensive observations it may become possible to improve the resolution by combining several nights' observations coherently on the assumption that the oscillations have a sufficiently long lifetime, although the problems of aliasing would then have to be dealt with. For solar oscillations such analyses were carried out by Deubner (1981), Claverie *et al.* (1981), and Bos and Hill (1983). Alternatively a continuous record could be obtained by combining data from several different observatories spaced around the Earth; and it may ultimately become possible to observe stars for extended periods of time from the South Pole, as was done by Grec *et al.* (1980) in the case of the Sun.

Further from the Main Sequence the periods become so long, and $\Delta\nu$ so small, that frequency resolved spectra would require months or even years of observation, and $\Delta\nu$ may eventually get so small that the modes appear to merge due to their finite lifetime. However it might still be possible to observe the oscillations as apparently irregular fluctuations in brightness or velocity, and from such observations to determine the overall distribution of power with frequency. In the range of stellar parameters covered by Table III the predicted maximum amplitudes are so large that the fluctuations should be seen with reasonable ease in ordinary photometric observations. In fact there may be some evidence that such fluctuations have been observed. In an extensive set of luminosity measurements made by Rufener, Maeder (1980) found luminosity fluctuations of up to $0\overset{m}{.}1$ among giants and supergiants. The amplitudes of these fluctuations, as a function of effective temperature, appears to have a minimum in the neighbourhood of the Cepheid instability strip, and this suggests that different mechanisms are at work among high- and low-temperature stars. If so, it seems possible that stochastically excited oscillations might be responsible for the fluctuations in the low-temperature region. Indeed the observations have the same trend, of increasing amplitude with increasing luminosity or decreasing effective temperature, as the results in Table III, and the observed and predicted amplitudes agree in order of magnitude. Radial velocity fluctuations of the order of 1 km s^{-1}, observed by Gun and Griffin (1979) in stars belonging to M3, might also be caused by stochastically excited oscillations. It is tempting to speculate that irregular and semiregular variability among red giants represents a further extension of this phenomenon; the oscillations of the true red variables are probably caused by linear instability. – Clearly further observations and more theoretical work is needed to clarify these possibilities.

5. Discussion

The physical model used in predicting the oscillation amplitudes, viz. equipartition between pulsational energy in one mode and kinetic energy in one 'resonating' convective eddy, is undoubtedly very rough. It is perhaps best regarded as a scaling law which, after being calibrated against the Sun, may be used to estimate how the amplitudes vary with stellar type. Thus the general trends in the neighbourhood of the Sun, i.e. the increase in amplitude with mass or age, are probably correct if the oscillations are indeed caused by stochastic excitation, whereas the predicted properties for models further from the Sun are far less reliable.

A major uncertainty is the use of traditional mixing length theory to describe the dynamics of the convective motion. Although the inferred time scales for solar convection are not in obvious conflict with observations of solar granulation, we clearly have almost no observational evidence for the distribution of energy among the convective eddies, and for other stars we have information about neither time scales nor energies. In particular it appears from anelastic modal calculations (Latour *et al.* 1981) that mixing length theory overestimates the convective flux in relatively hot stars with thin convection zones. If this is the case the convective energies are almost certainly overestimated as well, and the predicted amplitudes for the hotter stars in Table I would have to be reduced. Estimates of convective time scales and energies from more detailed calculations would provide useful information on this issue.

Even if the spectrum of convection had been known, we would still need a consistent scheme for calculating from this the surface amplitudes of the oscillations. It is encouraging that Gough and Poyet (Poyet, 1983) have made progress towards formulating such a scheme. Thus there is hope that a more reliable basis for making the amplitude estimates may soon become available.

Given the uncertainties in the theoretical results any observational test would clearly be highly valuable. Detection of stochastically excited oscillations in other main sequency stars and measurement of their amplitude would provide such a test, but has not yet been made. However as mentioned in Section 4 Maeder (1980) has found luminosity fluctuations in relatively cool supergiants, which agree roughly in magnitude and in the dependence on stellar luminosity and effective temperature with the results obtained in Section 3. The association of these fluctuations with oscillations has not been demonstrated. They could perhaps be caused by direct fluctuations in the convective flux (Schwarzschild, 1975), or by inhomogeneities in a stellar wind. Further observations, in particular of the relation between the luminosity fluctuations and possible fluctuations in radial velocity (which should be relatively easy to measure, if the estimates in Table III are realistic), may permit a choice between these different possibilities. Confirmation that the fluctuations are caused by oscillations would not only provide an immediate test of the results obtained here, but would also imply that this type of oscillations could be studied over a fairly large region of the HR diagram, with amplitudes large enough to allow traditional photometric techniques to be used.

6. Conclusion

From the results obtained here it seems reasonable to hope that 5 min oscillations of main sequence stars will be detected in a not too distant future, and that it will become possible to measure individual frequencies of such oscillations. The observational effort required is undoubtedly large; but the results should give direct information about the structure of Main Sequence stars and would therefore be very valuable in testing the theory of stellar evolution. The first quantity to be determined would probably be the mean frequency separation $\Delta\nu$ which, as pointed out in Section 3, to a large extent is a measure of the radius of the star. However with observations of sufficiently high frequency resolution to resolve the individual modes, it should be possible to assign values of l to the individual frequencies from the observed amplitudes and the distribution of modes, as has been done for the solar oscillations, (see e.g. Christensen-Dalsgaard, 1980); the frequencies would then give more detailed information about the structure of the star. Even higher resolution might enable detection of rotational splitting of the frequencies of modes with $l > 0$; this has apparently been achieved by Claverie *et al.* (1981) for the Sun, and would give information about the rotation rate of the star, possibly eventually even about its variation with position in the star.

It is worth pointing out that observations of this type of oscillations might also be useful in the study of stellar convection. The calculation of the oscillation amplitudes resulting from a given convective velocity field is probably considerably simpler than a direct computation of the convective velocities. Furthermore, although identification of individual frequencies may be difficult for stars that are not close to the Main Sequence, it may still be possible in such stars to determine the broad variation of power with frequency. Thus, if the general idea that these oscillations are excited stochastically by convection is correct, one may hope to be able to perform at least a limited inversion on the observed amplitudes, to get information about the properties of the convection. Such information, which can then potentially be obtained over a wide range of stellar parameters, would clearly be very useful in testing theories of stellar convection.

Acknowledgements

We are grateful to J. Andersen for the suggestion that led to the present work, and to T. M. Brown, D. O. Gough, M. Jura, A. Reiz, and J. Toomre for useful discussions. JC-D would like to thank Professor G. K. Batchelor for hospitality at DAMTP, Cambridge, where much of the paper was written.

References

Ando, H. and Osaki, Y.: 1975, *Publ. Astron. Soc. Japan* **27**, 581.

Ando, H. and Osaki, Y.: 1977, *Publ. Astron. Soc. Japan* **29**, 221.

Berthomieu, G., Cooper, A. J., Gough, D. O., Osaki, Y., Provost, J., and Rocca, A.: 1980, in H. A. Hill and W. Dziembowski (eds.), *Lecture Notes in Physics* **125**, Springer, Heidelberg, p. 307.

Bos, R. J. and Hill, H. A.: 1983, *Solar Phys.* **82**, 89 (this volume).

Böhm-Vitense, E.: 1958, *Z. Astrophys.* **46**, 108.

Christensen-Dalsgaard, J: 1977, Ph. D. dissertation, University of Cambridge.

Christensen-Dalsgaard, J: 1980, *Proc. 5th European Regional Meeting in Astronomy* (Institut d'Astrophysique, Liège).

Christensen-Dalsgaard, J.: 1982, *Monthly Notices Roy. Astron. Soc.* **199**, 735.

Christensen-Dalsgaard, J. and Frandsen, S.: 1983, *Solar Phys.* **82**, 165 (this volume).

Christensen-Dalsgaard, J. and Gough, D. O.: 1976, *Nature* **259**, 89.

Christensen-Dalsgaard, J. and Gough, D. O.: 1980a, in H. A. Hill and W. Dziembowski (eds.), *Lecture Notes in Physics* **125**, Springer, Heidelberg, p. 184.

Christensen-Dalsgaard, J. and Gough, D. O.: 1980b, *Nature* **288**, 544.

Christensen-Dalsgaard, J. and Gough, D. O.: 1981, *Astron. Astrophys.*, in press.

Christensen-Dalsgaard, J. and Gough, D. O.: 1982, *Monthly Notices Roy. Astron. Soc.* **198**, 141.

Claverie, A., Isaak, G. R., McLeod, C. P., van der Raay, H. B., and Roca Cortes, T.: 1981, *Nature* **293**, 443.

Cox, A. N. and Tabor, J. E.: 1976, *Astrophys. J. Suppl.* **31**, 271.

Deubner, F.-L.: 1981, *Nature* **290**, 682.

Dziembowski, W.: 1977, *Acta Astronomica* **27**, 203.

Dziembowski, W.: 1980, in H. A. Hill and W. Dziembowski (eds.), *Lecture Notes in Physics* **125**, Springer, Heidelberg, p. 22.

Goldreich, P. and Keeley, D. A.: 1977a, *Astrophys. J.* **211**, 934.

Goldreich, P. and Keeley, D. A.: 1977b, *Astrophys. J.* **212**, 243.
Gough, D. O.: 1977, *Astrophys. J.* **214**, 196.
Gough, D. O.: 1980, in H. A. Hill and W. Dziembowski (eds.), *Lecture Notes in Physics* **125**, Springer, Heidelberg, p. 273.
Grec, G., Fossat, E., and Pomerantz, M.: 1980, *Nature* **288**, 541.
Gunn, J. E. and Griffin, R. F.: 1979, *Astron. J.* **84**, 752.
Iben, I.: 1964, *Astrophys. J.* **140**, 1631.
Keeley, D. A.: 1977, *Proc. Symposium on Large-Scale Motions on the Sun*, Sacramento Peak Observatory, p. 24.
Keeley, D. A.: 1980, in D. Fischel, J. R. Lesh, and W. M. Sparks (eds.), *Proc. Conference on Current Problems in Stellar Pulsation Instabilities*, NASA Technical Memorandum 80625, p. 677.
Lamb, H.: 1909, *Proc. London Math. Soc.* **7**, 122.
Latour, J., Toomre, J., and Zahn, J.-P.: 1981, *Astrophys. J.* **248**, 1081.
Loumos, G. L. and Deeming, T. J.: 1978, *Astrophys. Space Sci.* **56**, 285.
Maeder, A.: 1980, *Astron. Astrophys.* **90**, 311.
Poyet, J. P.: 1983, *Solar Phys.* **82**, 267 (this volume).
Schwarzschild, M.: 1958, *Structure and Evolution of the Stars*, Princeton University Press, Princeton.
Schwarzschild, M.: 1975, *Astrophys. J.* **195**, 137.
Scuflaire, R., Gabriel, M., and Noels, A.: 1981, *Astron. Astrophys.* **99**, 39.
Shibahashi, H. and Osaki, Y.: 1982, *Publ. Astron. Soc. Japan* **33**, 713.
Traub, W. A., Mariska, J. T., and Carleton, N. P.: 1978, *Astrophys. J.* **223**, 583.
Unno, W. and Spiegel, E. A.: 1966, *Publ. Astron. Soc. Japan* **18**, 85.
Vandakurov, Yu. V.: 1967, *Astron. Zh.* **44**, 786.
Woodard, M. and Hudson, H.: 1983, *Solar Phys.* **82**, 67 (this volume).

HELIOSEISMOLOGY IN THE FUTURE*

(Invited Review)

R. M. BONNET

Laboratoire de Physique Stellaire et Planétaire, C.N.R.S., P.O. Box 10, 91370 Verrières le Buisson, France

Abstract. We review the observables of helioseismology that can contribute to our knowledge of the physical conditions in the solar interior. We discuss the limitations which presently prevent helioseismology from reaching its ultimate goal. We finally present a list of projects which either are already underway or that are planned for the near future, and we conclude by showing the crucial role that space observations may play in the future.

1. Introduction

Helioseismology is a recently born technique whose ultimate aim is to infer the physical conditions and the dynamical properties of the solar interior, starting from the top of the photosphere down to the centre. It derives from the techniques usually applied on Earth to study the internal structure of our planet through the properties of the vibrations induced by Earth quakes. As shown by Gough (1983), these techniques imply the mathematical inversion of an integral, whose results are model dependent. In this summary, we will not deal with this mathematical aspect that we assume to be sufficiently well mastered at present. We will on the contrary concentrate on the possible means of observation (and their limitations) that can be used for probing the solar interior, which is yet inaccesible to direct observation, except for the last two hundred outer kilometers and possibly the central core. By probing we mean inferring the density stratification, the chemical composition, the rotation and possibly studying the structure of large-scale convection (Gough and Toomre, 1983), and their possible variations with time.

2. The Observations Which Can be Used in Helioseismology

2.1. Neutrinos

We disregard the most direct messenger from the Sun's centre, the neutrino, because the matter of its detection has been dealt with extensively in other instance and would lead us much too far beyond the scope of this paper.

2.2. Routine observations of activity related manifestations and differential rotation

Helioseismology (in the broad sense) got a first observational support nearly 140 years ago at the time when the first systematic recordings of sunspot number and position were

* Proceedings of the 66th IAU Colloquium: *Problems in Solar and Stellar Oscillations*, held at the Crimean Astrophysical Observatory, U.S.S.R., 1–5 September, 1981.

Solar Physics **82** (1983) 487–493. 0038–0938/83/0822–0487$01.05.

undertaken. Indeed, the so-called 'butterfly diagram' which they serve to construct tells us something about phenomena which occur beneath the photosphere and about their relaxation time. More recently the high precision continuous measurement of the differential rotation by Howard and LaBonte (1981) and of its latitudinal variations indicates that a tight relationship exists between differential rotation and the solar cycle. This fact would not have been discovered without the careful and relentless work of dedicated observers.

These two examples are given here to illustrate the worthiness of routine observations which may well provide in the future some powerful clues to our understanding of solar subsurface phenomena.

However, in the past years, new techniques have developed or have reached a point of maturation that gives the observers and the theorists new means of investigation.

2.3. Radius variations

For example, one explanation offered to the neutrino deficiency is that the Sun's radius is shrinking at a substantial rate. Measurements of the solar diameter with high accuracy are the only possible way to check this explanation. Recently, Parkinson *et al.* (1980) have shown that combined data sets of the Mercury transit and total solar eclipse observations provide no solid basis for this assumption, at least over the past 250 years or so. However, there is some evidence that periodic changes of about 0.02% may have occurred on time scales of several tens of years.

2.4. Solar constant variations

The recent measurements, made from two artifical satellites, SMM and Nimbus 7, of the variations of the solar constant on time scales of days or weeks in association with the appearance of sunspots, may be indicative of modifications in the large scale convection that are deep seated in the convection zone. These measurements correspond to an accuracy of a few hundredths per cent and it is only through the use of space techniques and the continuous monitoring of the solar energy output that they have been made possible. A moderate spatial resolution at the disc surface would allow one to judge whether solar constant variations reflect similar variations in the luminosity or whether they are the result of time dependent inhomogeneities. The blocking by sunspots is not the only process that can cause luminosity variations. Such variations result from changes in the balance between thermal, gravitational and other forms of energy (Gough, 1980). Such changes modify the hydrostatic structure of the Sun and lead to variations both in luminosity and radius, that depend on the depth in the Sun of the perturbation that produces them. Simultaneous measurements of luminosity and radius are therefore of considerable interest.

2.5. Global oscillations

It is undoubtedly through the accurate measurements of periods and phases of dynamical oscillations that helioseismology can yield information about the solar interior, from the surface to the core. Provided the modes of oscillations can be identified, their frequencies can be used to diagnose the solar interior in a way similar to that which allows seismic waves to probe the interior of the Earth. The observations concern both velocity and luminosity oscillations.

The observations of each type fall into two groups: those that detect high-degree modes and require good spatial resolution (Deubner, 1975) and those that detect low-degree modes and concern integrated properties of the Sun (Claverie *et al.*, 1979; Grec *et al.*, 1980). The latter provide a nearly direct measure of conditions in the interior, in opposition to the first category which allows the sounding of only a fraction of the solar volume beneath the photosphere. Both categories have led to the conclusion that the standard model that fits the observations best has a relatively high helium content.

The amplitudes and relative phases of velocity and intensity oscillations depend on the reaction of the upper layer of the convection zone to oscillations with periods comparable with the eddy turnover time, and thereby they provide a valuable clue to our understanding of the solar convection zone. Some degree of spatial resolution is needed to provide direct information about the angular dependence of the modes.

In principle, rotational splitting of the nonaxisymetrical modes can provide us with information about the internal rotation κ of the Sun. If many modes are available the angular momentum distribution within the solar interior can be estimated, provided the modes originate from different portions of the solar interior. The condition required to obtain accurate measurements is that the oscillations maintain phase for several rotation periods and that they are observed continuously throughout that time.

3. The Present Limitations in Helioseismology

As stated in the introduction, helioseismology is in its infancy. However, we are already in a position where we can identify what are its observational limitations. With no pretention of being exhaustive, we isolate four areas in which present observations are now reaching their limits and where we think that considerable progress can be expected in the future.

3.1. Lack of Continuity

Ground-based as well as space-borne instruments are affected by the quasi-periodic eclipses of sunlight at night. This limitation is of no effect on routine observations that span large periods of time, but it affects substantially the analysis of shorter time periods. For example, the Fourier analysis of both the intensity (Deubner, 1981; Fröhlich, 1981) and the velocity oscillations suffers from side bands in the power spectrum, due to the quasi-periodicity of day/night cycles. These spurious signals can affect the detectability of frequencies which may be crucial for mode identification.

By observing from the South Pole, Grec *et al.* (1980) have already gone one step further by observing continuously for 5 concecutive days. The Birmingham group intends to observe from several ground based stations separated in longitude, and allowing in principle a continuous coverage.

From another point of view it is quite obvious that the dominant dips which appear in the records of the solar constant made from the SMM (Willson *et al.*, 1981) would not have been resolved by measurements performed once every month or every year, and had an isolated measurement been made during the period of such dips, the result would have been misleading.

3.2. The short duration of observations

The outcome of routinely collected data easily illustrate the importance of long duration observations. The excellent data of Grec *et al.* obtained from the South Pole also point out how important it is to have more than a few hours of continuous data. We have already mentioned the importance of long duration observations for the accurate measurement of the rotational splitting. It is necessary to follow the phase over several rotation periods for the effect to be measured precisely. Theory predicts that the low order modes which probe the whole solar interior, but have rather low amplitudes should maintain phase for several months, or longer.

The measurement of low order p modes quite certainly requires the use of spectrometers that are able to achieve a sensitivity of a few mm s^{-1}. Such performance would also be adequate to resolve f and low order g modes. Long observing times therefore appear as a necessity to help reducing the noise.

The measurement of periods like the 2^h40^m oscillations and longer, requires also that the observations be conducted for several months or years.

Talking of luminosity variations associated with changes in the hydrostatic structure of the convection zone and with the solar cycle, automatically implies that observations which would cover less than a substantial fraction of a solar cycle would be of poor value if any.

3.3. The Earth's atmosphere

Nearly all the observations analyzed in Section 2 are performed through the Earth's atmosphere.

Differential refraction, atmospheric oscillations and turbulence put strong limitations on the quality of measurements of the solar radius. The best sites offer a resolution of 0.3 arc sec, while the needs require an improvement of a factor 3 to 10, a goal yet impossible to reach from the ground.

The accurate photometric measurements of the solar constant and of white light oscillations are also strongly affected by the presence of the Earth's atmosphere and it is only through the use of space-borne instruments that it is possible to reach the accuracy of a few parts per million (Hudson, 1983), required to measure some of the modes of low order and to detect the minute variations in the solar total output.

The best measurements to date of the global velocity oscillations, by Grec *et al.* (1980), show that the low frequency region of the power spectrum where f and g modes are supposed to be found is very noisy. The source of this noise is either the Sun itself, or the instrument or the Earth's atmosphere. The precision of a few mm s^{-1} required to detect these modes is impossible to reach to day. By observing from space we can at least eliminate the atmospheric contribution and be able to substantially improve the solar signal.

4. Future Projects

We give now a summary list of projects that we are aware of and that may contribute to some progress in solar seismology. For the sake of coherency these projects are separated into two categories, i.e. ground-based and space borne. Again, we have no intention to be exhaustive and we apologize for any omission that may exist in this list.

4.1. Ground-based experiments

4.1.1. *Solar Radius Measurements*

Several groups are presently involved in developments in this area. Hill in the U.S.A. will measure the variations in the solar diameter through changes in the limb darkening at 6 positions around the solar circumference as defined by six 100 arc sec long slits placed tangentially to the Sun's limb.

The High Altitude Observatory is already involved in a monitoring programme.

Rösch and Yerle at Pic du Midi intend to continue their accurate measurements of the solar diameter through a comparison with a ULE rod calibre.

4.1.2. *Velocity Oscillations*

Two techniques are presently exploited which can help solving the continuity and long duration requirement.

The first one consists in a ring of stations located all around the world. This is the technique proposed by the Birmingham group. It has the drawback of multiplying the equipment and the number of observers. In addition, no one can never be sure that all stations do benefit from good weather conditions.

An alternative is to exploit the South Pole station that proved its efficiency with the pioneering observations of Grec *et al.* Several groups of observers are already developing equipments and planning their observations from this unique site: Stebbins and Harvey in the U.S.A., and the group from Nice University which intends to come back with a new equipment in 1982–1983.

Of course, like ground based observations, the success of this solution is severely weather-dependent and suffers from the presence of the atmosphere. This is the reason why the concept of a tethered balloon floating 5 km above the south polar cap has been considered in France. The severe requirement of knowing the relative velocity of the balloon with respect to the Sun with a precision of about 1 mm s^{-1} makes this project a very difficult one.

4.2. Space-borne experiments

In this category, only projects that are under study at this time can be quoted, and we do not know of any experiment yet in the planning stage. We list them according to their chronological entry on the stage:

(i) OGIS (Oscillations of the Global Intensity of the Sun) is a project which has been under discussion for more than three years between the Crimea Observatory and several eccentricity orbit and should measure the oscillations of the solar intensity in several bands of the visible and ultra-violet spectrum. The programme of this mission is still under question.

(ii) The 'Birmingham–Crimea' project consists in measuring velocity oscillations also from a Prognoz satellite. A feasibility study for the instrumentation has been undertaken in England but, as for the OGIS project, the mission is not yet programmed.

(iii) DISCO (Dual Investigation of the Solar Constant and Oscillations) is a project presently undergoing a Phase A study at the European Space Agency (E.S.A.). It is by far the best-studied space project among all those described here. It consists in measuring global velocity and intensity variations and oscillations from a spinning satellite located at the Lagrangian point L_1 which lies between the Sun and the Earth–Moon system. The lifetime of 2 years envisaged for DISCO and the use of a stable high resolution spectrometer should in principle provide the required accuracy of 1 mm s^{-1}. The rest of the instrumentation, consists of high accuracy radiometers, white light and broad band photometers, with limited (a few arc min) angular resolution capability, and of an extreme UV spectroheliograph providing a spatial resolution of 15 arc sec, sufficient to observe coronal holes and active regions. A detailed description of this mission can be found in the assessment study document ESA–SCI(81)6, November 1981. The decision to proceed after the completion of the phase A study will be taken at the beginning of 1983.

(iv) The Solar Internal Dynamics Mission (SIDM) is a NASA mission which was probably proposed before DISCO. However its state of definition is much less advanced and its concept not even defined yet. Its goals are similar to those of DISCO, except that it may offer more spatial resolution capability. If this is the case the SIDM would likely consist of a two or three axis stabilised spacecraft and therefore be of a higher degree of sophistication than DISCO.

(v) The Solar Diameter Measurement Instrument on Spacelab. This instrument has merely been suggested, and not yet formally proposed. It is an improved version of the Pic du Midi instrument and would consist of a 1 m × 10 cm telescope placed on Spacelab or a space station.

5. Conclusion

The future observations in helioseismology require progress in several directions. First, our discussion clearly indicates the role that space techniques can play in this context since they can resolve simultaneously all the three problems of continuity, long observing time, and perturbations by the Earth's atmosphere. This potentiality however has been discussed only recently, but there is little doubt that the most decisive progress in helioseismology will have to await the existence of a 'seismology observatory' or of instruments capable of measuring the parameters identified in Section 2.

We assume that the instrumentation itself has no intrinsic limitations. Those instruments which have already contributed substantially to the field, like the balloon radiometer of Frölich (1981), the ACRIM instrument on the SMM, the sodium optical resonance cell of Grec *et al.*, the potassium cell of the Birmingham group do not require a dramatic improvement in their performance in order to achieve the range of accuracy that is needed here.

More progress can be expected also in the near future in the area of data analysis, more specifically of ground based data, in particular for the detection of modes of low frequency, since it appears that this analysis has not always been conducted with enough precaution. Special care should be taken to eliminate all kinds of false periodicity effect of either instrumental or astronomical origin.

References

Claverie, A., Isaak, G., MacLeod, C., van der Ray, H., and Roca Cortes, T.: 1979, *Nature* **282**, 591.
Deubner, F. L.: 1975, *Astron. Astrophys.* **44**, 371.
Deubner, F. L.: 1981, *Nature* **290**, 682.
Frölich, C. and Wehlri, C.: 1981, in J. London (ed.), *Proc. IAMAP Symposium on Solar Constant*, Hamburg, 17–18 August, 1981.
Gough, D. O.: 1980, in S. Sofia (ed.), *Proc. Workshop on Solar Constant Variations*, NASA, November 1980.
Gough, D. O.: 1983, *Solar Phys.* **82**, 7 (this volume).
Gough, D. O. and Toomre, J.: 1983, *Solar Phys.* **82**, 401 (this volume).
Grec, G., Fossat, E., and Pomerantz, M.: 1980, *Nature* **288**, 541.
Howard, R. and LaBonte, B. J.: 1981, *Solar Phys.* **74**, 131.
Parkinson, J. H., Morrison, L. V., and Stephenon, F. R.: 1980, *Nature* **228**, 548.
Willson, R. C., Gulkis, S., Janssen, M., Hudson, H. S., and Chapman, G. A., 1981, *Science* **211**, 700.

TABLE OF CONTENTS

ARTICLES

TABLE OF CONTENTS

TABLE OF CONTENTS

Instrumentation for Astronomy with Large Optical Telescopes

Proceedings of IAU Colloquium No. 67, held in Zelenchuk, U.S.S.R., September 8–10, 1981

edited by C. M. HUMPHRIES

1982, xviii + 322 pp.
Cloth Dfl. 110,– / US $ 48.00 ISBN 90-277-1388-X
ASTROPHYSICS AND SPACE SCIENCE LIBRARY 92

At a time when recent advances have produced a dramatic increase in the detection efficiency of astronomical instrumentation (e.g. with the advent of charge coupled detectors and single photon counting systems), it would seem appropriate to review the progress which has been made. To this end, IAU Colloquium No. 67 was recently held in Zelenchuk, U.S.S.R. Topics discussed included spatial and spectral interferometry, conventional and new designs for spectrographs, image intensifiers and solid state detectors, large telescopes, and the efficient matching of telescope images and instrumentation. The spectral region considered included the infrared as well as the visible.

D. Reidel Publishing Company

P.O. Box 17, 3300 AA Dordrecht, The Netherlands
190 Old Derby St., Hingham, MA 02043, U.S.A.

Sun and Planetary System

Proceedings of the Sixth European Regional Meeting in Astronomy, held in Dubrovnik, Yugoslavia, 19–23 October, 1981

edited by
W. FRICKE and G. TELEKI

1982, xxii + 504 pp. + index
Cloth Dfl. 150,– / US $ 65.00 ISBN 90-277-1429-0
ASTROPHYSICS AND SPACE SCIENCE LIBRARY 96

This volume presents reports on recent research of the solar system. Under consideration are, the *Sun as a Star*, the Earth from the astronomical, geophysical and geodetic point of view, the physics of planets, comets, minor planets, satellites, the interplanetary medium, and the motions in the planetary system. A special part is devoted to the problems of astronomical refraction. Unprecedented progress has been made in recent years in all these fields by observers and theoreticians, and the subject has become attractive, not only to European astronomers, geophysicists, geodesists and physicists, but also to experts from non-European countries who took an active part in the *European Meeting.* The volume is unique in its combination of reviews of recent research on such entirely different problems, as they are offered by the various objects of the solar system.

D. Reidel Publishing Company
P.O. Box 17, 3300 AA Dordrecht, The Netherlands
190 Old Derby St., Hingham, MA 02043, U.S.A.

Solar Phenomena in Stars and Stellar Systems

Proceedings of the NATO Advanced Study Institute held at Bonas, France, August 25 – September 5, 1980

edited by
ROGER M. BONNET and ANDREA K. DUPREE

1981, x + 591 pp.
Cloth Dfl. 135,– / US $ 69.50 ISBN 90-277-1275-1
NATO ADVANCED STUDY INSTITUTES SERIES
C. Mathematical and Physical Sciences 68

The Sun is no longer an isolated astrophysical object but serves the role of representing the basic element of comparison to a large class of objects. This book reviews solar phenomena comprehensively and at each stage a general summation is attempted. Where necessary, as for instance in the case of solar flares, problems are also considered from a basic physics perspective.
The volume is organised into four sections, starting with an overview of the subject matter and then proceeding to a review of the physics of Stellar Interiors. Section 3 deals with the crucial aspect of existence and physics of chromospheres and coronae and how they relate to the presence of convective phenomena in stellar subsurface layers. The final section considers the question of variability. An extensive bibliography will be found throughout the various sections, to which the reader may refer for more detailed developments in different specialist areas.

Audience: Astronomers and other scientists who have an interest in variations in the earth's climatic conditions.

D. Reidel Publishing Company

P.O. Box 17, 3300 AA Dordrecht, The Netherlands
190 Old Derby St., Hingham, MA 02043, U.S.A.

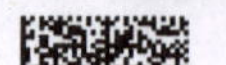